Safety Risk Assessment of Urban Underground Large Space Construction

Sheng-Xiang Lei · Zhong-Sheng Tan ·
Ming-Li Huang · Zheng-Quan Ding · Ke Lei

Safety Risk Assessment of Urban Underground Large Space Construction

人民交通出版社
China Communications Press

Springer

Sheng-Xiang Lei
China Railway Construction Corporation Limited
Beijing, China

Ming-Li Huang
School of Civil Engineering
Beijing Jiaotong University
Beijing, China

Ke Lei
School of Civil Engineering
Beijing Jiaotong University
Beijing, China

Zhong-Sheng Tan
School of Civil Engineering
Beijing Jiaotong University
Beijing, China

Zheng-Quan Ding
Department of Science and Technology Innovation
China Railway Construction Corporation Limited
Beijing, China

ISBN 978-981-97-7436-4 ISBN 978-981-97-7437-1 (eBook)
https://doi.org/10.1007/978-981-97-7437-1

Jointly published with China Communications Press Co., Ltd
The print edition is not for sale in China (Mainland). Customers from China (Mainland) please order the print book from: China Communications Press Co., Ltd.

Translation from the Chinese Simplified language edition: “城市地下大空间施工安全风险评估” by Sheng-Xiang Lei et al., © China Communications Press Co., Ltd 2021. Published by China Communications Press Co., Ltd. All Rights Reserved.

The original submitted manuscript has been translated into English. The translation was done using artificial intelligence. A subsequent revision was performed by the author(s) to further refine the work and to ensure that the translation is appropriate concerning content and scientific correctness. It may, however, read stylistically different from a conventional translation.

This Springer imprint is published by the registered company Springer Nature Singapore Pte Ltd.
The registered company address is: 152 Beach Road, #21-01/04 Gateway East, Singapore 189721, Singapore

Foreword

The development and utilization of underground space are important components of ecological civilization construction and represent an inevitable trend in the development of human society and cities. Urban underground space development and utilization is the most effective approach to address traffic congestion, land resource constraints, expand urban space, and alleviate environmental degradation. It is also a crucial initiative for achieving sustainable development, constructing a resource-efficient, and environmentally friendly society in human society and the economy.

In recent years, China's development of urban underground space, including underground transportation, underground commerce, comprehensive utility corridors, and municipal facilities, has seen rapid growth. The scale of construction is increasingly large, major projects are continuously increasing, technological levels are continually improving, and forward-thinking concepts are constantly being proposed. Simultaneously, new problems are emerging, and new challenges are being faced in the development and utilization of urban underground space and the technical support involved. Innovative approaches are needed to address these challenges. In response to scientific and key technical issues in underground engineering, systematic research and breakthroughs are essential for driving continuous innovation and development in urban underground space construction technology.

Against this backdrop, led by Chief Engineer Sheng-Xiang Lei of China Railway Construction Corporation Limited, the company relies on three national key research and development projects under the "Four Orientations": "Key Technology Research on Safe Construction in Urban Large Underground Spaces," "Key Technology Research and Demonstration of Comprehensive Monitoring for Urban Underground Infrastructure Operation," and "Research and Demonstration of Precision Detection Technology and Development Utilization in Urban Underground Space." Major scientific and key technical issues were identified and proposed, and systematic scientific research was conducted. This effort resulted in key technologies covering all aspects of urban underground and deep space development, including detection, planning and design, safe construction, intelligent monitoring, and smart operation and maintenance.

Based on the research outcomes and engineering practices, Chief Engineer Sheng-Xiang Lei organized the compilation of the "Key Technologies Series for Urban Underground Space Development and Utilization." This series not only reflects development philosophies but also provides explanations of key technologies and equipment applications. It showcases numerous breakthrough achievements, advanced practices, and typical engineering cases in China Railway Construction Corporation's field of urban underground space development and utilization. It is believed that this series will play a crucial and positive role in promoting the safe, orderly, and efficient development of urban underground space in China.

June 2021

Chen Xiangsheng
Dean of the School of Civil and Transportation Engineering at Shenzhen University
Academician of the Chinese Academy of Engineering
Beijing, China

Preface I

As China's urbanization rate continues to rise, urban land resources are becoming increasingly scarce, leading to a heightened emphasis on the development and utilization of urban underground space. In recent years, the development of urban underground space has shifted from isolated projects to block-based and networked developments. The engineering structures exhibit characteristics such as spatial multidimensionality, large scale, and structural complexity. The construction challenges, resource inputs, environmental impact, organizational complexity, and public concerns are distinct from conventional underground engineerings. Particularly during the construction phase, the inherent complexity of underground engineering, combined with large project scales and tight schedules, significantly increases the technological difficulty, resulting in heightened safety risks and frequent accidents.

To differentiate large-scale, complex, and high-risk underground engineerings from conventional ones, the concept of "urban large underground spaces" has been introduced. Currently, there is a lack of systematic research literature on the safety risks of urban large underground spaces. Therefore, a targeted and systematic study of construction safety risks in urban large underground spaces is imperative.

Due to their large scale, geological complexity, and intricate surrounding environments, urban large underground spaces exhibit characteristics such as diverse risk events, complex risk mechanisms, and numerous risk factors in their construction safety risks. Additionally, these spaces often serve as crucial nodes in the network of urban underground spaces or as landmark projects playing a pivotal role in urban development. Any safety incidents during construction could result in significant economic losses and extremely adverse social impacts. In the current context of deepening the overall national security outlook, the importance of workplace safety has reached an unprecedented level.

The construction of urban large underground spaces should prioritize safety management within the realm of engineering construction-related activities. By leveraging advanced management models and continually maturing technical systems, efforts should be maximized to control construction safety risks within acceptable limits. Thus, establishing a mature construction safety risk assessment system and forming a comprehensive construction safety risk control regime are critical for ensuring the secure, economic, and efficient development of urban large underground spaces.

This book represents a phase summary of the author's reflections and research findings on construction safety risk management in urban large underground space projects in recent years. Adhering to the principle of "safety first, life above all," and closely addressing the practical needs of frontline personnel engaged in urban large underground space construction, this book aims to address three core issues in construction safety risk management: the characteristics and analysis methods of construction safety risks in urban large underground spaces, the coupling risk formation and evolution mechanisms of construction risks in urban large underground spaces, and the evaluation methods and dynamic safety evaluation system for construction safety risks in urban large underground spaces.

The book delves into in-depth research around these three issues, covering:

1. Clarifying the concept of urban large underground spaces, defining their scope based on both engineering volume and environmental sensitivity, and providing a detailed summary of the types and characteristics of construction safety risks in urban large underground spaces.
2. Conducting theoretical research on the coupling and dynamic evolution of construction safety risks in urban large underground spaces, introducing the concepts of coupling effects and coupling coefficients, establishing a model for coupled evolution analysis of construction safety risks, and quantitatively describing the coupling effects and evolution patterns of risk factors.
3. Establishing an evaluation index system for construction safety risks in urban large underground spaces, determining quantitative standards for evaluation indicators, and providing a concrete and operational guide for the assessment of construction safety risks in urban large underground spaces.

The book aims to improve the quality of construction safety risk assessment in China's urban large underground spaces and provide theoretical and technical support for preventing major safety incidents in construction. It is co-authored by Sheng-Xiang Lei, Zhong-Sheng Tan, Ming-Li Huang, and Zheng-Quan Ding in eight chapters. The first two chapters (1 and 2) introduce the basic concept and characteristics of urban large underground spaces based on extensive research on the current status of domestic and international urban underground space development. They also present a summary and statistics of domestic urban large underground space cases. Chapter 3 compiles and analyzes 295 cases of underground space construction accidents worldwide, providing detailed analysis of several typical major construction safety accidents in urban underground spaces. Chapter 4 introduces the basic theory of construction safety risk assessment, summarizes the classification and characteristics of construction safety risks in urban large underground spaces, and proposes methods for analyzing construction safety risks. Chapter 5 focuses on the scientific problem of the coupling formation mechanism of construction safety risks in urban large underground spaces, presenting the concept of coupling effects and establishing a theoretical framework for quantitative research. Chapter 6 emphasizes the dynamic evolution mechanism of construction safety risks in urban large underground spaces, establishes predictive models, and summarizes the evolution patterns of safety risks during construction. Chapter 7 establishes a dynamic evaluation system for construction safety risks in urban large underground spaces, including assessment methods, indicator systems and quantitative standards, risk grading criteria and acceptance criteria, risk assessment procedures, and risk assessment systems. Chapter 8 provides a summary of the entire book.

Throughout the writing process, valuable opinions were provided by Zou Chunhua, Li Fengwei, Yang Huijun, and others. Assistance from members of the research team, including Wang Xiuying, Bai Mingzhou, Wang Weidong, Li Xiaoxue, Yao Hong'an, Tian Yuanyuan, Wang Hongkun, Ji Guodong, Guo Yajuan, Zheng Junfei, Yu Kun, Ding Yong, Huang Meng, Zhang Zhien, Gui Ting, Ren Shaoqiang, Huang Xin, Lin Zhiyuan, is sincerely acknowledged. Due to the limited expertise of the authors, there may be shortcomings in the book, and readers are encouraged to provide constructive feedback.

Beijing, China
June 2021

Sheng-Xiang Lei
Zhong-Sheng Tan
Ming-Li Huang
Zheng-Quan Ding
Ke Lei

Preface II

On March 5, 2017, General Secretary Xi Jinping emphasized during the deliberation of the Shanghai delegation at the Fifth Session of the Twelfth National People's Congress that urban management should be as meticulous as embroidery. China Railway Construction Corporation Limited (CRCC) deeply implemented the important instructions of the General Secretary and devoted all efforts to create the premier brand in urban underground space. Since 2018, CRCC has taken the lead in undertaking three national key research and development projects: "Key Technology Research on Safe Construction in Urban Large Underground Spaces," "Key Technology Research and Demonstration of Comprehensive Monitoring for Urban Underground Infrastructure Operation," and "Research and Demonstration of Precision Detection Technology and Development Utilization in Urban Underground Space." These projects are typical scientific research endeavors in the field of urban underground space during the 13th Five-Year Plan period.

In response, CRCC established a research team for urban underground space, fostering extensive collaboration among industry, academia, research, and application. They proposed a new construction philosophy of "human-oriented, green, resilient, intelligent, transparent, and legal" to promote the intensive and efficient development and utilization of urban underground space in China, aiming to build beautiful cities and create a better life.

In the field of urban underground space development, CRCC adheres to a problem-oriented, demand-oriented, and goal-oriented approach. Through theoretical innovation, technical research, patent layout, and demonstration applications, they have established comprehensive technical systems covering all elements of detection, planning and design, safe construction, intelligent monitoring, and smart operation and maintenance in urban large underground spaces, urban underground space network expansion, and deep space development. Numerous invention patents have been granted, forming a series of technical standards and methods that are crucial for solving pain points in traditional urban underground space development and utilization, addressing the hot issues of people's aspirations for a better life, and systematically enhancing the quality and safety of urban underground space construction and operation in China.

Based on the research outcomes, we have organized and compiled this series of "Key Technologies Series for Urban Underground Space Development and Utilization." The aim is to comprehensively introduce various scientific and technical issues related to the development and utilization of urban underground space from different perspectives, including development philosophies, planning and design, risk management, processes and methods, key technologies, and typical project cases. This series comprises a total of 8 volumes:

1. *Urban Underground Space Development and Utilization*
2. *Key Technologies for Networked Expansion and Renovation of Urban Underground Space*
3. *Analysis of Engineering Cases for Networked Expansion of Urban Underground Space*
4. *Risk Assessment of Construction Safety in Urban Large Underground Spaces*
5. *Safety Construction Technology of Prefabricated Integrated Structure for Pipe Roof*
6. *Investigation and Analysis of Underground Spaces in Japan*
7. *Research on Civil Defense Engineering Planning and Design for Urban Underground Space*
8. *Future Development Concepts of Urban Underground Space-Green, Human-Centric, Intelligent, Resilient, Networked.*

This series of books not only represents the summary of outcomes from major national research projects but also encapsulates the extensive practical experience of China Railway Construction Corporation (CRCC) in urban underground space engineering. We strive to connect theory with practice, summarizing, refining, and elevating our understanding through real-world applications. We sincerely hope that this series of books can serve as a reference for researchers, builders, and decision-makers engaged in the development and utilization of urban underground space. It is intended to be a valuable resource for students and educators in relevant disciplines at higher education institutions. The viewpoints presented in the series are the perspective of one organization, constrained by its expertise, and may not cover all aspects or may even contain inaccuracies. Constructive criticism and feedback from peers are earnestly welcomed to address any shortcomings.

Beijing, China
June 2021

Sheng-Xiang Lei

Contents

Chapter 1
Introduction

With the continuous acceleration of the process of urbanization in China, the accompanying "urban syndrome" is increasingly prominent. Issues such as crowded living spaces, traffic congestion, ecological imbalance, and environmental degradation are common, and the contradiction between the rapid expansion of urban space demand and the limited ground space is increasingly prominent. Urban expansion, increase of living space, rational use and optimization of land resources have become the key problems to be solved in the process of urbanization. Looking at the successful experiences given by developed countries in the world in solving these problems, demanding space, land, and resources from the underground has become one of the inevitable trends of modern urban development.

The 13th five-year Plan for the Development and Utilization of Urban Underground Space compiled by the Ministry of Housing and Urban–Rural Development of the people's Republic of China in May 2016 points out that the rational development and utilization of urban underground space is to optimize the urban spatial structure and management pattern. The necessary measures to alleviate the shortage of urban land resources are necessary to promote the transformation of cities from extension expansion to connotation promotion, improve urban environment and build livable cities. It has great significance to improve the comprehensive carrying capacity of the city. In December 2020, the Ministry of Housing and Urban–Rural Development of the people's Republic of China issued the guidance on strengthening the Construction of Urban Underground Municipal Infrastructure, which is helpful to strengthen the systematic construction of urban underground municipal infrastructure. Specific arrangements have been made to speed up the improvement of the standard of management system, make up for the shortcomings of planning and construction and safety management, promote the modernization of urban governance system and governance capacity, and improve the level of urban safety and comprehensive carrying capacity.

Under the high attention and policy support of the state, the process of planned and rational development of urban underground space is continuously advancing,

S.-X. Lei et al., *Safety Risk Assessment of Urban Underground Large Space Construction*, https://doi.org/10.1007/978-981-97-7437-1_1

and the construction scale of underground transportation networks, underground commercial complexes, underground public utilities, and underground civil defense engineering are all being carried out in an orderly manner. By the end of 2020, the total length of subway lines in operation in mainland China (excluding Hong Kong, Macao and Taiwan) reached 6280.8 km. The Shenzhen Futian comprehensive underground transportation hub, which was completed at the end of 2015, is the first fully underground railway station located in the city center in China. It is a three-dimensional integrated transportation hub integrating high-speed railways, intercity railways, subway transportation, buses and taxis. Through three-dimensional layered arrangement, it realizes the rapid transfer of high-speed trains and subways, with a total building area of 147,000 m^2. The underground intelligent parking garage on Changsha Du Zheng Street can provide 427 parking spaces, and the depth of the underground parking garage is 40 m. The public parking lot of Puyang Hospital of traditional Chinese Medicine in Henan Province adopts wellbore underground three-dimensional mechanical parking mode, and the area area of underground construction is up to 472.70 m^2. The development area of the underground space in the Shanghai Hongqiao underground commercial service area has reached 2.6 million m^2. Its blocks are connected by 20 underground passages and hubs to the underground passage of the National Convention and Exhibition Center (Shanghai), connecting all underground spaces, comparable to the Montreal Underground City in Canada. The above-mentioned underground space engineering practice cases further prove that the development and rational use of underground space has great practical significance for solving the contradiction between urban space demand and land resources and governing “urban syndrome”.

The current theme of urban underground space exploitation is developing in the direction of “overall planning, scientific layout, comprehensive functions, interconnection, intelligent integration, and multi-dimensional space”. At present, the new types of underground engineering, based on scientific research, have continuously made breakthroughs in terms of usage functions, spatial scale, structural form, etc., laying a solid foundation for further promoting the orderly development of urban underground space, building ecological cities, and guiding targeted research and development and industrial upgrading of related theories and technologies.

As the saying goes, “It is better to prepare for the rain than to dig a well before thirst.” The development of urban underground spaces cannot be separated from advanced top-level design and solid theoretical reserves. In 2014, General Secretary Xi Jinping first proposed the concept of overall national security at the Central National Security Committee. The overall national security concept clarifies the dialectical unity of development and security, that is to say, security is the premise of development, development is the guarantee of security, and development and security complement each other and cannot be neglected. In the development process of urban underground spaces, we must also deeply understand the important thoughts of the General Secretary’s overall national security concept, ensure safety in development, promote development in safety, and properly handle the relationship between development and security. Urban underground space is mostly original, leading, and challenging key urban project sand every stage of its life cycle will face great uncertainty and pose threats to safety. Therefore, the construction of urban underground spaces should follow the principle of "safety front", and safety considerations should be integrated into every link of construction. In the whole life cycle of urban underground space, the construction phase is undoubtedly the stage with the strongest uncertainty and the highest safety risk. Due to the large scale of the project and the complexity of the geology and surrounding environment, the construction difficulty of urban underground spaces is greatly increased compared to general underground engineering, and the construction safety risk is sharply increased. Improper treatment is very easy to cause catastrophic accidents, resulting in serious economic losses and social impacts. It can be said that construction safety issues are the absolute dominant factor affecting the overall safety situation of urban underground space construction.

In the final analysis, to solve the construction safety problem of urban underground space, it is necessary to innovate in design theory, construction technology, and construction management. But before this, we need to have a clear understanding of construction safety risks. Construction safety risk assessment is the basis for formulating construction plans and management strategies, and it occupies a pivotal position in construction safety management work. The construction safety risk of urban underground spaces also has unique characteristics different from general underground engineering, and it needs matching risk assessment theories and methods. This book systematically introduces the basic theory of construction risk assessment, sorts out the core issues that need to be resolved in the construction risk assessment of urban underground spaces, and conducts in-depth research around these issues, which could provide theoretical support for the construction safety risk assessment of urban underground spaces in our country, improve the quality of risk assessment work, and ensure the safe and efficient development of urban underground spaces.

1.1 Development Status of Urban Underground Space

1.1.1 Development Process and Main Achievements of Urban Underground Space

1. Foreign Urban Underground SpaceAs the birthplace of the Industrial Revolution, the UK is one of the earliest countries to make use of underground space. In 1861, the world's first underground comprehensive trench was built in London, introducing gas, water supply, sewage pipes, and residential pipelines into underground. The first subway line in the world was built in 1863 (Fig. 1.1), and the first underground postal transportation system was built in 1927. Japan, due to its small territory and dense population, built the world's first underground commercial street in Tokyo in 1934, opening the prelude to a comprehensive exploration of underground comprehensive development and utilization. With the comprehensive advancement of urban underground space development, underground space as an effective supplement to

urban surface space has been fully developed, including underground transportation, underground public utilities, underground municipal engineering, etc.

(1) Underground Transportation

The urban underground transportation system includes dynamic traffic (such as subways, underground expressways, underground pedestrian systems, etc.) and static traffic (such as underground parking lots). As an important measure to effectively alleviate urban surface traffic congestion, subway has become an inevitable choice for mega-cities in the world to ask for underground space. As of the end of 2016, the total mileage of subway operations in the world exceeded 6000 km, of which there were 52 large cities with more than 100 km, and the subway transportation accounted for 40–60% of the total urban transportation (Table 1.1).

The underground pedestrian system is laid out in conjunction with subway transfer points and underground commercial centers, which effectively solves the connection

Fig. 1.1 The world's first subway: London Metropolitan Subway. Image source from the Internet

Table 1.1 Lists the subway line statistics of some large foreign cities

City	Total length of netwoek(km)	Number of lines(lines)	Earliest opening date
New York	443.2	27	October, 1904
London	408.5	12	January, 1863
Moscow	313.1	12	May, 1935
Tokyo	304.5	13	December, 1927
Seoul	286.9	10	August, 1974
Madrid	284.4	13	October, 1919
Paris	220.6	16	July, 1900
Mexico City	201.7	11	September, 1969

problem between underground traffic, commerce, and surface space, and is widely used worldwide. The most representative ones are the Montreal underground pedestrian system in Canada and the Tokyo underground pedestrian system in Japan. The underground pedestrian system effectively connects subway stations, underground parking garages, underground shopping malls, and underground logistics warehouses, realizing the high integration of urban center resources, and is becoming an important part of the development of large city underground transportation.

With the rapid increase in the number of cars in cities, people's demand for parking space is becoming more and more vigorous, and the difficulty of urban parking has also become a common problem in the development of major cities around the world. Large-scale underground parking lots play a good role in alleviating this problem, such as France completed the planning of 41 dedicated underground parking lots in 1954, with more than 50,000 parking spaces; Japan planned underground public parking lots in large cities systems in the 1970s, which greatly relieved the pressure on surface parking space.

The inconvenience and high cost of underground parking have become important bottlenecks restricting its development. Countries around the world have made many explorations to solve the above problems, such as the construction of underground highways in central London, UK, and all underground parking lots are built on both sides of the highway. Through mechanical delivery, rapid parking is realized to improve parking efficiency. In addition, with the improvement of intelligence level, intelligent parking traffic systems (ITS), electronic parking fee systems (ETC), underground parking location guidance systems (DGS), etc. have been widely used in foreign parking fields, providing new ideas for the efficient development of underground parking systems.

(2) Underground public utilities

With the expansion of urban scale and the increase in intensification, many urban public service facilities (including underground commerce, underground cultural and sports entertainment facilities, underground science and education facilities, underground storage and logistics, etc.) have moved underground and have made great progress.

Underground commerce initially started in Japan and has been widely used around the world. At present, Tokyo, Japan has built 14 underground commercial streets, with a total area of 223,000 m^2; Nagoya has more than 20 places, with a total area of 169,000 m^2; and there are 26 underground streets in Japan with an area greater than 10,000 m^2. In addition, guided by underground commerce, a large number of multifunctional underground complexes covering transportation, commerce, culture and entertainment, and sports have appeared around the world. For example, the Montreal Eaton Center underground complex system in Canada (Figs. 1.2 and 1.3) includes subway stations, underground passages, underground public squares, underground commercial bodies, underground cultural and entertainment facilities, and is connected to 50 buildings above ground; the Le Halles underground complex in Paris, France is laid out in 4 layers, with a total construction area of more than 200,000 m^2, and through effective connection with public transportation, the organic integration of underground commerce and transportation is realized.

Furthermore, in order to fully utilize the advantages of constant temperature, humidity, sound insulation and no air pollution in underground space, more and more public service facilities are choosing to move underground, such as underground public libraries, underground museums, underground hospitals, underground research laboratories, underground fitness centers, underground warehouses, underground substations, etc. For example, Norway has fully utilized the advantages of caves and built the world's first Lillehammer Olympic underground stadium; in 2014, Singapore built the Jurong Island underground oil storage depot (Fig. 1.4), which cost 950 million Singapore dollars and is located 150 m below the seabed, making it the deepest underground public facility project in Singapore to date. Underground logistics systems have also been widely discussed in recent years, such as Germany, which since 1998, has built an underground logistics distribution system from the Ruhr industrial area. This system is about 80 km long and uses CargoCap to automatically load and unload containers, achieving efficient logistics distribution. But on the whole, under the influence of traditional concepts, the construction of underground public utilities still needs to be improved around the world.

(3) Underground Municipal Utilities

Underground municipal utilities serve the clean and efficient operation of the city, which belongs to the urban "Lizi" project, including underground municipal pipelines, common ditches (comprehensive pipe corridors) and so on. In the construction process of international metropolises, the construction of underground municipal utilities has always been highly valued and many classic masterpieces have been achieved. For example, the underground drainage system in the old town of Paris, France, has a total length of 2347 km, far exceeding the scale of the Paris metro, and still plays an efficient role after hundreds of years, which can be called the representative of the city's "conscience project" and provides an important foundation for the process of Paris as an international metropolis.

Since the twentieth century, with the increase in urban concentration, the requirements for underground municipal utilities have become higher and higher, and the municipal pipe network system (such as water supply and drainage, sewage, electricity, communication, heating, gas, etc.) has been concentrated in the same ditch to improve the efficiency and convenience of public utility management, which is called "common ditch". In the 1920s, Tokyo built the first underground common ditch under the Kudan regional arterial road in the center of the city, and concentrated municipal utilities such as electricity, telephone lines, water supply, and gas pipes in a single pipe ditch. In 1963, the Japanese government promulgated the "Law on Special Measures for the Construction of Common Ditches", which required the construction of common ditches to accommodate a variety of municipal public facilities under the main arterial roads with large traffic flow and vehicle congestion, and provided legal protection.

At present, Japan has built a total of 2057 km of underground common trenches in nearly 80 cities including Tokyo, Osaka, Nagoya, Yokohama, and Fukuoka, playing

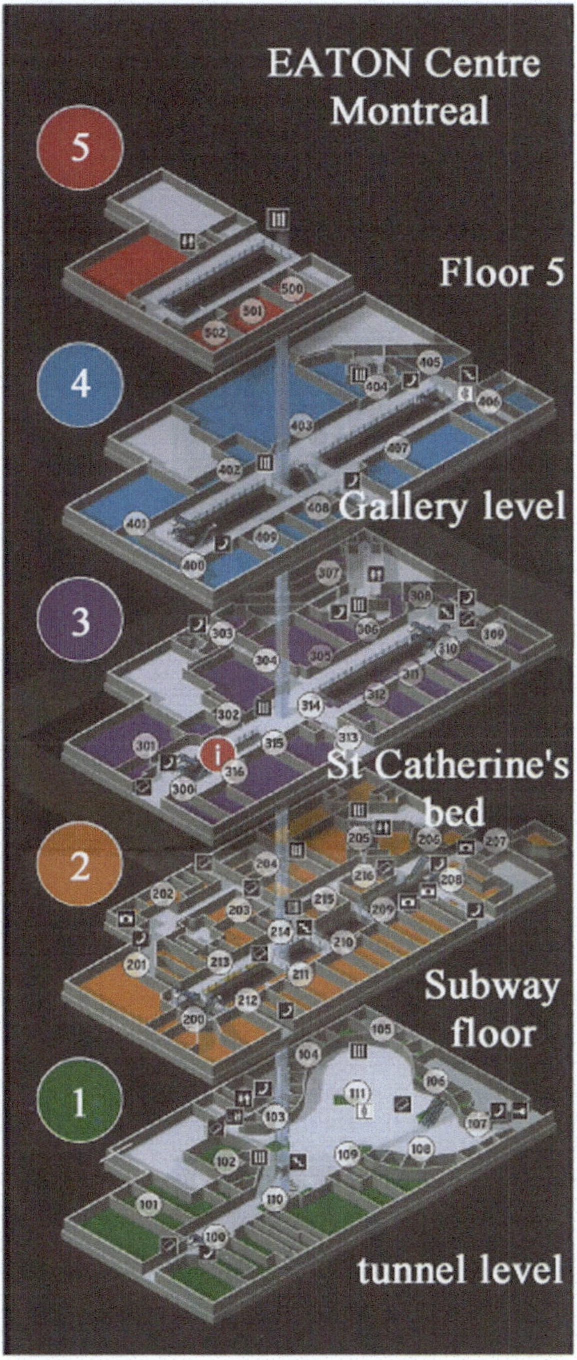

Fig. 1.2 Montreal Eaton Underground City Layout. Image source from the Internet

Fig. 1.3 Interior view of Montreal Eaton Underground City. Image sourced from the internet

Fig. 1.4 Singapore Jurong Island underground oil storage depot. Image sourced from the internet

an important role in the modern and scientific construction and development of Japanese cities. In addition, in recent years, Japan has organically combined the waste classification collection system with the construction of common trenches, improving the effective service level and service life of the common trenches.

In addition, developed countries such as the United Kingdom, France, Germany, Spain, the United States, Singapore, and Russia have also paid great attention to the construction of underground integrated pipe gallery projects in the process of large city construction, and have completed many effective works. In recent years, with the advancement of the internet and information technology, modern information technologies [such as artificial intelligence, remote sensing, building information modeling (BIM), geographic information system (GIS), augmented reality/virtual reality technology (AR/VR) technology, etc.] are gradually being used for the whole process monitoring, early warning, and regulation of pipe gallery operations, more comprehensively ensuring their efficient operation.

(4) Underground civil defense engineering

As an important protection project to resist the destruction and destruction with certain weapon effect and protect the safety of people's lives and property in wartime, underground civil defense engineering runs through the whole process of human modern urban construction. After the First World War, some Western countries paid great attention to the combination and utilization of civil defense projects and urban underground space. For example, Stockholm, Sweden, began to fully construct urban air defense shelters in 1938 to ensure that urban residents could quickly enter protective facilities. The completed "three-defense" Stockholm underground hospital, with a total area of 4000 m^2, can protect 3000 people or accommodate 150 beds during wartime, and can simultaneously play the dual effects of civil defense and emergency rescue; its completed high-speed train tunnel connecting the city center and the airport, which is 40 km long, can achieve rapid evacuation of the city population during wartime, and is a typical civil defense project for both peacetime and wartime. At present, the construction of urban civil defense projects in countries such as the United States, Russia, the United Kingdom, France, and Germany has reached a considerable scale. Among them, Russia has built multiple shock-resistant underground civil defense engineering in major central cities and important industrial areas; Norway, taking advantage of its natural cave features, has built multiple cave-type shelters in large cities including Oslo, which can effectively ensure the safety of residents during wartime.

The advancement of new types of war weapons (such as ground-penetrating missiles, precision-guided weapons, etc.) poses new challenges to traditional civil defense projects. Civil defense projects located in the shallow underground can no longer meet the real needs of modern warfare, and the development towards "deep ground, special civil defense construction" has become a global issue for urban civil defense projects.

2. Urban Underground Space in China

In the early days of the founding of New China, the use of urban underground space began with the construction of civil air defense projects. According to the guiding ideology of "digging deep holes, storing grain widely, not seeking hegemony, preparing for war and famine for the people" proposed by Chairman Mao Zedong in the early 1970s, cities across the country began a relatively extensive construction of people's air defense tunnels. Later, according to the evolution of the actual situation, economic construction became the main theme of society.

In 1978, the central government proposed the civil defense construction policy of "comprehensive planning, focusing on key points, combining peace and war, and quality first", but the use of urban underground space at this stage was only limited to civil defense construction. At the national symposium on the combination of civil defense construction and urban construction in 1986, the idea of "combining peace and war" for civil defense projects was proposed. Civil defense projects should be combined with urban construction, which pointed out the direction for the development of urban underground space utilization. With the rapid development of urban construction in our country, in order to alleviate the increasingly serious traffic congestion problem, major cities have carried out vigorous subway construction, which has laid a solid technical foundation for the development of urban underground space utilization. Since entering the twenty-first century, the problem of "big city disease" tends to be obvious, and the demand for the comprehensive development and utilization of urban underground space has reached an unprecedented height. New underground space utilization facilities, including underground complex, underground hub, underground block and so on, continue to emerge, which opens a new chapter for the comprehensive utilization of urban underground space in our country. (1) Underground Transportation.

The first subway in our country was the Beijing Subway built in 1965, and the Shanghai Subway, which opened in 1993, is currently the largest and longest subway system in the world. By December 2020, 45 cities in China have opened subways, and perfect subway rail transit networks have been built in Beijing, Shanghai, Guangzhou, Shenzhen and other first-tier cities. Nanjing, Chongqing, Wuhan, Chengdu and other cities have completed the basic rail transit network, while urban rail transit backbone lines such as Nantong, Shijiazhuang and Lanzhou have been basically completed. To raise the overall level of China's subway rail transit to a new level. The rapid development of China's rail transit has also driven the development of large underground transportation hubs. The Beijing West Railway Station underground transportation hub integrates railway stations, subways, buses, parking lots, and commercial plazas, effectively alleviating the previously congested situation at the West Railway Station.

The Shenzhen Futian comprehensive underground transportation hub, completed at the end of 2015, is the first fully underground railway station located in the city center in China. It integrates high-speed railways, intercity railways, subway transportation, buses, and taxis into a three-dimensional transfer hub. With a total construction area of 1.47 million m^2, it is currently the largest underground transportation hub project in Asia (Fig. 1.5). In order to better preserve the original features of lakes, rivers and hills in the urban pattern, some big cities choose underground city tunnel crossing to complete, and also put forward new ideas for the development and utilization of urban underground transportation. For example, the Wuhan East Lake Tunnel, with a total length of about 10.6 km, passes under the East Lake Scenic Area and is currently the longest lake tunnel in the city. Similar projects include the Hangzhou West Lake Tunnel, Nanjing Xuanwu Lake Tunnel, Yangzhou Slender West Lake Tunnel, etc. For example, the mountain city of Chongqing effectively connects the main urban area through the Huayan Tunnel, the Two Rivers Tunnel, etc., achieving interconnection of urban functions; the Hangzhou Purple Tunnel, with a total length of 13.9 km, is the longest city tunnel in the country so far.

The difficulty of urban parking is becoming a common problem in the development of large cities in China. The construction of underground parking garages is also in full swing in major cities. For example, the Changsha Du Zheng Street underground intelligent parking garage can provide 427 parking spaces and adopts

Fig. 1.5 Shenzhen Futian underground comprehensive transportation hub. Image source from the Internet

"intelligent parking, App reservation payment" and other modes to achieve convenient and fast parking. In addition, the underground parking garage is as deep as 40 m; the public parking lot project of the Henan Puyang Traditional Chinese Medicine Hospital adopts the well-type underground three-dimensional mechanical parking method, with an underground construction area of 472.70 m^2, actively exploring the use of underground static traffic space. In addition, the depth of the underground parking garage is 40 m, and the public parking lot of Henan Puyang traditional Chinese Medicine Hospital adopts wellbore underground three-dimensional mechanical parking, with an underground construction area of 472.70 m^2, actively explore the utilization of underground static traffic space. In addition, the "sharing economy" model that has sprung up in recent years in China has also made a preliminary development in the field of urban parking, and the efficient use of underground parking spaces through the "Internet + parking" model is becoming another important measure to improve parking difficulties.

(2) Underground public utilities

With the rapid development of China's underground transportation engineering, underground public utilities have also been rapidly improved in the twenty-first century. At present, China has built more than 200 underground complexes with an area of over 10,000 m^2, among which, the underground space development area of the Shanghai Hongqiao underground commercial service area reaches 2.6 million m^2 (Fig. 1.6). The blocks are connected by 20. The underground passage and hub connect the National Convention and Exhibition Center (Shanghai) underground passage, linking all underground spaces, comparable to the Montreal Underground City in Canada. Beijing's Wangfujing combines subway stations and underground malls through the form of underground streets, effectively combining pedestrian systems, sunken gardens, and other space transformation forms to build a three-dimensional underground commercial system in Wangfujing. Our country's underground public utilities (such as underground museums, underground hospitals, underground laboratories, underground entertainment centers, underground warehouses, etc.) are also actively exploring. For example, the Shaanxi Hanyangling Underground Museum is developed and protected in situ, fully integrating modern technology, and intuitively presenting a magnificent underground kingdom. The constant temperature and humidity of the underground also provide innate conditions for medical and health facilities. According to the "Hangzhou Underground Space Development and Utilization Special Plan (2012–2020)", Hangzhou will build a certain scale of underground medical space in the future; in addition, there are a small number of reports on underground research laboratories, underground entertainment and fitness centers, and underground storage centers developed from early residential basements and civil defense projects. However, overall, there is still a large room for development in the comprehensive use of underground space for public service facilities in our country.

(3) Underground municipal public works

Every summer and autumn, many cities in our country often experience "urban waterlogging", and there is a trend of spreading; in addition, due to the impact of pipeline management and repeated excavation and maintenance issues, "road zippers" have always been a stubborn problem in our country's urban construction. The two major issues are always touching the sensitive nerves of urban builders. Therefore, our country has also accelerated the pace of urban underground pipeline engineering construction and improvement.

In 2015, our country issued the "Guiding Opinions on Promoting the Construction of Urban Underground Comprehensive Pipe Galleries" (Guobanfa [2015] No. 61), pointing out: by 2020, a batch of internationally advanced underground comprehensive pipe galleries will be built, significantly improving the "road zipper" problem, enhancing pipeline safety levels and disaster prevention and mitigation capabilities, gradually eliminating spider-web overhead lines on main streets, and improving the

Fig. 1.6 Shanghai Hongqiao underground commercial service area. Image source from the Internet

urban ground landscape. In addition, in 2015, the Ministry of Housing and Urban–Rural Development issued the "Technical Specifications for Urban Comprehensive Pipe Gallery Engineering" (GB 50838–2015), which stipulated the technical requirements and standards of the laying and installation of urban engineering pipelines that contain water supply, drainage, rainwater, sewage, reclaimed water, electricity, communications, gas, and heat, etc., providing technical support for the construction of our country's comprehensive pipe gallery projects. Subsequently, 36 large and medium-sized cities across the country have successively launched pilot projects of underground comprehensive pipe galleries, providing important references for subsequent urban underground comprehensive pipe gallery construction. By the end of 2016, the total mileage of urban underground comprehensive pipe gallery construction that had already started in our country had reached 2005 km, and it was still maintaining rapid development. For example, the underground comprehensive pipe gallery under construction in Tongzhou Caoyuan South Street and Yirui East Road in Beijing, with a total length of 3.8 km, is currently the most comprehensive underground pipe gallery under construction in Beijing; the main and branch pipe gallery under construction in Xi'an, with a total length of 73.13 km, is the largest underground comprehensive pipe gallery project in our country. In addition, some problems that have gradually emerged in the rapid progress of our country's underground comprehensive pipe gallery construction, such as: ① the need for further refinement of related standards and norms for pipe galleries (such as pipeline system planning, layout in the gallery, pipe material and diameter, etc.); ② the connection problems between pipe gallery planning and municipal pipeline planning ③ the compatibility issue of multiple pipelines entering the gallery; ④ the coordination issue of special pipeline elevation and pipe gallery elevation. With the continuous development of our country's underground comprehensive pipe gallery construction technology, the above problems are expected to be gradually resolved.

(4) Underground civil defense engineering

In the early days of New China, the construction of civil defense projects lacked overall planning and design, adopting a mass line approach called "create as you go, design as you go, build as you go". This approach was disjointed from the overall layout and urban development, leading to the waste of many underground spaces. Subsequently, the policy of "combining peace and war" was fully implemented, through the idea of coordinated development of urban construction and civil defense projects. However, there was still a lack of systematic understanding of the development and utilization of urban underground space at that time, and serious problems still existed in the coordinated planning and design of "combining peace and war". Entering the twenty-first century, with the increasing demand for the development of urban underground space, the idea of overall planning and coordinated development of underground space has gradually gained acceptance. This has led to the three-dimensional design of urban traffic hubs, important infrastructure, underground parking lots, underground shopping malls, etc., and the conduct of special anti-weapon damage and war refuge survival probability assessment analysis, truly reflecting the dual role of civil defense engineering. For example, in 2002, the underground civil defense project of the South Square of Shanghai Railway Station, with an area of over 22,000 m^2, was put into use. It can park 561 cars during normal times, effectively alleviating the parking difficulty in the railway station area. By the end of 2016, the area of civil defense projects developed and utilized in combination with peace and war nationwide had exceeded 100 million square meters, with an annual output value of more than 30 billion yuan, a total profit of more than 12 billion yuan, and providing about 7 million jobs. In addition, in recent years, civil defense engineering has shifted from simply preventing "war disasters" to preventing "man-made disasters (including war disasters)" and "natural disasters", strengthening its disaster prevention breadth and depth, and further expanding the core functions of civil defense engineering such as emergency refuge and mobile evacuation. At the same time, with the increase in the comprehensive utilization of urban underground space in our country, the discussion on the diversified integration planning led by large—scale underground complexes has become a new proposition, that is, the civil defense function is integrated into the diversified underground complexes, to realize the strong civil defense function in the development of urban underground space, and to ensure the multi-function of refuge, rescue and mobile evacuation in wartime.

But in all, the peacetime use of civil defense engineering in our country started relatively late, the use form is relatively single, and the types of underground facilities are relatively few. Compared with foreign countries, there is a large gap in terms of technical scale, economy, and social and environmental benefits.

1.1.2 Main Problems in the Development of Urban Underground Space

Based on the above analysis of the development status at home and abroad, the development and utilization of urban underground space has made significant progress, but for our country, it is still in the early stage of its development, and there are many scientific problems that need to be solved urgently. Compared with the development status of developed countries abroad, it can be summarized in detail as follows.

(1) The development and utilization of underground space in various regions and cities across the country is unbalanced, and the understanding of the utilization of underground space is insufficient. First of all, there is a huge gap in the development of urban underground space between the eastern and western regions. According to statistics from Chen Zhilong and others, the total investment in underground space development in the eastern provinces in 2015 was 1.43 times that of the west, and the development of underground space in the east has turned to the construction of public service facilities such as underground municipal engineering, while the west is still dominated by the development of underground rail transit. Secondly, the overall planning, development scale, construction and operation level, and public awareness of underground space

in second and third-tier cities are still low, and the current use of underground space is still limited to single-line planning around underground transportation, civil defense facilities, and other facilities. Underground space is a typical non-renewable resource, once developed and utilized, it will not be easy to reuse, and the later resource recovery and remedial protection will cost a high price. Therefore, establishing forward-thinking thinking and overall planning and design is an important proposition for the future use of urban underground space.

(2) The utilization form of underground space is single, the development layout is scattered, and the comprehensive management mechanism is immature. At present, the development and utilization of urban underground space involves civil defense, construction, municipal, environmental protection, electric power, transportation, communications and many other departments. Since the state has not clearly defined a comprehensive management agency to manage it uniformly in the system, and there is no unified planning for the use of underground space between various departments, therefore, there is a lack of unified coordination mechanism among various departments in development and management, and they are in a state of fragmentation and self-fighting, forming a situation where multiple management and no one manages coexist. For example, the construction process of subways often does not take into account municipal pipelines, communication cables, and other functions causing the development and utilization of urban underground space to become a free, short-term behavior lacking long-term planning. In addition, due to the concealment of underground engineering, their related information is not as easy to obtain as above-ground projects. And the data information of different underground spaces is scattered in multiple departments, and there is no clear specific department responsible for the integration of information resources. For a long time, there has been a lack of data sharing, poor communication, inconsistent statistical calibers and standard systems, and the progress of underground space surveys is not smooth, which brings many obstacles to the planning, construction, and management of the development and utilization of underground space.

At present, many cities have made beneficial explorations in the management of the development and utilization of underground space, and have adopted management models dominated by planning management, construction management, civil defense management, safety management, and coordination management, but none of them have achieved unified management. As a result, the construction of underground space is chaotic, and multiple management does not manage. The whole process of planning, design, construction, and completion acceptance of underground space development and utilization projects, as well as all aspects of material and equipment quality safety, do not have a clear subject for full-process and all-round supervision, resulting in blind construction and frequent quality safety accidents. Therefore, orderly control of the development and utilization activities of urban underground space resources, rational layout and overall arrangement of the construction of various underground space functional facilities, is the basic premise for the future development and utilization of urban underground space.

(3) The development of underground space lacks overall unified planning, and lacks layered and coordinated above and below ground. In recent years, although many cities in our country have compiled various types of underground space plans, covering different levels such as overall planning, detailed planning, and urban design, they are still in the exploratory stage as a whole, and the compilation of urban underground space planning generally lags behind construction and development practice.

① The planning organization is not clear, the planning system is not clear, the planning system is not perfect, there is a lack of unified norms, key elements control failure, and then there is poor connectivity of underground space, lack of connection and continuity between adjacent projects, and sporadic, scattered, isolated development and other problems, leading to a general lack of systematic problems in the development and utilization of underground space.
② At present, the development of urban underground space is still concentrated on shallow development (<50 m), and there is a lack of layered planning thinking for different underground structures, resulting in underground transportation, municipal pipelines, and communication cables appearing at the same depth, which often leads to engineering accidents. Therefore, according to the core development functions of different underground structures, the actual layout depth geological requirements, etc., carry out different vertical layered planning research, consider different level accommodation objects and specific development requirements, properly consider the phased development of deep underground space, will be one of the research topics that must be considered in the later development of underground space.
③ As a complete organic entity that links the surface and underground of a city, it is necessary to coordinate the development and utilization of above and below ground spaces. However, due to the limitations of previous urban planning and design, the development of underground space passively adapts to the actual needs of surface buildings, bringing many practical difficulties. In addition, the separately compiled underground space planning is disconnected from the surface planning, and the positioning and layout of underground functions are not closely related to the surface location conditions, land use functions, and structural forms. There is a lack of spatial coordination and connection, and even contradictions, which is also one of the reasons for the poor implementation effect of underground space planning. In planning new areas (such as Xiong'an New Area, Shenshan Cooperation Zone), and new city development zones, it is necessary to establish a concept of linked planning and design.

(4) There is a serious lag in the policies and legislation for the development and utilization of urban underground space. Although China has issued many laws, regulations, government regulations, and normative documents on the development and utilization of urban underground space, many legislative issues have not been resolved:

① There is no authoritative national legislation. The highest legislative body of the country has not formulated special laws or administrative regulations of the State Council on the development and utilization of urban underground space. The "Regulations on the Management of Urban Underground Space Development and Utilization" by the Ministry of Construction is just an administrative regulation of a government department, lacking sufficient permission setting authority, and its legal effect is relatively low. The current version was revised on January 26, 2011, some provisions no longer meet the needs of development, and due to its non-specific provisions and incomplete content, it is difficult to effectively implement in practice.
② The basic ownership issue has not been resolved. The "Property Law of the People's Republic of China" only stays at the macro level, and does not essentially establish the concept of underground space rights and the form of property rights, and does not establish the acquisition method, transfer, mortgage method and ownership management of underground space rights.
③ No special technical standards have been formulated. The standards for underground space construction are not uniform, there are no professional design, construction, and acceptance standards, and they are only implemented in accordance with relevant technical standards.
④ No specific preferential policies have been proposed. The state has given corresponding preferences to civil defense projects, but there are no specific preferential policies for other underground space development and utilization in terms of investment and financing, pricing, taxation and so on.

(5) The level of science and technology is relatively low

China has achieved the localization of shield construction and tunnel boring machines (TBM), and has introduced and digested a variety of advanced materials. However, compared with foreign countries, there are still many gaps, a large number of key equipment and materials depend on imports, the research on mechanisms and performance is not sufficient, and the degree of mechanization and automation is low. There is a lack of construction machinery types, lack of large-scale construction equipment, low product quality, such as cutting continuous wall equipment, micro-tunnel shield construction, water-containing fine sand grouting drilling machine, as well as automatic monitoring technology in informatized construction, remote control and global positioning system (GPS) positioning technology in micro-tunnel excavation, pre-cutting groove and pre-cutting block technology, tunnel excavation and lining synchronous construction (ECL) technology, application of digital excavation technology, etc. At the same time, the localization ability of construction equipment and key materials is low, the life of building materials is short, and the construction technology of large-section and full-section is backward.

(6) The survey methods are relatively backward

In terms of surveying technology, there is still a certain gap between our country and the Western developed countries, mainly manifested in the mechanical physical means of surveying, such as drilling machinery, soil sampling equipment, and testing analysis methods which are relatively backward, automatic survey data processing, urban geographic geological information database and other software technologies which are still in the initial stage. Further research should be conducted on the surveying methods and evaluation methods of regional underground geotechnical geology, such as underground buildings at different depths, underground pipelines, distribution of ancient relics, complex geological sections of linear underground engineering, and modern surveying techniques of underground seepage fields.

1.2 Current Status of Underground Space Risk Assessment Research

1.2.1 Current Status of Underground Space Risk Assessment Method Research

A lot of research work has been carried out by scholars at home and abroad in the field of underground engineering risk management system. Risk analysis originated in the early twentieth century, with representative figures such as EINSTEIN H. H from the United States who wrote many valuable documents. His main contribution was to point out the characteristics and principles of tunnel engineering risk analysis, such as "Geological model for a tunnel cost model" (EINSTEIN, 1974), "Risk and risk analysis in rock engineering" (EINSTEIN, 1996), "Decision Aids for Tunneling" (EINSTEIN, 1998). STUZK (1996) analyzed various complex factors and technologies in large-scale underground engineering construction, pointing out that wrong decisions in underground engineering will lead to unnecessary risks, so it is necessary to study its decision-making process and propose practical methods of risk analysis.

KAMPMANN (1998) used the Copenhagen Metro project as a research object, applied risk assessment technology, proposed more than 10 types of risks existing in metro construction, a total of more than 40 risk factors, and discussed how to fully record the adopted risk assessment process to provide defensible, traceable records, and established a set of risk type tables. Professor J REILLY (2000) analyzed the causes of tunnel construction accidents, proposed that "the construction process of tunnel engineering is a comprehensive process of risk management and risk sharing", identified potential safety risk factors in the tunnel construction process, summarized 4 types of risks, namely risks that cause the tunnel construction period to be extended, risks that cause casualties, risks that cause changes in the construction project budget cost, and risks that cause construction requirements to be unmet. This provides a theoretical reference for risk decision-makers in future tunnel construction. SMITH (2005) faced a variety of geotechnical types in the construction process of Asian subways, and summarized 15 types of risks including 33 risk factors based on practical experience. MOLAG M (2006) provided applicable methods for hazard identification during the design and operation maintenance phase of the tunnel, but did not discuss the safety risk identification method during the construction phase.

SEJNOHA et al. (2009) proposed to use Fault Tree Analysis (FTA) and Event Tree Analysis (ETA) for risk quantitative assessment of tunnel excavation process in the construction of the New Of a Tunnel in the Czech Republic, but only identified risk events and 3 main causes in the paper, and the number of risks identified was too small. BU-QAMMAZ (2009) studied the influence relationship between risk factors related to transnational construction projects, used network analysis method to derive the relative priority of risk factors and applied it in bidding decision-making. This method can help decision-makers estimate the risk level so that they can rank alternative projects according to the risk level, and can define appropriate contingency values after making bidding decisions.

The International Tunneling Association published the "Tunnel Risk Management Guide" in 2004, providing theoretical guidance on risk identification and assessment for tunnel and underground engineering practitioners. In the same year, the "Underground Engineering Risk Management Joint Specification" was introduced in the UK. The specification comprehensively analyzes the risk analysis methods for underground engineering and the factors to consider during application. Its purpose is to provide a series of theoretical guidance and norms for risk management in underground engineering. As can be inferred from the above, foreign research on risk analysis theory and application in subway tunnels and underground engineering is quite mature. AL-BAHAR proposed in 1988 that risk classification should be based on risk source factors, with a total of 6 risk categories: force majeure risk, financial and economic risk, natural risk, political and environmental risk, design analysis, and construction risk. CHOI HYUN-HO simplified it in 2004 to four types: force majeure analysis, financial and political risk, design-related risk, and construction-related risk. British scholar TAH JHM classified risks into internal and external risks from the perspective of management resources and controllability. Internal risks refer to the risks of the internal systems of the parties involved in the project, and external risks refer to the risks caused by the external environment. The Project Management Association of the United States, in order to standardize risk classification, has set the following three methods for risk classification: one is to classify according to the knowability of the risk, which can be divided into known risks, known unknown risks, and unknown risks; the second is to classify according to the impact of the risk on the project, which can be divided into scope risk, quality risk, schedule risk, and cost risk; the third is to classify according to the nature of the risk, which can be divided into insurable risk and business risk. FORTEZA FRANCISCO J (2017) has achieved many results by studying and analyzing risks from the perspective of accidents. He summarized the types of accidents that may occur during subway construction and formulated a large number of risk measures. EMMANUEL CHIDIEBERE EZE (2018) and others emphasized in their study of underground space safety accidents that risks should be studied and analyzed from the perspective of accidents, and the results of the analysis should be used to guide the safety operations of underground spaces. In comparison, research in the field of risk in China is relatively backward. In recent years, with the large-scale development and utilization of urban underground space, risk analysis theory research has attracted great attention from domestic scholars and has made some progress. The pioneers in the field of risk analysis in China mainly include Professor Guo Zhongwei from Tsinghua University. His "Risk Analysis and Decision Making" written in 1987 detailed foreign risk analysis theories and comprehensively reviewed domestic and foreign research results, which still has great reference value today. In addition, Professor Ding Shizhao from Tongji University (2003) analyzed and summarized various risk factors at various stages of subway construction in China, and deeply explored the pros and cons of various risk evaluation models during the subway construction phase. The team of Academician Ding Lieyun from Huazhong University of Science and Technology has successively completed several scientific research achievements such as "Automatic Identification and Visualization Safety Warning System for Subway Construction Risks" (2008), "Research on Safety Risk Identification and Warning Technology for Subway Construction" (2009), "Research on Safety Risk Control for Subway Construction in Complex Environments" (2009), etc. They have authored many papers on underground engineering safety risks, including "Research on Shield Construction Risk of Wuhan Yangtze River Tunnel Project" (2007), "Research on Safety Evaluation Standards for Subway Construction" (2011), "Real-time Sensing and Warning Research on Safety Risks of Yangtze River Subway Liaison Channel Construction" (2013), etc. The subway construction risk automatic identification and visualization safety warning system they developed has been successfully applied in more than 80 subway construction projects in four cities: Wuhan, Shenyang, Zhengzhou, and Kunming, and nine subway lines (including the first subway tunnel crossing the Yangtze River). The team of Professor Huang Hongwei from Tongji University has carried out a large amount of work in the field of underground engineering safety risk research. Representative literature includes "Progress in Risk Management Research in Tunnel and Underground Engineering Construction" (2006), "Research on Risk Management of Subway Projects throughout the Life Cycle" (2006), "Current Status and Progress of Engineering Risk Analysis and Insurance Research" (2010), "Monitoring and Warning Methods for Civil Engineering Risk Visualization" (2015), "Progress in Fine Sensing and Control Research on Safety Risks of Long and Large Tunnel Engineering Structures" (2020), etc. Professor Huang Hongwei's team has conducted systematic research on various issues such as underground engineering risk management, risk assessment, risk control, and risk warning. The research objects cover various types of underground engineering such as urban underground rail transit, long and large mountain tunnels, and shield tunnels. They have developed a comprehensive and practical underground engineering project risk management software, which has been promoted and applied in many projects. The software uses a risk database as the backend data support, takes risk indicators as evaluation standards, realizes the basic process of tunnel risk identification, risk assessment and decision-making, and risk tracking management, and establishes a relatively complete risk management system. Liao Weiquan and Deng Siqian (2004) identified the risk factors affecting the construction progress in "Risk Analysis of Construction Progress", and used statistical data and decision tree methods to estimate and evaluate the duration risk, and found the biggest risk source. Yuan Yong (2005) analyzed the main risk factors in the waterproofing of construction projects and explained the concept of waterproofing risk grading. Li Feng (2007) proposed graded risk prevention measures based on the specific situation of construction and the interference of surrounding factors. Regarding subway station construction, Li Bing from Beijing Jiaotong University (2007) analyzed the construction accidents of subway stations, summarized the construction safety risk factors with each participating entity as the target, used the fuzzy hierarchy analysis method for risk assessment, and proposed preventive measures. However, only a qualitative description of the risk source was carried out, and no classification and quantitative research was conducted on the importance of various risk sources. Zhou Hongbo and others (2006) used the Work Breakdown Structure (WBS) to decompose the work structure of a super-long rail transit foundation pit project in Shanghai, used the fault tree method to identify risk events and risk factors, applied the integrated integrated risk assessment method for risk assessment, and obtained the risk levels of various types of foundation pits. However, the identification of risk factor layers is relatively simple. Chen Taihong and others (2008) summarized the Nanjing Subway 2. The risk units of the first phase of the Line 1 station pit project (such as deep pit excavation risk, internal support system stability, impact of underground continuous wall joints, etc.) have listed the risk triggering factors for four main types of risks, applied the risk index method for risk assessment and proposed pre-control measures, but did not establish a specific correspondence between the risk triggering factors and the basic risks.

Wang Huawei of Southwest Jiaotong University (2008) identified risk events such as object strikes, lifting injuries, falls from heights, and collapses, listed the main hazards of subway station construction, used the accident tree method to analyze the risk of crane hooking, and used the Analytic Hierarchy Process and fuzzy evaluation method to assess the safety status of the construction site, but the analysis of the hazards was not detailed. Zhou Hongbo et al. (2009) introduced the "Work Breakdown Structure and Risk Breakdown Structure" (WBSRBS) to couple the work breakdown structure and risk source breakdown structure of subway pit engineering, and explained the corresponding risk factors or risk events, conducted sensitivity analysis of risk factors, and proposed corresponding preventive measures. Chen Zhong of Southwest Jiaotong University (2009) qualitatively listed the risk environment and risk factors of Chengdu subway shield tunnel construction, summarized 10 risk events, and used the $R = P \times C$ (R represents risk, P represents the probability of risk events, C represents risk loss) method to rank construction risks, but did not establish the relationship between risk and risk factors. Zheng Zhibin (2009) summarized the main risk factors of tunnel construction through expert survey method, used the $R = P \times C$ method to assess construction risks, and used FLAC3D software to simulate the impact of tunnel excavation on nearby bridges and existing subway lines. Zhang Bo (2009) used the basic theory of risk management and the method of combining with subway shield construction practice, used Venn diagrams to identify risk factors in the process of subway shield construction, and applied the improved Fuzzy Analytic Hierarchy Process (FAHP) model to conduct risk assessment of the underground diameter line project from Beijing Railway Hub Beijing

Station to Beijing West Station, verified the practicality of the model, and proposed corresponding risk control measures.

Hou Yanjuan et al. (2009) divided recent Beijing subway construction safety accidents into 5 types according to the main causes of accidents: collapse caused by excessive deformation of strata, sudden disasters of poor geological bodies, construction-induced damage to underground pipelines, poor construction management, and accidents caused by construction equipment and operational technical errors. For each type of accident, corresponding prevention and control measures and plans were formulated, but the risk factors of the accidents were not systematically summarized. Ren Qiang (2010) in his doctoral thesis, considering the impact of Beijing's hydrological conditions and special geological conditions on subway construction, comprehensively and systematically studied and analyzed various risk factors in the process of tunnel construction of Beijing Subway Line 4, established a qualitative or semi-quantitative risk analysis model, and proposed control measures to reduce the risk occurrence rate. Zhai Zhigang (2011) studied the risks of subway construction from two aspects: the risks generated by the safety of subway construction itself and the risks caused by surrounding environmental factors.

Wu Xiangguo (2012) believes that subway risk assessment relies too much on the subjective opinions of industry insiders and is not efficient enough, and has developed an automatic recognition system, which has the advantages of accurate, fast and intelligent risk recognition; Yang Qianhui (2013) proposed to build a WBS-RBS with the background of safety risk management of subway construction projects. The matrix has enhanced the identification of risk factors during subway construction. He Meili (2012) introduced the theory of information entropy into the evaluation method to reduce the influence of human factors on the weight of risk indicators. At the same time, the theory of uncertain measurement was introduced, and an uncertain measurement model was established by combining the two theories, effectively solving many uncertainties in the risk evaluation of underground engineering. Wang (2018) took the shallow buried tunnel of urban subway as the main research object, proposed a dynamic risk rating method for shallow buried tunnel construction based on Bayesian network, proposed an index scale based on triangular fuzzy number, and applied it to the network hierarchy analysis method, established a general method for tunnel engineering risk analysis, and applied it to the engineering application of crossing Beijing railway line in the section of Wuluju-Cishou Temple Station of Beijing Subway Line 6.

From the above content, it can be seen that the application and academic research of urban underground space construction risk have been popular in developed countries such as Europe and America for a long time. However, in terms of urban underground space construction risk research, evaluation and safe construction, although foreign research started earlier, there are still many deficiencies and shortcomings due to the complex factors and unknown factors of urban underground space engineering construction. Compared with foreign countries, the research on safety management and risk assessment of underground urban space construction in China is relatively backward, and the risk assessment research on subway construction risk is even more lacking, and the corresponding application research is also very few. However, with the large-scale development of underground space in China in recent years and the vigorous development of urban rail transit, the field of subway construction has received widespread attention, and scientific researchers from major design institutes, universities and construction enterprises have carried out related explorations and research in recent years.

1.2.2 Current Status of Risk Coupling Mechanism and Theoretical Research

Urban shallow buried large-span tunneling projects, deep large foundation pits, adjacent construction projects and other urban large underground space construction, with long construction cycles, multiple steps, high geological environment uncertainty, complex engineering structures, numerous construction risk factors interwoven with each other, and significant coupling effects between risk factors.

Risk coupling theory has been studied abroad for a long time and is still a hot topic. The fields studied by foreign experts and scholars include aviation, coal mining, enterprises, ecology, construction engineering, subway construction, etc., and the research content involves risk collision models, system dynamics models, human–environment risk coupling system models, etc. GOMATHI S and others proposed a method to describe the concept of interactive coupling based on software development research, and used model examples to verify the feasibility of the method. SAHIN O and others, based on Australia's water supply and demand problems, conducted a study on a series of risk coupling factors affecting water supply problems by establishing a system dynamics model, which helps decision-makers to formulate effective and sustainable water supply strategic plans. MERCAT-ROMMENS C and others formed a decision support tool, which is also a risk management tool, based on the coupling of Multiple Criteria Decision Aids (MCDA) and Geographic Information System (GIS). YODZIS M and others proposed a human–environment risk coupling system model to reflect and evaluate the impact of water pollution on fisheries and human health.

Domestic research on risk coupling has also emerged in recent years, mostly focusing on related fields such as coal mining, ecology, aviation, and subway construction, with varying degrees of complexity in various model algorithms. The fields studied by domestic experts and scholars are similar to those abroad. They have conducted certain research on the classification of risk coupling, the causes, coupling mechanisms, development trends, and decoupling methods after risk occurrence. The coupling evolution of urban large underground spaces should consider comprehensive factors such as people, machines, environment, and management, and couple major risk factors, and conduct dynamic risk assessment according to the construction process. However, there is still little research on the theory and methods of risk coupling in urban large underground spaces in China. The underground large space structure, adopting new construction methods and new structural forms, still has many blind spots in the evolution process and impact range of disasters, and there is also a great construction risk. Wu Xiangguo first applied the N-K model to the statistical probability of subway accident coupling, and later research on coupling simulation in subway construction gradually increased. Zhang Fuqing constructed a regional economic industrial ecological coupling evaluation model and its indicator system. Xu Kui used the finite element method to explore the deformation rules of existing subway stations and the stations themselves using Beijing subway measured data as an example. Hu Xingjun and others constructed a construction project safety risk coupling degree model from the perspective of human factors. Xu Tao proposed a subway construction safety human-object risk coupling system dynamics model and gave risk control suggestions. Chen Mengjie used fuzzy entropy theory to solve the weights of various subway construction safety risks, and calculated the coupling value between the four major first-level risk factors of "people, machines, environment, management" with weights and utility functions. Qiao Wanguan calculated the risk coupling degree of coal mine safety production system based on the coupling mechanism of coal mine production safety risk using the coupling degree model. Ren Zhen established a subway station pit construction risk coupling index system, combined with risk degree, numerical simulation, and coupling degree model to assess the evolution trend of subway deep pit construction risk coupling. Wang Hui constructed a subway construction safety system risk index system that includes "human, mechanical, environmental, management" four major factors, and established a coupling degree model. In summary, existing research mainly studies the coupling effects of risk factors from a management perspective. These theories inevitably have many problems when applied to the complex mechanical process of urban large underground space construction. The coupling of urban large underground space construction risks is not only logical coupling, probability coupling, but also mechanical, physical level coupling. For this coupling effect, no one has yet proposed a clear definition, and there is a lack of related leading research. Therefore, the study of urban large underground space construction risk coupling needs to start from actual accident cases, extract the conceptual core of risk coupling from the accident context, and construct a set of risk coupling analysis methods suitable for urban large underground space construction.

1.2.3 Risk Evolution Mechanism and Theoretical Research Status

Movement is the essence of things, and so is risk. The evolution of risk is not a new topic, but an inherent attribute of risk. However, due to the complexity of the concept of risk itself, people's attention and research on the issue of risk evolution are still in the early stages. Due to different scholars' understanding of the issue of risk evolution and different academic fields, there are significant differences in the theories and methods of studying the issue of risk evolution. In recent years, scholars at home and abroad have conducted a series of related studies on the mechanism of risk evolution and have achieved fruitful research results. Wang Chengtang et al. (2019) used the fault tree method to analyze risks, constructed a polymorphic Bayesian network to calculate the probability of pit collapse risk, and dynamically assessed the risk during the pit construction process based on Bayesian network forward reasoning. Xu Tian (2018) used the WBS-RBS method to construct a risk assessment system for deep foundation pit construction. Li Yicheng (2019) and others introduced the concept of dynamic weight and created a new set of dynamic risk assessment methods. Wang Yongsheng (2011) proposed a risk management theoretical framework from the perspective of "group" based on complex system theory. Wang Fan (2013) constructed a feedforward control model for subway shield construction and established a dynamic evolution model of subway construction safety risk (DRAFTs) by integrating various sub-models. You Pengfei (2013) used the theory and method of system dynamics to systematically analyze the subway tunnel construction process and established a system dynamics model from the mechanical and energy perspectives. Zhao Xianli (2013) constructed a mathematical model of airport flight area risk evolution and a risk evolution topological structure model based on the complex network model, and proposed a risk chain-breaking control scheme. Jiang Xin (2015) used the Cobb–Douglas function to describe construction risk and constructed a system dynamics (SD) model of the evolution of construction risk in engineering project groups. Liu Qing (2016) constructed a system dynamics (SD) model of the evolution of ship navigation risk in the Three Gorges Dam area by analyzing the causal relationship and influence function between the navigation risk of the Three Gorges Dam and its influencing factors. Meng Xiangkun (2017) proposed a semi-quantitative risk evolution evaluation method for pipeline system leakage evolution based on complex network theory. Qin Pan (2018) established a risk evolution theoretical relationship model based on the Structural Equation Model (SEM) method and studied the evolution law of ship navigation safety risk in the Three Gorges Dam area.

In summary, although the principles and methods of studying risk evolution are diverse, almost all researchers use abstract models constructed based on system science and management science to carry out their research, and there is little involvement in the real, physical mechanisms of risk generation and evolution. This leads to the fact that although most models seem to be very perfect in terms of structure and algorithm, they lack the support of objective principles and data, and face many difficulties in practical application.

Chapter 2
Definition of Urban Large Underground Space

2.1 Basic Concept of Urban Large Underground Space

2.1.1 Proposal of the Concept of Urban Large Underground Space

1. New characteristics of urban underground space

In recent years, with the advancement of engineering technology and the enhancement of urban functional requirements, the development of urban underground space has begun to move towards integration, diversification, and integration, giving birth to more and more distinctive, powerful, and high-quality new urban underground space projects. These projects mainly have the following characteristics.

(1) Traditional underground space often exist as appendages of above-ground structures, such as underground garages, underground malls, etc. Although such underground space partially have their own independent functions, they fundamentally serve above-ground buildings. Nowadays, more and more urban facilities are beginning to place underground structures in their primary positions, even using underground space as the main functional areas. Today's urban underground space has become an important part of urban space together with above-ground buildings, and its functions are also developing towards diversification and integration.
(2) The difficulty and cost of developing underground space are much higher than those of above-ground space, so traditional underground space generally have limited volume and are relatively narrow. Some underground space exist only to perform certain specific functions and cannot serve as human activity space. Nowadays, with the development of underground engineering technology, the difficulty of developing underground space is gradually decreasing, and the construction of large-scale underground projects has become possible. In addition, with the gradual exhaustion of available space above ground in cities, urban

S.-X. Lei et al., *Safety Risk Assessment of Urban Underground Large Space Construction*, https://doi.org/10.1007/978-981-97-7437-1_2

residents' activity space are beginning to move underground, more and more large-scale underground projects are appearing.

(3) The characteristics of traditional underground space determine that they are mostly "separate and self-governing" independent units. Nowadays, urban underground space is gradually showing a trend of integration and connectivity. Seamless transfers between underground rail transit have basically been achieved, and the exits of rail transit are increasingly directly connected to commercial facilities, new residential communities have begun to adopt integrated underground structures. It can be said that today's urban underground space are developing in the direction of being accessible in all directions and functionally integrated.

(4) Since it is easier to develop the shallow underground space, most projects were built within 30 m deep under the ground surface in the early stages of urban underground space development, and the underground space was spread out in a planar manner. Nowadays, the city's shallow underground space has become increasingly tight space, the construction depth of new underground space is constantly increasing, the super-deep foundation pits and undercrossing projects frequently appear. At many transfer nodes of underground rail transit, stations and stations, lines and lines have formed overlapping three-dimensional space networks vertically. Driven by the demand for functional diversification, spatial expansion, and network structure, the development of urban underground space will inevitably explore deeper strata in the future.

2. New problems in the development of urban underground space

New characteristics and new demands often come with new challenges. With the continuous enlargement, deepening, and complexity of urban underground space, safety issues in their construction process are increasingly highlighted. Numerous engineering practices have shown that the structural forms and construction methods used by large-volume, deep underground space are vastly different from those of traditional small-volume underground space, and the difficulty of engineering construction and the impact on the surrounding environment increase rapidly with the increase in scale.

This book collects data from 86 open excavation projects and 85 undercut projects in China, and statistically analyzes their spatial volume and construction risk levels, as shown in Fig. 2.1. From the figure, it can be seen that with the increase in spatial volume, the average number of construction risk sources per project shows a nonlinear increase trend, and the risk level gradually changes from Level III to mainly Level I and II. It is not difficult to find that the "quantitative change" in the scale of urban underground space leads to a "qualitative change" in construction difficulty, and the "qualitative change" in construction difficulty further leads to a "sudden change" in construction risk level and characteristics. This puts higher demands on the concept, methods, and system of urban underground space construction safety risk management, and requires more targeted research.

Therefore, based on extensive research on urban underground space in China, combined with the author's experience in many urban underground space engineering

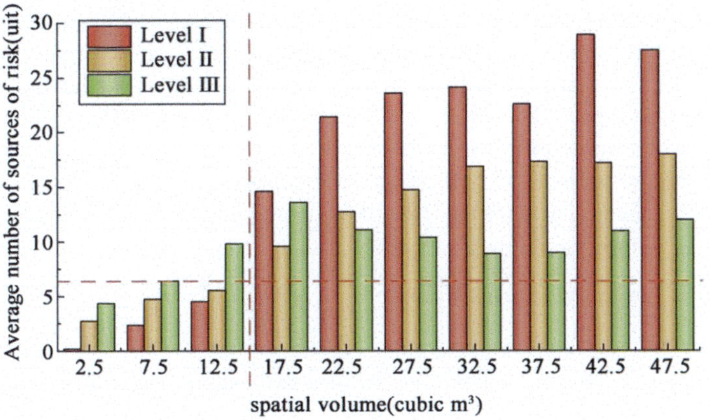

Fig. 2.1 Urban underground space construction risk source statistical results

practices and discussions with many industry experts, the concept of "urban large underground space" is proposed.

2.1.2 Definition of Urban Large Underground Space

Urban large underground space refers to the area below the surface within the administrative region of a city, within the depth range of controllable engineering risk and cost, built to meet specific production, living and disaster prevention needs, with a large structural span (18 m or above) or a certain scale (150,000 m^3 or above) of underground single large space or networked underground space.

2.1.3 Classification of Urban Large Underground Space

Based on the project scale, construction method, connectivity between underground space, importance and proximity of surrounding environmental facilities, urban large underground space are divided into monolithic large underground space by open excavation method, and networked expansion underground large space.

1. Monolithic Large Underground Space by Open Excavation Method

In principle, deep foundation pit projects that meet the following conditions belong to monolithic large underground space by open excavation method.

(1) Underground subway station projects of four floors or more constructed by open excavation method or underground projects of equivalent scale.

(2) Underground subway station projects of three floors constructed by open excavation method or underground projects of equivalent scale, and there are important or above environmental facilities within the secondary impact area.
(3) Underground projects with an excavation depth exceeding 30 m and an excavation volume exceeding 150,000 m^3.
(4) Underground projects with an excavation depth exceeding 25 m, an excavation volume exceeding 100,000 m^3, and there are important or above environmental facilities within the secondary impact area.

2. Monolithic Large Underground Space by Undercut Method

The number of structural layers, span, cross-sectional size, etc. of monolithic large underground space by undercut method projects should meet the following conditions.

(1) Underground subway station projects of four floors or more constructed by undercut method or underground projects of equivalent scale.
(2) Underground subway station projects of three floors constructed by undercut method or underground projects of equivalent scale, and there are important or above environmental facilities within the secondary impact area.
(3) Underground single-span subway station main projects constructed by undercut method or underground projects of equivalent scale, and there are important or above environmental facilities within the secondary impact area.
(4) Underground projects with an excavation cross-section exceeding 500 m^2, and there are significant or more important environmental facilities within the secondary impact area of the underground project.
(5) The excavation section is larger than 300 m^2, and there are important or more important environmental facilities within the secondary impact area of the underground project.
(6) The excavation width of a single span is more than 18 m, and there are important or more important environmental facilities within the main impact area of the underground project.

3. Large Underground Space by Networked Extension Space

Networked expansion projects refer to the requirements for underground space networking. On the basis of existing underground space, through connection, expansion, reconstruction, addition, and construction of underground space, the underground space are organically linked with each other and the surface, forming a networked underground space that is connected in plan and interconnected up and down. The expansion methods of urban underground space are divided into five types: adjacent addition, connection, vertical layering, small to large expansion, and multi-dimensional expansion.

In principle, networked expansion projects that meet the following conditions belong to urban large underground space projects.

(1) For the underground space expanded by vertical layering, the completed project is not less than 3 floors and the height is not less than 20 m of the subway station project or underground project of the same scale.
(2) For the underground space expanded by small to large, the volume of the single space after completion is more than 1 times the original volume and the total volume is not less than the standard 3-layer subway station underground project or underground project of the same scale.
(3) For the underground space expanded by adjacent addition, connection, and multi-dimensional expansion, the volume of the space after completion is not less than 3 times the standard single-layer subway station underground project or underground project of the same scale.

2.1.4 Classification of Impact Areas and Importance of Environmental Facilities for Urban Large Underground Space Projects

1. Division of impact areas for urban large underground space projects

The impact area of urban large underground space projects refers to the strata and urban areas that may be significantly disturbed during the construction process. In most cases, the construction of urban large underground space will form a passive earth pressure area and potential slip surface within a certain range around it. The strata and surface buildings (structures) within this range will produce additional displacement during the construction process. According to the theory of earth pressure, the size of this range is mainly related to the excavation depth and stratum characteristics. For simplicity, the impact areas of urban large underground space projects are divided into three levels in terms of plan range, as shown in Table 2.1.

Table 2.1 The impact areas of urban large underground space projects

Construction method	Approach (main impact area)	Secondary approach (secondary impact area)	General approach (general impact area)
Open excavation method	Within 0.7H (include) of pit perimeter	0.7 H–1.0H (include) of pit perimeter	Outside 1.0H of pit perimeter
Undercut method	Directly above and within 0.7H of the tunnel perimeter	0.70.7H–1.0H (include) of tunnel perimeter	Outside 1.0H of tunnel perimeter

Note H—the distance from the excavation depth of the foundation pit or the invert of the tunnel to the ground

2. Classification of the importance of surrounding facilities

The importance of surrounding environmental facilities is determined comprehensively based on the type, function, usage nature, characteristics, scale, etc. of the environmental facilities, and is divided into three levels: extremely important, important, and relatively important, as shown in Table 2.2.

Table 2.2 The levels scale of environmental facilities' importance

Level	Basic conditions	Basis for amendment
Extremely important	(1) Existing rail lines and railway (2) State-level protection of cultural relics and ancient buildings (3) The iconic buildings of the national cities (4) Airport Runway and Apron, etc.	When the following situations occur, an upgrade by one level is permissible (1) Environmental objects have special protection requirements (2) Environmental objects underneath new urban rail structures (3) Rivers and lakes are hydraulically connected to groundwater (4) In the vicinity of rivers and lakes where seasonal and differential water levels exist and where construction is likely to take place during the flood season etc.
Important	(1) Municipal-level protection of cultural relics and ancient buildings (2) Recent outstanding buildings, important industrial buildings, high-rise or super-high-rise civil buildings with more than 10 floors, important underground structures (3) Gas mains or natural gas mains with a diameter of more than 0.6 m, municipal heat trunks, rainwater and sewage pipe mains (4) Continuous box beams for main bridges of viaducts and overpasses at traffic nodes (5) Urban expressway, highway (6) 500 kV and above high voltage lines (7) Important rivers and lakes	
Relatively important	(1) Relatively important industrial buildings (2) Recent outstanding buildings, important industrial buildings, high-rise or super-high-rise civil buildings with more than 10 floors, important underground structures (3) Urban primary roads, secondary roads (4) 110–500 kV high voltage lines (5) Relatively important rivers and lakes	

2.2 Statistical Analysis of Large Underground Space in Chinese Cities

2.2.1 Open Excavation of Single Large Underground Space

This book has surveyed 86 cases of underground projects constructed by open excavation in 20 cities across the country, of which 71 are large underground space projects. The information of each project is shown in Table 2.3.

2.2.2 Monolithic Large Underground Space by Undercut Method Space

This book has collected data on 85 cases of underground projects constructed by tunneling in more than 10 cities across the country, of which 26 are large underground space projects. The information of each project is shown in Table 2.4.

2.2.3 Networked Expansion of Underground Large Space

This book has collected data on 30 underground engineering projects in 15 cities across the country, of which 24 are examples of networked expansion of large space. The information for each project can be found in Table 2.5.

2.3 Urban Large Underground Space Cases

2.3.1 Wuhan Optics Valley Square Project

1. Project Overview

(1) Project Introduction

The Wuhan Optics Valley Square Project, located under the existing Optics Valley Square in Wuhan's East Lake High-tech Zone, is a comprehensive project that combines rail transit engineering, municipal engineering, and underground public space. The Optics Valley Plaza Complex has set a model for urban renewal and comprehensive development of large underground space. The colorful super-large underground plaza and the rhythmically brilliant "star river" have become a new landmark in Wuhan. The completed Optics Valley Plaza Complex is shown in Fig. 2.2.

Table 2.3 Summary of urban large underground space projects constructed by open excavation

S. No.	Area	Name of station	Number of basement floors	Sizes (m)
1	Beijing	Wangjing Xi (W) Station of Beijing Subway Line 13, Line 15, Line 17	Four basement floors	Length 228 Width 27.7
2	Beijing	Sanyuanqiao Station of Beijing Subway Line 10	–	Length 149.5 Width 23.4 Depth 37.1
3	Beijing	Jiaomen Xi (W) Station of Beijing Subway Line 10	–	Length 86.9 Width 28.2
4	Beijing	Shaoyaoju Station of Beijing Subway Line 10	–	Length 183.4 Width 20.7 Depth 18
5	Beijing	Yongdingmen Wai Station of Beijing Subway Line 8, Line 14	–	Length 139.2 Width 24.7 Depth 32
6	Chengdu	Yushuang Road Station of Chengdu Metro Line 6	Three basement floors	Length 206 Width 22
7	Chengdu	Niuwangmiao Station of Chengdu Metro Line 6	Three basement floors	Length 339.7 Width 22.5 Depth 25.3
8	Guangzhou	Shiliugang Station of Guangzhou Metro Line 18	Three basement floors	Length 567 Width 34
9	Guangzhou	Guangzhou East Railway Station of Guangzhou Metro Line 18	Five basement floors	Length 622.52 Width 26.2
10	Guangzhou	Nancun Wanbo Station of Guangzhou Metro Line 18	Four basement floors	Length 296 Width 35
11	Guangzhou	Pizhou xi(W) Station of Guangzhou Metro Line 18	Three basement floors	Length 439 Width 34.1
12	Guangzhou	Baiertan Station of Guangzhou Metro Line 22	Five basement floors	Unexcavated
13	Guangzhou	Chencunbei Station of Guangzhou Metro Line 7	Three basement floors	Length 188 Width 23
14	Guangzhou	Guangzhou South Railway Station of Guangzhou Metro Line 7	Four basement floors	Length 592.6 Width 25.54 Depth 32
15	Guangzhou	Nanchong Station of Guangzhou Metro Line 7	Three basement floors	Length 176.2 Width 20.1 Depth 23.83
16	Guangzhou	Shaxi Station of Guangzhou Metro Line 18	Three basement floors	Length 465.35 width 34.1 Depth 30

(continued)

Table 2.3 (continued)

S. No.	Area	Name of station	Number of basement floors	Sizes (m)
17	Guangzhou	Panyu Square Station Complex of Guangzhou Metro Line 3, Line 17, Line 18, Line 22	Five basement floors	Length 540 Width 52
18	Kunming	TANGZIXIANG Station of Kunming Rail Transit Line 6	Three basement floors	Length 465.9 Width 22.9 Depth 23
19	Luoyang	Shanghai Market Station of Luoyang Metro Line 1	Two basement floors	Length 275.6 Width 21.2 Depth 26.6
20	Luoyang	Wangcheng Park Station of Luoyang Metro Line 1	Three basement floors	Length 169 Width 24.3 Depth 23.2
21	Luoyang	Zhou Wangcheng Square Station of Luoyang Metro Line 1	Three basement floors	Length 216 Width 21.4 Depth 23.56
22	Luoyang	Qingniangong Station of Luoyang Metro Line 1	Three basement floors	
23	Luoyang	Jiefang Road station of Luoyang Metro Line 1	Three basement floors	Length 315 Width 23
24	Luoyang	Wuhan Road Station of Luoyang Metro Line 1	Three basement floors	Length 160.9 Width 21.9 Depth 25
25	Luoyang	Jichang Road Station of Luoyang Metro Line 2	Three basement floors	Length 223.3 Width 22.1 Depth 33.3
26	Luoyang	Peony Square Station of Luoyang Metro Line 1	Three basement floors	Length 649 Width 84 Depth 20.7
27	Nanjing	Gulou Station of Nanjing Metro Line 1, Line 4	Three basement floors	Length 211.03 Width 49.6
28	Nanjing	Yuantong Station of Nanjing Metro Line 2,Line 10	Three basement floors	Length 219.282 Width 21.2 Depth 19.2
29	Nanjing	Taifenglu Station of Nanjing Metro Line S8	Three basement floors	Length 523.2 Width 22.1 Depth 18.94
30	Nanjing	Jimingsi Station of Nanjing Metro Line 3,Line 4	Three basement floors	Length 178.26 Width 22.3 Depth 19.4
31	Qingdao	Laoshan six middle school Station of Qingdao Metro Line 4	Three basement floors	Length 165 Width 19.5 Depth 28.0

(continued)

Table 2.3 (continued)

S. No.	Area	Name of station	Number of basement floors	Sizes (m)
32	Qingdao	Haipo Bridge Station of Qingdao Metro Line 1,Line 4	Open excavate five floors, concealed excavate two floors	Length 192(57 of open excavation)
33	Xiamen	Hongkeng Station of Xiamen Metro Line 3	One basement floor, two above ground	Length 540.8 Width 42.9
34	Shanghai	Zhenru Station of Shanghai Metro Line 11,Line 14	Three basement floors	Length 383.1 Width 20.44
35	Shanghai	Xujiahui Station of Shanghai Metro Line 11	Four basement floors	Length 204.8 width 21.6
36	Shanghai	Hanzhong Road Station of Shanghai Metro Line 13	Five basement floors	Length 204 Width 21.5
37	Shanghai	Shanghaixi Railway Station of Shanghai Metro Line 11	Three basement floors	Length 98 Width 31
38	Shanghai	People's Square Station of Shanghai Metro Line 1,Line 2,Line 8	Three basement floors	–
39	Shanghai	Century Avenue Station of Shanghai Metro Line 4	Three basement floors	–
40	Shanghai	Basement of Bank of Shanghai Headquarters	Three basement floors	Length 112.5 Width 104
41	Shanghai	Shanghai Expo Transformer Foundation	Four basement floors	Cylinder diameter 130
42	Shanghai	Shanghai Hongqiao Lot 08 D13 Project Foundation	Three Basement floors	Length 313 Width 179
43	Shanghai	Shanghai International Aviation Service Center Project Foundation	Three basement floors	Length 210 Width 180
44	Shanghai	Shanghai Tower Foundation	Five basement floors	Length 200 Width 200
45	Shenzhen	Longyuan Station of Shenzhen Metro Line 16	Two basement floors	Length 192.48 Width 22.4
46	Shenzhen	Laojie Station of Shenzhen Metro Line 1,Line 3	Four basement Floors	Length 129.97 Width 18.75 Depth 20.65
47	Shenzhen	Grand Theater Station of Shenzhen Metro Line 2	Three basement floors	–
48	Shenzhen	Futian Railway Station of China Railway Shenzhen	Three basement floors	Length 1023 Width 78.86 Depth 32
49	Shenzhen	Houhai Station of Shenzhen Metro Line 2,Line 11	Three basement floors	Length 497.8 Width 24.3

(continued)

Table 2.3 (continued)

S. No.	Area	Name of station	Number of basement floors	Sizes (m)
50	Shenzhen	Huaqiang North Station of Shenzhen Metro Line 2,Line 7	Three basement floors	–
51	Shenzhen	Civic Center Station of Shenzhen Metro Line 4	Three basement floors	Length 162.3 Width 22 Depth 16.69
52	Shenzhen	Yinhu Station of Shenzhen Metro Line 6,Line 9	–	Length 313.8 Width 20.4 Depth 32
53	Shenzhen	Tai'an Station of Shenzhen Metro Line 5,Line 7	Three basement floors	Length 621.6 Width 20.4 Depth 32
54	Shenzhen	Shenzhen World South Station of Shenzhen Metro Line 12,Line 20	Two basement floors	Length 364.3 Width 43.3 Depth 15
55	Tianjin	Jintanglu Station of Tianjin Rail Transit Line 5	Two basement floors	Length 176 Width 27.1
56	Tianjin	Liuweilu Station of Tianjin Rail Transit Line 4	Four basement floors	Length 196.2 Width 21.3 Depth 26.86
57	Tianjin	Lujiazui Financial Plaza Station of Tianjin Rail Transit Line 1	Four basement floors	Length 348 Width 154
58	Tianjin	Jingjianglu Station of Tianjin Rail Transit Line 2, Line 5	Three basement floors	Length 504 Width 19.5
59	Tianjin	Xinjiayuan east Station of Tianjin Metro Line B1	One basement floor, one above ground	Length 454.0 Width 21.2 Depth 40.3
60	Tianjin	Siyuandao Station of Tianjin Rail Transit Line 5	Three basement floors	Foundation diameter 827,240
61	Urumqi	Nianzigou Station of Urumqi Metro Line 2	Three basement floors	Length 246.6 Width 23.1
62	Wuhan	Tangjiadun Station of Wuhan Metro Line 1	Three basement floors	Length 227.8 Width 30
63	Wuhan	Sanyang Road Station of Wuhan Metro Line 1,Line 7	Two basement floors	Length 273 Width 26–80
64	Wuhan	Tiandi Yifang Jinjiang Foundation	Three basement Floors	Circumference 1051, average width 90
65	Wuhan	Wuhan Yangtze River Shipping Center	Three basement floors	Length 210 Width 170
66	Wuhan	Xianggang Road Station of Wuhan Metro Line 3, Line 6,Line 7	Three basement floors	Length 290 Width 25.1

(continued)

Table 2.3 (continued)

S. No.	Area	Name of station	Number of basement floors	Sizes (m)
67	Wuhan	Wuhan Greenland Center	Five basement floors	Length 304 Width 121
68	Wuhan	Optics Valley Square Complex	Four basement floors	Diameter about 200 m
69	Xi'an	Mutasi Station of Xi'an Metro Line 6	Three basement floors	Length 189 Width 23 Depth 23.94–22.95
70	Xi'an	Dayanta Station of Xi'an Metro Line 3, Line 4	–	Length 213.05 Width 24.8 Depth 20.63
71	Zhengzhou	Nanyang Road Station of Zhengzhou Metro Line 5	Three basement floors	Length 202 Width 23

The Optics Valley Square is a three-story underground structure with a construction area of about 160,000 m^2 and an earthwork excavation volume of 1.8 million m^2, making it the largest urban underground complex in Asia. It includes:

① Subway engineering: The section from Optics Valley Plaza Station to Luoxiong Road Station on Line 2 South Extension, and Luoxiong Road Station; Optics Valley Plaza Station on Line 9; Optics Valley Plaza Station on Line 11 and its synchronously constructed sections; Transfer passage at Optics Valley Plaza Station on Line 2.

② Municipal supporting projects: Luoyu Road, Lushan Road municipal road tunnels.

③ Underground public space: The underground circular space in the disc area and the underground public space within the construction range of the subway station, subway section, and municipal tunnel are comprehensively utilized.

The locations and overall plan of each line of the Optics Valley Plaza Complex Project are shown in Fig. 2.3.

(2) Structure and Functional Layout

The Optics Valley Plaza Complex Project adopts a three-story multi-span frame structure underground, with a soil cover thickness of 1.0–1.7 m, and the planned ground level is about 1 m higher than the existing ground level. The underground first floor, second floor, and part of the third floor (Line 11), with the bottom plate buried depths of about 14 m, 21 m, and 33 m respectively, are constructed by open excavation method and bored pile enclosure. The function distribution of each floor is as follows.

① Mezzanine of the first underground floor: Lushan Road municipal tunnel, Line 9 platform (140 m long).

Table 2.4 Summary of urban large underground space projects constructed by concealed excavation

S. No.	Area	Name of station	Transfer	Number of basement floors	Sizes (m)
1	Beijing	Panjiayuanxi Station of Beijing Subway Line 17	–	Two basement floors	Length 265.6 Width 14 Height 22.25
2	Beijing	Gongzhufen Station of Beijing Subway Line 10	Line 1	Three basement floors	Length 202 Width 13.45 Height 16.5
3	Beijing	Xidan Station of Beijing Subway Line 4	Line 1	Two basement floors	Length 222.3 Width 22.7
4	Beijing	Dongdan Station of Beijing Subway Line 5	Line 1	Two basement floors	Length 204.4 Width 23.08
5	Beijing	Guomao Station of Beijing Subway Line 10	Line 11	Three basement floors	Length 133.2 Height 12.57
6	Beijing	Gulou Dajie Station of Beijing Subway Line 8	Line 2	Three basement floors	Length 242.2 Height 28
7	Beijing	Yonghegong Lama Temple Station of Beijing Subway Line 5	Line 2	Three basement floors	Length 121.20
8	Beijing	Chaoyang Men Station of Beijing Subway Line 6	Line 2	Two basement floors	Length 188 Width 22.35 Height 23
9	Beijing	Chongwen Men Station of Beijing Subway Line 5	Line 2	Three basement floors	Length 202.9 Width 14 Height 21.8
10	Beijing	Haidian Huangzhuang Station of Beijing Subway Line 4	Line 10	Two basement floors	Length 216.70 Width 23.10
11	Beijing	Ping'anli Station of Beijing Subway Line 4	Line 6	Two basement floors	Length 225.45 Width 25.1 Height 21.15
12	Beijing	Dongsi Station of Beijing Subway Line 5	Line 6	Two basement floors	Length 197 Width 26.2
13	Beijing	Hujialou Station of Beijing Subway Line 10	Line 6	Two basement floors	–

(continued)

Table 2.4 (continued)

S. No.	Area	Name of station	Transfer	Number of basement floors	Sizes (m)
14	Beijing	Shuangjing Station of Beijing Subway Line 10	Line 7	Three basement floors	Length 237.6 Width 23.1 Height 16.15
15	Beijing	Puhuangyu Station of Beijing Subway Line 5	Line 14	Three basement floors	Length 146 Width 22.6 Height 16.3
16	Beijing	Aolinpike Gongyuan Station of Beijing Subway Line 8	Line 15	Three basement floors	Length 205.5 Width 23.3
17	Beijing	Ping'anli Station of Beijing Subway Line 4	Line 6	–	Length 224.5 Width 25.1 Height 21.37
18	Nanjing	Nanjingnan Railway Station of Nanjing Metro Line 1	Line 3	Three basement floors	Length 282.86 Width 24 Height 29
19	Beijing	Dongdaqiao Station of Beijing Subway Line 17	Line 6, Line 28	Two basement floors	Spans 29
20	Beijing	Erligou Station of Beijing Subway Line 16	Line 6	Three basement floors	Length 303 Width 19.8 Height 8.57–9.67
21	Beijing	Niujie Station of Beijing Subway Line 19	–	Three basement floors	Length 239 Width 23.3 Height 23.49
22	Qingdao	Neimenggu Rd Station of Qingdao Metro Line 4	–	Two basement floors	Height 16.18
23	Qingdao	Changle Rd Station of Qingdao Metro Line 4	Line 5	Two basement floors	Width 25.52 Height 19.886 Depth 20.8–24.9
24	Shenzhen	Gongbei Tunnel	–	Two basement floors	Width 18.8 Height 21.0
25	Taiyuan	Yingze Avenue Underpassing Train Station Project	–	One basement floor	Length 462 Width 18.2 Height 10.5
26	Chongqing	Hongqihegou Station of Chongqing Rail Transit Line 3	Line 6	Five basement floors	Width 25.55 Height 32.83

Table 2.5 Summary of network underground large space development projects

S. No.	Area	Name of station	Case of proximity	Number of basement floors	Sizes (m)	Specific method
1	Beijing	Pingguoyuan Station of Beijing Subway Line 6	Line 1, Line 6, Line S1	Three basement floors	Length 324.4 Width 23.5 Height 26.8	PBA, open excavation
2	Beijing	Suzhou Jie Station of Beijing Subway Line 16	Line 10	Two basement floors, island platform	Length 260.65 Width 23.5 Height 17.12	PBA
3	Beijing	Chaoyang Park Station of Beijing Subway Line 3	Line 14, Line 3	Two basement floors	Length 295.6 Width 23.1 Height 25.4	PBA
4	Beijing	No.8 and No.9 assembly tunnel shield access shaft of Beijing Daxing International Airport Express	–	–	–	–
5	Chengdu	Jinfu Station of Chengdu Metro Line 6	The subway station is located between the Long Jiang Mechanical and electrical city and the existing underpass tunnel	Two basement floors	Length 311.3 Width 20	Half cover excavation method
6	Chengdu	Chengdu University of TCM and Sichuan Provincial People's Hospital Station of Chengdu Metro Line 5	Line 5, Line 2, Line 4	Two basement floors, one-sided platform	Length 284.3 Width 8.9–39.15	Open excavation

(continued)

Table 2.5 (continued)

S. No.	Area	Name of station	Case of proximity	Number of basement floors	Sizes (m)	Specific method
7	Chengdu	Chengdu Metro Line 10 Airport 2 Station Renovation Project	–	–	–	Undercut
8	Guangzhou	Nancun Wanbo Station of Guangzhou Metro Line 18	–	Four basement floors	Length 289 Width 42.1 Height 40.7	Open excavation, concealed surface method
9	Guangzhou	Donghu Station of Guangzhou Metro Line 10	Line 10, Line 12	Five basement floors	Length 199.8 Width 54.2 Height 40.2	Open excavation
10	Guangzhou	Guangzhou Metro Panyu Square Undercut Tunnel Project	Triple interchange between Line 18, Line 22 and access lines	–	Length 259.287 Sectional area 160 m^2	Undercut
11	Jinan	Renjia Zhuang Station of Jinan Metro Line 2	–	Two basement floors	Length 210.1 Width 19.3 Height 16.5–18.0	Open excavation
12	Lanzhou	Dongfanghong Square Station of Lanzhou Rail Transmit Line 1	–	–	–	Open excavation
13	Qingdao	Qingdao Metro Line 1 Qingdao Station Renovation Project	Line 1, Line3	–	–	–
14	Qingdao	Kaifeng Road Station of Qingdao Metro Line 1	–	Two basement floors	Length 218.3	Undercut

(continued)

Table 2.5 (continued)

S. No.	Area	Name of station	Case of proximity	Number of basement floors	Sizes (m)	Specific method
15	Qingdao	Jiangsu Road Station of Qingdao Metro Line 1	Line 1, Line4	Two basement floors	Length 183.2 Width 20.9 Height 17.26	Undercut
16	Shanghai	Shanghai Railway Station North Square Comprehensive Transportation Hub Project	–	–	Length 138 Width 69.4 Height 6.05	Open excavation
17	Wuhan	Wuhan Railway Station	Line 4, Line 10, Line 19, Line 20	Three basement floors	Length 241.5 Width 39.8	Open excavation
18	Wuhan	Jiyuqiao Station of Wuhan Metro Line 5	Line 2	–	Length 315 Width 21.3 Height 17.5	Open excavation
19	Xi'an	NANSHAOMEN Station of Xi'an Metro Line 5	Line 2, Line 5	Three basement floors	Length 178.75 Width 23.1 Height 22.95	Open excavation
20	Xi'an	Xi'an Metro Line 2 Weiyi Street No. 1 and No. 2 duct reconstruction	–	–	–	–
21	Changchun	Yuanjiadian Station Assembly Subway Station of Changchun Metro Line 2	–	Two basement floors	Length 310 Width 19.7 Height 20	Open excavation
22	Zhengzhou	CONVENTION and EXHIBITION CENTER Station of Zhengzhou Metro Line 4	Line 1	Three basement floors	Length 167.02 Width 23.1	Open excavation

(continued)

Table 2.5 (continued)

S. No.	Area	Name of station	Case of proximity	Number of basement floors	Sizes (m)	Specific method
23	Chongqing	Shapingba Railway Hub Comprehensive Renovation Project	–	–	Length 560 Width 125	Open excavation
24	Chongqing	Chongqing West Railway Station Comprehensive Transportation Hub	Connecting to Phase I in the west and Xinfeng Middle Road in the east	Three basement floors (four basement floors a few)	Length 700 Width 140	Open excavation

Fig. 2.2 The completed Optics valley plaza complex

② First underground floor: Circular hall, realizing subway transfer and pedestrian functions, connecting various station lines and surrounding commercial underground space.

③ Second underground floor: Luoyu Road municipal tunnel, Line 2 South Extension section.

④ Third underground floor: Line 11 platform (186 m long, 14 m wide).

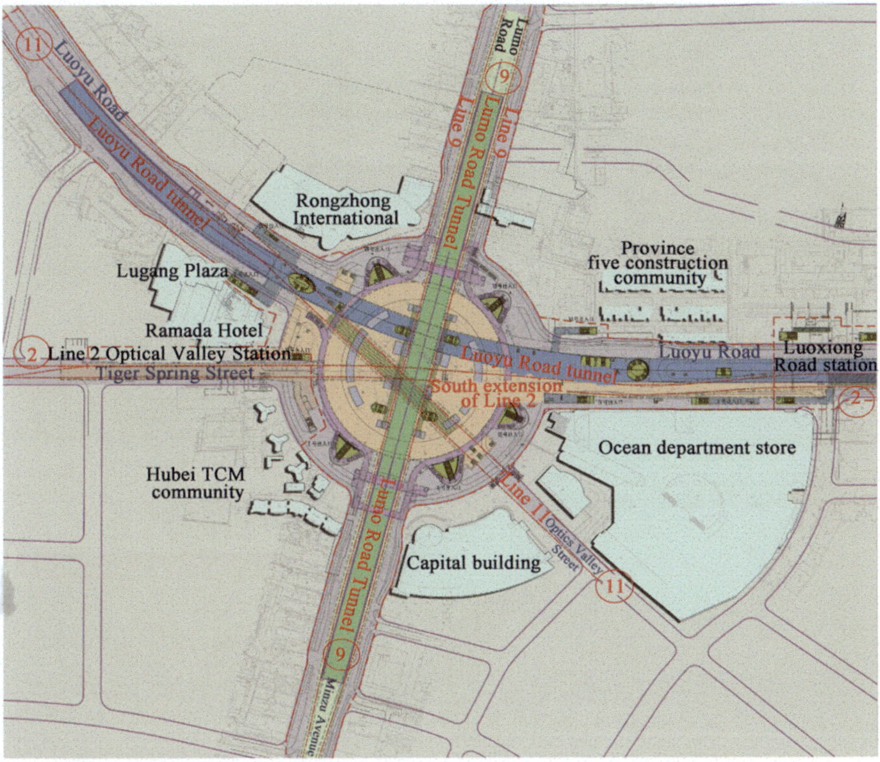

Fig. 2.3 Locations and overall plan of each line of Optics valley square project

The structure of the underground space of the Optics Valley Plaza Complex Project is shown in Fig. 2.4.

(3) Brief Introduction of Main Components

The components of the Optics Valley Plaza Complex are shown in Fig. 2.5.

① Luoxiong Road Station

Luoxiong Road Station is an underground three-layer island platform station, with a length of 250.30 m and a width of 34.35–36.20 m. It is constructed with a reinforced concrete box frame structure, and the station's floor is buried at a depth of 22.47–29.20 m.

② Luoyu Road Tunnel

The Luoyu Road Tunnel starts near the Lulane Post Office in the west and ends on the east side of Huazhong University of Science and Technology in the east. It is 1270 m long, with a red line width of 60 m. The start and end points of the project are connected to the existing Luoyu Road. The tunnel is a two-way six-lane double-hole structure, constructed with a reinforced concrete frame structure.

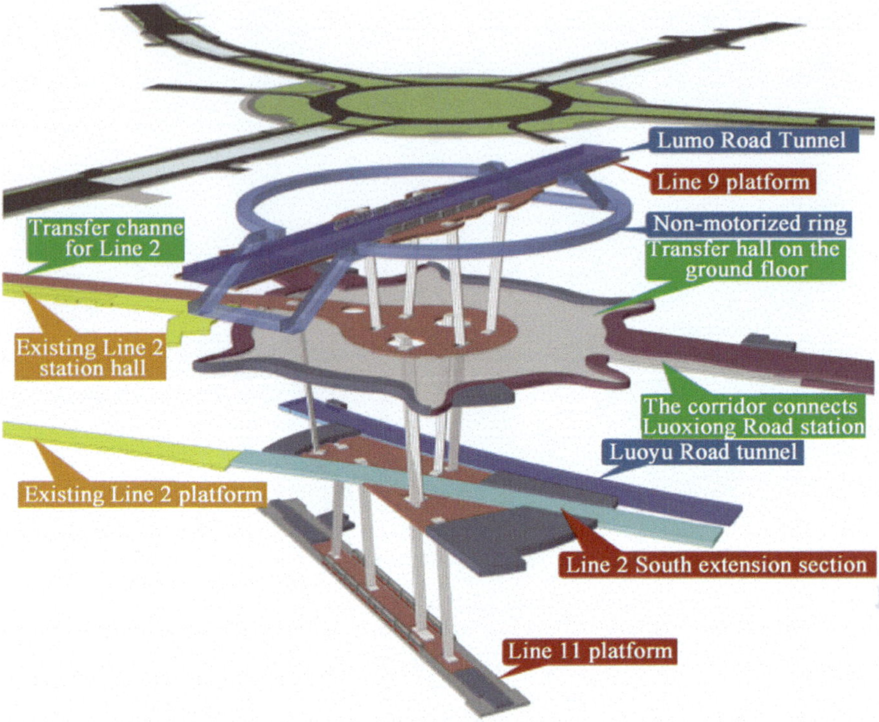

Fig. 2.4 Schematic diagram of the underground space structure of the Optics valley square

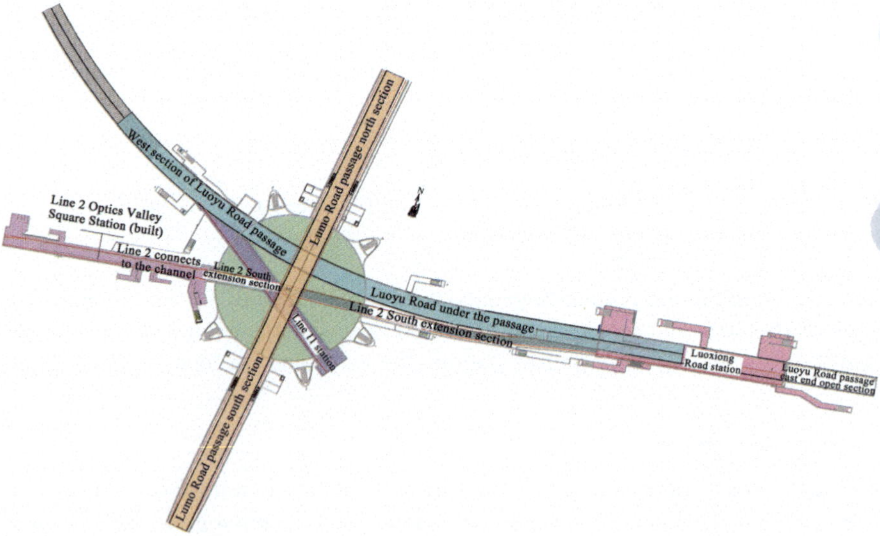

Fig. 2.5 Components of the Optics valley square

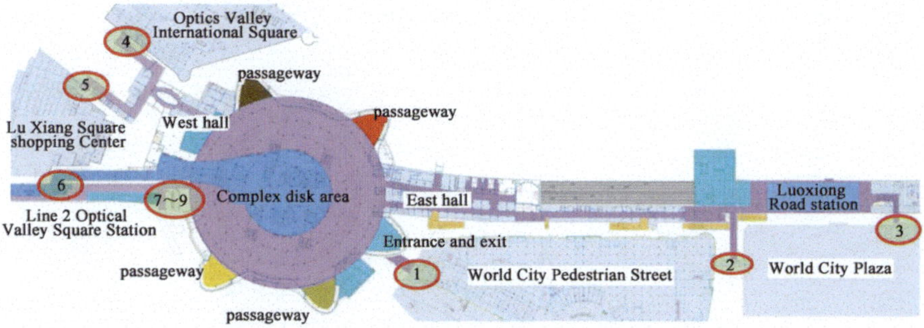

Fig. 2.6 Schematic diagram of the connection and transfer between the complex and the surrounding underground space. 1—Complex ↔ World City Pedestrian Street; 2—Luoxiong Road Station ↔ World City Pedestrian Street; 3—Luoxiong Road Station ↔ World City Square; 4—West Hall ↔ Optics Valley International Square; 5—West Hall ↔ Lulane Square Shopping Center; 6—Complex ↔ Line 2 Optics Valley Square Station (transfer); 7—Complex ↔ Line 2 Optics Valley Square Station (entrance/exit); 8—Complex (Line 2 right line) ↔ Line 2 Optics Valley Square Station; 9—Complex (Line 2 left line) ↔ Line 2 Optics Valley Square Station

③ Lumo Road Tunnel

The Lumo Road Tunnel, outside the Optics Valley Square, has a southern section length of 221.77 m and a northern section length of 258.23 m. The southern section corresponds to an open section length of 190 and 220 m. It is a two-way six-lane double-hole structure, constructed with a reinforced concrete frame structure.

④ Connectivity and Transfers

The Optics Valley Square, Line 2 South Extension, and Luoxiong Road Station are connected to the surrounding existing underground space, commercial squares, and existing stations at 9 points, as shown in Fig. 2.6.

(4) Surrounding Environmental Conditions

① Traffic Conditions

There are four main roads connecting the complex with the square: Lumo Road, Minzu Avenue, Luoyu Road and Luoyu East Road, as well as Huquan Street and Optics Valley Street, forming a six-road intersection. At the same time, non-motorized vehicles and pedestrians do not form a complete connectivity system, with intertwined traffic and pedestrian flow, resulting in high traffic pressure.

② Surrounding Building Conditions

Surrounding Optics Valley Square are Luxiang Square Shopping Center, Ramada Hotel, Optics Valley International Square, Optics Valley World City, Optics Valley Capital Building, Optics Valley Square, etc., all with independent underground space. The surrounding environment of the Optics Valley Square Disc is shown in Fig. 2.7.

Fig. 2.7 Optics valley square disc surrounding environment

2. Project Features

(1) Integration of Internal and External Environments

To create a comfortable and beautiful underground space environment, the main project of Optics Valley Square uses a large-span structure to enhance the spatial effect, in the spatial layout, it solves the problem of traffic routes dividing the underground space through methods such as underground station halls elevated platforms, double-height halls, etc. This greatly enhances the sense of space and effectively avoids the oppressive feeling of large space with low clearance; through large sunken squares, skylights, large atriums, and daylighting skylight of 2500 m^2 bring natural light into the underground, creating a blend of indoor and outdoor space, filled with sunlight, green plants, and a comfortable and pleasant underground space. The effect of the first ground floor lobby of the Optics Valley Square is shown in Fig. 2.8.

(2) Integrated Up and Down Structure

The arrangement of the main structure of the project is reasonably laid out according to the main passenger flow direction. The column grid combines the passenger flow direction and adopts new arrangement forms such as radial and circumferential to avoid blocking the circumferential pedestrian flow. The column grid adopts a large-span structure, especially in the densely populated core area, to improve the spatial effect and comfort through the large span. The structural beam system is optimized in combination with the direction of the main equipment pipeline, and various new types of force systems are selected according to the project conditions, the main beams and secondary beams are flexibly arranged, and large indoor pipelines are

Fig. 2.8 Optics valley square underground first floor lobby renderings

arranged in the main beam grid to save the net height of the space. The structural system of the Optics Valley Square complex is shown in Fig. 2.9.

(3) Landscape Guidance Integration

The direction of large underground space is difficult to identify. For clear guidance, a clear guidance system is achieved through integrated underground decoration, ground landscape, and guidance system. For the decoration of entrances and exits of each quadrant around and the corresponding ground square landscape, they are designed with different themes and colors according to the business characteristics of the plot, and the clear guidance and rapid diversion of the underground space are achieved through innovative means such as color mark guidance. The recognition is clear, the visual effect is good, greatly improving the recognition and sense of direction of the space, and the bright colors inject vitality into the hall, creating a beautiful

a) b)

Fig. 2.9 Optics valley square structural system

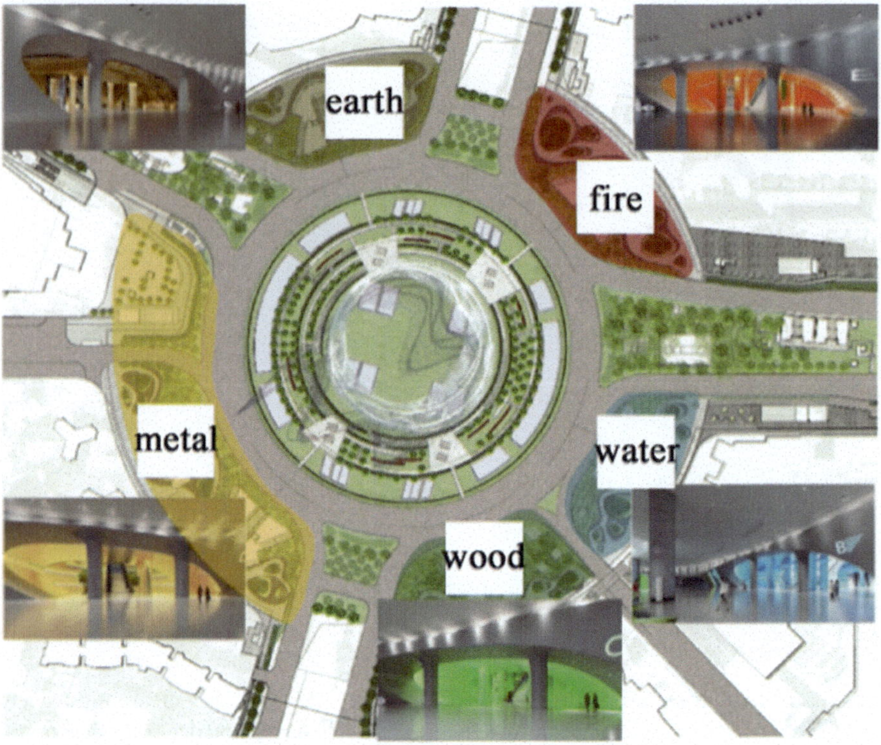

Fig. 2.10 Optics valley square color mark theme landscape and guidance

underground space atmosphere. The color mark theme landscape and guidance of the Optics Valley Square complex are shown in Fig. 2.10.

3. Project Risks and Difficulties

(1) Solving the Traffic Bottleneck of Multi-line Radial City Center

Optics Valley Square gathers all types of traffic from 6 roads, with serious mixed traffic of people and vehicles, frequent traffic jams, high saturation of existing surrounding roads, daily traffic volume of Optics Valley roundabout area reaches 13,000 vehicles, passenger flow of 230,000–260,000 people, peak traffic volume can reach 19,000 vehicles, passenger flow can reach 300,000 people, traffic is extremely busy. How to achieve smooth passage of motor vehicles, non-motor vehicles, subways, and pedestrians in the limited space of the city center node where multiple roads converge, solve the traffic bottleneck of the city center node, efficiently use scarce land resources, optimize the traffic function of the city center node, intensive resource utilization, and interconnect with the underground space of surrounding plots, is a major difficulty in project construction.

(2) Safety of Large-scale Irregular Deep Foundation Pit and Complex Structure Construction

The 200 m diameter circular foundation pit where six roads and five lines (two municipal tunnel lines + three subway lines) intersect, has a staggered internal plan, vertical staggered layers, variable depth, and complex spatial relationship of the structure, and the load transfer and stress state present complex spatial characteristics; the foundation pit of the subway and municipal road extension part, the overall plan is irregular, and the depth varies in different areas; the multiple entrance and exit foundation pits hanging around the disc, have complex shapes. In the city center traffic node with high-rise buildings, congested ground traffic, dense and crisscross pipelines, and complex environmental conditions, centralized implementation of a large-scale, complex structure of the complex project, the construction difficulty, and safety guarantee difficulty are very high.

(3) Limited Space Site with Large Traffic Impact

The project is located in the center of the Optical Valley Turntable, surrounded by dense businesses, radially connected six city roads. The roads are highly saturated and traffic is extremely busy. Due to the construction site occupying most of the vehicular and pedestrian roads, a detailed traffic relief plan must be formulated before construction to ensure normal vehicle and pedestrian traffic during construction. Secondly, during the construction process, it is necessary to strictly control the settlement deformation of the ground roads around the foundation pit to prevent the interruption of surface traffic caused by construction safety or the negative social impact caused by long-term congestion. In addition, the construction site is close to the existing Optics Valley Square Station of Metro Line 2, and it is necessary to ensure the normal operation of Line 2 and the opening of the southern extension of Line 2 on time during the construction period. In summary, construction has a significant impact on traffic, on-site traffic relief is difficult, and the requirements for surface settlement control are high.

(4) There are significant safety risks in phased and expansion construction

Due to the impact of pipeline relocation and traffic relief, the entrance and exit, sunken courtyard structure and municipal tunnel structure around the circular underground space need to be built in the circular core part and implemented after part of it is put into use, involving the expansion problem on the basis of the existing underground structure. The already operational Metro Line 2 station needs to expand the transfer connection channel, and it is necessary to solve the problems of modification of the existing project, node processing, and construction impact. Therefore, the impact on the existing main structure during the expansion construction, the stability of the expansion foundation pit support system, and the treatment of the structural seams between the expansion structure and the existing structure are the difficulties to be solved.

(5) There are many connection points

The underground space of the surrounding buildings of the complex, such as Luxiang Square Shopping Center, World City Pedestrian Street, World City Square, and Optics Valley International Square, are independent and not effectively connected. The Optics Valley Square complex has a total of 9 connection points to realize the interconnection of underground space. During the construction process, there are high requirements for construction quality such as partial demolition of existing retaining piles, opening of existing main structures, and connection of old and new structures, and there are risks such as local stress concentration of existing main structures, differential settlement, and leakage of interface structures.

2.3.2 Chongqing West Station Integrated Hub Phase II Project

1. Project Overview

(1) Project Introduction

Chongqing West Station Railway Integrated Transportation Hub is a passenger integrated transportation hub that mainly focuses on railway passenger transportation and integrates urban rail (ring line, line 5, line 12), taxis, and social vehicles. Its main function is to achieve zero-distance relief between various modes of transportation. The Chongqing West Station Railway Integrated Transportation Hub mainly consists of railway station buildings, station yards, and local supporting integrated transportation hubs. The local supporting integrated transportation hub project is divided into two phases. The first phase of the project was completed and put into use in January 2018. The second phase of the project started in August 2018 and was completed at the end of 2020. The under-construction integrated transportation hub Phase II project is shown in Fig. 2.11.

The Chongqing West Station Integrated Transportation Hub Phase II Project is located underground in the square in front of Chongqing West Station, west of the Phase I project, and east of Xinfeng Middle Road. The project is about 700 m long from south to north and about 140 m deep from east to west. The main structure is divided into three floors (partially four floors, for the transfer passage of Line 12), among which the three underground floors are the Chongqing West Station subway station, Line 5, ring line track section and B2 municipal traffic underpass and other structures, the second underground floor and the first underground floor are transportation hubs and municipal projects, all of which are underground projects and are constructed by open excavation. The plan position relationship of the project is shown in Fig. 2.12, and the spatial position relationship of the project is shown in Fig. 2.13.

The construction content of the second phase of the Chongqing West Station Integrated Transportation Hub includes subway engineering, hub engineering foundation

Fig. 2.11 Under-construction integrated transportation hub Phase II project

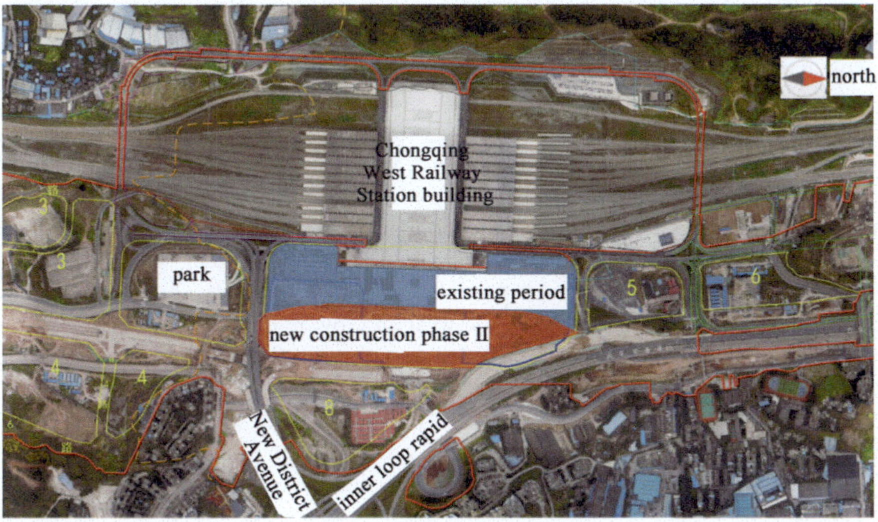

Fig. 2.12 Plan position relationship diagram of the project

and main structure construction, and civil and general installation construction engineering with a total construction area of 195850 m^2. The transfer passage of Line 12 located on the fourth underground floor has a construction area of 1625.4 m^2; the total construction area of the third underground floor is 33214 m^2, of which the subway station construction area is 23605 m^2, the construction area of the B2 underpass is

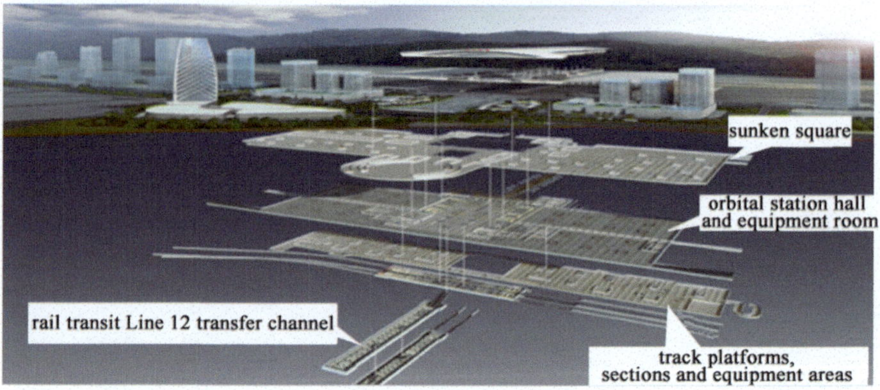

Fig. 2.13 Spatial position relationship diagram of the project

2609.2 m^2, the total length of the open- excavation section structure of Line 5 and the ring line is about 920 m (four lines), and the area is about 7000 m^2 (this part is not included in the total construction area); the second underground floor is the municipal subway station hall paid and unpaid area, hub social vehicle parking lot, municipal taxi waiting area, with a height of 5.2 m, and a construction area of 85900 m^2; the first underground floor is the hub supporting area, mainly with a height of 5.0 m, and a construction area of 79700 m^2.

(2) Surrounding Environment

The project is located in Shapingba District and Jiulongpo District of Chongqing City. To its east is the underpass of Xinfeng Middle Road, with the Southwest Pharmaceutical Logistics Company, Huayan Putian Logistics Market, and residential areas on the surface; to the west is the first phase of the hub project, with Fengzhong Road running north–south through the entire second phase of the hub construction site. The terrain is generally high in the west and low in the east, the site terrain is relatively flat, mostly various types of construction sites, the surface is covered with varying thickness of artificial fill and silty clay, and the site is partially characterized by exposed sloping bedrock. The original ground elevation is 298.38–321.46 m, with a relative height difference of 23.1 m, the terrain slope angle is generally 3°–8°, relatively flat overall; the slope of the local slope is steep, with a slope angle of 15°–40°.

2. Key Difficulties of the Project

(1) Large-scale construction of the first and second phase projects to interface the old and new structures

The interface plane division of the first and second phases of the Chongqing West Station Integrated Transportation Hub project is bounded by the construction seam between the Q axis and the R axis, and the vertical plane division is bounded by the construction seam that vertically penetrates from the bottom plate of the third underground floor to the ground floor (±0.000) bottom plate. The total length of the new and old structure connection treatment is about 2750 m, and the position of the operation surface boundary line is shown in Fig. 2.14. The first phase of the project was completed and delivered in January 2018. After more than 20 months of interface retention, the preliminary construction preparations for the second phase of the project were ready, and the construction of the main structure expansion began.

(2) Construction of large-span beam slab structure with unbonded prestress

The maximum span of the subway transfer hall roof of this project is 24 m, using unbonded prestressed tension construction, and the roof beam uses bonded prestressed tension construction, which requires high specialization in construction. The completed large-span beam slab structure is shown in Fig. 2.15.

(3) Construction of large volume concrete

The thickness of the subway station floor slab in this project is 1 m, and the height and width of the subway conversion beam reach 2.5 m and 1.8 m respectively. The maximum thickness of the subway station wall is 1.0 m. It is particularly important to take effective measures to ensure the quality of large volume concrete during the construction process.

2.3.3 Tianjin Subway Line 5 Siyuan Road Station Underground Space Construction Project

1. Project Overview

The underground space construction project of Siyuan Road Station on Tianjin Subway Line 5 is located at the intersection of Siyuan Road, Qunfang Road, Baiyang Road and Hongxing Road in Hebei District. The construction project is distributed on the east and west sides of the main structure of the existing Siyuan Road subway station (completed in 2014, a double-layer island-style station), which is a three-layer underground space structure with a total construction area of 60,307 m^2, of which the first underground layer is a motor vehicle parking garage, the second underground layer is a shopping mall, and the third underground layer is a motor vehicle parking garage and equipment room. The elevation of the second underground layer is consistent with the elevation of the existing subway station hall, and two connecting passages are set up in the subway station hall to achieve equal height connection between the constructed underground space and the subway station hall; in the constructed underground project and the development on the cover of Siyuan Road Station, the ground building is a three-story skirt building and 6 high-rise buildings. The constructed underground project is shown in Figs. 2.16 and 2.17.

The pit of the construction project is irregular shape, with the east side pit along Siyuan Road Station being 191.08 m long, the widest in the vertical direction is 107.57 m, and the pit excavation area is about 12,500 m^2; the west side pit along

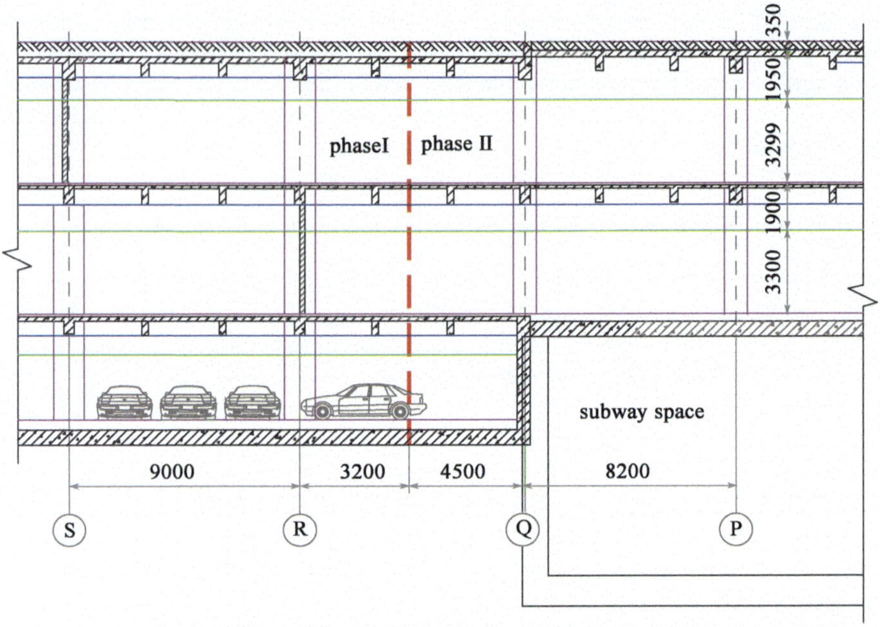

a) working surface boundary profile diagram (dimensional unit: mm)

b)Operational Surface Boundary Elevation

Fig. 2.14 Position of the boundary line between the first and second phases

Fig. 2.15 Completed large-span beam slab structure

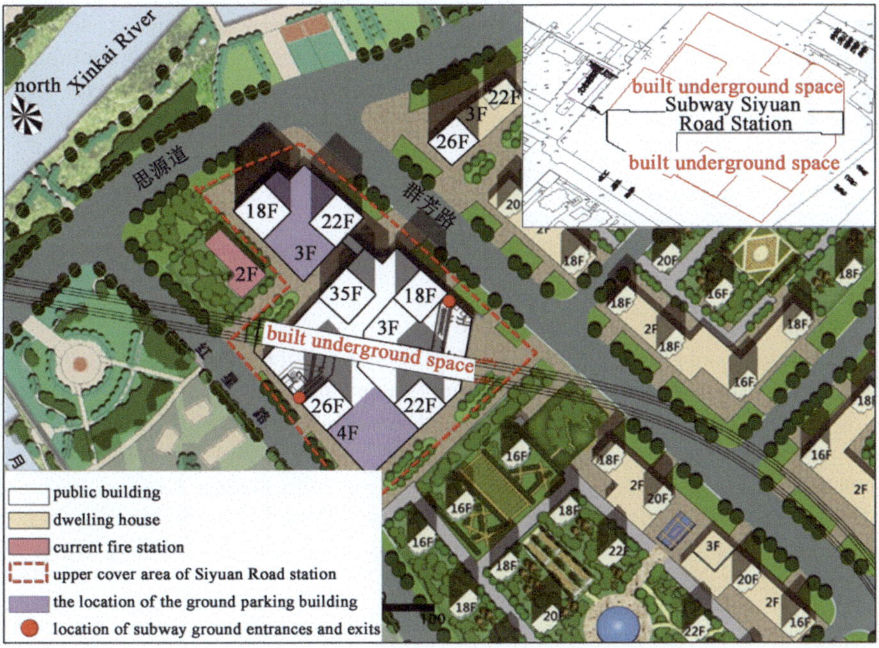

Fig. 2.16 Plan schematic of Siyuan road station construction of underground space project

Siyuan Road Station is the longest at 145.113 m, the widest in the vertical direction is 95.778 m, and the excavation area is about 8500 m^2. The average depth of the pit is 16 m, the retaining structure adopts the form of underground continuous wall plus internal support, the wall thickness is 800 mm, the depth of the underground

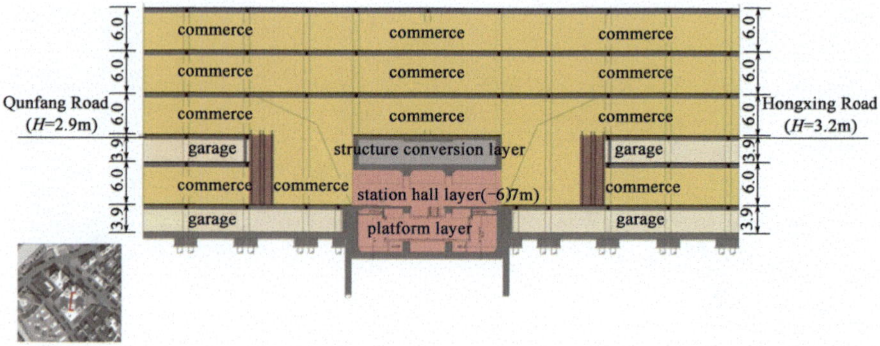

Fig. 2.17 Section schematic of Siyuan road station construction of underground space project (unit of measurement: m)

continuous wall is 32 m, a reinforced concrete crown beam is set at the top of the wall, and the internal support is three concrete supports; in the deeper excavation area of the skirt building elevator pit, three-axis cement soil mixing piles are set, with an effective pile length of 9 m. The cold joint between the underground continuous wall of the constructed project and the original subway station is sealed with high-pressure rotary jet piles. The plan and section of the pit support structure are shown in Figs. 2.18 and 2.19.

2. Key Difficulties of the Project

(1) Super large and super deep irregular pit ring support system and excavation construction

This project adopts the support method of underground continuous wall + ring internal support, and the main support adopts concrete support with a section of 700 × 1200 mm (first row) and 900 × 1200 mm (second and third rows). The excavation of the deep foundation pit close to the existing subway station at zero distance is likely to cause disturbance to the main structure of the existing subway station, especially when the excavation is deeper on both sides of the existing station, it is easy to cause deformation of the main structure of the existing station; if the excavation is not symmetrical on both sides, it is easy to cause bias pressure on the existing structure, and serious damage to the structure.

(2) Large area pit dewatering in high pressure water stratum

Dewatering of the foundation pit is a prerequisite and guarantee for the successful excavation of the pit. The groundwater type of this project is the Quaternary pore pressure diving, which is stored in the second continental layer. The groundwater in the lower silt sand and silty soil layers is under pressure. The base of the foundation pit is 0.5–1.0 m above the first layer of confined aquifer, and the water head height of the first layer of confined aquifer is 0.2 m (2.2 m from the original ground surface). The dewatering of the confined water stratum is difficult; the foundation pit has a large area

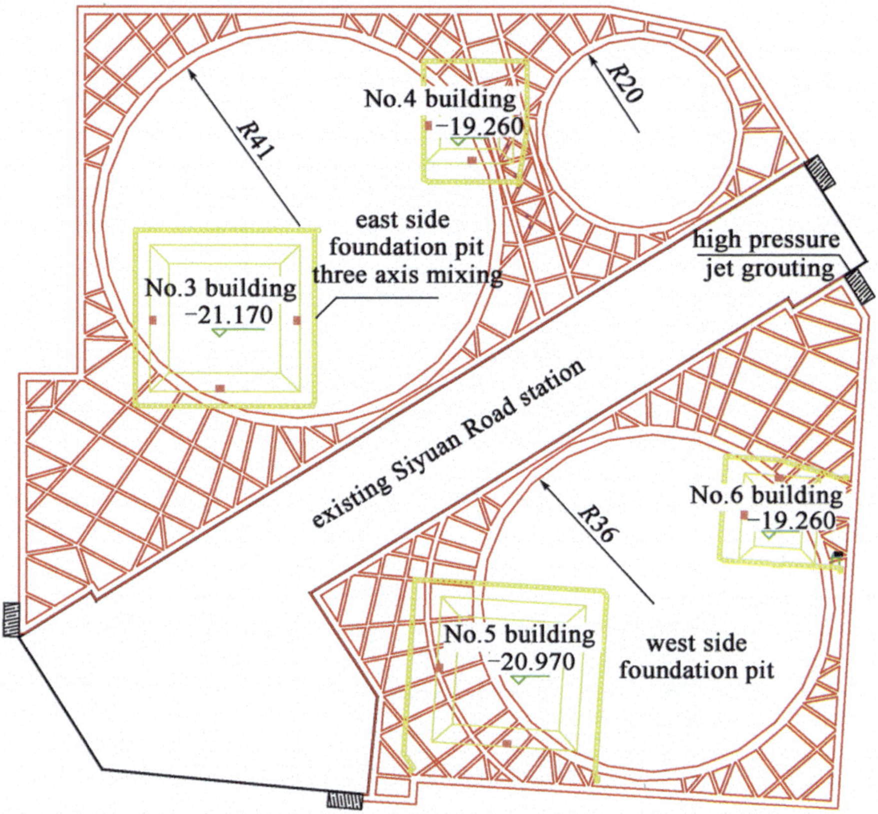

Fig. 2.18 Plan of the foundation pit of Siyuan road station construction project (dimension unit: m; elevation unit: m)

and uneven local depth. If the graded precipitation and on-demand precipitation are not well controlled, it is easy to cause water and sand gushing or uneven settlement of surrounding structures; to ensure the safety of the existing subway station structure, it is particularly important to choose a reasonable dewatering method.

(3) Treatment of cold joints in new and old underground diaphragm walls

The construction time of the existing structure's underground diaphragm wall was in 2012, and the construction time of the retaining structure of the construction project was in 2016. The time span is large, and the intersection of the new and old underground diaphragm walls is located at the yang corner of the existing structure. This part of the existing underground diaphragm wall has bulging phenomenon. The connection between the new and old structures is prone to form cold joints, leading to leakage during excavation, which may cause water and sand gushing, seriously threatening the safety of the existing subway operation, and even causing structural

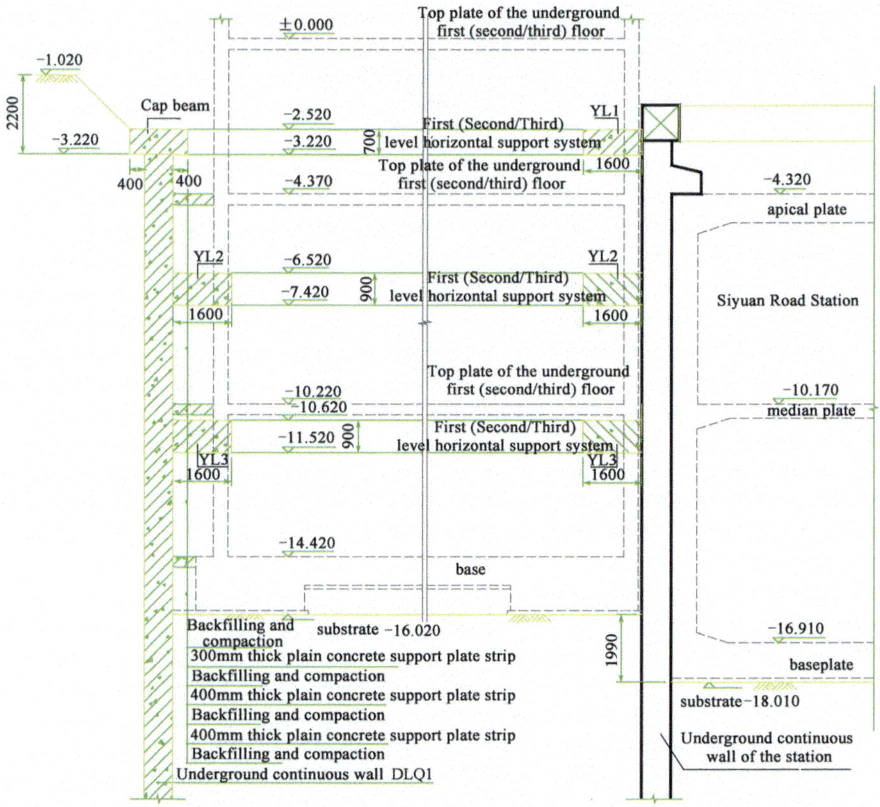

Fig. 2.19 Section of the foundation pit of Siyuan road station construction project (dimension unit: mm; elevation unit: m)

damage. How to deal with cold joints and ensure no leakage is one of the key points of this project.

2.3.4 Chongqing Shapingba Railway Hub Comprehensive Renovation Project

1. Project Overview

(1) Project Introduction

The Chongqing Shapingba Railway Hub Comprehensive Renovation Project is located in the central area of Shapingba, Chongqing. It is the first domestic case in China to develop the high-speed railway station yard into an urban complex. (Fig. 2.20). The project fully utilizes the city's underground space, the upper space of the high-speed railway station yard, and the city's road space to create a modern urban core comprehensive transportation hub integrating high-speed rail, urban rail transit, road traffic, pedestrian traffic, as well as a large-scale urban complex integrating high-end commerce, hotels, office buildings, and apartments. Through the transfer channel in the hub, it realizes convenient transfer between high-speed rail, urban rail transit, urban ground traffic, and urban pedestrian system, and is evaluated by industry insiders as the third-generation high-speed railway station. This project is of great significance to the improvement of traffic in the core area of Shapingba, the upgrade of the business circle, the expansion of the core area, the promotion of the city's image, and the enhancement of the local economy. It is also a model of intensive land use and has pioneering exploration and typical demonstration significance. The project started on April 28, 2013, and was completed on August 31, 2018.

The Shapingba Railway Hub Comprehensive Renovation Project covers the station area of Shapingba Railway Station, railway living office land, and urban construction land. It is bordered by the city's main roads, Zhandong Road and Zhanxi Road, to the north, and the planned east–west connection roads on the east and west sides, and the planned city main road, ZhannanRoad, to the south. The Metro Line 9 is located in the underground of Zhandong Road and Zhanxi Road, and the Metro Loop Line is located under Tianchen Road. Through the transfer channel in the hub, it realizes convenient transfer between the railway passenger dedicated line, urban rail transit, urban ground traffic, and urban pedestrian system. The overall plan of the project is shown in Fig. 2.21, and the longitudinal section is shown in Fig. 2.22.

Fig. 2.20 Three-dimensional effect diagram of Chongqing Shapingba railway hub comprehensive renovation project

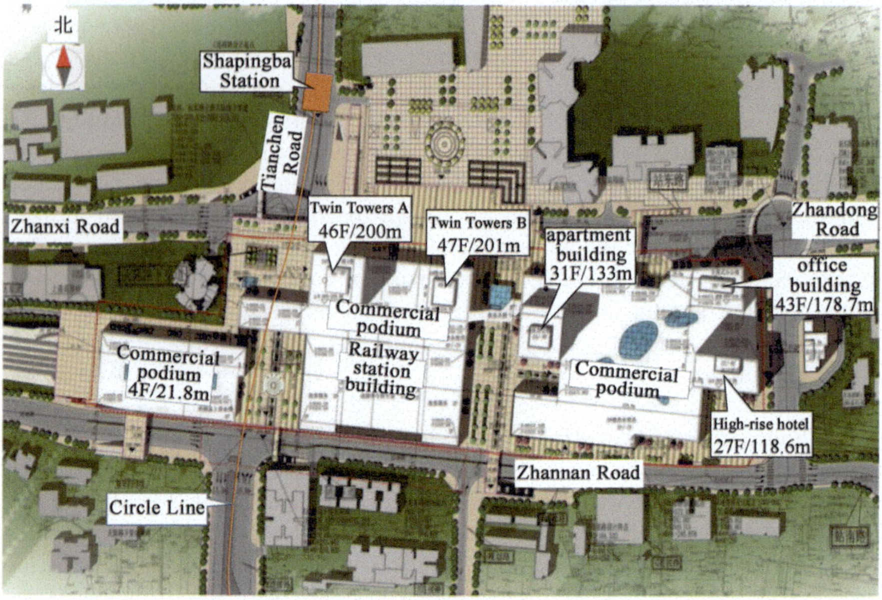

Fig. 2.21 General layout diagram of the project

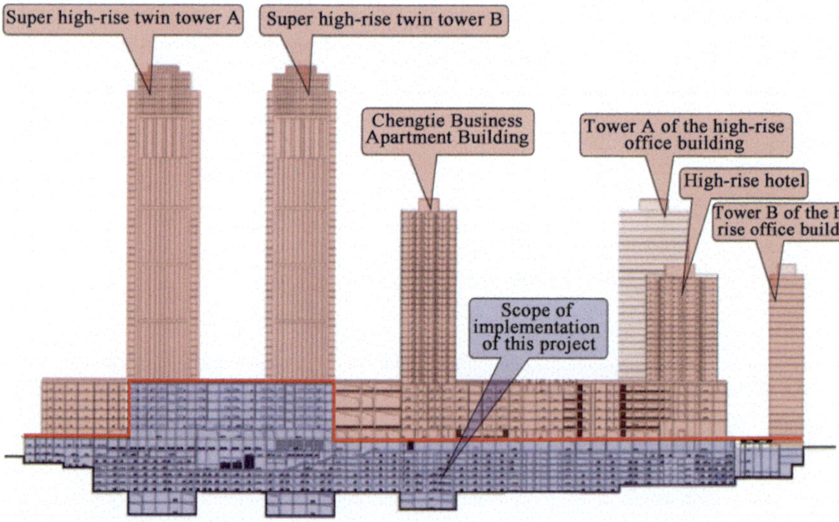

Fig. 2.22 Longitudinal section of the project

(2) Functional Layout

The main content of the comprehensive renovation project of the Shapingba railway hub is divided into four parts according to the three-dimensional space interface.

① Shapingba station of Chengdu-Chongqing passenger dedicated line and related projects: including the construction of new passenger station buildings, railway station covers, Shapingba station yard renovation, etc., with a construction area of 13974 m^2; Construction above railway station yard (including the cover frame structure and the cover bridge structure) is 816 m long from east to west, 71.5 m wide from south to north, and the construction area is 49095 m^2; A new passenger yard is built at the existing Shapingba station, with a line length of 1.319 km, and the passenger yard is arranged with 3 platforms and 7 lines.

② Urban road projects within the land use range: The project derived 8 sections of reconstructed or newly built urban roads, of which 3 sections were reconstructed, namely the section from Zhandong Road to Zhanxi Road, Tianchen Road North Section, Tianchen Road South Section; 5 new roads, namely the underpass from Zhandong Road to Zhanxi Road, Tianchen Road underpass, East Connection Road, West Connection Road and Station South Road, with a total length of 4.239 km for reconstructed and newly built roads.

The layout of urban roads is shown in Fig. 2.23.

③ Transportation hub complex: including the transfer hall and entrance and exit passages between the passenger station and urban transportation; new bus stations, taxi stations, underground garages, and the upper cover property development supporting project. The transfer system covers the internal traffic organization of the hub, the transfer between Chengyu Passenger Dedicated and urban

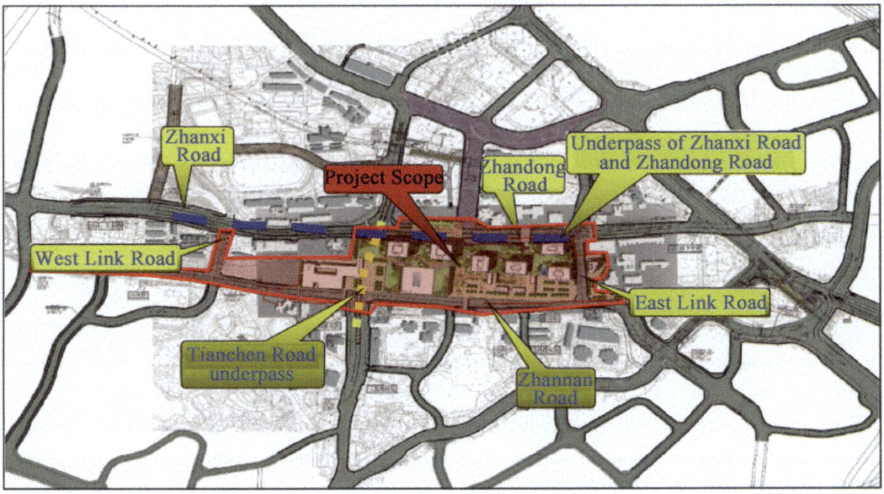

Fig. 2.23 Urban road layout diagram

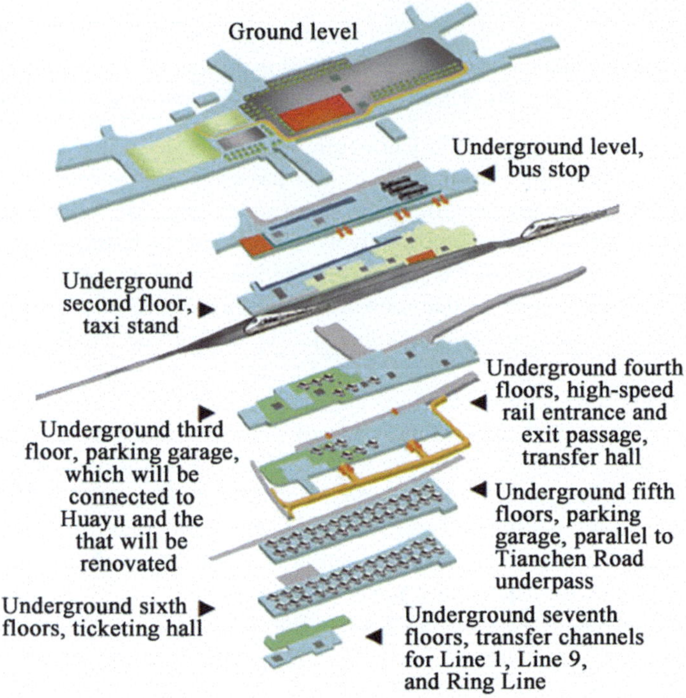

Fig. 2.24 Hub facade layout diagram

rail transit, urban buses, taxis and other transportation, as well as the pedestrian system, etc. The layout of the hub facade is shown in Fig. 2.24.

④ Urban rail transit project: The synchronized constructed urban rail transit project includes the section from Shapingba Station to Xiaolongkan Station on Line 9, Shapingba Station on Line 9, the shared No. 2 wind pavilion group project of the ring line and Line 9, the No. 2 entrance and exit of Shapingba Station on the ring line, the entrance and exit from the station hall of Line 9 to the ground, and the open excavation section of the No. 3 entrance and exit of Shapingba Station on the ring line. The layout of urban rail transit is shown in Fig. 2.25.

(3) Surrounding environment

The project is located in the core area of Shapingba District, Chongqing City, with a very complex surrounding environment, surrounded by buildings and densely populated.

From east to west, the main buildings on the north side are several 34-story commercial and residential buildings and commercial blocks of Huayu Plaza, Liyuan Hotel, and Wengda Ping'an Building, all of which are frame structures or frame shear structures. The deep foundation pit boundary is about 50 m away from the high-rise building at the main entrance of Huayu Plaza, with an excavation depth of 35.4 m;

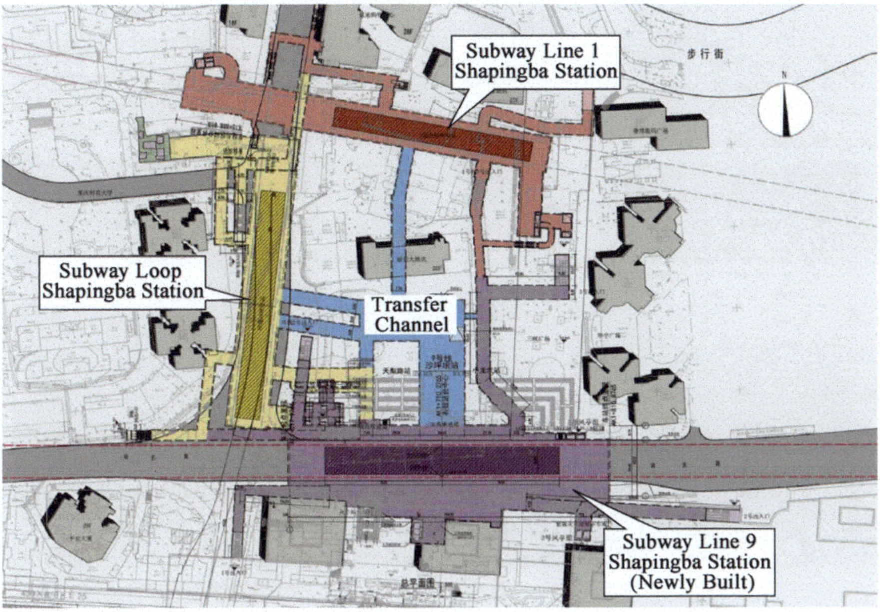

Fig. 2.25 Urban rail transit layout diagram

the minimum distance between the boundary of the deep foundation pit and the high-rise commercial and residential building (commercial block) of Huayu Plaza is about 16 m, and the foundation pit excavation depth is 34.8 m; the boundary of the deep foundation pit is 57–85 m away from the 18-story Liyuan Hotel, with a foundation pit excavation depth of 34.8 m; the excavation area boundary is 30–50 m away from the underground water pump room, underground staff canteen and underground parking lot of Liyuan Hotel; the minimum distance between the excavation area boundary and Wengda Ping'an Building is about 14 m, the foundation pit forms two levels of steps, the length of the foundation pit slope is 56 m, and the depth of the foundation pit is 8.5–17.5 m.

The main buildings on the south side from east to west are Xuelin Jiayuan A, B buildings, Honghuayuan, 10-story building in Radio Village, Edward Hospital, and the 32-story frame-shear structure commercial and residential building above. The surrounding environment of the project is shown in Fig. 2.26.

2. Engineering difficulties

(1) The surrounding environment of the super-large deep foundation pit is complex, and the blasting construction is difficult

The foundation pit of this project is 125 m wide from south to north, 560 m long from east to west, with an excavation area of about 56,000 m^2, and the excavation depth of the foundation pit is 4–45 m, with a total excavation volume of about 1.36 million m^3, of which the stone accounts for more than 82%. The slope of the foundation pit

a)The original appearance of the existing Shapingba Station building

b)Western environment

c)Northern environment

d)Southern environment

Fig. 2.26 Surrounding environment of the project

is only 10 m from the basement of the 32-story Huayu Building, 95 m from Edward Hospital, the vibration speed must be controlled below 0.5 cm/s, and the shallow hole controlled blasting of the deep foundation pit stone in the complex urban environment is one of the difficulties of the foundation pit project. At the same time, because the project is located in the core business area of Three Gorges Square, with many high-rise buildings, adjacent to schools and hospitals, and more than 500,000 residents around, how to minimize construction noise and dust is also a problem.

(2) The surrounding buildings are dense, and the foundation pit support is difficult and risky

There are many high-rise and super high-rise buildings around the foundation pit, and the requirements for pit deformation and risk control are extremely high. Affected by the surrounding environment and geological conditions, the forms of foundation pit support are diverse, mainly balance weight retaining wall, plate ribbed anchor rod retaining wall, and sprayed anchor net slope protection. Among them, the plate ribbed anchor rod retaining wall is a vertical cast-in-place concrete structure, with a maximum wall height of 31.35 m for the single-sided wall without steps. Anchor rods are set at the rib column, and the maximum length of the rod is 25 m, composed of 1–7 HRB400 grade steel bars with a diameter of 32 mm, and the maximum drilling diameter is 250 mm. The drilling equipment needs to be specially made, and the anchor rod construction is another difficulty of the foundation pit project.

(3) The surrounding environment of the renovated road is complex, and the construction and coordination are difficult

① East section of Zhannan Road

The east section of Zhannan Road is 524 m long, starting from Tianchen Road and ending near Building A of Xuelin Jiayuan. The main project contents include E, F, G, H underground connecting roads, 3 pedestrian evacuation passages, Zhannan Road, bridges, new integrated pipe network, as well as existing pipe network relocation and modification and related supporting projects. Among them, the E connecting road connects the 1st and 2nd vehicle passages to the fourth underground floor of the hub, and connects the F connecting road to the surface of Zhannan Road. H, G connecting roads connect to the fifth underground floor of the hub. The newly built rainwater pipeline uses D1800 mm fiberglass sand sandwich pipe, and the sewage pipeline uses D400–600 mm fiberglass sand sandwich pipe. The excavation depth of the rain and sewage pipeline is 3–8 m.

The existing rain and sewage combined culvert and sewage pipe in the road range, with a maximum burial depth of 7.5 m; D500 mm water supply pipe, with a burial depth of about 1 m; 10 kV red platform, Red pump power line passes above the ground. The above pipelines conflict with the E connecting road, the newly built rain and sewage pipelines, and the plane of the Zhan Nan Road bridge pier, affecting the construction of the new project, requiring relocation and modification, and temporary transition measures need to be taken during construction.

After the completion of Zhan Nan Road, the road height is 1–5.8 m, the bridge height is 5.8–8.1 m, and the minimum distance from the edge to the existing houses is 3–5.7 m. It affects the surrounding road traffic, the business of the community's commercial facade, the lighting of residents, etc., and the construction and coordination are difficult.

② East section of Zhan Dong Road

The eastern section of Zhandong Road is a renovated road, with a total length of 280 m. To the north of the renovated road are the Lider Building, Huayu Branch Road, several commercial and residential buildings in Huayu Square and its commercial district, the No.1 temporary road, and within the road section there are two exits from the underground garage of Huayu Square, one entrance and exit of Carrefour supermarket, and one entrance and exit of the residential area. The deep foundation pit retaining wall is about 9 m away from Building C of Huayu Building, with an excavation depth of 34 m; the minimum distance between the retaining wall of the deep foundation pit and Building E of Huayu Building and its commercial district is about 7 m, with an excavation depth of 21 m. To the south of the renovated road is the newly built East Connection Road and the second phase of the foundation pit.

The main contents of the road renovation project include: mechanical excavation of about 200,000 m^3 of earth and stone; the remaining section and upper structure of Shapingba Station of Metro Line 9, about 61.1 m long, constructed by open excavation method, the structural space from bottom to top includes the platform layer of Shapingba Station of Line 9, the station hall layer, the underpass from

Zhanxi Road to Zhandong Road, property development, Zhandong Road surface and cover structure; the underpass and upper structure of Zhandong Road, 238.9 m long, the structure from bottom to top includes the underpass from Zhanxi Road to Zhandong Road, the garage layer and Zhandong Road. The double-layer double-span rectangular frame structure section is 191.9 m long, constructed by open excavation method; the straight middle wall composite continuous arch tunnel section is 47 m long, constructed by undercut method, with a tunnel depth of about 7 m.

The road renovation project involves existing pipelines including water supply, electricity, communication, rain and sewage pipelines, etc., which need to be relocated and renovated in sync with the project. The existing Zhandong Road mainly carries vehicles coming from Zhanxi Road, Tianchen Road South and North sections, with busy traffic, and this road section involves 1 bus route, 5 bus dispatch rooms, and 2 bus parking bays. Except for the foundation pit retaining wall and pipeline relocation which can be implemented before traffic conversion, the rest of the project needs to be implemented after the third stage of traffic conversion.

2.3.5 Comprehensive Renovation Project of Lanzhou Metro Dongfanghong Square Hub Station

1. Project Overview

(1) Project Introduction

The comprehensive renovation project of Lanzhou Metro Dongfanghong Square Hub Station, starting from Pingliang Road in the east to Jinchang Road in the west, from the Chairman's Review Stand in the north to the South Square Road in the south. The project mainly includes the renovation and expansion of the underground space of Dongfanghong Square, the construction of municipal supporting facilities around the square and pipeline relocation, and the square's ground green space, pavement, and the upgrading of lighting. The total land use of the project is 120,900 m^2; the total building area is 58,300 m^2, of which 3500 m^2 is above ground and 54,800 m^2 is underground; 750 new parking space are added.

The comprehensive project of Dongfanghong Square Hub Station is a typical case of revitalizing and upgrading the existing underground space based on the traffic function of the subway station, and also a typical case of upgrading and renovating the ground pavement, greening and beautification, and landscape lighting in combination with underground space development. After the completion of the project, the underground commercial area seamlessly connects with the subway station, and the underground parking lot significantly improves the parking difficulty in the core area; at the same time, it solves the problems of chaotic exterior facades of surrounding buildings, old and aging square ground pavement, and above-ground buildings affecting the view, becoming a high-quality "city reception hall". The comparison before and after the completion of the Dongfanghong Square Hub Station comprehensive project is shown in Fig. 2.27.

a)Original appearance before renovation

b)The effect of renovation (Architectural Design and Research Institute of Tongji University)

Fig. 2.27 Comparison before and after the completion of the Dongfanghong square hub station complex project

(2) Surrounding Environment and Functional Layout

The underground expansion of the Dongfanghong Square adopts an inverted U-shaped layout, all of which is constructed using the open excavation method. The northern A area is a two-story underground structure, adjacent to Qingyang Road and the interchange station of Metro Lines 1 and 2. The southern central area is the operating Dongfanghong underground commercial and entertainment city, which is connected to the existing underground level of the commercial and entertainment city. The C area is a three-story (partially two-story) underground structure, with the west side closely adjacent to the Dongfanghong underground commercial city, and the east side closely adjacent to Guofang Department Store (one underground level, eight above-ground levels). The B area is a three-story (partially two-story) underground structure, with the east side closely adjacent to the Dongfanghong underground commercial city, and the west side closely adjacent to Lanzhou Stadium. The plan of the underground space of the Dongfanghong Square Hub Station Comprehensive Remediation Project is shown in Fig. 2.28.

The interchange station of Metro Lines 1 and 2 is a two-story three-span structure with a structural height of 14.74 m, a soil cover thickness of about 3 m, open excavation construction, a foundation pit depth of 17.24 mm, and a φ1000@750 drilling bitten pile enclosure structure. The Dongfanghong underground commercial and entertainment city is a one-story underground structure with a soil cover thickness of 1.5 m, open excavation construction, a structural bottom plate depth of about 6.2 m, and slope excavation.

According to the arrangement of architectural functions and the relationship of basement levels, the underground expansion plan is divided into three areas, with garages on the east and west sides, and the one to three-story underground structures serving as civil defense. Among them, the height of the third underground level is 4.15 m, the height of the two underground levels is 3.5 m, the height of the first underground level is 4.25 m, and the soil cover on the ground is about 1.4 m. The north side is a two-story underground structure, with the height of the two underground levels being 4.70 m, the architectural function is arranged as a garage, and it also serves as civil defense. The height of the first underground level is 6.40

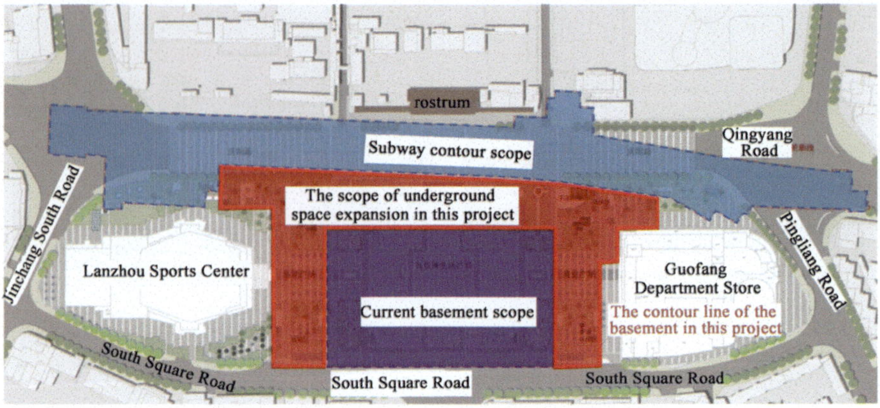

Fig. 2.28 Schematic diagram of the underground space plan of the Dongfanghong square hub station comprehensive remediation project

m, the architectural function is arranged as commercial, sunken plaza activity area, equipment room, etc., and the soil cover on the ground is about 2.2 m. The plan of the expanded underground first floor space is shown in Fig. 2.29, and the section of the expanded underground space is shown in Fig. 2.30.

The newly built underground level is connected to the subway station at 4 places on the north side; it is connected to the existing underground space through 6 passages in the east, west, and north directions; the east side of the C area underground level is connected to the Guofang Department Store underground commercial area; the west side of the B area reserves the condition for underground space expansion. After the completion of this project, the total building area of the underground space (excluding

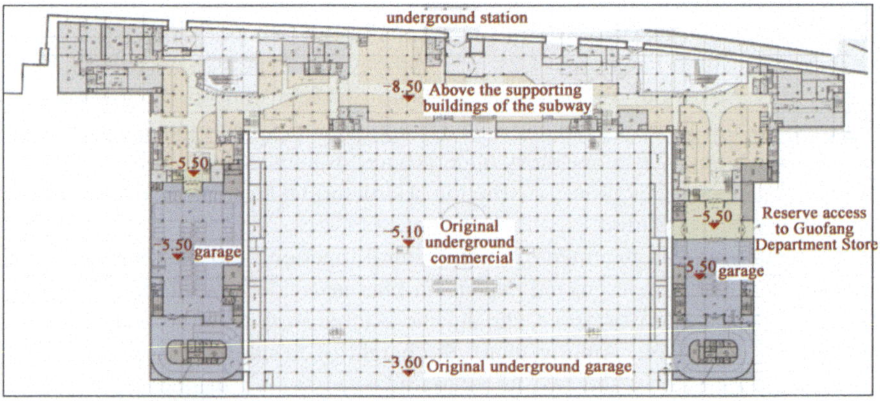

Fig. 2.29 Schematic diagram of the layout of the expanded underground level (Tongji University Architectural Design Research Institute) (elevation unit: m)

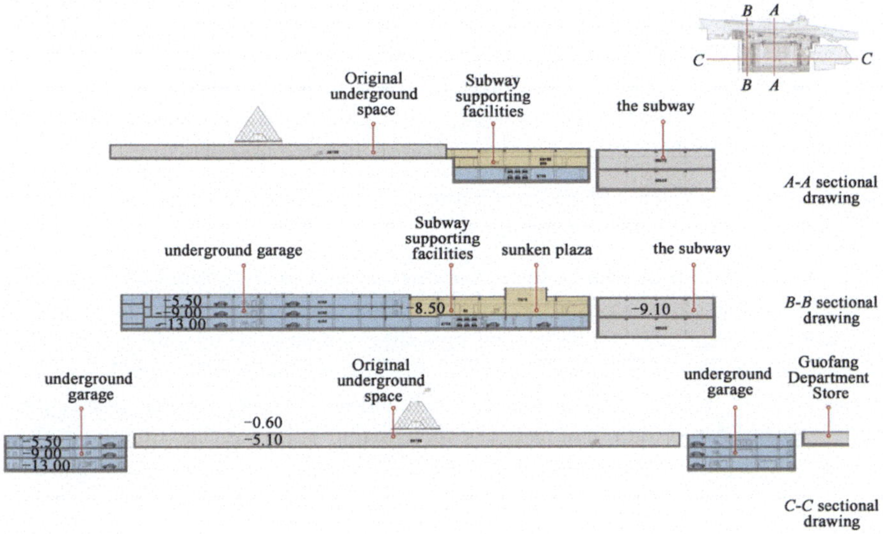

Fig. 2.30 Section of the expanded underground space (Tongji University Architectural Design Research Institute) (elevation unit: m)

the subway station) is 53.7 thousand m^2, and it forms a networked underground space system with the surrounding existing underground structures.

2. Major difficulties of the project

(1) The surrounding environment of the foundation pit is complex and the geological conditions are poor, posing a high safety risk

The construction area of the Lanzhou Metro Dongfanghong Square Hub Station Complex Project is surrounded by existing subway stations, underground commercial streets, and large surface malls, making the surrounding environment complex. The stratum contains a large amount of red sandstone, which greatly reduces its strength and self-stability after disturbance, and the groundwater level is high. If the foundation pit construction is not handled properly, it can easily induce deformation or even collapse, posing a great construction safety risk.

(2) Control of settlement and deformation of existing underground structures at zero distance

The Dongfanghong Square Hub Station complex project involves a foundation pit excavation depth of 14.3 m for the underground space, with the operating subway station sharing a contiguous wall for adjacent construction. The existing underground commercial and entertainment city structure bottom plate has a burial depth of 6.2 m, with the new underground project being constructed nearby. During the excavation of the deep foundation pit, under the unloading conditions of the subway station's unilateral soil body, lateral deformation or bulging settlement can easily occur. However, the subway operating line requires that the track deformation cannot exceed 3 mm.

The existing commercial and entertainment city underground structure is relatively shallow and has certain structural defects. Uneven settlement can cause the structural conditions to deteriorate, exacerbating cracking or water leakage, which may affect normal use. Therefore, it is necessary to strictly control the deformation of the foundation pit and the settlement of neighboring underground structures.

(3) Reinforcement and interface treatment of existing underground structure connection and docking openings

According to the functional requirements of the new underground structure, the underground first floor needs to be connected and docked with the existing underground commercial and entertainment city structure and the operating subway at multiple points. The existing underground structure was built earlier and has certain disease defects. The opening of the side wall and the docking construction require reinforcement, demolition, and repair of the existing structure. The treatment of its interface structure and waterproof nodes is one of the major difficulties of the project.

2.3.6 Beijing Subway Line 4 Xuanwumen Station Additional Transfer Passage Project

1. Project Overview

(1) Project Introduction

The Xuanwumen Station of the Beijing Subway is located at the intersection of Xuanwumen East and West Streets and Xuanwumen Inner and Outer Streets in the central old city area within the Second Ring of Beijing. It is a transfer station for Subway Lines 4 and 2. The Line 2 station is arranged east–west and was built in the early 1970s; the Line 4 station was opened in 2009, arranged north–south, and crosses under the Line 2 station in a cross shape. The Line 2 and Line 4 stations are connected by two opposite "E" shaped passages for transfer. Before the renovation, the transfer passage was for one-way use, as shown in Fig. 2.31a. Shortly after the opening of the Line 4 subway in 2009, there was insufficient transfer capacity between the two lines, and the transfer passage was severely crowded during peak hours in the morning and evening, posing a significant safety risk. After research, it was decided to implement the Xuanwumen Station Additional Transfer Passage Project.

The main objectives of this project are to increase transfer capacity, eliminate safety hazards, improve passenger experience, and enhance urban spatial quality.

The Xuanwumen Station Additional Transfer Passage Project is divided into five parts according to the plan layout: northwest quadrant, southwest quadrant, northeast north quadrant, northeast east quadrant, and southeast quadrant. Underground ticket inspection halls are added in each quadrant and related system equipment is renovated. Two entrances and exits are newly built in the northeast quadrant to solve the problem of no entrances and exits in the northeast direction of the original Line 4. New transfer passages in the northwest, northeast, and southwest are built

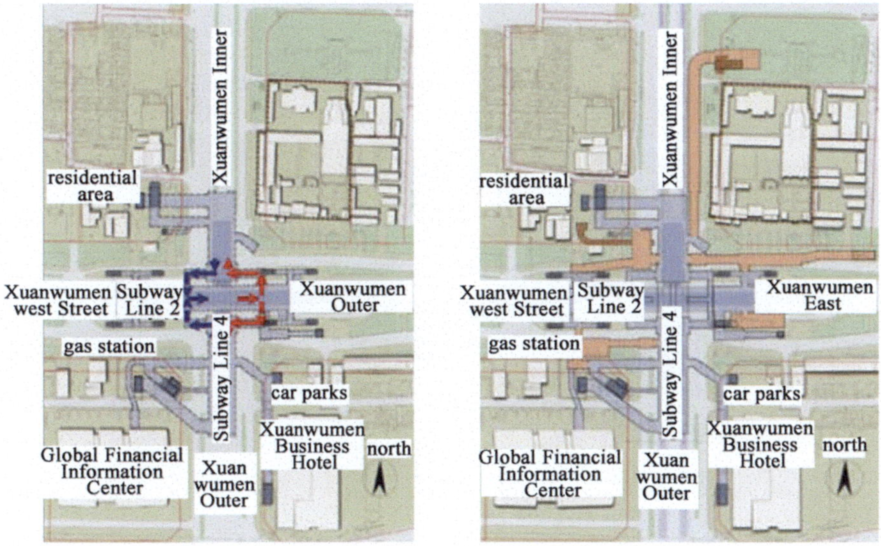

a)Status quo plane b)new plan

Fig. 2.31 Schematic diagram of the additional transfer channel project at Xuanwumen Station

to realize the transfer from Line 4 to Line 2, and the original "E" shaped passage is only used for transfer from Line 2 to Line 4. In the southeast quadrant, the original Line 2 entrance and exit are demolished, and a combined entrance and exit are built with the newly built ticket inspection hall. Through expansion, the transfer capacity between Line 2 and Line 4 is improved, facilitating passenger entry and exit from the station and solving congestion problems. The station layout after expansion is shown in Fig. 2.31b.

(2) Structural Form and Construction Method

The northwest and southwest quadrants add vertical shafts and horizontal passages, and excavate underground excavation transfer halls and transfer passages.

In the north of the northeast quadrant, on the east side of Xuanwumen Inner Street and the south side of Dafang Hutong, the ticket inspection hall is constructed by open excavation, the entrance and exit passages are constructed through open excavation foundation pits, by undercut method, and the existing space is used to connect to Line 4.

The entrance and ticket inspection hall are arranged longitudinally along the auxiliary road on the north side of Xuanwumen East Street in the northeast quadrant, constructed using the open- excavation method, and connected to the station hall of Line 4 and the entrance of Line 2 from the west end of the open- excavation section.

The original entrance of Line 2 in the southeast quadrant is demolished, and a new ticket inspection hall and entrance are constructed using the open- excavation method.

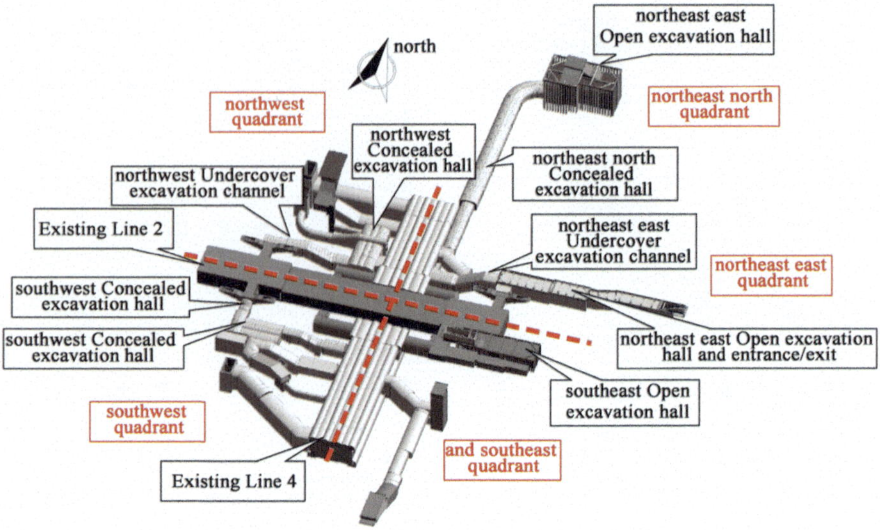

Fig. 2.32 Schematic diagram of the structural form of the additional transfer channel project at Xuanwumen Station

The structural forms and construction methods of each direction are shown in Fig. 2.32 and Table 2.6.

(3) Surrounding environment

There are several important buildings near Xuanwumen Station, including the South Church of the Catholic Patriotic Association (a cultural relic protection unit), Changan Kindergarten in Xicheng District, Fanxing Drama Village, Global Financial Information Center Building, Xinhua News Agency, Yuexiu Hotel, Xuanwumen Business Hotel and Zhuangsheng Chongguang Department Store, as shown in Fig. 2.33. The underground pipelines near the station are dense, including several large-scale important pipelines such as a 2000 × 2350 mm power tunnel, a 2600 × 1500 mm thermal tunnel, a D1800 mm water supply pipe, a D1250 mm sewage pipe, a 1800 × 1500 mm rainwater pipe trench, and a D500 mm high-pressure gas pipe.

2. Key difficulties of the project

(1) Construction of various forms of connection

The new transfer channel project connects with the existing subway lines 2 and 4 at a total of 10 locations, each with different structures and interface forms. The largest opening section is 7.2 × 4.85 m (width × height), and the deformation of the existing structure is required to be no more than 2 mm. How to reduce vibration, control deformation, protect existing structures, ensure construction safety, control dust, and ensure the normal operation of existing lines is the key to the connection construction. The existing Line 2 was built a long time ago, and the waterproof materials are different from the new structure waterproof materials, and the performance is quite

Table 2.6 Engineering construction method and structural form site area

	Structural form	Construction method	Scale	Note
Northwest quadrant	Underground hall: single-layer double-span arch-top straight wall frame section structure	Undercut middle hole method	The excavation width is 15.5 m and the length is about 32.4 m	Add shaft cross channel alone
Channel	Transfer channel: single span arch-top straight wall frame section structure	CRD by under-cutting method	The excavation width is 7.9 m and the length is about 53 m	Excavation by using underground excavation hall
Southwest quadrant	Underground hall: single-layer double-span arch-top straight wall frame section structure	Undercut middle hole method	The excavation width is 15.1 m and the length is about 32.1 m	Add shaft cross channel alone
Channel	Transfer channel: single span arch-top straight wall frame section structure	CRD by under-cutting method	The excavation width is 6.2–7.7 m and the length is about 46 m	Excavation by using underground excavation hall
Northeast Quadrant north	Underground hall: double-layer double-span rectangular frame section structure	Open excavation method	Foundation pit size: 37.3 m (length) × 21.8 m (width) × 17.1 m (depth)	
Channel	Transfer channel: single span arch-top straight wall frame section structure	CRD by under-cutting method	The excavation width is 7.4–10.1 m and the length is about 139.9 m	Excavation by open excavation section
Northeast Quadrant east	Underground hall and entrance: single-layer single-span rectangular frame section structure (local second floor)	Open excavation method	Foundation pit size: 105.3 m (length) × 7.2–12.1 m (width) × 2.4–11.4 m (depth)	
Channel	Transfer channel: single span arch-top straight wall frame section structure	CRD by under-cutting method	The excavation width is 7.9–10.1 m and the length is about 41.5 m	Excavation by open excavation section
Southeast quadrant	Underground hall and entrance: single-layer rectangle frame section structure	Open excavation method	Foundation pit size: 54.9 m (length) × 5.9–12.2 m (width) × 2.4–8.0 m (depth)	

a)the South Church of the Catholic Patriotic Association b)Changan Kindergarten c)Fanxing Drama Village d)Xinhua News Agency

e)Global Financial Information Center Building f)Xuanwumen Business Hotel

Fig. 2.33 Buildings around Xuanwumen station

different. At the same time, due to the lap of new and old structure waterproofing and other issues, ensuring the waterproof quality of the interface is also the key to the connection construction.

(2) Protection of surrounding buildings (structures)

In the northeast quadrant of the excavation entrance parallels side through the South Church, with the foundation of the South Church located above the entrance, with a horizontal clearance of 15.0–17.5 m, and it runs parallel to the existing Line 4 tunnel for a distance of 131 m, partially close to it.

The construction shaft in the northwest quadrant is close to the existing Line 4 northwest air duct on the north side, with a clearance of 8.05 m; it is close to the existing Line 4 E entrance on the east side, with a clearance of 5.45 m, and the bottom of the shaft is 10.32 m lower than the entrance floor; there are 2 early residential buildings in the impact area on the west side of the shaft, with a clearance of 7.3 m. The northwest quadrant connects to the underground excavation channel of Line 2, which crosses Xuanwumen West Street auxiliary road, with a soil cover thickness of 2.5 m, and is closely adjacent to the existing Line 2 entrance.

The construction shaft in the southwest quadrant is close to the existing Line 4 H entrance, with the bottom of the shaft 4.9 m lower than entering and exiting the bottom plate; the large-span underground ticket inspection hall and the underground excavation channel connecting line 4 are adjacent to the entrance and exit of line 4 H, with clearances of 1.85 m and 8.25 m respectively. The deformation of the main body and interval structure of the station is required to be no more than 2 mm, and the deformation of the auxiliary structure is required to be no more than 3 mm. The deformation control requirements during construction are high; there is a high risk to the protection of cultural relics.

(3) Deformation control of important municipal pipelines under large-span underground excavation chambers

The underground ticket halls in the northwest and southwest quadrants are single-layer double-span arch roof straight wall structures, with shallow burial, large span, dense and old important pipelines above, and some pipelines have a settlement allowance of only 10 mm. Among them, the ticket hall in the southwest quadrant is close to the flat roof straight wall structure, with a span of 14.9 m, a height of 7.22 m, a top soil cover of 5.95 m, and the excavation surface is mainly located in the fine sand layer. The construction process of shallow buried large-span concealed excavation is complex, the difficulty of deformation control is high, and the safety risk is high.

(4) Construction of Shallow Buried Large Slope Undercover Excavation Channel with Upward Excavation

The Line 2 transfer passage in the northwest and southwest quadrants uses the CRD method for 30° upward excavation, with shallow soil cover, passing under municipal roads and pipelines, and some pipelines have a settlement allowance of only 10 mm. Among them, the Line 2 transfer passage in the northwest quadrant is 7.6 m wide and 8.87 m high, with a minimum soil cover thickness of 2.5 m, and it passes through a D1250 mm sewage pipe on its side with a horizontal clearance of only 0.9 m. How to ensure the safety of large slope upward excavation and control ground settlement and pipeline deformation is a challenge for this project.

2.3.7 *Beijing Subway Line 6 West Extension Project, Pingguoyuan Station*

1. Project Overview

(1) Introduction to Beijing Pingguoyuan Transportation Hub

Beijing Pingguoyuan Transportation Hub is a large-scale integrated transportation hub that combines rail transit, regular buses, rapid transit buses, and P + R parking. Subway Line 1, Line 6, and Maglev S1 Line all transfer here, and passengers traveling by various modes of transportation will achieve efficient and convenient three-dimensional transfers within the hub. Among them, the Pingguoyuan Station of the Line 6 West Extension Project is an important part of it, crossing the core position of the transportation hub from east to west, connecting the Subway Line 1 and regular buses on the north side, and the S1 Line and rapid transit buses on the south side. The underground transfer passage centered on the Line 6 transfer hall will provide the

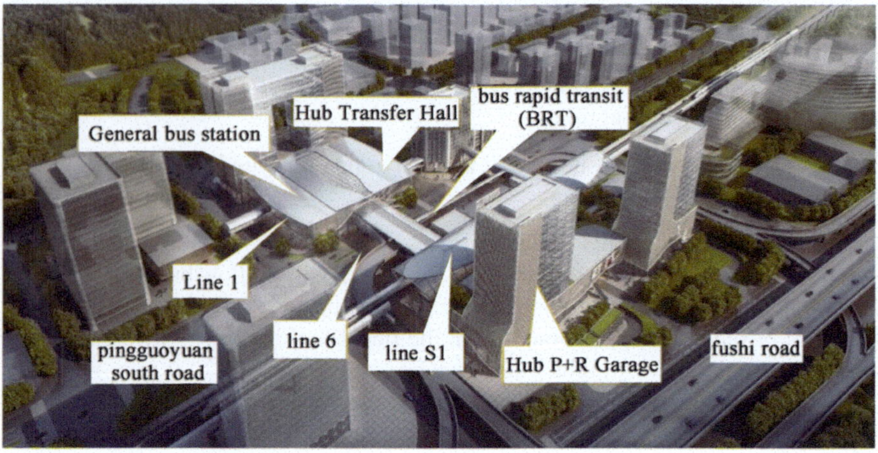

Fig. 2.34 Effect of Pingguoyuan transportation hub

most convenient transfer path for passenger traffic on the 3 railways; the expanded and renovated Subway Line 1 station and the underground space on the north side of the transportation hub are integrated, and the underground cross-street passage and station entrances and exits built simultaneously with Line 6, the ground city buses, underground rail transit, elevated rapid transit buses, and underground parking lots are seamlessly connected; the entrances and exits of Line 6 provide shuttle conditions for the commercial and business underground space on both the north and south sides. The effect of the Pingguoyuan Transportation Hub is shown in Fig. 2.34.

Controlled by factors such as ground traffic, Subway Line 1 station, underground pipelines, and construction sites, it is difficult for the Line 6 station to be constructed by open excavation. After comprehensive analysis, it is determined to adopt a method mainly based on under-cutting method. Considering the position of the Line 6 station in the integrated hub and the need to enhance the function of the transfer station, a through-type station hall double-layer building scheme is selected, and some sections are designed as three layers, arranging a convenient transfer hall. Combined with the construction period of Line 6 and the layout of the construction shaft, the construction planning requires the station structure to be connected as soon as possible to ensure that Line 6 is opened on schedule. The plan and section of the Pingguoyuan Transportation Hub are shown in Figs. 2.35 and 2.36 respectively.

(2) Introduction to Line 6 Pingguoyuan Station

The main structure of Line 6 Pingguoyuan Station is 324.4 m long and is divided into 5 sections.

① Standard sections at the east and west ends. The lengths are 145.5 m and 51.7 m respectively; the cross-section is a double-layer triple-arch structure, 23.3 m wide, 17.14 m high, with a top slab soil cover thickness of about 10 m, and a bottom slab burial depth of 26.8 m. Within the construction impact range,

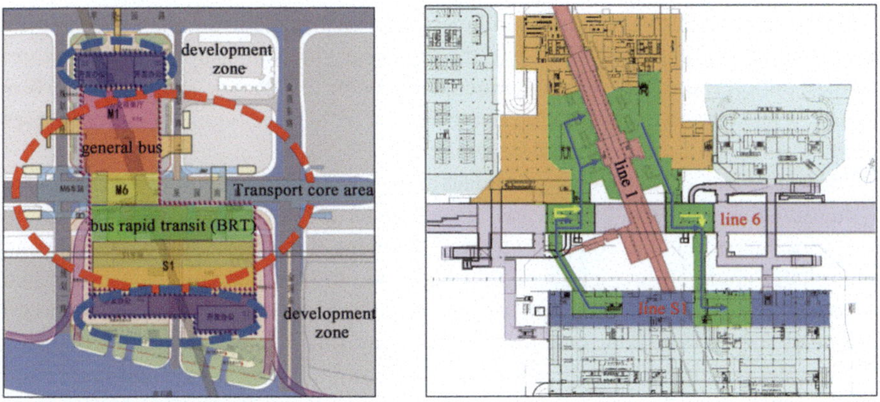

a) b)

Fig. 2.35 Plan of Pingguoyuan transportation hub

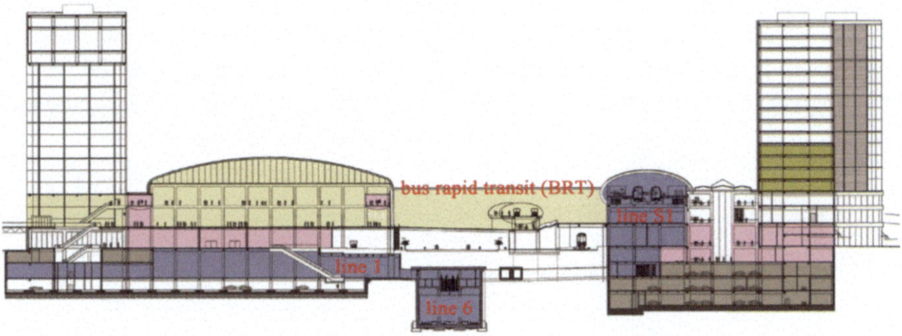

Fig. 2.36 Section of Pingguoyuan transportation hub

the stratum is mainly composed of cobblestones, with a large distribution of large diameters. Due to the constraints of existing construction equipment, it is decided to use the 8-guide hole PBA method for construction, and the side piles and middle columns are completed by manual drilling (the same below).

② Middle section. The total length is 52.4 m, obliquely crossing at 70° under the existing subway line 1 station, using a two-layer three-span box frame structure, with a maximum excavation width of 27.8 m, an excavation height of 16.24 m, a top slab burial depth of 11.8 m, and a bottom slab burial depth of about 27.0 m.

③ Three-layer section of the station. The three-layer section is divided into east and west parts, each section is 37.4 m long, located between the undercrossing section and the two standard sections at both ends, the cross-section is a three-layer three-span structure, 23.1 m wide, 22.50 m high, soil cover thickness 4.31–4.83 m, bottom slab burial depth about 27.5 m.

The second and third underground layers are constructed using the under-cutting method, with a height of 14.82 m, and the temporary top slab is a flat roof structure; the first underground layer is constructed by open excavation after the line 6 is opened, with an open excavation pit depth of 12.2 m, and the retaining structure uses retaining piles + internal support system.

The station has a total of 4 entrances and exits, 2 ventilation pavilions, 2 emergency evacuation channels, and 1 safety exit; two transfer channels are set up on the north side of the transfer hall to connect with the basement of the north plot of the hub, and two transfer channels are set up on the south side to transfer with the S1 line; two pedestrian overpasses are implemented at the same time on the top of the standard section of the station. The layout of the subway line 6 Pingguoyuan station, the relative relationship between the subway line 6 Pingguoyuan station and line 1, and the cross-section of the main structure of the subway line 6 Pingguoyuan station are shown in Figs. 2.37, 2.38 and 2.39.

(3) Surrounding environment

The subway line 6 Pingguoyuan station is located on the east side of the intersection of Pingguoyuan South Road and Fushi Road, set up along the north side of Pingguoyuan South Road from east to west, the station angle with the existing subway line 1 Pingguoyuan station is about 70°, and it crosses between the proposed Pingguoyuan

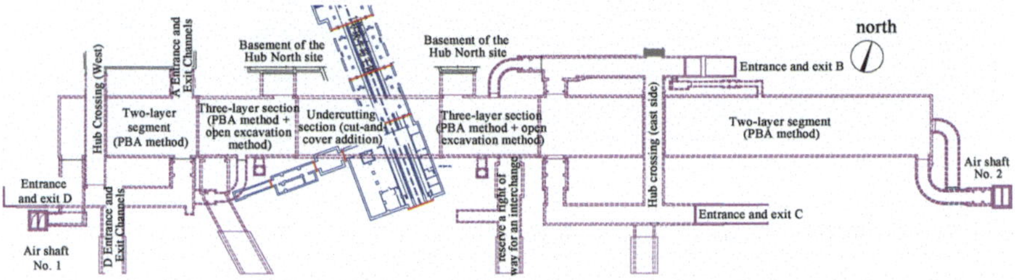

Fig. 2.37 Schematic diagram of the layout of Subway Line 6 Pingguoyuan station

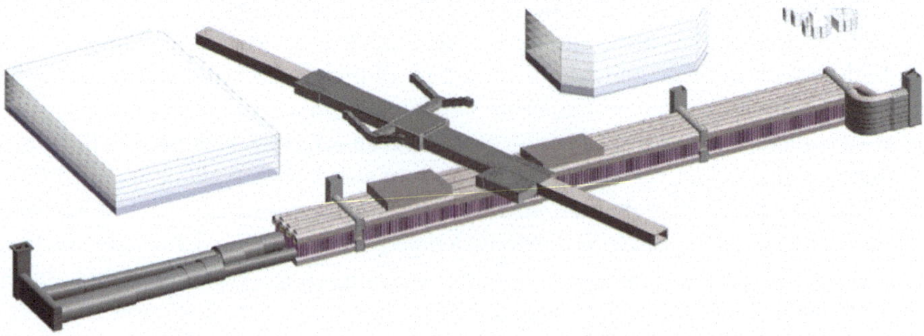

Fig. 2.38 The relative relationship between subway Line 6 Pingguoyuan station and Line 1

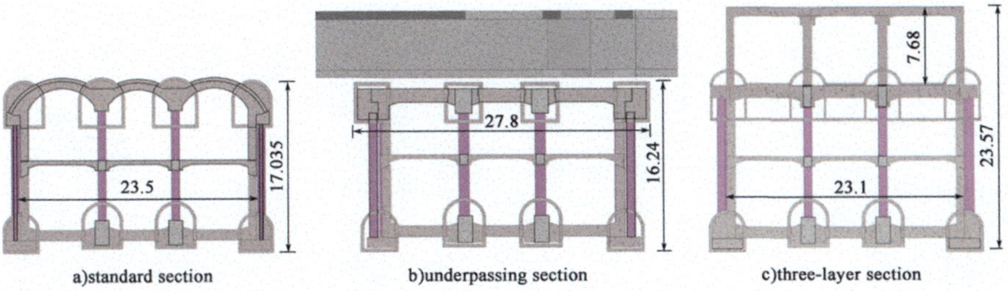

Fig. 2.39 Schematic diagram of the cross-section of the main structure of subway Line 6 Pingguoyuan station (unit of measurement: m)

hub south and north plots. The newly built station crosses under or is close to the existing subway, important pipelines, and underground passages in many places.

On the south side of the station location, Pingguoyuan South Road, there are 3–5 story shops (a total of 4 buildings, respectively 5 story, 3–4 story, 5 story and 3 story) and low-rise bungalows; to the south of the shops is the east–west Jingmen Railway, to the south of the Jingmen Railway is the proposed S1 line Pingguoyuan station (elevated station), the west section of the south side of the station is a bus station in use; the west section of the north side of Pingguoyuan South Road is a park, the middle section is the under-construction Pingguoyuan hub J plot building (basement depth 19.2 m, the upper structure is far from the station), the subway line 1 Pingguoyuan station and its affiliated structures, the proposed Pingguoyuan hub north plot, the east section near Pingguoyuan South Road is the under-construction Pingguoyuan hub business district I plot building (basement depth 19.4 m), further east is an 18-story residential building.

During the construction of Line 6 at the Pingguoyuan subway station, multiple pits and buildings within the surrounding comprehensive transportation hub were under construction at the same time. Due to the limited site, there was mutual interference; many bus routes pass through Pingguoyuan South Road, which is busy with traffic. The environment around the Pingguoyuan station is shown in Fig. 2.40.

2. Key difficulties of the project

(1) High precision control of settlement during underpass construction, strict control of settlement of existing Line 1 subway.

The top floor of the new Line 6 Pingguoyuan subway station closely underpasses the main structure of the Line 1 Pingguoyuan subway station, with a vertical clearance of only 0.27–0.50 m between the new and existing structures. During the excavation process, the dynamic load of the train greatly disturbs the excavated soil of the underpass section; the Line 1 subway station was built in 1966, and the subway operating company requires the settlement control of Line 1 to be within 3 mm, which not only ensures the safe operation of the existing line but also ensures the safety of the structure construction. Any carelessness could potentially affect the

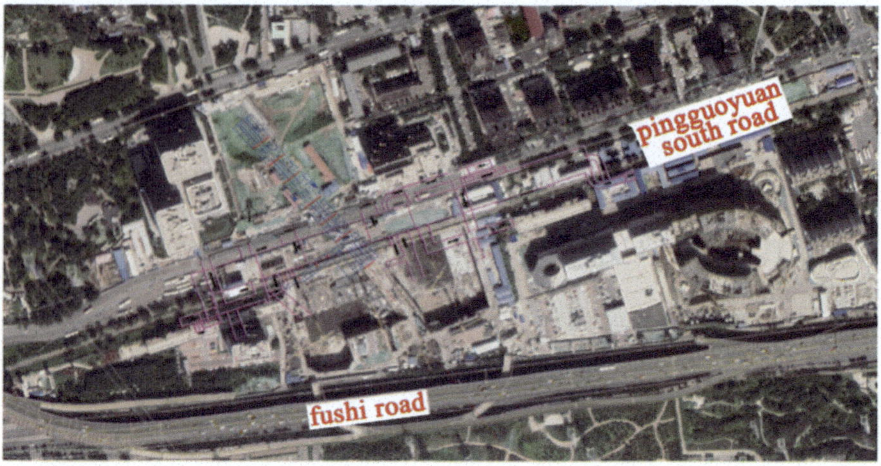

Fig. 2.40 Schematic diagram of the environment around Pingguoyuan station

normal operation of the subway, making the underpass of the operating station a major risk source for the project construction.

The Line 6 subway station simultaneously underpasses a certain passage affiliated with the Line 1 subway station, the affiliated structure of the station entrance and exit, multiple rain and sewage, water supply pipelines, and laterally passes through nearby ground buildings, involving 8 first-level risk sources. The construction safety and quality control of the underpass section have become the key difficulties of this project. The main risk sources of the underpass section are shown in Fig. 2.41, and the details are shown in Table 2.7.

(2) Construction of the pit for the additional layer on the operating station structure.

Due to factors such as the project schedule and implementation conditions, this project is constructed in two stages. The first stage uses the PBA method to complete the underground second and third layer structures. After the Line 6 subway is opened, the additional layer is excavated on its basis to construct the underground first layer (two transfer halls). The excavation size of the pit is 38.8×26.9 m, and the excavation depth is 12.2 m. Because the pit excavation is carried out above the operating station structure, some of the retaining piles are in the form of "hanging piles", and the additional layer excavation pit is close to the main structure of the south end of the Line 1 Pingguoyuan station and the military passage structure, involving 4 types of first-level risk sources. During the construction process, the control of the deformation of the retaining structure of the open- excavation foundation pit, the protection of the nearby Line 1 subway structure, the protection of the structure of the operating Line 6 subway below, and the control of deformation (bulging) are another focus of this project. The relationship between the cover layer foundation pit and its surroundings is shown in Fig. 2.42, and the main risk sources of the cover layer section are shown in Fig. 2.43, with details in Table 2.8.

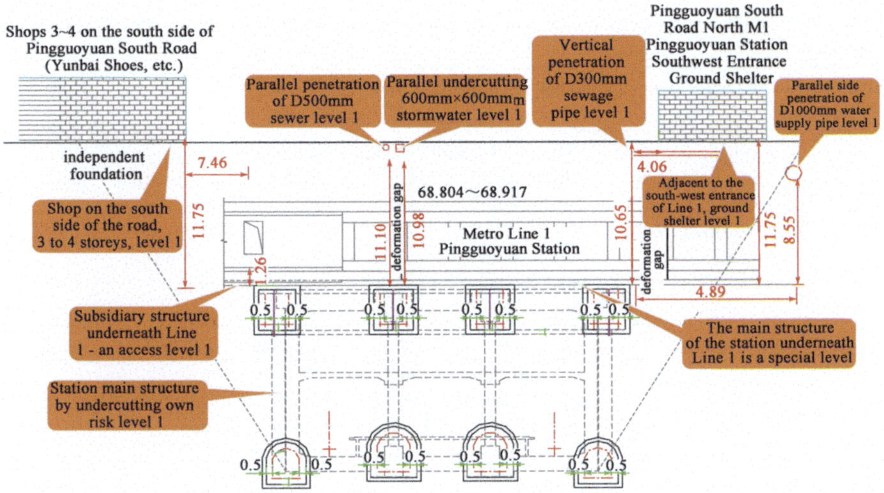

Fig. 2.41 Schematic diagram of the main risk sources of the underpass section (unit of measurement: m)

2.3.8 *Beijing Subway Line 17 Dongdaqiao Station*

1. Project Overview

(1) Project Introduction

The Dongdaqiao Station on Beijing Subway Line 17 is a transfer station, with a T-shaped passage for transfer with the existing Dongdaqiao Station on Subway Line 6, and conditions for transfer with the planned Line 28 are reserved. Dongdaqiao Station is located at the intersection of Dongdaqiao Road, Workers' Stadium East Road, Chaoyangmen Outer Street, and Chaoyang North Road. The station has a total of 4 entrances and exits, 3 safety exits, 2 transfer passages, and 2 sets of wind pavilions. The 4 entrances and exits are located in the four quadrants on both sides of Dongdaqiao Road and Gongti East Road. The layout of Dongdaqiao Station is shown in Fig. 2.44.

The newly built Dongdaqiao Station is an island station with a total length of 336.8 m and the maximum depth of the station floor is 34.102 m. The station axis direction is divided into three parts: the standard section on the south side is 250.6 m, the standard section on the north side is 49.7 m, and the middle underpass section is 36.5 m. Among them, the standard section is located on both sides of the existing tunnel of Subway Line 6 (middle island type); the underpass section of Subway Line 6 is divided into 2 single-hole tunnels (side platforms), with a net spacing of 5.2 m soil isolation, ensuring the safety of the existing Subway Line 6 tunnel. The relative position of the newly built station and the existing Subway Line 6 interval tunnel is shown in Fig. 2.45.

Table 2.7 Main risk sources for construction of downward layer addition in close-packed concealed excavation

S. No.	Name of the risk source	Description of the underlying risk profile	Risk level
1	Metro Line 6 stations are closely aligned with Metro Line 1 stations down to the next level	The main station of Metro Line 6 adopts a two-story three-span box frame structure for the downward additional layer section, which is constructed based on the PBA method. The maximum excavation width is 27.8 m, the excavation height is 16.54 m, and the depth of base plate is about 27.0 m	Special level
2	Station undercutting main structure	Flat-top PBA method concealed excavation construction, length 52.4 m, the whole section is located in the pebble layer, the maximum excavation width is 27.8 m, the excavation height is more than 16.5 m, and the thickness of the overlying loose soil is about 11.8 m	First level
3	Station concealed excavation main body under the subway line 1 station accessory structure-military corridor	Affected subsidiary a channel section has two, a section width 8.2 m, height 4 m; another affected section width 7.55 m, height 4.4 m, are open excavation construction. Subway line 6 station body under a channel, vertical minimum clearance 1.26 m	First level
4	Station main body downward added layer section adjacent to the metro Line 1 station southwest entrance ground shelter	Horizontal distance between the main structure of the additional floor and the existing entrance structure 4.60 m, vertical distance 11.75 m	First level
5	The main body of the station is adjacent to the stores on the 3rd–4th floors of the south side of Pingguoyuan South Road	The neighboring stores were built in 1994, frame structure, 3–4 floors above ground, no basement, independent foundation, the horizontal distance between the main structure of the station and the building is 7.46 m, and the vertical distance is about 11.75 m	First level
6	ϕ1000 water main on the side of the station body	Minimum distance from the pipeline to the roof of the structure 8.55 m, horizontal distance 4.89 m	First level

The standard section of the newly built station is a double-layer double-column three-span structure, 24.5 m wide and 17.85 m high, constructed using the PBA method. The standard cross-section of the station is shown in Fig. 2.46.

The middle underpass section is two single-layer single-hole tunnels with a net spacing of 5.2, 9.9 m wide and 9.52 m high, crossing the existing Subway Line 6 interval tunnel at an angle of 86°. The minimum net distance of the underpass is only 2.148 m, and the Center Cross Diagram Method (CRD method) is used for construction. The cross-sectional dimensions of the underpass station structure are shown

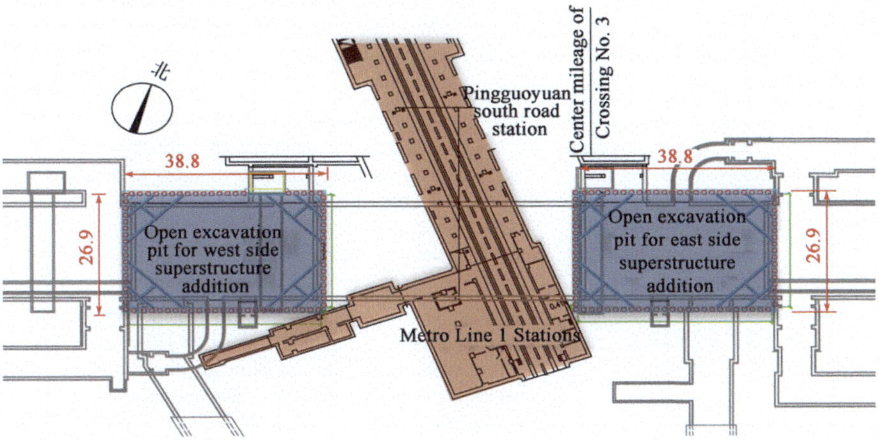

Fig. 2.42 Schematic diagram of the relationship between the cover layer foundation pit and its surroundings (unit of measurement: m)

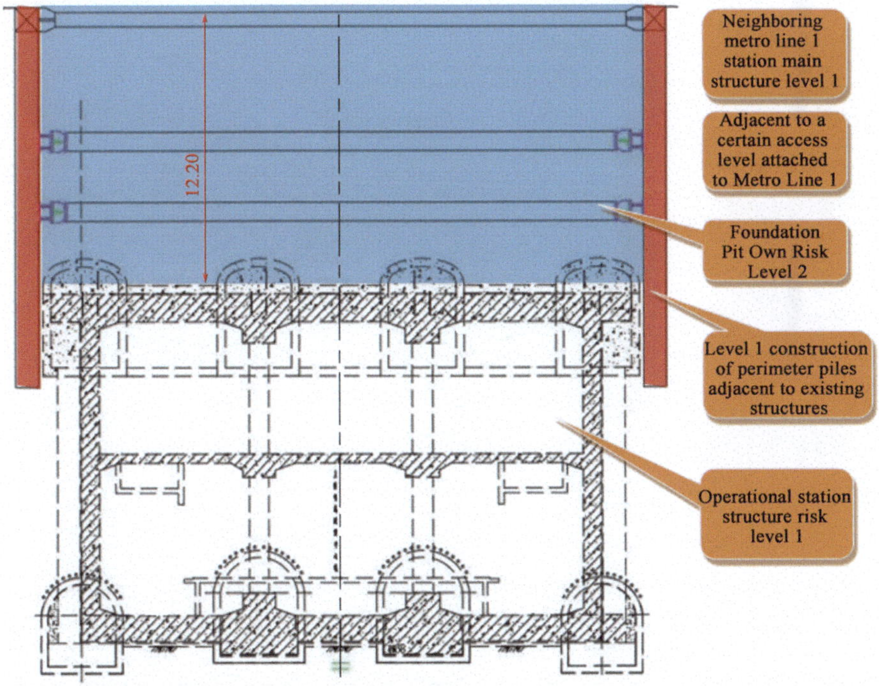

Fig. 2.43 Schematic diagram of the main risk sources of the cover layer section

Table 2.8 Main risk sources for construction of open excavation superstructure additions

S. No.	Name of the risk source	Description of the underlying risk profile	Risk level
1	Open excavation foundation pit for the west side transfer hall	The pit is 33.8 m in length, 26.9 m in width and 12.2 m in depth, and adopts the form of enclosure of piles and internal support; the enclosing piles at the east and west ends fall on the top arch of the station; the stratum of the pit is mainly pebble layer, and the pit is adjacent to the military access road, with a minimum clear distance of 2.3 m	First level
2	Open-excavation foundation pit for the east side transfer hall	The pit is 33.8 m long, 26.9 m wide and about 12.2 m deep, adopting the enclosure form of piles and internal support; the enclosure piles at the east and west ends fall on the top arch of the station; the stratum of the pit is mainly pebble layer; the pit is adjacent to the main body of Metro Line 1, with a minimum clear distance of 1.5 m	First level

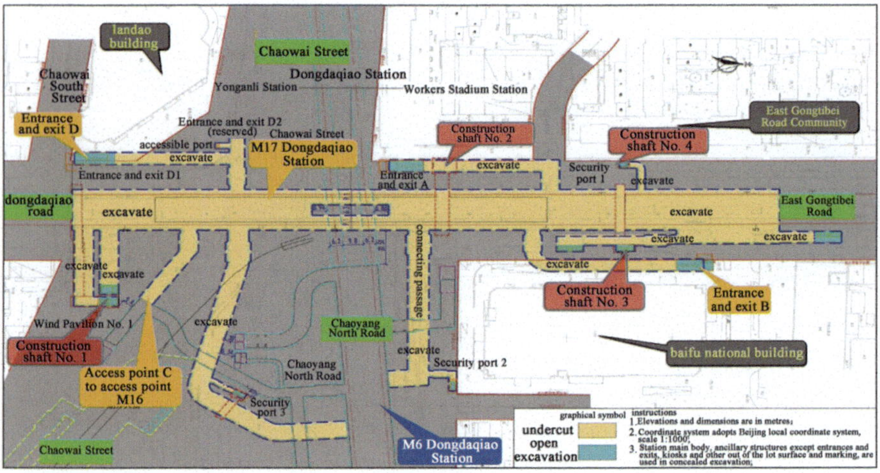

Fig. 2.44 Dongdaqiao station layout

in Fig. 2.47. During the construction period, a tubewell dewatering programme was used to carry out waterless undercutting operations. (2) Surrounding Environment.

In the northwest quadrant of Dongdaqiao Station on Subway Line 17, there is a residential area on Workers' Stadium East Road, the Blue Island Building in the southwest quadrant, the Baifu National Building and a bus station in the northeast quadrant, and the Dongdaqiao East Community in the southeast quadrant. The buildings are far from the underpass section (horizontal distance greater than 40 m), and the impact on construction is small; the surrounding environment of Dongdaqiao

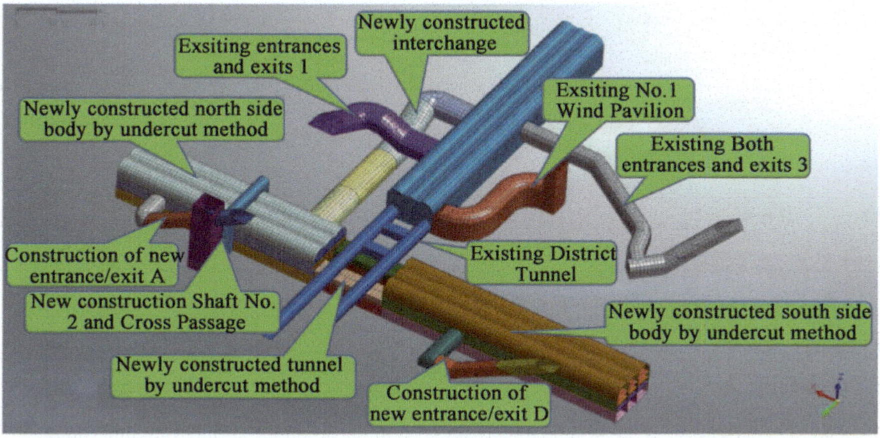

Fig. 2.45 Schematic diagram of the relative relationship between Dongdaqiao station on subway Line 17 and the eastward section of subway Line 6

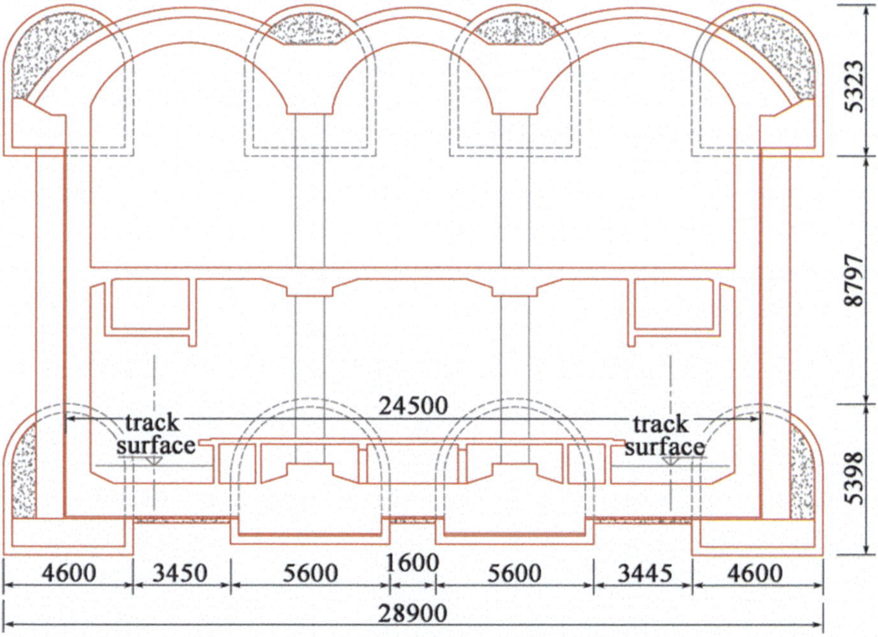

Fig. 2.46 Standard cross-section of the station (dimension unit: mm)

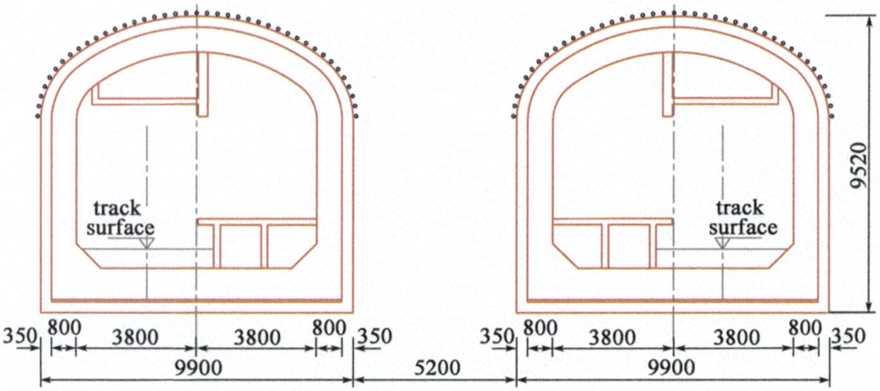

Fig. 2.47 Cross-section of the middle underpass section of the station (dimension unit: mm)

Station is shown in Fig. 2.48. Chaoyangmen Outer Street and Dongdaqiao Road, Workers' Stadium East Road have many bus lines, and the traffic is busy.

There are many municipal pipelines within the construction impact range of the newly built station, mainly including: 2000 × 2350 mm power tunnel; 4500 × 2800 mm thermal ditch, 2600 × 2300 mm thermal ditch, 1600 × 950 mm thermal ditch; D1050 mm sewage pipe, D1350–1550 mm variable diameter sewage pipe; 4500 × 3000 mm rainwater ditch; D1750 mm water pipe, D600 mm water pipe; 2 D500 mm sub-high pressure natural gas pipes, D300 mm low pressure gas, D400 mm

Fig. 2.48 Schematic diagram of the surrounding environment of Dongdaqiao station

low pressure gas, D406 mm sub-high pressure gas, D508 mm medium pressure gas pipeline total 4.

2. Engineering Difficulties

(1) Close Construction

The standard section of the newly built Dongdaqiao Station on Subway Line 17 is close to the structure of the eastward tunnel section of Subway Line 6, and the minimum horizontal net distance between the two is 3.48 m.

(2) Multiple Tunnels and Multiple Disturbances

Due to the construction of new station tunnels, cross tunnels, bored piles, and arch construction, the surrounding soil of the existing tunnel has been disturbed multiple times, causing deformation of the existing tunnel (Figs. 2.49 and 2.50). The deformation control of the existing operational tunnel is extremely strict. Reducing the impact of the excavation of the new station tunnel on the existing tunnel and ensuring the safe operation of the existing tunnel is a key focus of this project.

(3) Underpass construction

The middle underpass section of the new station consists of two single-layer single-tunnel tunnels with a clear distance of 5.2 m. The tunnel clearance is small, and there is significant interference during construction. At the same time, the two station tunnels pass under the existing subway line 6 tunnel at an angle of 86°. The minimum

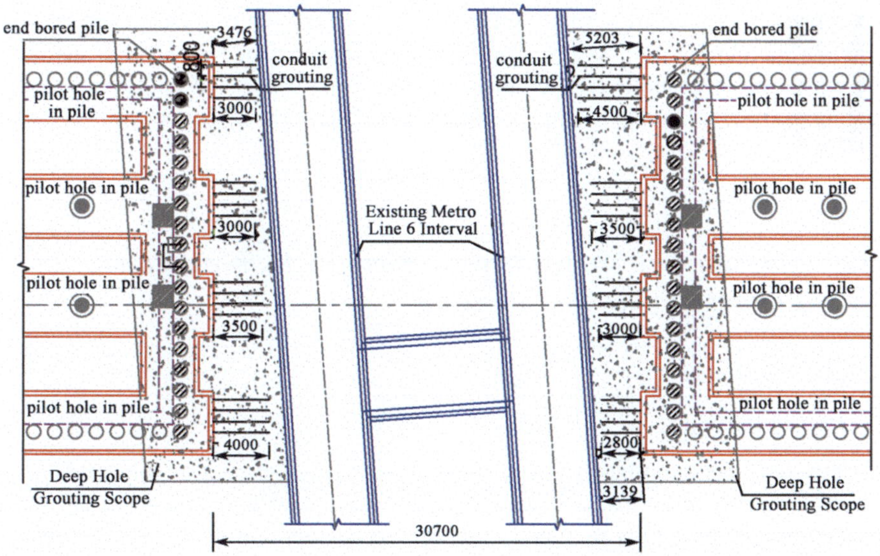

Fig. 2.49 PBA method station structure and subway line 6 tunnel plan relationship (unit of measurement: mm)

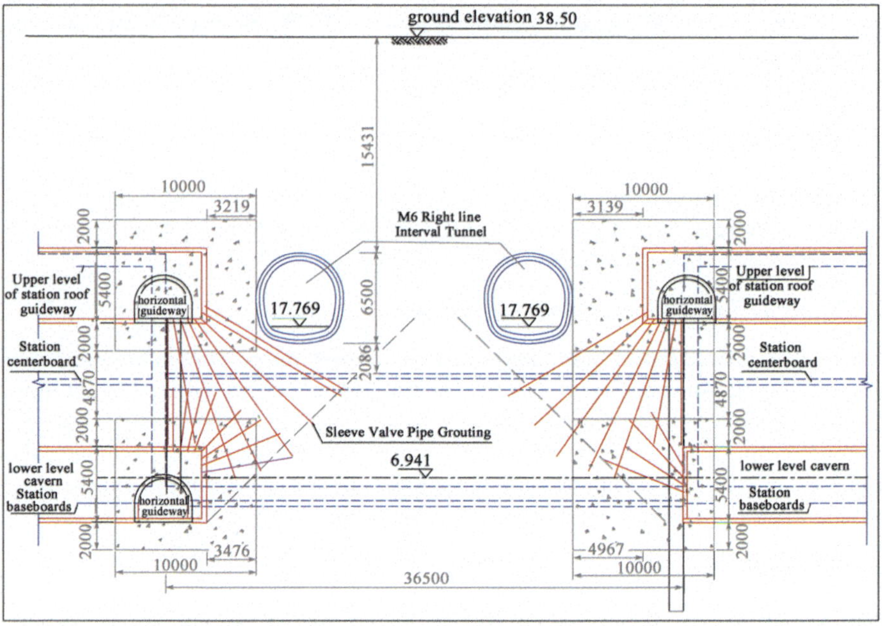

Fig. 2.50 PBA method station structure and subway line 6 tunnel cross-section relationship (unit of measurement: mm; elevation unit: m)

vertical clearance is only 2.15 m, which is far less than 1 times the tunnel diameter. The construction process is very likely to cause deformation of the existing tunnel, seriously affecting traffic safety. The project environmental risk level is special. Therefore, the close underpass of the separated small-clearance tunnel under the existing operational tunnel is a major difficulty in this project, as shown in Figs. 2.51 and 2.52.

2.3.9 Taiyuan City Yingze Street Underpass Railway Station Tunnel Project

1. Project Overview

(1) Project Introduction

The Yingze Street underpass railway station construction project in Taiyuan City is a controlling project for the eastward extension of Yingze Street and one of the main channels for Taiyuan City to expand eastward. Yingze Street splits into two lanes, up and down, in front of Taiyuan Railway Station, passing under the station from the north and south ends of the rain shed column. Taiyuan Railway Station is a century-old station, a special class railway station, connecting the Shitai passenger line, the

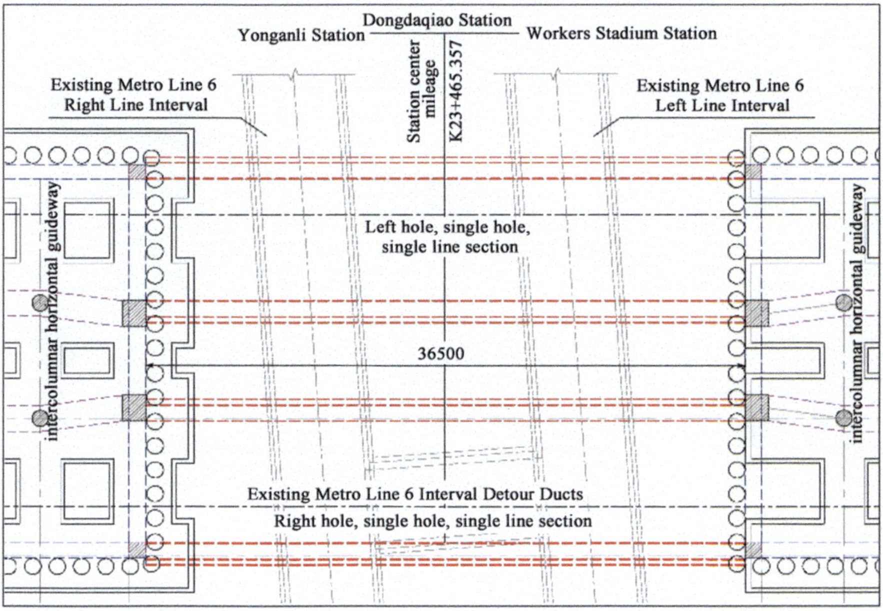

Fig. 2.51 Station single tunnel section and existing tunnel plan position relationship (unit of measurement: mm)

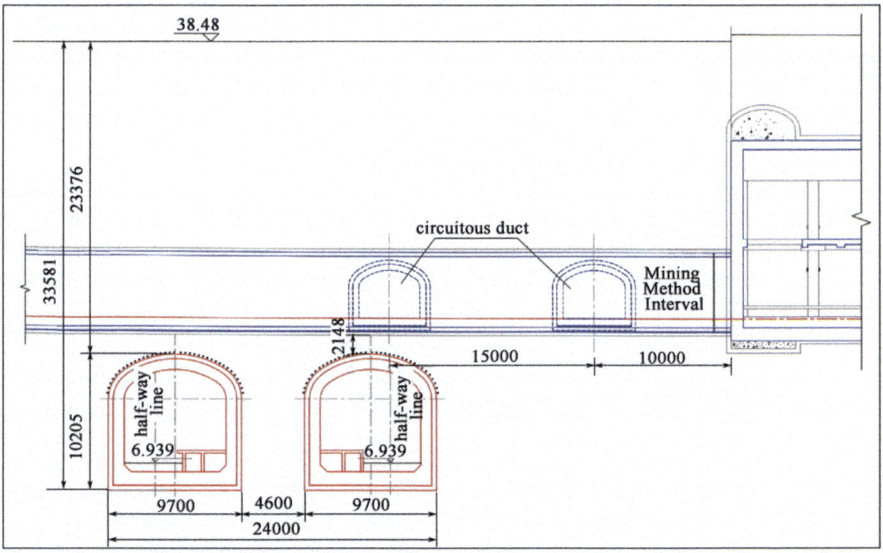

Fig. 2.52 Station single tunnel section and existing tunnel cross-section relationship (unit of measurement: mm; elevation unit: m)

Fig. 2.53 Project location schematic

Daxi high-speed line, the north–south Tongpu railway, the Shitai railway, etc., with more than 140 trains departing for all parts of the country every day. Therefore, the construction of the tunnel cannot affect the normal operation of Taiyuan Station. The main project of the underpass railway station is 2 1 × 15 m vehicle tunnels, with a total length of 463 m, of which the pipe -roof construction section is 210.1 m long (the north side vehicle tunnel pipe -roof section is 102.5 m long, and the south side vehicle tunnel pipe-roof section is 107.6 m long). The project location is shown in Fig. 2.53.

(2) Support Structure and Construction Method

The underpass section of this project uses 20 steel pipes with a diameter of 2000 mm and a wall thickness of 20 mm as the main support structure, with a pipe spacing of 165–235 mm. The full width of the pipe- roof section structure is 18.2 m, the full height is 10.5 m, the design speed is 30 km/h in the near term and 50 km/h in the long term. The cross-section of the pipe-roof section is shown in Fig. 2.54.

This project adopted the pioneering "new pipe-roof method" in the construction of the underpass section. The new pipe-roof method is a brand new undercut method. Its main construction steps are (Fig. 2.55):

① Outside the clearance profile of the tunnel, use a pipe jacking machine to jack large-diameter steel pipes into the soil to form a circle of steel pipe-roofs.
② Perform layering and segmentation for steel pipe cutting and inter-pipe welding within the steel pipe, forming a continuous, closed lattice steel structure.
③ Tie steel bars and pour concrete inside the steel pipe to form a high-rigidity integrated permanent support structure.
④ Excavate the earthwork under the protection of the support structure, and finally complete all underground structures.

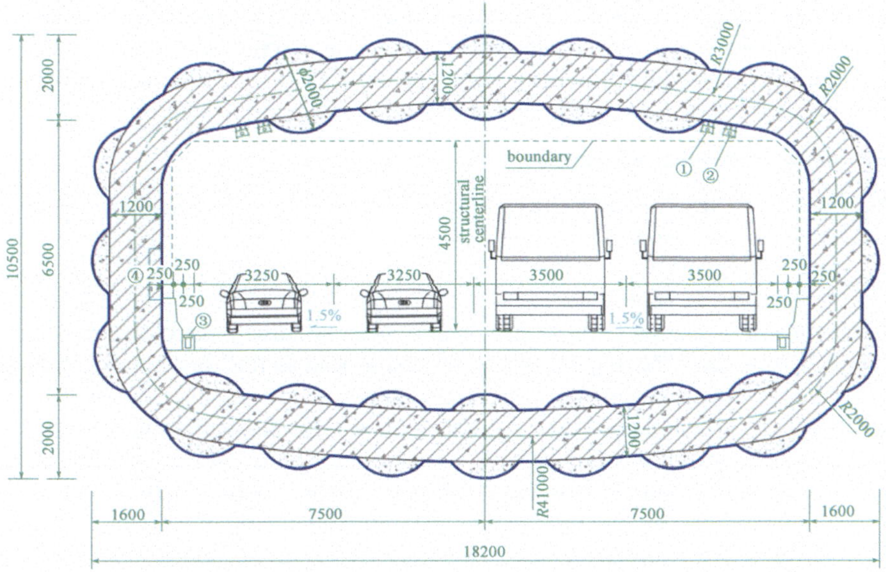

Fig. 2.54 Pipe-roof section cross-section diagram (unit of measurement: mm)

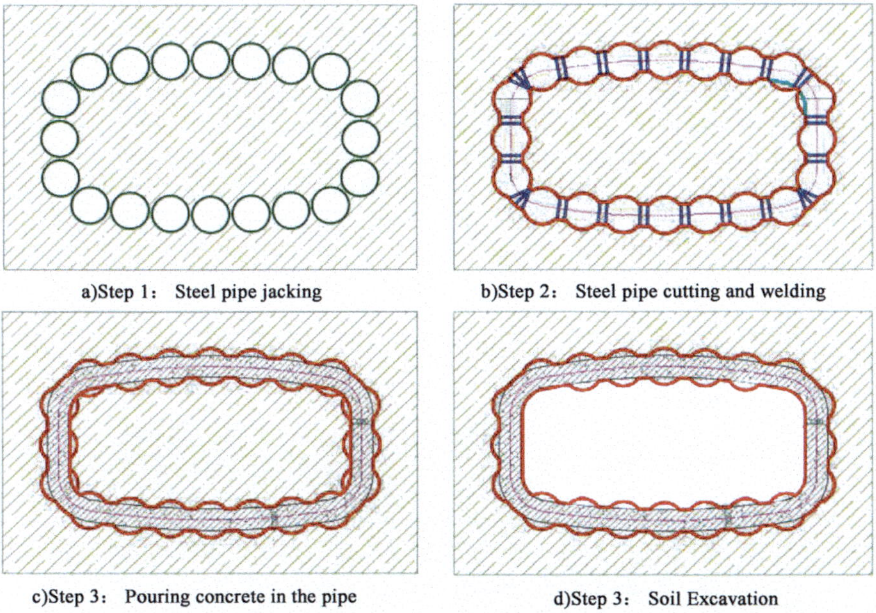

Fig. 2.55 Schematic diagram of the construction steps of the new pipe-roof method

The new pipe-roof method proposes the concept of solving geotechnical problems in a structural way, realizing the integration of temporary support and permanent main structure, adopting a disruptive construction process of structure first and excavation later, which is a brand new integrated design form of support structure.

(3) Surrounding environment

The passage crosses Taiyuan Railway Station, and the existing buildings (structures) around the passage are dense, mainly including:

① Railway

Taiyuan Railway Station is the intersection of several railway lines such as Shitai Passenger Line, Daxi Passenger Line, North–South Tongpu Railway, Shitai Railway, etc. There are 4 platforms in the station, 10 main lines and arrival lines, among which 1, 3, 4, 7–9 are arrival lines (P60 track, concrete wide sleeper, ballast thickness 0.35 m), 2, 10 are Daxi lines (P60 track, III type sleeper, ballast thickness 0.5 m), 5, 6 are Shitai passenger lines (P60 track, 5 is III type sleeper, ballast thickness 0.5 m; 6 is concrete wide sleeper, ballast thickness 0.35 m). The minimum distance from the top of the tunnel structure to the top of the rail is 3.5–3.9 m. The longitudinal section of the passage is shown in Fig. 2.56.

② Canopy foundation

After the renovation of Taiyuan Railway Station in 2008, a column-free canopy was adopted on the platform, with a diameter of 560 mm for the double-column steel structure. Under the column are 4 prestressed concrete pipe piles with a diameter of 500 mm, a pile length of 23 m, a single pile bearing capacity characteristic value of 770 kN, and a test pile loading single pile vertical limit bearing capacity of 1540 kN. The canopy foundations affected by the construction of the passage are mainly the 2nd to 3rd groups (north passage) and the 10th to 11th groups (south passage). The

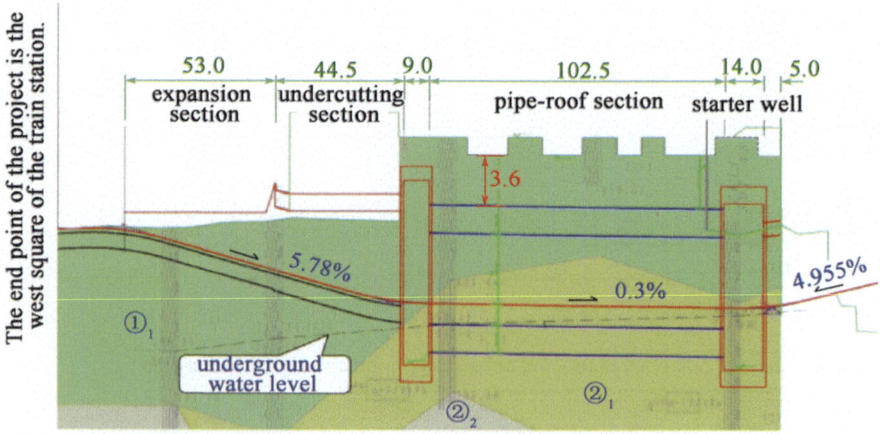

Fig. 2.56 Longitudinal section of the passage (unit of measurement: m)

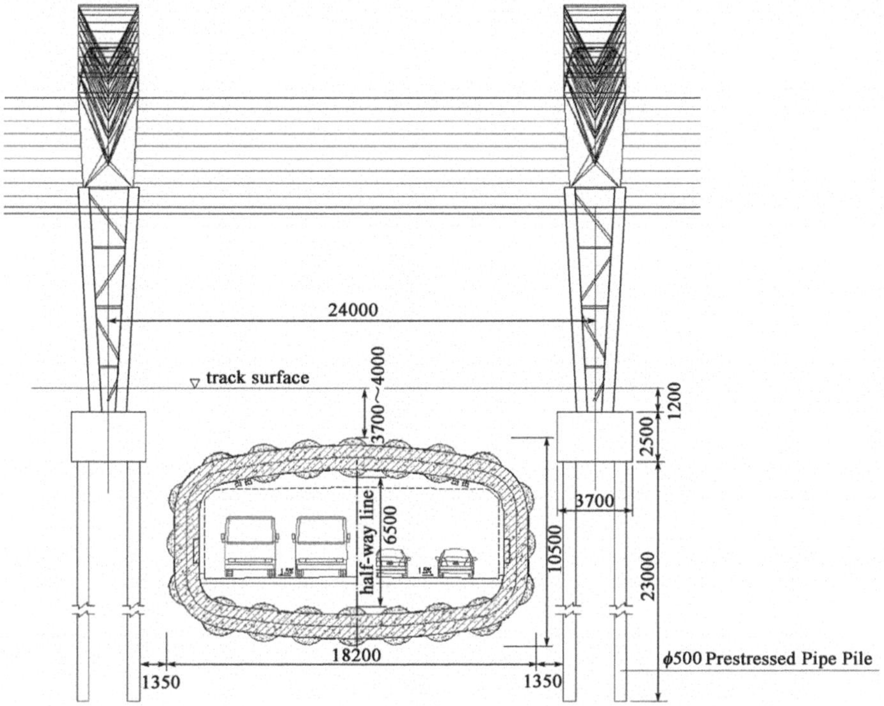

Fig. 2.57 Relationship between the canopy column and the cross-section of the underpass (unit of measurement: mm)

minimum net distance from the pipe-roof structure to the pile foundation under the column is 1.35 m. The relationship between the canopy column and the cross-section of the underpass is shown in Fig. 2.57.

③ Abandoned canopy foundation

Before the renovation of Taiyuan Railway Station, the platform was equipped with a reinforced concrete canopy. The foundation of the canopy column used a ϕ1200 mm bored pile with a pile depth of 10.0–13.8 m. During the renovation construction, all pile foundations were abandoned and left in the stratum.

④ Luggage tunnel

The south passage needs to pass under the existing luggage tunnel entrance and exit opening section. The entrance and exit opening section of the tunnel is excavated by slope, reinforced concrete structure, the thickness of the structure side wall and bottom plate is 45 cm. The minimum net distance from the pipe-roof to the luggage tunnel structure is 1.22 m. The relationship between the underpass and the luggage tunnel section is shown in Fig. 2.58.

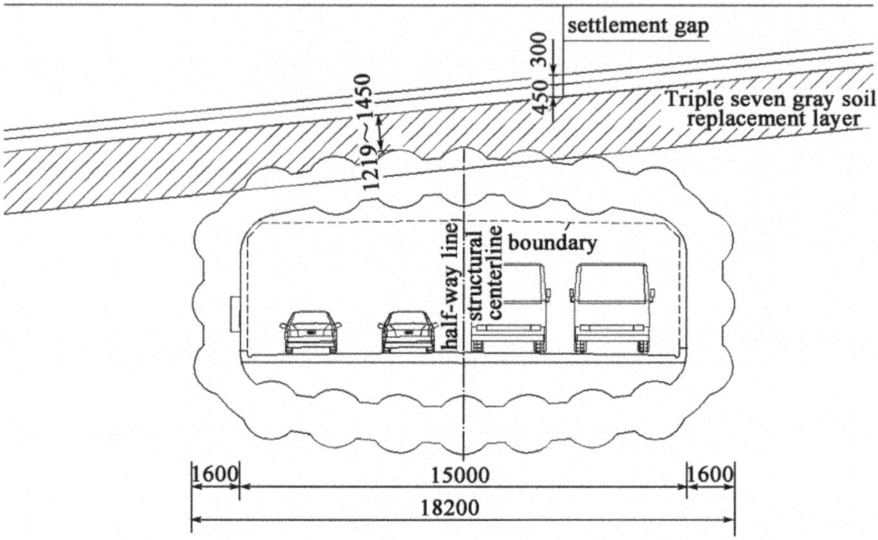

Fig. 2.58 Relationship between the underpass and the luggage tunnel section (unit of measurement: mm)

⑤ Station yard drainage culvert

According to existing data, after the longitudinal drainage ditch in the station collects rainwater, it is discharged into the drainage culverts in the south and north ends perpendicular to the direction of the track. The southern end drainage culvert is located between the 11th and 12th groups of rain shed columns, with the deepest measurement of the inspection wellbeing 6.2 m (at the platform); the northern end drainage culvert is located between the 2nd and 3rd groups of rain shed columns, with the deepest measurement of the inspection wellbeing 4.6 m (at the platform), and the culvert interferes with the northward underpass. The culvert on the north side is a reinforced concrete pipe with a diameter of 1.05 m. The distance from the top of the pipe-roof to the bottom of the culvert is only 250 mm. Before the pipe jacking construction, the culvert on the north side needs to be relocated. The relationship between the station yard drainage culvert and the pipe-roof structure cross-section is shown in Fig. 2.59.

⑥ Contact network and communication, signal, power cables and pipelines. The first platform has communication, signal cables, water supply pipes, drainage pipes, drainage ditches and other pipelines, which must be relocated and protected during on-site construction.

Between the 2nd and 3rd groups of rain shed columns, there are 2 existing contact network columns within the construction impact range of the north tunnel pipe-roof section. There are 3 contact network columns on the track corresponding to the axis of the 2nd group of rain shed columns, and 5 contact network columns on the track corresponding to the axis of the 11th group of rain shed columns, which are close

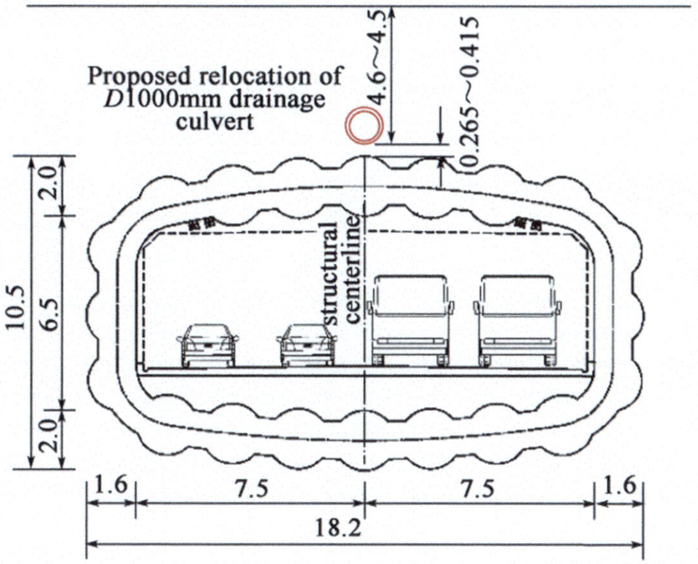

Fig. 2.59 Station yard drainage culvert and pipe-roof structure cross-section relationship (unit of measurement: m)

to the pipe-roof section structure, with the minimum net distance on the plane being 1.93 m.

⑦ Before the station renovation, the old contact network column foundation, according to the preliminary survey results, there may be old contact network foundations and reinforced concrete foundations within the underpass range, with the foundation buried depth being 3–4 m.

⑧ Railway operating buildings around the working well.

According to the site survey, the main station buildings within the range of the south line vehicle passage and pedestrian passage include the Taiyuan Railway Station Office Building and the Taiyuan Railway Housing Construction Section Taiyuan Housing Construction Heating Workshop; the main station buildings within the range of the north line vehicle passage include the Taiyuan Station North Line Package Room and the China Railway Express Business Hall. All these buildings need to be relocated before the construction of the project.

2. Engineering difficulties

(1) The construction method is novel, the process is complex, and there are few similar engineering cases that can be referenced. The underpass project of Taiyuan Railway Station adopts the pipe-roof pre-construction method, which has few domestic cases, and it is the first time to go under a railway station.

The key issue of the pipe-roof pre-construction method is the burial depth of the steel pipe. According to the design scheme, the actual overburden thickness of the pipe-roof is only 2.7 m. At the same time, the tunneling process of the pipe-roof pre-construction method is complex, with pipe jacking, mud jacket replacement, steel pipe cutting, steel plate welding, support column arrangement, and main body structure casting interlocking and mutually restrictive, making the construction difficult and technically demanding. The construction site is shown in Fig. 2.60.

(2) High construction risk of operating railway lines and nearby operating lines

According to the relevant regulations of the railway operation department, part of this project is carried out within the station, including pipeline relocation and line reinforcement, which is considered construction on operating lines; some projects such as pipe-roof working well construction, open excavation and buried section construction, and pipe-roof pre-construction are considered construction near operating lines. Given the important status of Taiyuan Railway Station, the safety of the railway operating lines must be ensured first during construction, which poses a high construction risk.

(3) Strict control of ground settlement

The tunnel passes under 4 platforms, 10 regular and passenger dedicated lines, and according to the "High-speed Railway Ballasted Track Line Maintenance Rules (Trial)", the maximum settlement of the subgrade is 10 mm. How to control ground settlement is the key to this project.

(4) There are many obstacles in the pipe-roof crossing area, making the crossing difficult

Fig. 2.60 Construction site of the underpass project at Yingze Street in Taiyuan City passing under the railway station

According to the survey, there are many obstacles within the station area that have been identified to affect the construction of the pipe-roof. Some obstacles can be clearly investigated for their specific location and depth, such as the drainage culvert on the north side; some obstacles can determine the depth and size but not the specific location, such as the old rain shed foundation, old contact network foundation; there are also some obstacles that cannot determine the depth and location, such as sleepers, steel rails and other discarded building materials that may be buried in the subgrade during station modification, and optical cables whose location cannot be determined.

During the pipe-roof jacking process, if small obstacles are encountered, these obstacles have poor stability underground. If the cutterhead is forced to excavate, the obstacles may roll in different directions in the soil, causing impact on the existing railway or the foundation of the existing rain shed; if large obstacles such as old contact network foundations, reinforced concrete culverts are encountered, special measures need to be taken. How the jacking machine deals with obstacles is the biggest problem faced by this project.

(5) The difficulty of construction coordination is high

Taiyuan Railway Station has undergone multiple renovations and expansions, with many buildings (structures) and pipelines, complex layout, and missing foundation data for some buildings (structures) and pipelines; construction needs to avoid rain shed foundations, existing luggage tunnels, contact network foundations, etc. Some pipelines and buildings (structures) need to be relocated, some need to be reinforced or protected, and the relocation and protection construction involves many railway departments, making coordination difficult and construction risk high.

(6) Limited construction conditions, low construction efficiency

The tunnel construction technology of the pipe-roof pre-construction method is complex, the jacking of the jacking machine, the replacement of the mud sleeve, the cutting of the steel pipe, the welding of the steel plate, the arrangement of the support column, and the casting of the main structure are interlocked and mutually restrictive. At the same time, the working environment is small, the working conditions are poor, the construction efficiency is low, and large-scale parallel operations cannot be carried out.

2.3.10 Hong Kong-Zhuhai-Macao Bridge Gongbei Tunnel Project

1. Project Overview

(1) Project Introduction

The Hong Kong-Zhuhai-Macao Bridge Zhuhai Connection Line Project is an important part of the Hong Kong-Zhuhai-Macao Bridge. It is a solution to the land passenger and cargo transportation requirements between the Hong Kong Special Administrative Region and the mainland (especially the west bank of the Pearl River) and the Macao Special Administrative Region. It is a new land transportation channel connecting the east and west banks of the Pearl River in the Greater Pearl River Delta Region, radiating the Pan-Pearl River Delta region; it is a key project connecting the sea bridge and tunnel main project of the Hong Kong-Zhuhai-Macao Bridge, improving the "Pearl River Delta Ring" of the national highway network and the "Pearl River Delta Outer Ring Expressway" of the Guangdong Province highway network.

The construction of this project is of great significance for improving the comprehensive transportation system and highway network of the country and Guangdong, Hong Kong and Macao, strengthening the economic and social ties between the west bank of the Pearl River and Hong Kong, improving the investment environment of the west bank of the Pearl River, enhancing the comprehensive competitiveness of the Pearl River Delta, maintaining the continuous prosperity and stability of Hong Kong and Macao, and promoting the coordinated economic and social development of the two banks of the Pearl River. The route generally runs from east to west.

The Gongbei Tunnel of the Zhuhai Link of the Hong Kong-Zhuhai-Macau Bridge is a key project of the Zhuhai Link. The tunnel is a double-deck six-lane tunnel in both directions, passing under the restricted area of the Gongbei Port, with two working wells set up (one on each of the east and west sides). The geological conditions of the project are complex and the surrounding environment is sensitive. The plan and cross-section of the tunnel are shown in Figs. 2.61 and 2.62 respectively.

(2) Support system and excavation method

The tunnel uses a pipe-roof for advanced support, which consists of 36 ϕ1620 mm steel pipes, half of which are integrated with the initial support; the method of freezing in sections is used for water stopping between the pipe roofs, with the minimum

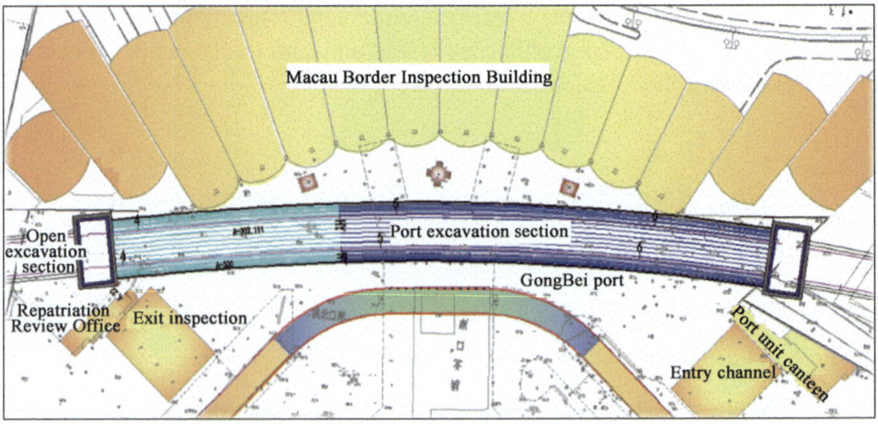

Fig. 2.61 Plan of the Gongbei tunnel

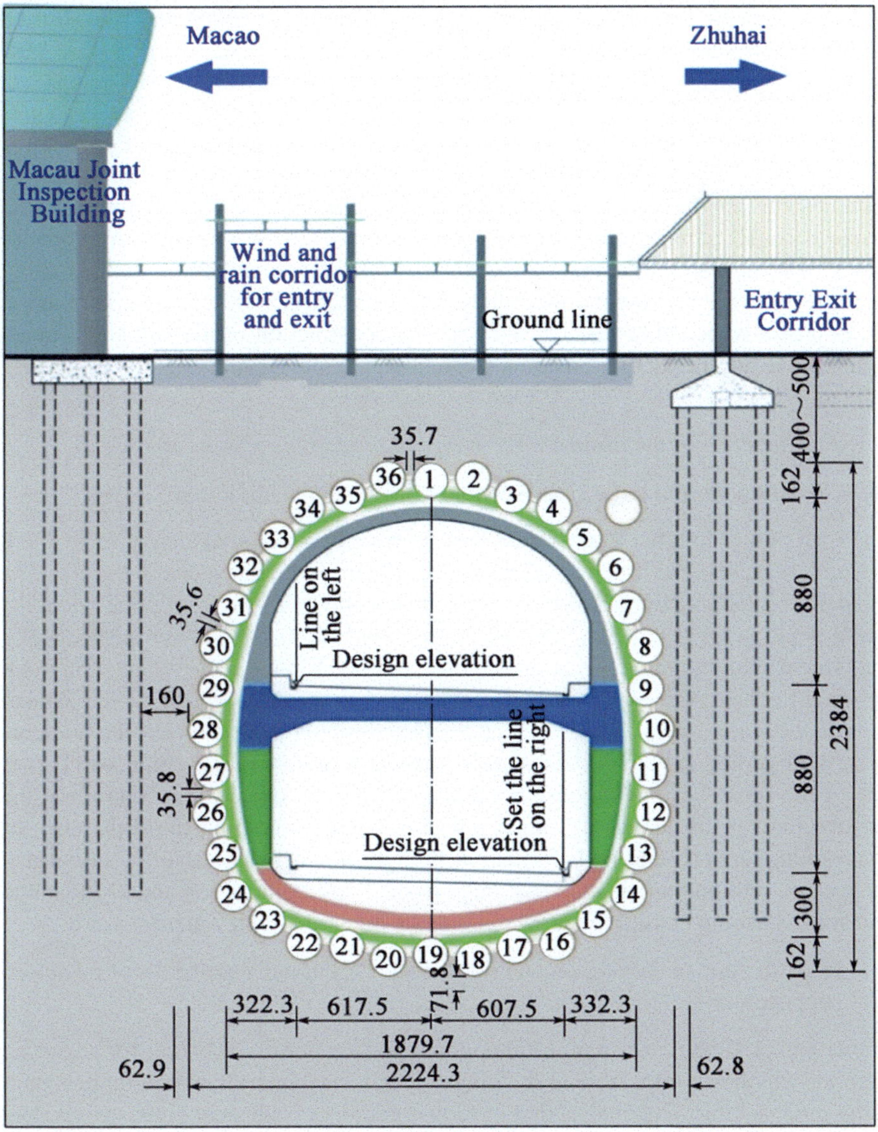

Fig. 2.62 Cross-section of the Gongbei tunnel (unit of measurement: cm)

thickness of the frozen soil curtain designed to be 2 m and the maximum thickness not exceeding 2.6 m; the multi-step method is used for step-by-step excavation, with an excavation width of 18.8 m; the excavation height is 21.0 m, and the excavation cross-sectional area is 336.8 m^2; three times compound lining support; the thickness of the soil cover on the top of the pipe roof is 4–5 m. The tunnel is located on a transition curve and a circular curve with a radius of curvature of 885.852–906.298

m, with a longitudinal slope of 0.35%. The sequence of step-by-step excavation of the tunnel is shown in Fig. 2.63.

(3) Surrounding environmental conditions

The tunnel passes under a narrow strip of land between the Gongbei Port and the Macau Border Gate, close to the Macau Border Inspection Building. The minimum distance between the tunnel outline and the building is only 1.46 m, and the minimum distance from the tunnel pipe-roof to the pile foundation of the building's rain corridor is only 0.46 m. The port handles an average of 10,000 vehicles entering and exiting the country every day, with a total daily flow of about 300,000 people. Due to the heavy customs clearance tasks at the port, the construction of the Gongbei Tunnel cannot have any impact on customs clearance, making the construction extremely difficult.

2. Key difficulties of the project

(1) Construction of a large cross-section tunnel under shallow sea

The span of the Gongbei Tunnel is 18.8 m, with a cross-sectional area of up to 336.8 m^2, and the actual construction area exceeds 400 m^2, which is close to the size of a standard basketball court. However, compared to the large cross-sectional area, the burial depth is less than 5 m, and the cover span ratio is only about 0.25, making it a typical shallow tunnel. In addition, the Gongbei Tunnel is the connection point between the sea excavation section and the land excavation section of the Zhuhai Link of the Hong Kong-Zhuhai-Macau Bridge, close to the coastline of the Gongbei Port, with geological conditions of soft and water-rich strata near the sea. This is one of the most extreme adverse conditions for tunnel excavation. Therefore, the construction of the Gongbei Tunnel not only has to overcome the difficulties of excavation, support, and arch formation brought about by the shallow burial and large cross-section, but also face the huge risks of soil liquefaction and sudden mud and water gushing caused by large amounts of water and high water pressure.

(2) Construction of large-diameter long-distance curved pipe-roof for advanced support

The Gongbei Tunnel uses a pipe canopy with a diameter of 1620 mm for advanced support, which is comparable to the diameter of some water diversion tunnels built using pipe jacking. This means that the construction of each steel pipe is equivalent to building a small diameter tunnel. In addition, the Gongbei Tunnel has a dark excavation section with a curved structure that is 255 m long, and the line shape is a combination curve. The length of the pipe canopy is currently the longest in the country, and the combination form of the pipe canopy is a first in the country and rare in the world. Under the current technical level, the construction accuracy of long-distance curved pipe jacking can be basically controlled when the line shape is a circular curve, but there is no mature solution for the accuracy control of the combination curve. Because the distance between adjacent pipe jacks in the Gongbei Tunnel is only 355–358 mm, if the accuracy of the pipe jacking is not properly

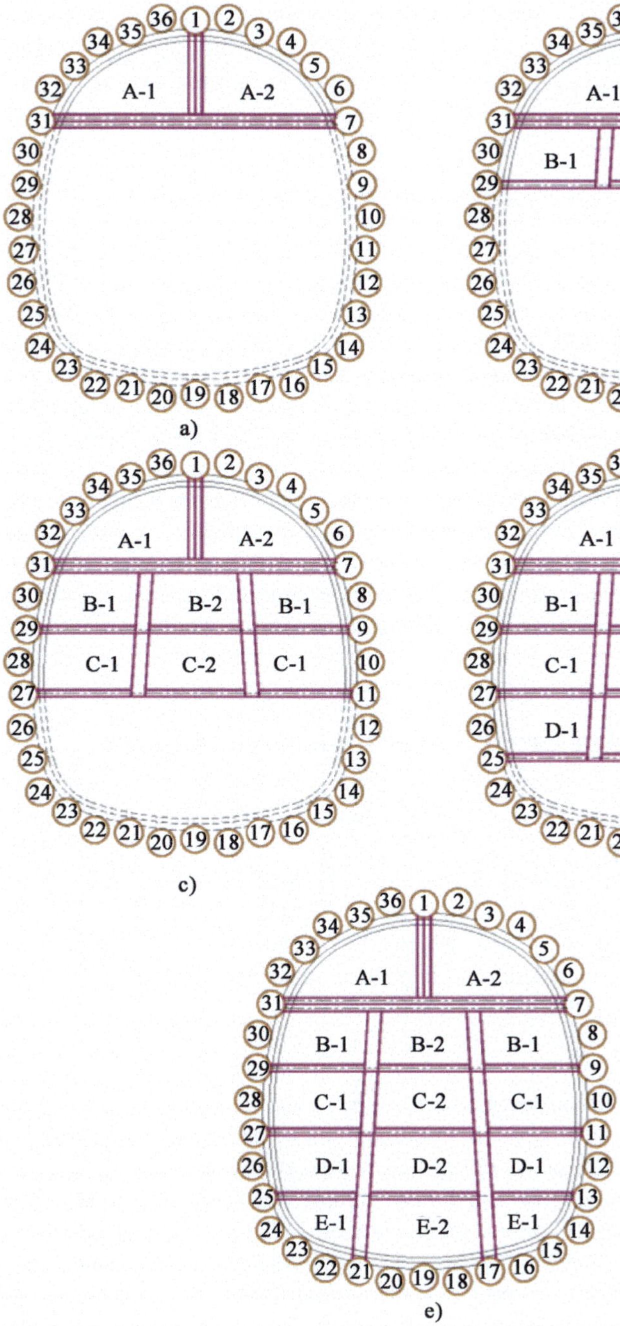

Fig. 2.63 Dark excavation section partial excavation sequence diagram

controlled, it will at least affect the subsequent freezing reinforcement and water stop effect, or it will occupy the position of the adjacent pipe jacking, making construction impossible. How to control the accuracy of the large diameter long-distance curved pipe canopy has become a major problem that needs to be solved in the construction of the Gongbei Tunnel.

(3) Large-scale pipe canopy freezing method reinforcement

The Gongbei Tunnel is the first large-scale project in the country to actually apply the pipe canopy freezing method. Because the tunnel is curved, conventional horizontal drilling cannot be implemented, so it is necessary to arrange the freezing pipe inside the pipe canopy steel pipe; at the same time, to ensure sufficient rigidity, after the pipe canopy is jacked, concrete will be poured into half of the steel pipes to form solid steel pipes, which are alternately arranged with hollow steel pipes. Under this condition, whether the freezing method can form a sufficiently thick frozen soilcurtain is a question that needs to be verified. In addition, under the dual requirements of construction water stop and settlement control, it is necessary to prevent water leakage caused by insufficient freezing, and to prevent frost heave caused by excessive freezing. This requires that the thickness of the frozen soilcurtain cannot be too thin or too thick, and it must remain stable under repeated thermal disturbances during excavation. This poses a great challenge to freezing technology.

2.4 Main Features of Urban Large Underground Space

According to the definition, urban large underground space include but are not limited to new construction projects and existing underground facility expansion projects for transportation, municipal, public management and services, commerce, logistics warehousing, disaster prevention and mitigation, etc. These projects are generally highly socially concerned key projects, which are of great significance for improving the quality of life of urban residents and increasing resident satisfaction. They have high social value, and their main features are as follows.

1. Vastness of the spatial scale

The primary feature of urban large underground space is their large vastness of the spatial scale, which directly leads to the amount of earthwork excavation, the depth of foundation pit excavation, or the area of dark excavation section far greater than general underground engineerings. For example: the construction area of the Wuhan Optics Valley Square Complex project is about 160,000 m^2, and the earthwork excavation volume exceeds 1.8 million m^3; the maximum depth of the foundation pit excavation of the Chongqing Shapingba Railway Hub Comprehensive Renovation Project reaches 45 m, and the earthwork excavation volume is about 1.36 million m^3; the actual construction section area of the single tunnel dark excavation section of the Hong Kong-Zhuhai-Macao Bridge Gongbei Tunnel exceeds 400 m^2, which is approximately the size of a standard basketball court.

As is well known, as the vastness of the spatial scale of underground structures expands, the load they bear will significantly increase, and the requirements for excavation and support methods and processes will also be markedly different; in addition, the expansion of the vastness of the spatial scale of underground structures will also expand the range of environmental impacts during the construction process, thereby making their planning and design difficulties incomparable to those of small-scale underground structures. It can be said that it is precisely because of the significant differences in construction difficulty, construction investment, and risk that it becomes extremely necessary to distinguish urban large underground space from general underground structures. Therefore, vastness of the spatial scale is the most prominent feature of urban large underground space.

2. Integrated functionality

Due to the increasing scarcity of urban above-ground space, future urban development will inevitably extend more and more to underground space. Urban large underground space, as a natural product of this trend, will inevitably also undertake more traditional urban functions. Wuhan Optics Valley Square not only serves as the transfer station for Wuhan Metro Lines 2, 9, and 11, but also integrates the functions of Lumo Road Tunnel, Luoyu Road Tunnel, and non-motorized vehicle passages; Tianjin Metro Line 5 Siyuan Road Station, after reconstruction, has formed a combination of underground rail transit, underground commerce, and underground parking, effectively utilizing the commercial added value brought by rail transit; after the transformation of the railway hub in Shapingba District, Chongqing, a modern urban core comprehensive transportation hub integrating high-speed rail, urban rail transit, road traffic, and pedestrian traffic has been formed, and it is directly connected with existing high-end commerce, hotels, office buildings, apartments, etc., forming a functional unity. It can be foreseen that the development of future urban large underground space will inevitably further develop in the direction of functional synthesis, in addition to traditional transportation, office, and commercial functions, it will also undertake diversified functions such as civil defense, medical care, and urban communication, energy and environmental control centers (Fig. 2.64).

Fig. 2.64 Diversified functions of urban large underground space

3. Unified overall planning

Compared with above-ground space, the development history of urban underground space is shorter, which determines that there are not many successful experiences in underground space development to learn from, and it is necessary to "cross the river by feeling the stones". In addition, our cities are in a period of rapid development, accompanied by the people's increasing demand for quality of life, the scale of urban underground space construction is rapidly expanding, making many historical problems in the early development of underground space gradually exposed. Due to limitations in technical level, outdated planning and design theories, and fragmentation of urban management functions, the early development of urban underground space has a certain degree of disorder and blindness, leading to a lack of foresight and integrity in urban underground space planning, complex and interwoven pipelines, poor connectivity and integration of buildings, artificially creating many hidden dangers in future new construction projects.

The large underground space in cities has emerged alongside the intensive development of cities, and it is also a concentrated reflection of the efficient use and comprehensive development of urban resources. Its prominent features are unified planning, comprehensive development, future-oriented, top–bottom coordination, complementary functions, supporting construction, and people-oriented. For example, in the comprehensive utilization plan of the underground space along the east–west axis of the Xiong'an New Area in Hebei, the first underground layer is filled with underground space throughout the area, mainly underground pedestrian passages, connecting all blocks, including crossing peripheral roads, connecting with adjacent blocks, and setting up underground businesses along pedestrian roads as needed, forming a connected underground city, as shown in Fig. 2.65. The pedestrian system is multi-layered. In addition to the two-layer viewing platform set up by the main city trunk roads combined with the buildings on both sides, the ground of the People's Axis is for landscape, cycling, jogging, and sightseeing, and a longitudinal pedestrian system is set up on the first underground layer, connecting with the rail transit transfer hall and the plots on both sides. Considering the control of scale and necessity, each side of the longitudinal direction within the 200-m red line is 60 m wide, and a transverse contact channel is set up along the longitudinal direction every 500 m. Consider setting up centralized business and public service facilities within the pedestrian system. The second underground layer is a comprehensive pipe gallery layer, integrating municipal pipelines such as electricity, communications, gas, heat, and water supply and drainage, not only avoiding the disorderly layout of underground pipelines causing disorderly division and waste of underground space, but also facilitating the operation and maintenance of pipelines. The third and fourth underground layers are intelligent logistics routes and rail transit layers, undertaking the function of networked and efficient connection of regional transportation. The functions of each layer are clearly divided and well coordinated, and the function of the underground space is coordinated with the urban spatial structure, supported by the rail transit network, forming a "one axis, five hearts, multiple points" urban underground space layout structure.

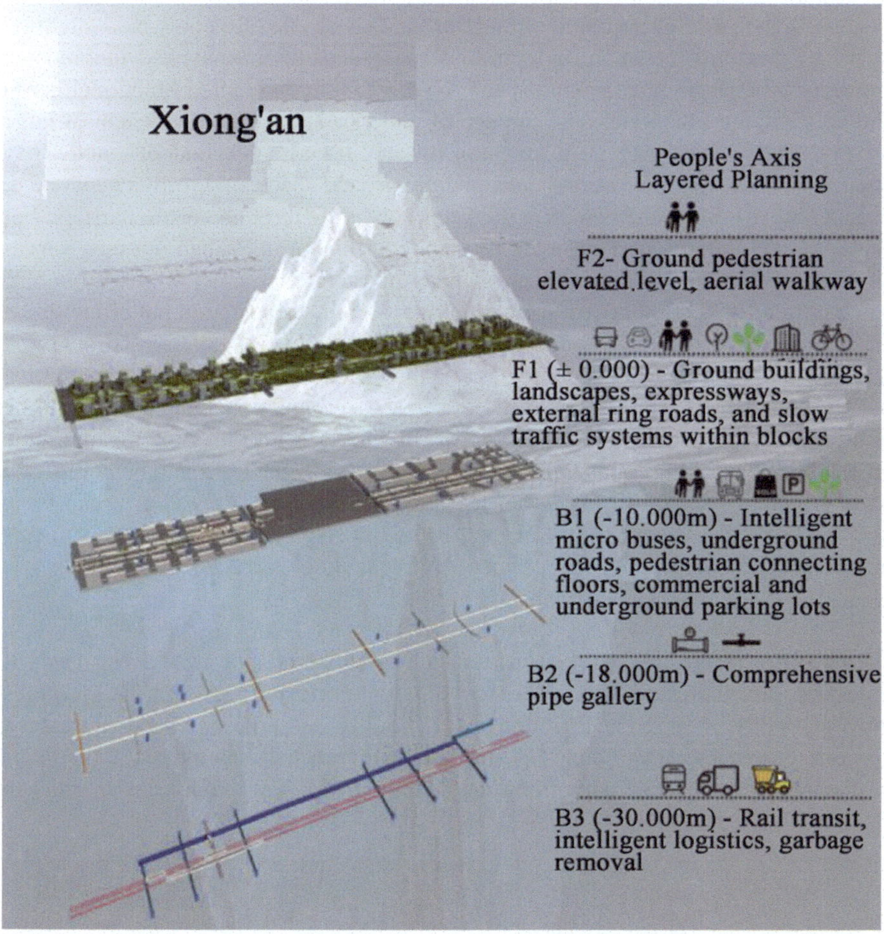

Fig. 2.65 Comprehensive utilization plan of underground space along the east–west axis of Xiong'an new area

It can be seen that the unified planning of urban large underground space can effectively strengthen the urban functions of the central area, centrally solve a series of contradictions such as land shortage, space congestion, traffic congestion, and environmental degradation in urban ground space planning and construction, avoid the waste of underground space resources and low development benefits caused by isolated development, single function, and lack of facilities in the underground space, and fully play the advantages of functional agglomeration of underground buildings.

4. Networked and scalable

The inherent characteristics of underground space determine that it is inevitably a "easy to divide, difficult to combine, easy to block, difficult to exchange" spatial system. In the past development of underground space, we often only paid attention to the connectivity between underground space and the ground, and ignored the connection between underground space. To meet more diversified functional needs and enhance the functional convenience of underground space, the design of large urban underground space must abandon the traditional "independent" mode, pay attention to the interconnection between underground space, and form a networked, expandable overall space, which is the key to breaking through spatial barriers. For example, the underground complex of Guanggu Square in Wuhan not only serves as a transportation node for multiple subway lines and municipal projects, but also has multiple connection nodes, forming a one-stop interconnection pattern with the surrounding existing underground space, commercial squares, existing stations, etc.; Dongdaqiao Station of Beijing Subway is a three-line transfer node of Subway Line 6, Line 17 and the planned Line 28. After the overall completion, Dongdaqiao Station will form a huge underground space network, where all lines can interchange with each other in both directions without obstacles, and all functions can be completed underground, as shown in Fig. 2.66. It can be said that the "large" of urban large underground space is not only reflected in its huge spatial volume, but also in its huge scale and convenient interconnection of spatial network structure.

5. High Quality

Urban large underground space serves the high-quality production and living needs of the people, and therefore must adhere to the "people-oriented" concept. This requires that the quality of large underground space should meet the user's needs for safety, comfort, efficiency, and greenness. Due to the comprehensive nature of the functions of underground space, people often spend a long time below the ground. In order to

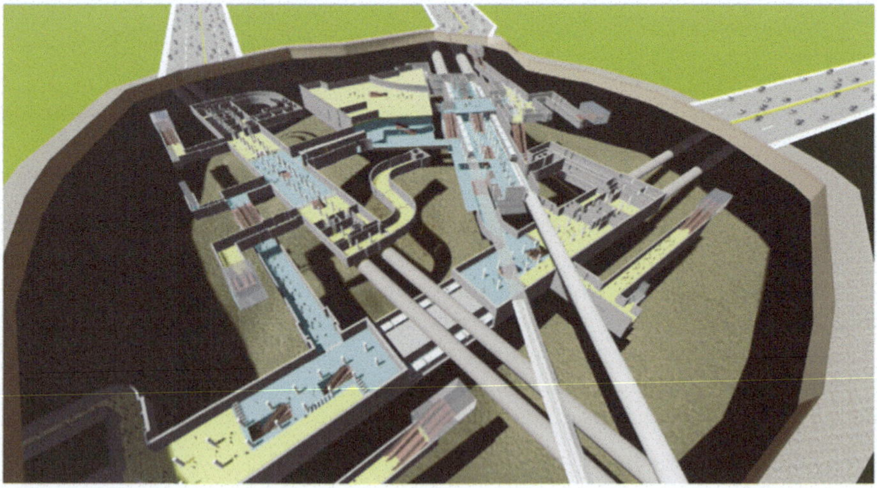

Fig. 2.66 Future spatial layout of Dongdaqiao station

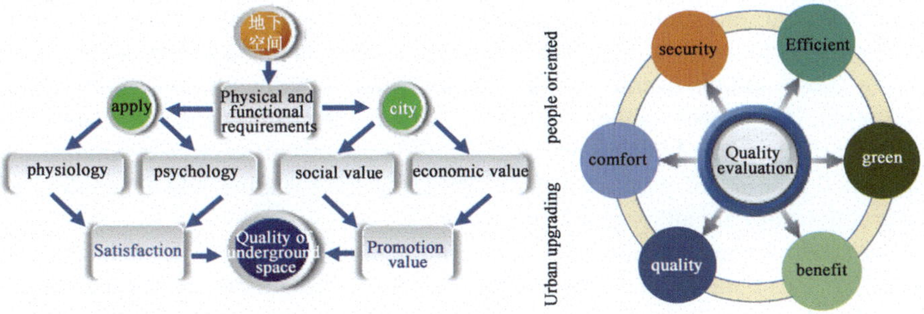

Fig. 2.67 Quality requirements of urban large underground space

reduce the impact of the underground environment on the physiological and psychological aspects of humans, great attention should be paid to the improvement of the quality of underground space in the design, including creating a good air environment, light environment, sound environment, visual environment, and other high-quality indoor environments. In addition, through intelligent means, comprehensive monitoring, intelligent guidance, terminal visualization, and information release systems, urban large underground space can enhance the functional carrying capacity of local space elements, meeting the comprehensive needs of safety, function, economy, and quality. The quality requirements of urban large underground space are shown in Fig. 2.67.

6. Construction Complexity

The construction of urban large underground space can play an important role in enhancing the city's image, the value of urban functions, and the people's living standards, and has great economic and social value. However, the price that comes with it is the high construction difficulty and complexity of urban large underground space. The construction complexity of urban large underground space is reflected in the following aspects: the construction of new underground space is huge, making it difficult for the entire space to be formed at once, requiring multi-step excavation, which increases the complexity of construction; if the new underground space is connected to the existing underground space to form a large underground space, this requires different forms of breakthrough construction on the existing buildings, causing great disturbance to the existing structures, and may affect the original force transmission system of the existing structures, in addition, the connection reliability and waterproofing at the connection interface need to be paid attention to, and the construction process requirements are high; urban large underground space often have complex, irregular spatial geometries, making their structural systems often special, requiring the use of new methods and processes to achieve. For example, the Wuhan Optics Valley Square complex project is a circular foundation pit with a diameter of 200 m, and the internal plane is divided into different shapes, vertical depth is staggered, and the structural spatial relationship is complex. For such a unique project, existing mature construction plans are not applicable, and finally, by dividing the pit into south, north, and middle areas (Fig. 2.68), each area is separately constructed with a protective structure, a net frame support structure is set up (Fig. 2.69), and the overall structure is completed by layered step-by-step retreat excavation. In this process, the erection of supports, the excavation and transportation of earthwork, the blasting and removal of temporary protective structures, the partitioned pouring of permanent structures, and the planning and scheduling of construction sites are all extremely complex.

The construction foundation pit of the underground space project connected to the Siyuan Road Station of Tianjin Metro Line 5 is an irregularly shaped foundation pit with a length and width of about 100 m each, and the diagonal direction is separated by the existing Siyuan Road Station, so a very special partitioned irregular support + ring-shaped internal support structure is used, and after excavation, it also

Fig. 2.68 Division of construction areas in Optics valley square

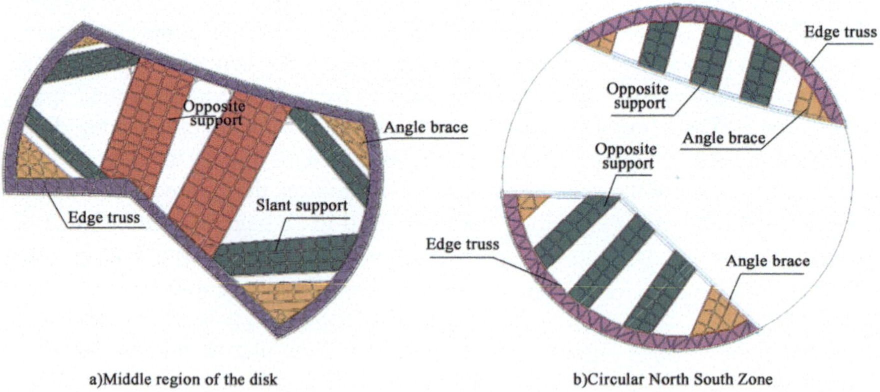

Fig. 2.69 Support arrangement in the foundation pit of Optics valley square

has to deal with the floating problem caused by the unloading of the excavation of the existing Siyuan Road Station, the processing steps are very tedious and time-consuming. In fact, each urban large underground space project is unique, with its own unique features and advantages, and also faces unknown difficulties and challenges in construction.

7. High Risk

High risk is an inevitable result of the above characteristics of urban large underground space. The high risk of urban large underground space runs through its planning, design, construction, and operation lifecycle.

(1) Planning Stage

The development of urban large underground space is irreversible. Once such a massive underground structure is built, it will semi-permanently affect the surrounding environment. Whether it's renovation, expansion, or rebuilding, all will incur substantial costs. Improper planning can lead to many difficult-to-handle legacy issues. Therefore, the planning of urban large underground space must be done with great caution. On one hand, the relationship between the structure and the surrounding above and below ground environment should be properly considered. On the other hand, for some layers where the development function is not clearly defined, there should be a mindset of "planning for white space", leaving room for further development in the future, and avoiding as much as possible issues like "zero-distance close contact", "destructive removal for expansion", and "avoidance of irregular structures".

(2) Design Stage

Due to the uniqueness and innovation in the structural forms and construction methods of urban large underground space, their design often lacks precedents to follow, and many past design methods may no longer be applicable. In addition, the structural performance requirements and functional quality requirements of urban large underground space are higher than those of general underground space, and the reliability and rationality of the design scheme are extremely demanding. Therefore, there is a high risk in the design stage of urban large underground space, requiring designers to have a solid theoretical foundation, advanced design methods and tools, a thorough grasp of the site conditions, and a comprehensive design scheme demonstration.

(3) Construction Stage

The high risk of the construction stage of urban large underground space is reflected in the following three aspects: ① Deep and large foundation pits, underground structures with large sections of undercuts, and complex expansion structures have relatively special and strict requirements for excavation methods, structural construction processes, and auxiliary measures, and rely heavily on the professionalism of the construction team; ② Under the disturbance of large-scale underground space construction, the urban environment is relatively fragile, and all kinds of existing buildings, roads and bridges, municipal pipelines, and even natural terrain and landforms are potential risk sources during construction; ③ The lengthy construction process, numerous construction steps, and massive personnel and machinery input make the organization and management of the construction process extremely complex, and any management oversight in any link can potentially trigger and catalyze accidents.

(4) Operation Stage

Urban large underground space are carriers of key urban functions and are underground distribution centers for urban residents and important materials. If they become paralyzed during operation, it can lead to problems such as traffic interruption, water cut-off, power outage, etc., causing huge economic losses, and in severe cases, endangering the lives of a large number of people. Due to the relative closure of underground space, coupled with the vast spatial range and complex network topology of urban underground space, ensuring the safety, orderly, and rapid evacuation of people in the event of a hazard is also a major challenge. In addition, urban underground space have complex engineering structures and a variety of mechanical and electrical equipment, making the maintenance and inspection during operation a very complicated and easily overlooked task.

Chapter 3
Analysis of Safety Accidents in Urban Underground Space Construction

3.1 Statistical Analysis of Accident Patterns

Risk includes three elements: risk source, risk factor, and risk event, among which the risk event is the core of the three elements. Risk events and accidents are two concepts of the same coin; a risk event is a potential accident, and an accident is the final transformation of a risk event. Therefore, accident analysis is the starting point of risk research and the basis for analyzing the formation mechanism, influencing factors, and consequences of risk.

This chapter investigates 295 safety accidents in urban underground space construction and several major foreign underground space construction accidents (excluding accidents that occurred during the operation phase and non-urban underground space projects) in more than twenty cities nationwide, including Beijing, Shanghai, Shenzhen, and Guangzhou, from 2004 to 2019. The main sources are data from the State Administration of Work Safety (abolished in March 2018), the Ministry of Housing and Urban–Rural Development of the People's Republic of China, as well as publicly published literature and news media reports, etc. The analysis of the year of the accident occurrence, geological area, accident type, risk source direction, accident occurrence mechanism, etc., lays the foundation for the study of safety risks in urban large underground space construction.

3.1.1 *Statistical Analysis of Accident Types*

Referring to the accident categories in the "Classification of Employee Injury Accidents" (GB6441—1986), and combining the characteristics of risk accidents in urban underground space construction, the types of urban underground space construction accidents are divided into collapse, falling from height, water and soil inrush, object strike, mechanical injury, damage to the surrounding environment, lifting injury,

S.-X. Lei et al., *Safety Risk Assessment of Urban Underground Large Space Construction*, https://doi.org/10.1007/978-981-97-7437-1_3

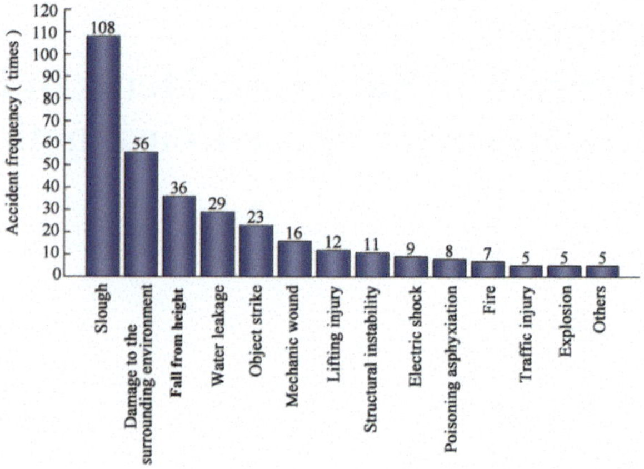

Fig. 3.1 Bar chart of accidents that occurred from 2004 to 2019 by accident types

vehicle injury, fire, electric shock, explosion, poisoning and suffocation, structural deformation and destruction, and others. Collapse includes open pit collapse and tunnel collapse; damage to the surrounding environment refers to accidents caused by damage to surrounding buildings (structures), various types of pipelines, and roads during the construction of urban underground spaces; others refer to low-probability non-typical accidents that do not belong to the above types, such as landslides in the pit, etc. It should be pointed out that a single accident can have multiple types of accidents, such as damage to surrounding pipelines caused by construction, which then leads to water leakage, and finally causes pit collapse. This accident includes damage to the surrounding environment and water and soil inrush, as well as collapse. The 295 accidents that occurred from 2004 to 2019 are statistically analyzed according to the above accident types, as shown in Fig. 3.1.

According to statistical results, among various types of safety accidents caused by urban underground space construction, collapse accidents occur most frequently, totaling 108 times, accounting for 33% of all accidents; followed by damage to the surrounding environment, accounting for 17%; then falls from height, water seepage, object strikes, and mechanical injuries, accounting for 11%, 9%, 7%, and 5% respectively; the remaining types of accidents account for about 20% in total. On the other hand, from the perspective of the objects and scale of the accident's impact, the above types of accidents can be divided into two categories: ① large-scale accidents that seriously interfere with the construction process, including collapse, damage to the surrounding environment, water seepage, structural instability, fire, and explosion; ② independent accidents that cause harm to individual construction workers, including falls from height, object strikes, mechanical injuries, etc. Among all the accidents counted, there were 216 of the first category, accounting for 65%, and just the two most common types of accidents, collapse and damage to the surrounding environment, accounted for as much as 50%. It can be seen that, unlike general industrial production safety accidents, in urban underground space construction safety accidents, major accidents with serious consequences and a large impact range dominate. This phenomenon indicates that the construction of urban underground spaces has a high degree of uncertainty. Even if the preliminary planning, survey, design, and construction management are all very perfect, it is impossible to ensure that the construction process is foolproof.

3.1.2 *Statistical Analysis of Geological Conditions in the Accident Occurrence Area*

By analyzing the geological structure, stratum characteristics, topography and geomorphology, hydrological and meteorological conditions, and other underground engineering data of different cities, the country is divided into five types of areas: soft soil areas, flood deposit soil layer areas, loess areas, expansive soil areas, and bedrock or complex geological unit areas. The soft soil area mainly includes Ningbo, Hangzhou, Shanghai, Wuxi, Tianjin, and other cities, with marine sediment and river–lake silt mud and silty soil as the main strata; the flood deposit soil layer area mainly includes Beijing, Shijiazhuang, Zhengzhou, Changchun, Shenyang, Chengdu, Hohhot, and other cities, with sandy soil, pebbles, loess, and clay as the main strata; the loess area mainly includes Xi'an, Lanzhou, Taiyuan, and other cities, with aeolian loess as the main stratum; the expansive soil area mainly includes Nanning, Hefei, and other cities; the bedrock or complex geological unit area includes Guangzhou, Shenzhen, Chongqing, Dalian, Qingdao, Nanjing, Jinan, Urumqi, and other cities, the accidents are classified by geological area as shown in Fig. 3.2.

As can be seen from Fig. 3.2, the area with bedrock or complex geological units has the largest proportion of total accidents, with 109 accidents, accounting for about 42%.

Affected by the depth of urban underground space development, the shallow bedrock is strongly weathered, the integrity of the rock mass is poor, the rock strength is not high, the rock properties of the bedrock in each city are different, the integrity

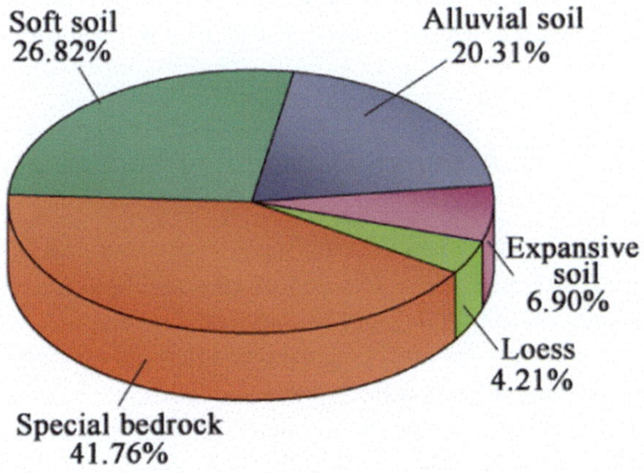

Fig. 3.2 Accident occurrence area pie chart

of the rock mass is different, and the engineering problems brought about are also different. For example, the slate in Dalian will disintegrate when it encounters water, and some people liken the construction of urban underground space in this situation to drilling holes in tofu dregs, which shows its strength drops sharply after encountering water.

In addition, Dalian also has limestone, and the sudden surge of karst water in limestone can bring disaster to the construction of urban underground space. Next is the soft soil area with 70 accidents, accounting for 27%. This is because the mechanical properties of the soil in the soft soil area are poor, and the groundwater is abundant, there are multiple layers of groundwater in the stratum, among which the confined water has a prominent impact on the project.

There were 53 accidents in the flood deposit soil layer region, accounting for 20% of the total. The prominent feature of the northern region is that due to water scarcity, the groundwater level is generally buried very deep, the soil particles vary from coarse to fine, there are large differences in permeability, and the mechanical indicators vary greatly. For example, the floating stone stratum in the western part of Beijing is a good foundation for general building construction; however, for the construction of urban underground spaces, it brings a series of construction difficulties such as difficulty in hole and pile formation, and wear and tear of shield construction tools.

There were 18 and 11 accidents in the expansive soil and loess regions, accounting for 7% and 4% respectively. The number of accident cases in the loess and expansive soil regions is relatively small, which is related to the slow economic development of the region, fewer civil engineering projects, and insufficient statistical cases. The verticality and stability of the soil layer in the loess region are good, which is very beneficial for engineering construction, but loess has the characteristic of collapsing when it encounters water, which can easily lead to instability of urban underground spaces and slopes. In addition, the ground cracks in the loess region will affect the structure of the underground space; in the expansive soil region, the expansive soil has the dual properties of water absorption expansion and water loss shrinkage. When the excavation surface encounters atmospheric precipitation, it is easy to expand, which reduces the shear strength and causes significant changes in the physical and mechanical properties, thereby affecting the stability of the foundation soil.

3.1.3 Risk Source Distribution Statistical Analysis

The evolution of risks during the construction process of urban underground spaces into accidents is inevitably cumulative, and risks exist in all stages of the construction project, including the survey stage, design stage, construction stage, etc. The hidden risks in the survey and design stages often only become apparent during the construction stage, which is reflected in the occurrence of construction accidents. Therefore, by analyzing and statistically determining the direction of accident risk sources, the causes of accidents can be effectively revealed. Due to objective conditions, some

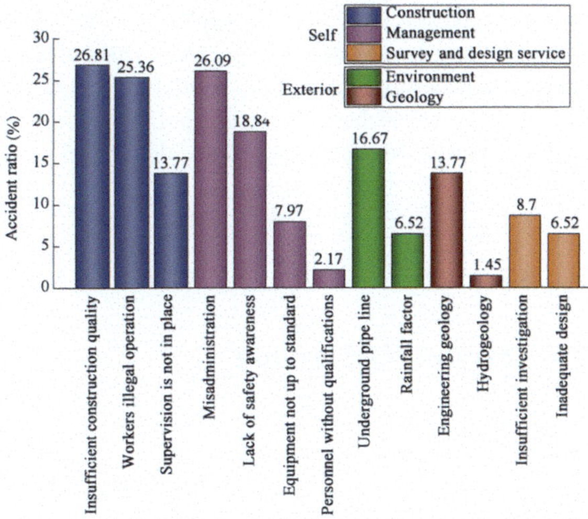

Fig. 3.3 Bar chart of accident cause classification statistics

accident causes are missing. After excluding the missing accidents, the statistics of risk source distributions are shown in Fig. 3.3.

As can be seen from Fig. 3.3, the main causes of accidents in the construction of urban underground spaces are divided into two categories: project inherent reasons and external factors. The project inherent reasons mainly include survey design, management, and construction. The risk sources of survey design mainly include insufficient survey and design; the risk sources of management mainly include insufficient safety awareness, improper management, substandard equipment, unqualified personnel, inadequate supervision, workers violating regulations, and substandard construction quality, etc.; external factors mainly include engineering geology, hydrogeology, underground pipelines, and rainfall factors, etc. From the perspective of risk source direction, the risk sources with the highest proportions are substandard construction quality, improper management, workers violating regulations, insufficient safety awareness, and the impact of underground pipelines, accounting for 26.81%, 26.09%, 25.36%, 18.84%, and 16.67% respectively. It should be pointed out that an accident may be caused by one risk factor or multiple risk factors, so the total risk sources counted here are greater than 1.

Overall statistical data shows that in the past urban underground space construction projects, construction safety accidents were relatively frequent, and often accompanied by casualties, the consequences of which were very serious. urban large underground space projects, as the largest volume, most complex construction process, and the type of underground project with the greatest impact on the surrounding environment and public opinion, the possibility and consequences of construction safety risks are only more severe than ordinary underground spaces. Therefore, in line with the principle of "safety first", urban large underground space projects must adopt more scientific, comprehensive, and effective targeted risk control measures. However, there are many types of construction safety risks in urban underground space projects, the causes are complex, and the risk occurrence mechanisms are extremely complex, which brings great difficulties to the formulation of corresponding risk control measures. To clarify the mechanism of major construction safety risks in urban large underground spaces, this section selects representative major accidents from the surveyed accident cases for further detailed analysis.

3.2 Analysis of a Subway Station Pit Collapse Accident

3.2.1 Project Background

1. Project Overview

The pit where the accident occurred is the main open-cut construction pit of a subway station. The main structure of the station has a construction area of approximately 36,082.5 m^2, it is a two-story underground structure, with a total length of about 934.5 m, a standard section width of 20.5 m, a 12 m wide island platform station, and a maximum burial depth of about 17.7 m. The accident pit is 106 m long, 20.5 m wide, the bottom plate is buried 16 m deep, the continuous wall pit bottom is about 17 m deep into the soil, and the top plate of the main structure of the station is covered with 1.8 m of soil. The excavation depth of the standard cross-section pit is 16.28 m, and it is protected by an 800 mm thick underground reinforced concrete continuous wall structure, with a wall length of about 33 m. Four steel pipe inner supports are set at 1.56, 5.26, 9.06, and 12.76 m below the ground. After the pit is sealed, when the permanent structure is constructed, one support is added at 10.56 m after removing the bottom support (12.76 m), and one support is added at 6.46 m after removing the support at 9.06 m.

2. Engineering Geological Conditions

The engineering characteristics of the soil layers within the pit range are described from top to bottom according to the stratigraphic sequence.

(1) ①$_2$ Fill layer (Q_{43}^{ml}).
(2) ②$_2$ Clayey silt (Q_{43}^{al}).
(3) ③$_2$ Silty clay, coastal and bay phase (Q_{42}^{m}).
(4) ⑥$_1$ Silty clayey silt, shallow sea, drowned valley phase (Q_{41}^{m}).
(5) ⑧$_2$ Silty clay interlayered with silty sand, lacustrine marsh phase sediment (Q_{32}^{l+k}): Saturated, soft plastic state, thin-layered, contains organic matter, humus.

Locally dominated by fine sand, contains a small amount of shell fragments. Natural water content 33.9%, porosity ratio 1.040, average cone tip resistance 1.46 MPa. Locally distributed, layer thickness 0.70–20.10 m. Below 40–50 m is strongly weathered to moderately weathered silty sandstone.

a) b)

Fig. 3.4 Collapse scene photo

3.2.2 Accident Process

On November 15, 2008, the road surface on the west side of the accident pit sank significantly, and the base of the pit became unstable, causing the continuous wall on the west side of the pit to break, the pit to collapse, and the collapse length was about 75 m. A large amount of flowing water from the river on the east side, the sewage under the road on the west side, and the broken water pipe immediately rushed into the pit, with a water depth of up to 9 m. The collapse scene is shown in Fig. 3.4.

3.2.3 Causes of the Accident

1. Overestimating the bearing capacity of the base

The original design was to reinforce the pit bottom 3 m below the pit bottom elevation with cement mixing piles to form a grid. However, this pit bottom reinforcement measure was cancelled before construction, believing that dewatering in the pit can achieve the purpose of reinforcing the base soil, which is a major mistake.

2. Irregular excavation

The original design of the pit was divided into 6 working sections (each about 25 m) for excavation, with steel pipe supports set at a height of 0.5 m below the support during excavation, and the bottom was sealed in stages. However, before the accident, 3 of the working sections of the pit had been excavated to near the bottom of the pit and the fourth layer of steel pipe support had not been installed, and the excavation progress was ahead of the support erection and bottom sealing.

3. Misuse of design parameter

When the design unit calculated the anti-sliding moment of the foundation soil, it used the saturated weight γ_{sat} of the foundation soil. However, if the foundation soil is normally consolidated clay, the vertical effective consolidation pressure it has received in the long geological history is $\gamma' H$ (γ' is the buoyant weight of the foundation soil), so the anti-sliding moment must be calculated with the buoyant weight of the foundation soil. The use of saturated weight for calculation resulted in a lower actual safety factor in the design.

4. Highly sensitive soft soil

The main soils involved in the project, ④$_2$ silty clay and ⑥$_1$ silty silty clay, are both under-consolidated soils and belong to highly sensitive soft soils. Table 3.1 shows the sensitivity test results conducted by a design institute after the accident. It can be seen that the undisturbed soil sensitivities of the main soil layers ④$_2$ and ⑥$_1$ are 6.6 and 9.1, respectively, which are already highly sensitive and extremely sensitive soft soils. Field tests show that the minimum static cone tip resistance of the disturbed soil in the pit is only 35% of the original soil.

For this type of soft soil, sufficient attention should be paid to the selection of strength and construction arrangements. However, before the accident, due to the renovation of nearby roads, all vehicles nearby passed through the road on the west side of the pit, causing obvious disturbance to the highly sensitive soft soil.

The cause and effect of the accident are shown in Fig. 3.5. The basic objects (boxes) in the accident cause and effect diagram represent an event, and the directed connections between these objects indicate their cause and effect relationship. In the Fig., the orange box represents the key cause of the accident, called the key disaster-causing factor; the orange arrow represents the key cause and effect chain of the formation and development of the accident, called the key disaster-causing path.

From the figure, it can be seen that the pit collapse accident is not a simple, isolated accidental event, but a major safety accident formed by the influence of many adverse factors, triggered by some small adverse events, and then developed and expanded step by step. When erroneous behaviors (such as non-standard excavation) and adverse objective conditions (long-term rainfall, highly sensitive soft soil) occur at the same time, their adverse effects quickly amplify, developing along the key disaster-causing path into more serious safety hazards (unfavorable stress on the retaining structure, decrease in soil performance), and then develop into accident signs (cracks in the underground continuous wall, sinking of the road surface). The

Table 3.1 Unconfined strength and sensitivity of soft soil layers in the field

Stratigraphic sequence	Unconfined compressive strength of peripheral soil q_u/ sensitivity S_t	Unconfined compressive strength of disturbance zone outside the pit q_u/ sensitivity S_t	Unconfined compressive strength of soil in pit q_u/ sensitivity S_t
④$_2$	47.9/6.6	37.5/4.8	
④$_3$	58.2/3.8	38.4/2.4	
⑥$_1$	51.9/9.1	40.3/6.4	45.0/7.5
⑥$_2$	60.5/11.5	47.5/8.0	54.2/8.0
⑧$_1$	76.0/8.0	61.2/6.8	29.5/3.7

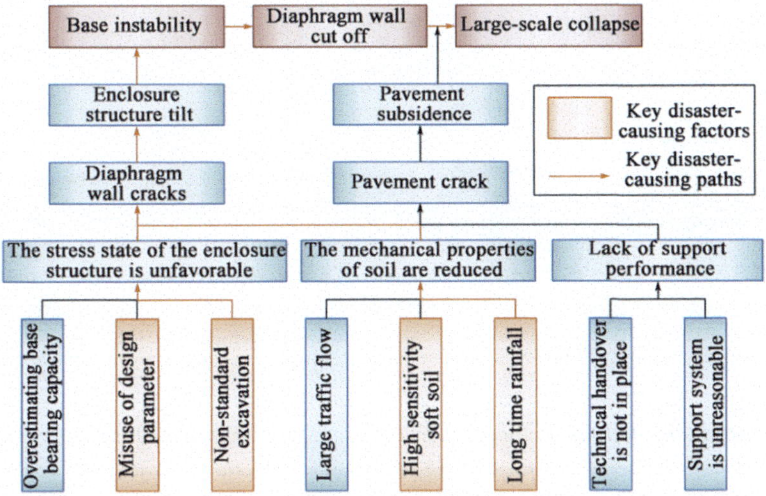

Fig. 3.5 Cause and effect diagram of a subway station pit collapse accident

construction party, knowing the existence of accident hazards, failed to take effective countermeasures and carried out reckless construction, which ultimately led to the occurrence of this serious accident.

3.3 Analysis of a Subway Station Tunnel Collapse Accident

3.3.1 Project Background

A certain subway line (the yellow line in Fig. 3.6) is 12.5 km long, connecting the city center and the western urban area, with four transfer stations.

The construction of the tunnel in this section is divided into three segments. Segment 1 uses Earth Pressure Balance Shield for excavation, Segment 2 uses the

Fig. 3.6 Subway line map

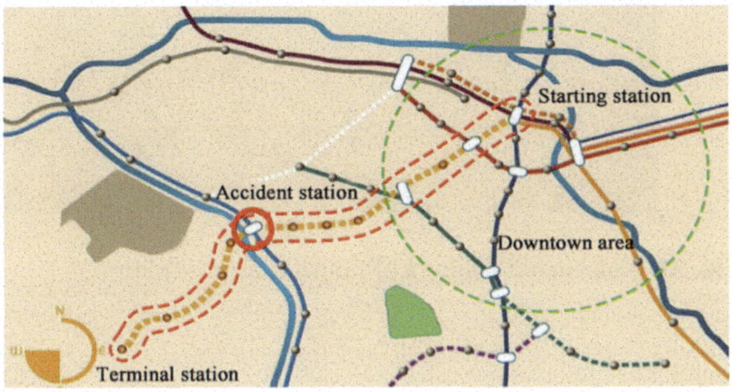

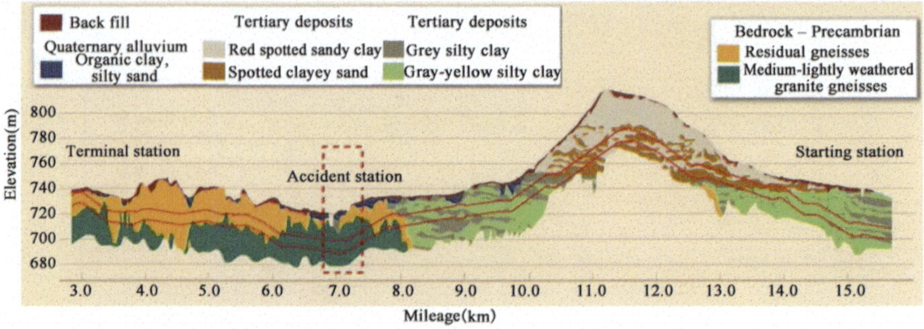

Fig. 3.7 Geological conditions along the tunnel

New Austrian Tunneling method for excavation, and Segment 3 uses the Cut-and-Cover method. All subway stations are built using the Cut-and-Cover method or the New Austrian Tunneling method. Figure 3.7 shows the geological conditions along the line, with the red box indicating the station where the accident occurred.

The accident station section is mainly located in the moderately weathered to slightly weathered granite gneiss stratum. The station is built using the New Austrian Tunneling method, including a large diameter vertical shaft, two platform tunnels (18.6 × 14.2 × 46 m), and two transfer passages. The station has side platforms and a double-line tunnel (diameter 9.6 m). The plan position and three-dimensional schematic of the tunnel are shown in Figs. 3.8 and 3.9.

The typical cross-section of the platform tunnel is shown in Fig. 3.10. The tunnel arch lining support uses shotcrete (350 mm thick, later increased to 580 mm) + grid beams spaced at 830 mm, while the sidewalls use steel fiber reinforced shotcrete (150 mm thick). The invert only has a thin layer of shotcrete (70 mm), with no structural function. During step-by-step excavation, if necessary, steel anchor bolts will be added according to the actual situation.

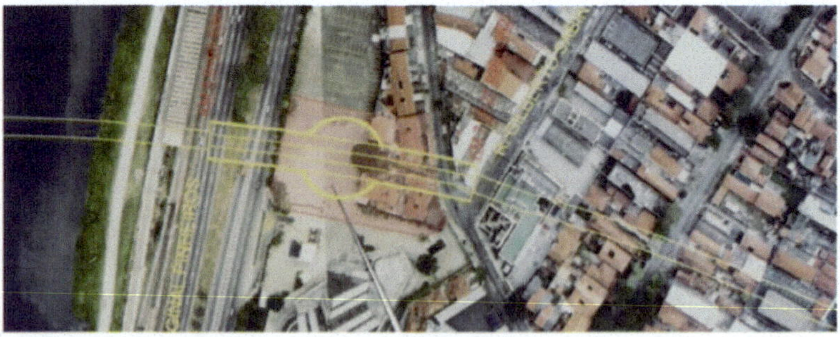

Fig. 3.8 Schematic diagram of tunnel plan position

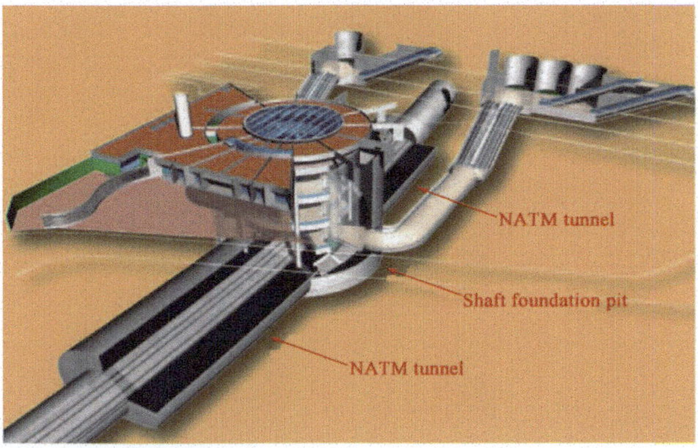

Fig. 3.9 Three-dimensional schematic of the tunnel

Fig. 3.10 Cross-section of the tunnel (unit of measurement: mm)

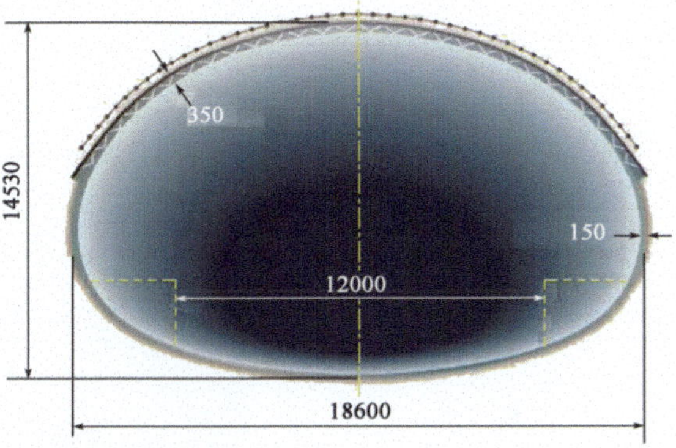

3.3.2 Accident Process

On the afternoon of January 12, 2007, a 40 m section of the accident station tunnel suddenly collapsed, followed by the collapse of the nearby station vertical shaft foundation pit, with nearly half of the completed vertical shaft support structure instantly sinking into the ground. Figure 3.11 shows the situation before the tunnel collapse, at this time the steps in the tunnel of the nearby vertical shaft foundation pit had not yet begun construction. Between the tower crane and the tree is a nearby street. Between mileage 7080 m (vertical shaft foundation pit sidewall) and mileage 7120 m (edge after collapse), it roughly indicates the 40 m long station tunnel collapse range.

Fig. 3.11 Bird's eye view of the station vertical shaft

Fig. 3.12 Collapse site condition

Figure 3.12 shows the situation after the collapse, about 15,000–20,000 tons of jointed, leaf-shaped highly weathered rock ridge fell rapidly by nearly 10 m after the collapse, and the huge air suction generated sucked the victims on the nearby ground into a deeper place. The speed of the collapse was so fast that it caused an air explosion in the tunnel under construction, and one escaping worker was even blown over.

3.3.3 Cause of the Accident

1. The difference between the stratum distribution revealed by geological exploration and the actual existence

The average depth of the tunnel vault is 21 m below the ground surface. Geological borehole (No. 8704) drilled near the center of the tunnel revealed that the absolute elevation of the rock surface is 706 m, which is exactly the same as the average rock surface elevation revealed in the other four nearby boreholes. Figure 3.13 shows the anticipated geological cross-section based on the results of the geological survey. Because the rock surface elevations revealed by the boreholes are similar, it was initially thought that the rock surface in the section should be roughly horizontal.

However, after the tunnel collapsed, a unique, steep rock ridge was discovered in the direction of the tunnel axis as the excavation of the collapsed rock and soil body deepened. According to on-site back-analysis, this rock ridge is 10–13 m higher than the surrounding rock surface and extends tens of meters along the tunnel and tunnel (Figs. 3.14 and 3.15). Because borehole 8704 is located exactly in the low concave position between the two large rock ridges, it just missed the protruding rock ridge and revealed the rock surface elevation that is the same as the average rock surface elevation (706 m) of the nearest five boreholes, thus creating an illusion for the engineers.

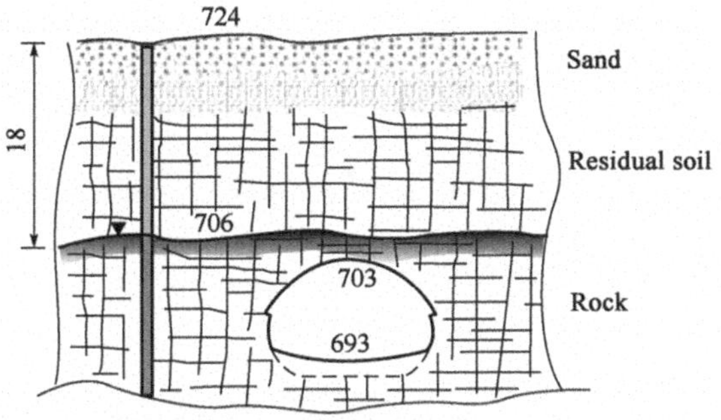

Fig. 3.13 Estimated rock surface elevation

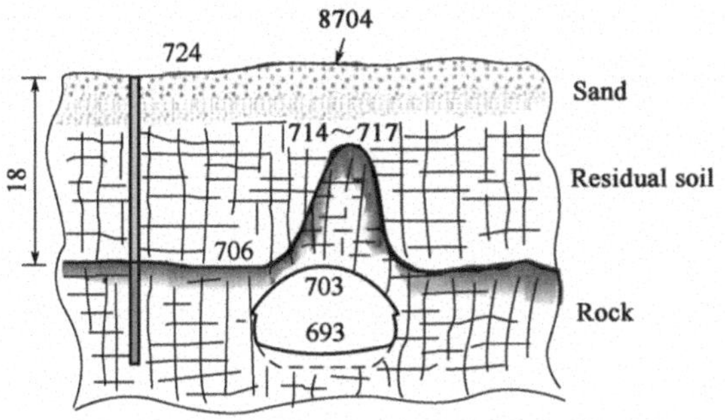

Fig. 3.14 Back-analyzed rock surface elevation based on excavation results (cross-section)

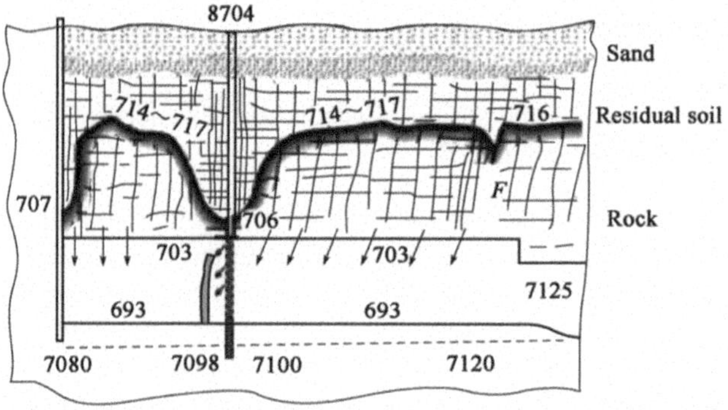

Fig. 3.15 Back-analyzed rock surface elevation based on excavation results (longitudinal section)

2. Use of open support

The tunnel adopted the relatively rare open half-section support, using the "elephant legs" set on both sides of the rock as the support for the arch, as shown in Fig. 3.16.

Some scholars have found through numerical simulation that under the site's surrounding rock conditions, the total vertical load applied to the upper part of the support structure can reach an astonishing 20,000 kN. Under these conditions, not only will the support structure quickly yield and soften shortly after the excavation face passes, but the rock "elephant legs" supporting the support structure will also develop a large number of cracks and enter an unsafe state (Fig. 3.17). Although theoretically, no existing support form is sufficient to resist such a huge load in this project, some believe that the use of a closed support structure can effectively prolong the destruction process, make the precursors of accidents more obvious, and increase the time for the construction party to react, thereby reducing the losses caused by accidents.

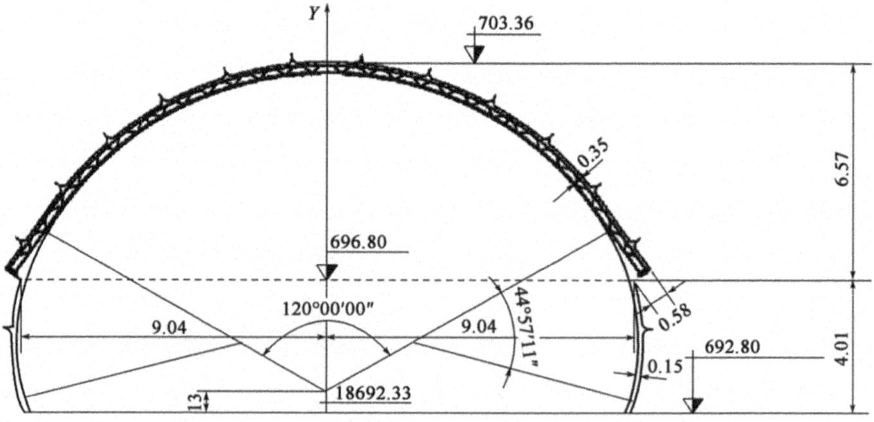

Fig. 3.16 Schematic diagram of arch arrangement (unit of measurement: m)

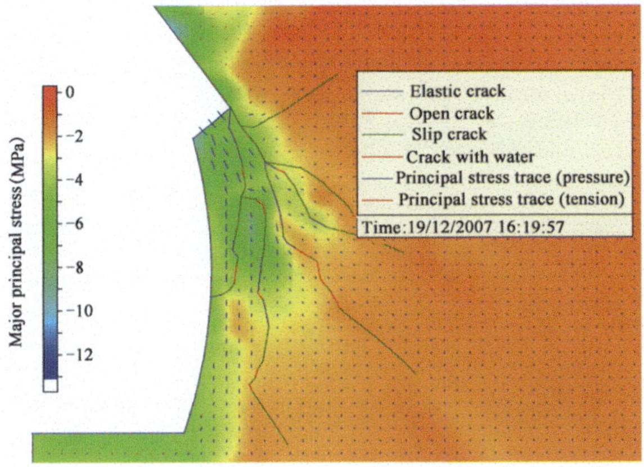

Fig. 3.17 Cracks in the rock "elephant legs"

3. Cancellation of pre-excavation grouting

In the period before the accident occurred, geological engineers discovered more and more surrounding rocks of grade III (RMR = 44–48, RMR refers to the rock mechanics classification index value proposed by the South African Council for Scientific and Industrial Research) in the middle of the tunnel (Fig. 3.18). Figure 3.19 shows the grading of rock quality on the tunnel cross-section, with the two figures showing the changes in rock quality on the cross-section at different times with excavation. Clearly, along the length of the tunnel, the distribution pattern of rock quality grades is almost the same, that is, the core area of grade III is surrounded by poorer quality grade IV rock (RMR = 34–36) (A/B/A pattern).

Due to the gradually improving quality of the rock mass, the construction party cancelled the pre-excavation grouting. However, in reality, this is evidence of the appearance of differential weathering in the rock layers.

Fig. 3.18 Excavation face showing signs of a "core area"

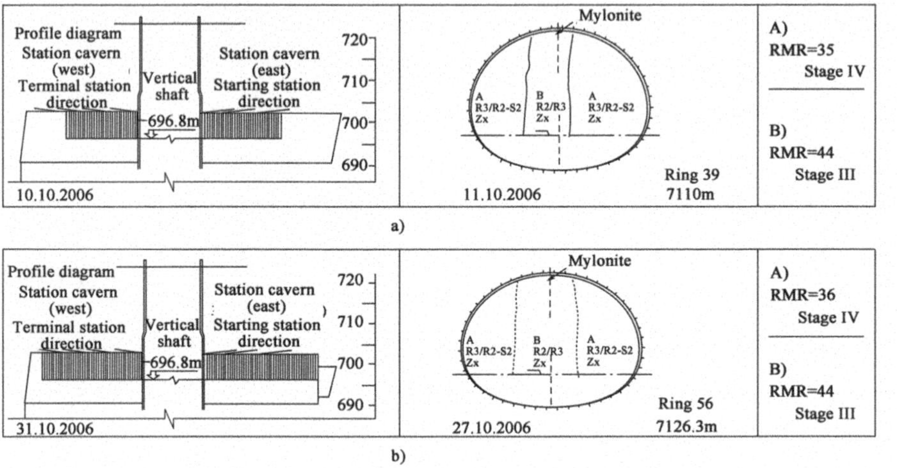

Fig. 3.19 Changes in the quality grade of the excavated rock face

4. Adverse terrain in nearby areas

In this project, a rock fault (Fig. 3.20) crosses the tunnel at an almost vertical angle, which would normally be the most advantageous. However, at the tunnel location 20 m below, this fault is difficult to distinguish from the smooth rock joint (fracture) area, resulting in it not receiving enough attention. Additionally, a rainwater pipe built 30 years ago happens to cross this fault, and the diameter of the pipe cross-section decreases from 1000 to 700 mm before crossing the fault, meaning that the water-crossing cross-sectional area is reduced by 50%. In the weeks before the accident, the rainfall was exceptionally heavy. It can be speculated that under the high water pressure and the fault displacement caused by tunnel excavation, this rainwater pipe developed leaks, exacerbating the softening of the weathered soil body around the rock ridge.

According to the accident data, the cause and effect of this subway station collapse accident can be sorted out as shown in Fig. 3.21.

From the above analysis, it can be seen that the key link in this accident is the high load brought by the rock ridge formed by differential weathering. Firstly, due to insufficient drilling density, the position of the key borehole indirectly led to a misjudgment of the rock face height, placing the tunnel in such a dangerous position. Secondly, due to insufficient understanding of the mechanism of differential weathering of rocks, the signs of the existence of the core area revealed on the face of the tunnel did not receive enough attention, and even the advance reinforcement was cancelled as a result. In addition, the open support used in this tunnel is also one of the important reasons for the aggravation of the accident. Generally speaking, such a large-scale collapse would first show signs of severe deformation of the tunnel, destruction of the support structure, etc., but the open support structure does not have

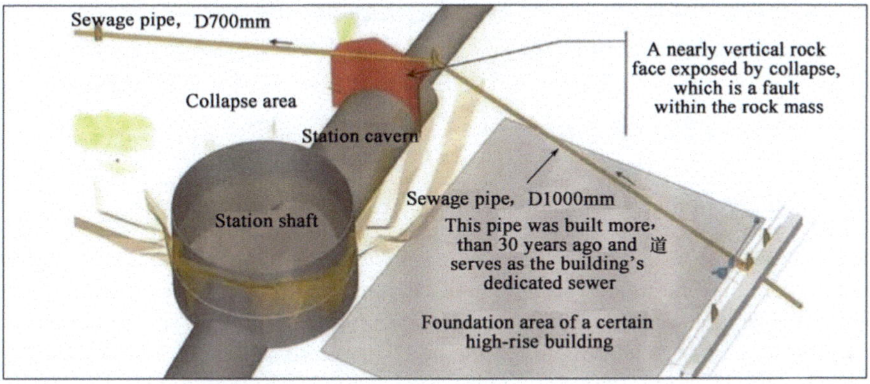

Fig. 3.20 Discontinuous surface near the collapse area

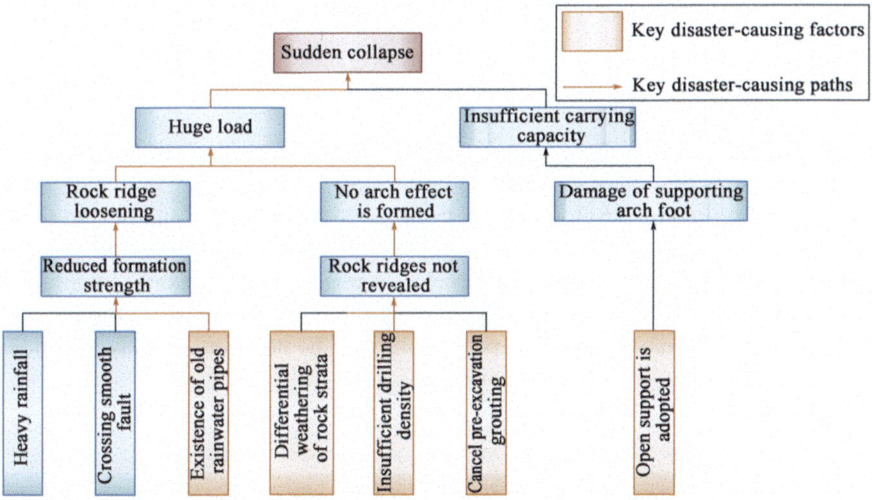

Fig. 3.21 Cause and effect diagram of a subway station tunnel collapse accident

enough ductile deformation capacity, and the huge load directly destroys the bearing part of the support structure, causing the support structure to lose overall stability. On the other hand, the nearby existing faults and old rainwater pipes also became catalysts for this accident, and the absence of corresponding monitoring projects made the accumulation and development of these deep-seated adverse factors hidden. These adverse factors, when existing alone, may not lead to such a serious accident, but when these conditions are present at the same time, the occurrence of the accident is almost inevitable, and any remedial measures are difficult to reverse. This accident has very distinct characteristics and important warning significance.

3.4 Analysis of a Major Water Leakage Collapse Accident in a Shield Tunnel

3.4.1 Project Background

1. Accident section location and surrounding environmental conditions

The tunnel where the accident occurred is a subway section tunnel, running from east to west, a double-line tunnel, located under the main road of the city, passing under an overpass and a small road bridge. The single-line length of the section tunnel is about 1932 m, constructed by shield method. There are a large number of various types of pipelines laid within 5 m depth below the ground along the section, which have a certain impact on tunnel construction. The plan of the section tunnel location is shown in Fig. 3.22.

2. Engineering geological conditions

The main traversing area of the accident tunnel is mostly loose rock and soil, with low bearing capacity and poor self-stability, and the overall engineering geological conditions are very poor.

The accident tunnel is approximately 30.5 m deep, from top to bottom it consists of man-made fill, silty clay, silt, silty clay, silty sand, medium sand, rounded gravel, and strongly weathered silty sandstone. Prior to the incident, the middle and lower parts of the right line shield machine were at the junction of medium sand and silty sand, the top of the shield tunnel was a layer of silty clay, the bottom of the tunnel was a weak layer of silty sand, and the front of the shield head was at the junction of silty sand and medium sand. The permeability of the medium sand layer is in the medium to strong range, containing artesian water.

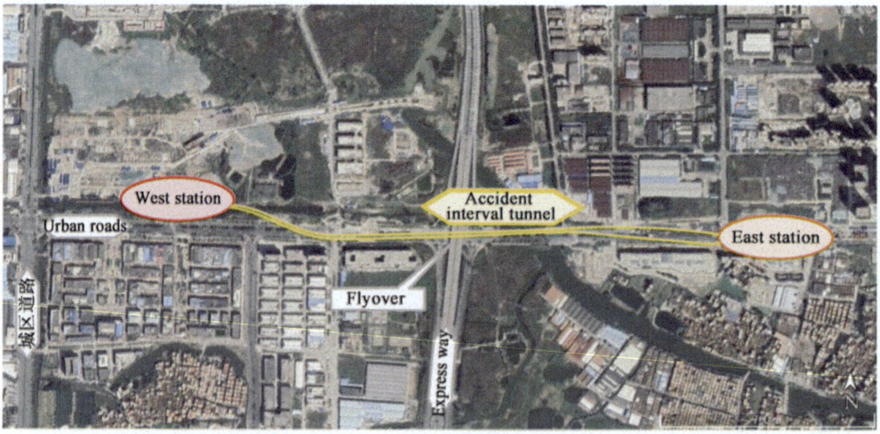

Fig. 3.22 Accident tunnel location plan

3. Hydrological conditions of the project

The accident tunnel passes under a river, which has water all year round and is abundant. The groundwater mainly consists of Quaternary loose layer pore water and bedrock fissure water, with pore water divided into phreatic water in the upper sticky soil layer and artesian water in the lower sand and gravel layer. The sand and gravel layers are widely distributed and abundant in water.

4. Tunnel design overview

The left line of the accident tunnel is 1929.702 m long, the right line is 1931.976 m long, the section line longitudinal section is a V-shaped slope, with a maximum slope of 27‰, the line depth is 15.58–33.83 m, and the tunnel top cover soil thickness is 10.29–28.54 m. The shield segment adopts a 6-part plan, 1 top seal block, 2 adjacent blocks, 3 standard blocks, and the lining ring is assembled with staggered joints.

5. Shield machine overview

The accident tunnel uses 2 earth pressure balance shield machines with a diameter of 6980 mm for construction.

(1) Shield tail seal structure

Shield machines generally use a shield tail seal method of filling sealant between multiple wire brushes, each wire brush is generally composed of a spring steel sheet, wire, and tail end steel plate (Fig. 3.23). Shield tail seal method used in the construction shield machine is tail brush + sealant, using a structure of 2 wire brushes + 1 steel plate wire brush + 1 check plate, shield tail grease is injected by a pneumatic grease pump installed in the rear support system, using injection pressure and injection volume dual control method. The maximum pressure resistance of shield tail seal design is 1.0 MPa.

(2) Use of shield tail sealant

From February 4, 2018 to February 7, 2018, the right line shield machine used about 10 barrels of shield tail grease, excavated a total of 30 rings (each ring 1.5 m, the same below), with an average usage of about 3 rings/barrel. During the excavation of rings 875–905, the average grease pressure was above 2.0 MPa.

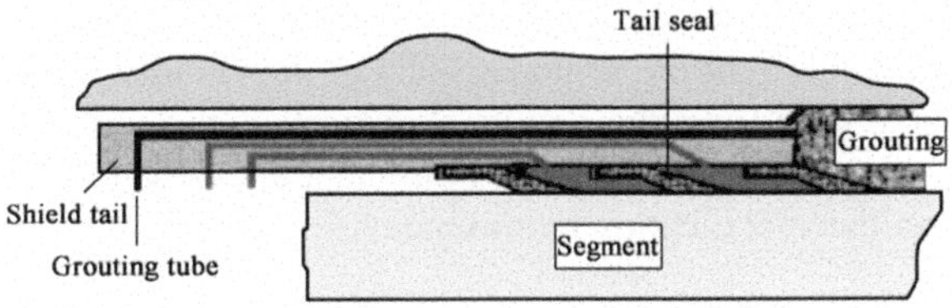

Fig. 3.23 Shield tail seal principle schematic

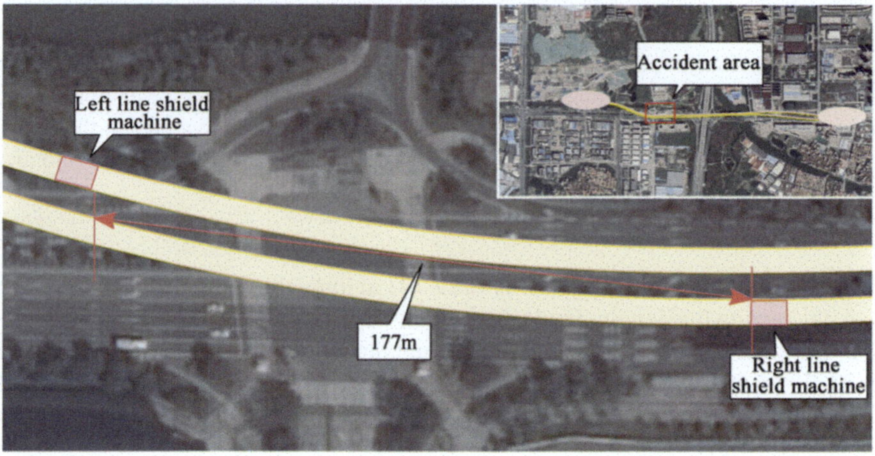

Fig. 3.24 Planar position relationship of left and right line shield machines at the time of the accident

6. Construction progress

The right line shield machine of the accident tunnel started on May 10, 2017, and as of February 7, 2018, the right line had completed 904 rings of construction, 1356 m. At the time of the accident, the left line shield had excavated to the 1028th ring, and the distance between the left and right line shield machines was about 177 m. The planar position relationship of the left and right line shield machines at the time of the accident is shown in Fig. 3.24.

7. Supplementary inspection, testing and other situations

The investigation team commissioned a quality and safety inspection company to test the quality of the right line segment. The test results showed that the concrete strength, rebar diameter, spacing, protective layer thickness, and segment thickness all met the relevant specification requirements. The team also commissioned a geological engineering survey institute to conduct supplementary drilling surveys around the accident collapse area. No new adverse geological conditions were found, and the hydraulic connection of the upper aquifer near the accident site in the fourth series was weak, with no direct hydraulic connection to the confined water. The investigation team also sent the segment bolts to the Quality Supervision and Inspection Center for testing, and the results showed that the tensile strength of the bolts met the requirements. After expert compliance review, the survey and design of the project met the specification requirements.

Fig. 3.25 Curve of abnormal pressure increase in the 905th ring soil

3.4.2 *Accident Process*

On February 7, 2018, during the construction of the 905th ring of the right line, the pressure in the shield machine's soil bin suddenly increased by about 40 kPa, from 233 to 270 kPa (Fig. 3.25). The pitch angle at the rear of the shield body began to increase, the tail of the shield began to sink, and at the same time, the first segment (A2 block) that had just been assembled inside shield tail suddenly erupted upwards near the right side (approximately 6 o'clock position of the shield tail). The construction personnel took emergency plugging measures but were unsuccessful. The slurry quickly rose in the tunnel, and the vertical deviation of shield tail of the shield machine continued to increase.

About 2 h after the mud and sand started to gush, a large area of collapse appeared on the ground above the shield machine. The large amount of mud and sand that suddenly gushed out in the tunnel pushed the shield machine trolley to slide back about 700 m. The accident eventually led to a city surface collapse of about 65 m east–west, about 81 m north–south, 6–8 m deep, and an area of about 4192 m^2, with a collapse volume of nearly 25,000 m^3. Photos of the collapse area are shown in Fig. 3.26.

3.4.3 *Cause of the Accident*

1. Poor geological conditions

(1) The tunnel section of the accident is located in silty soil and sand layer. The soil is loose, with low bearing capacity, poor self-stability, and prone to collapse. The silty sand layer is a liquefied soil, with continuous distribution, wide range, large burial depth, strong permeability, abundant water, and the upper silty soil forms a relative water barrier layer. The groundwater in the lower sand layer is under pressure, and the hydrogeological conditions are poor.

Fig. 3.26 Aerial photo of the ground collapse area

(2) At the time of the accident, the shield machine was just at the junction of silty sand and medium sand. The shield machine itself was in the weak silty sand layer. The longer the shield machine stays at this junction, the greater the risk of water pressure imbalance and breaking the seepage prevention system. The permeability of the medium sand and the round gravel layer below it is stronger than that of the silty sand layer, and the water is abundant and under pressure. This is a very unfavorable combination. Once the silty sand layer gushes water, the silty sand under the shield machine will be lost with the gushing water, which is very likely to cause pipe gushing and loss of silty sand.

2. Performance degradation of shield tail sealing device

(1) Before the accident, the right line shield machine had excavated about 1.36 km in total. Tail brush was worn, and the water-stopping performance of shield tail seal decreased. During the assembly of the segments, the gap at shield tail was in a disadvantageous state of being larger at the bottom and smaller at the top, and it was easy for slurry and water to leak from the bottom of the shield tail.
(2) The shield tunneling machine is carrying out segment assembly work. When the segment assembly machine hoists the second segment of the 905th ring, the external load at the tail of the shield increases, and the pressure in the soil bin suddenly rises by about 40 kPa, which is detrimental to the sealing of the shield tail. These factors cause the water pressure resistance of shield tail seal to decrease during use, leading to shield tail seal being punctured by external pressure.

3. Delayed evacuation

After the occurrence of ground inrush, a large amount of mud and sand has accumulated in the tunnel, and shield tail of the shield tunneling machine has obviously sunk. The laser guidance system can no longer detect the vertical deviation of the shield tail. These phenomena indicate that the tunnel is in a dangerous state. The construction personnel injected emergency plugging grease into shield tail seal and

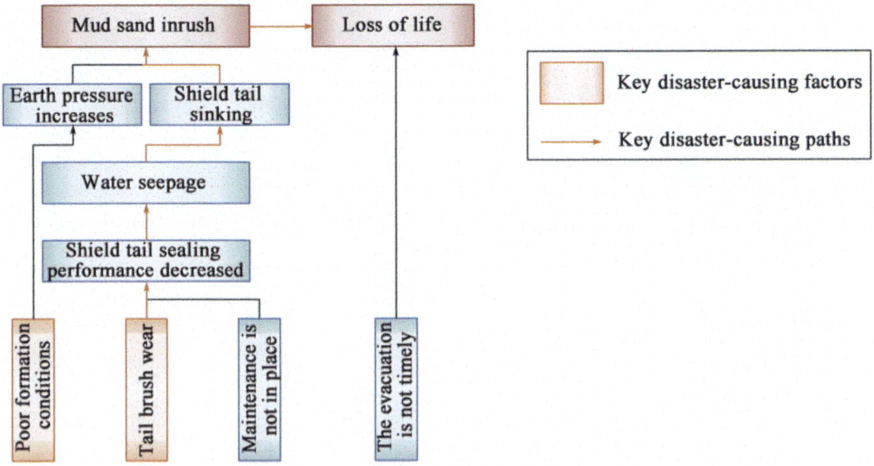

Fig. 3.27 Cause and effect diagram of a shield tunnel water leakage and collapse accident

threw sandbags to counter pressure at the leakage point of the shield tail. However, the phenomenon of water leakage and mud gushing at shield tail continued, indicating that the emergency measures were difficult to effectively control the danger. In this case, repair efforts should be immediately abandoned and the tunnel evacuated.

According to the accident data, the cause and effect of the accident can be sorted out as shown in Fig. 3.27. The direct cause of the accident was the unfavorable hydrogeological conditions and the wear caused by the overuse of shield tail brush. There was a clear evolution process in the occurrence of the accident. The wear of shield tail brush led to a decrease in the sealing of the shield tail, the sinking of the shield tunneling machine, and the further enlargement of the gap in the upper shield shell, developing into water seepage. At this point, the occurrence of the accident was hard to avoid. Subsequently, due to the failure to abandon the rescue and evacuate in time, casualties that could have been avoided were caused under the impact of the sudden gushing of mud and sand.

3.5 Analysis of a Subway Section Open-Cut Foundation Pit Collapse Accident

3.5.1 Project Background

1. Project Overview

The accident pit is a section of open-cut tunnel of a certain subway loop line. The total length of this section is about 2.8 km, including underground three-layer stations

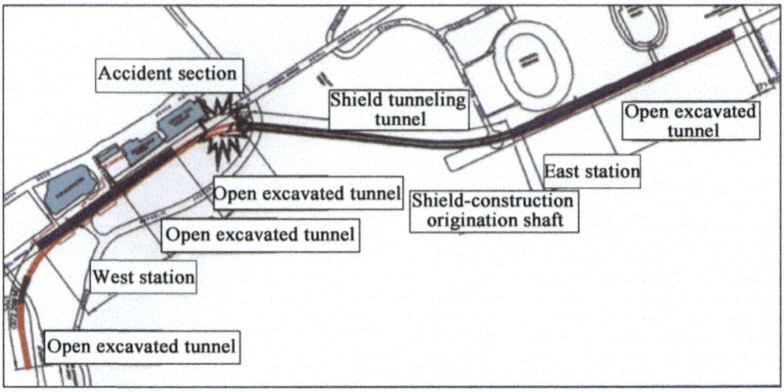

Fig. 3.28 Overall layout of the accident section

on both sides and the interval tunnel. There are 800 m of the interval tunnel under a certain river, which is constructed by shield tunneling. The shield starting shaft is located at the end of the east station, and the receiving shaft is located on the west side of the river, which is a circular working shaft with a diameter of 34 m. The other sections are constructed by open-cut method. Figure 3.28 is the overall layout of the accident section.

This section of the line is adjacent to a certain expressway, and on the north side of the expressway is the parallel Beach Road. The entire area south of Beach Road was formed by two rounds of sea enclosure. As shown in Fig. 3.29, the land between Beach Road and the expressway was formed by sea enclosure in the 1940s; the north side of the expressway was formed by sea enclosure in the 1970s, which means that the accident section has only settled and consolidated for 40 years, and it is a young land.

The accident pit was dug to a maximum depth of about 33 m and a width of about 20 m. The retaining system is a 0.8 m thick underground continuous wall, with ten steel supports from top to bottom, and column piles set in the middle. Between the ninth and tenth supports, there is a 1.6 m jet grouting reinforcement layer as a hidden support, and at the bottom of the final excavation surface, there is a 3.0 m thick jet grouting reinforcement layer as the bottom plate during the construction phase. The project uses open excavation construction method, Fig. 3.30 is the plan of the accident pit, where M2 and M3 are the sections where the accident occurred. Figure 3.31 is a cross-sectional view of the accident pit support.

2. Geological conditions of the project

The distribution of strata from top to bottom is as follows: a 4–5 m thick fill layer; 5–30 m is a marine mud layer, with a thin alluvial layer in the middle of the marine mud layer, and the bottom of the marine mud layer is relatively hard old sediment. The underground continuous wall generally needs to be inserted into the old sediment 3–5 m. In this accident section, this layer of old sediment is not horizontally stratified, and

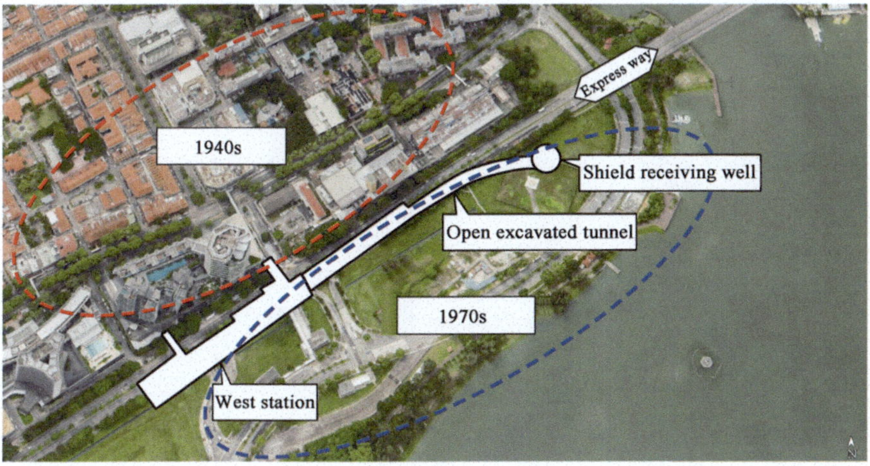

Fig. 3.29 Schematic diagram of the geological age of the accident section

Fig. 3.30 Accident pit plan

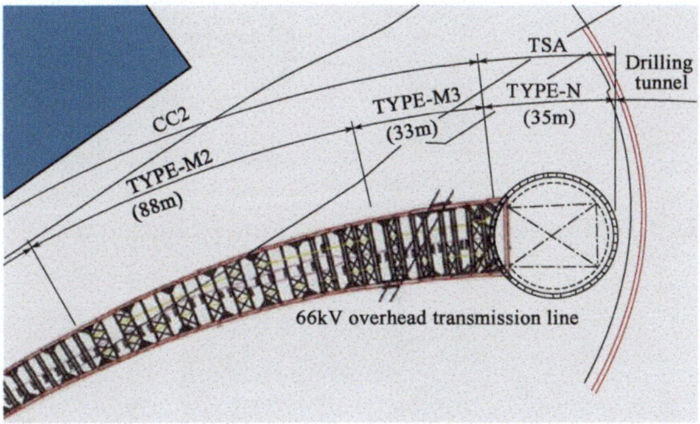

it is relatively shallow on the side of the expressway. Judging from the distribution of soil layers, the entire excavation is carried out within the marine mud layer, which is generally soft, under-consolidated, and has poor permeability, which also leads to significant time–space effects of the deformation of the retaining system during the excavation of the pit.

3.5.2 Accident Process

On April 20, 2004, the site engineer heard two sounds during the site inspection and found that the reinforce channel steel and flange of the north side retaining girder yielded (Fig. 3.32a), and 10 min later, the south side retaining girder yielded (Fig. 3.32b). After discussion, the construction party decided to pour a 200 mm thick

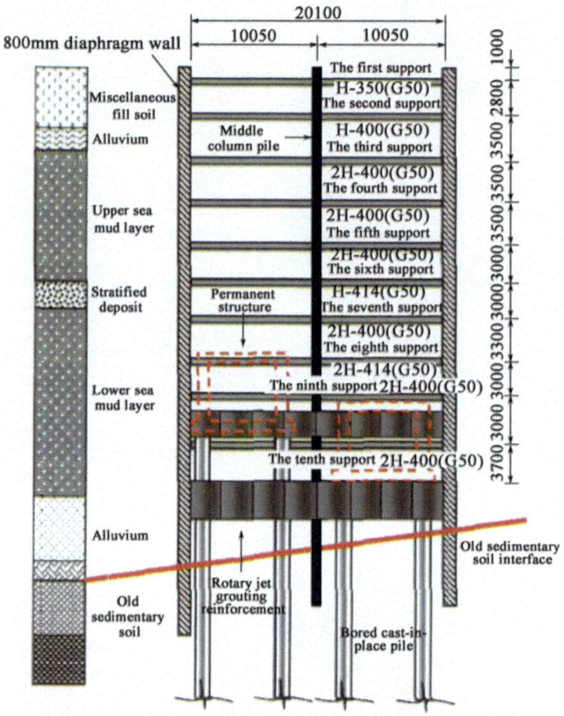

Fig. 3.31 Cross-sectional view of accident pit support (unit of measurement: mm)

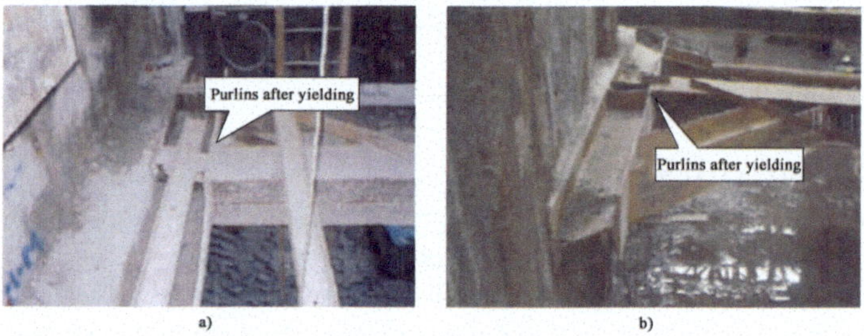

Fig. 3.32 Steel retaining girder yield situation

plain concrete cushion layer as temporary support on the excavation surface, and the concrete poured on the upper part of the retaining girder as reinforcement for the upper part of the retaining girder. The monitoring data at this time showed that the deformation of the underground continuous wall developed from 349.81 mm on April 17 to 440.55 mm on April 20, an increase of 90 mm in three days (Fig. 3.33a); the axial force of the ninth support began to decrease, while the axial force of the eighth support correspondingly increased (Fig. 3.33b).

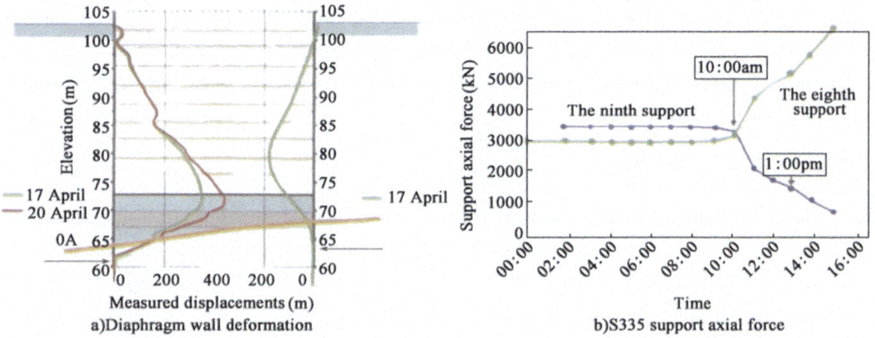

Fig. 3.33 Changes in the deformation of the underground continuous wall and the axial force of the support before the accident

a) b)

Fig. 3.34 The pit after the accident

In the afternoon of the same day, a worker found that the retaining girder and support of the ninth support had completely separated, and then the supports began to fail one after another, leading to the overall collapse of the pit. The entire collapsed section is about 100 m long, and during the collapse process, a gas pipeline broke and caught fire, a 66 kV cable was damaged causing a power outage in some areas, etc. The accident scene photos are shown in Fig. 3.34.

3.5.3 Causes of the Accident

1. Newly deposited marine silt soil

The accident pit is located by the sea, and the stratum is marine silt soil. This type of stratum has low deformation modulus and shear strength, which is a very unfavorable geological condition; one side of the pit is even a newly deposited stratum formed by land reclamation, with a short consolidation history and loose soil.

Building a pit in such a stratum already faces high risks, and has high requirements for the rigidity and strength of the pit support system.

2. Incorrect calculation parameters

(1) In the finite element simulation analysis of soil-structure interaction, the parameters under drainage conditions are used for saturated soft soil, which makes the soil–water load in the analysis model lower than the actual situation, thereby underestimating the deformation and internal force of the retaining system, leading to an unsafe design of the retaining system.
(2) In the detailed design of the connection between the support and the retaining beam, the suspension length of local buckling instability was incorrectly used, which greatly overestimated the limit load of structural buckling instability.

3. Insufficient strength of support connection nodes

(1) Using channel steel instead of stiffening plates greatly reduces the ductility of the system without significantly increasing the ultimate bearing capacity.
(2) In the field construction, some of the figure-eight braces of the supports were omitted, which added insult to injury to the inherently insufficient support system, leading to a complete collapse of the system in just a few hours.

4. Technical errors exist in the back analysis

The inversion analysis based on the deformation measurement data of the underground continuous wall simply adopted the method of reducing parameters to fit the curve, without seeking the physical meaning and mechanical explanation of the reduced parameters, and did not evaluate the overall safety of the entire retaining system.

The cause and effect of the accident are shown in Fig. 3.35. The main reasons for the accident can be divided into two aspects: on the objective side, the stratum near the pit is quite special, with weak marine mud on one side and short-term artificial land on the other, making the stability of the pit a great challenge; on the human side, due to incorrect design parameters, the structural design load is underestimated, and the neglect of the strength of the support connection nodes during construction has weakened the stability of the support. It can be said that under the combined effects of the above factors, the pit was in an extremely unsafe state during construction, and the construction party failed to detect the problems presented by the monitoring data in time, leading to a chain of destruction of the pit support system, further inducing secondary disasters such as widespread power outages and gas pipe explosions.

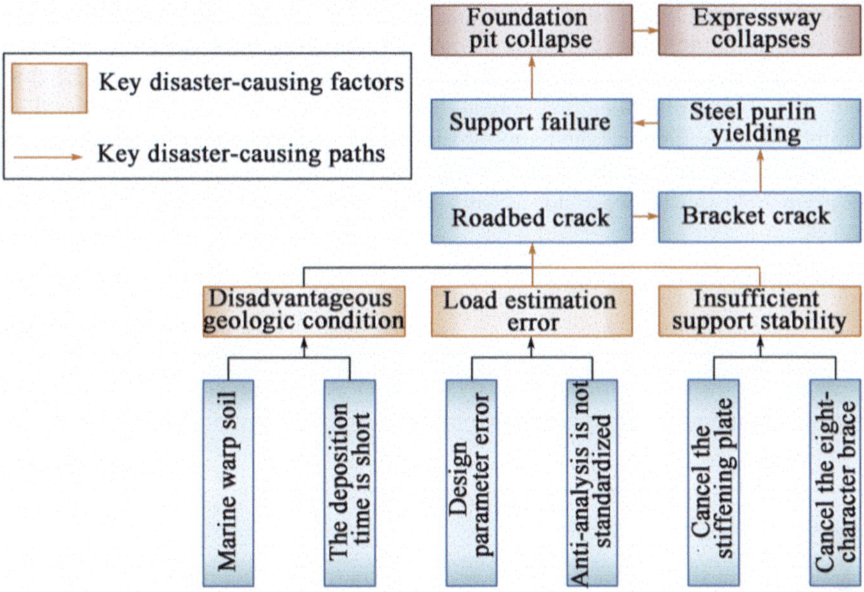

Fig. 3.35 Cause and effect diagram of the collapse accident of a certain subway section open pit

3.6 Analysis of the Water Leakage and Collapse Accident in a Cross-River Tunnel of a Certain Subway Section

3.6.1 Project Background

1. Project Overview

The accident tunnel is the cross-river section of a certain subway section tunnel. The total length of the upline of this section is about 1998 m, the total length of the downline is about 1982 m, the total length of the cross-river section is 440 m, and the maximum burial depth of the tunnel bottom is 37.35 m. The tunnel is constructed by shield method, and there is a middle air shaft on the side of the river.

The accident occurred near the air shaft. The original design plan was to first construct the upper half of the air shaft by open excavation, then cross the air shaft by shield method, and reinforce it by vertical freezing method, and construct the passage connecting the tunnel and the air shaft by mining method; there is a bypass channel under the air shaft, which is constructed by mining method combined with horizontal freezing.

2. Surrounding Environment

The accident area is located between the main urban road and the river flood wall, 53 m away from the flood wall, which is a 6.7 m high L-shaped reinforced concrete wall.

To the west, there is a wholesale market and a commercial building, to the north, there is a restaurant, two commercial buildings (5-story brick-concrete structure) and the approach bridge of the cross-river bridge, to the south, there is a 22-story residential community, local tax bureau, bank building and a commercial building. Various pipelines such as water supply, electricity, gas, communication, cable, and rain and sewage are laid above and below the main roads around the accident area. The main road on the west side is very busy, with dozens of bus routes passing through.

3. Hydrogeological conditions

(1) Geological conditions

According to the geotechnical investigation report, the physical and mechanical properties of the main strata are shown in Table 3.2.

(2) Hydrological conditions

① Groundwater type

The survey data shows that the groundwater along the line mainly includes phreatic water in the shallow cohesive soil and silty soil layers, and confined water in the deep silty soil and sandy soil layers. The 7th layer is the first confined aquifer in this area, and the 9th layer is the second confined aquifer in this area. The first and second confined aquifers in the site are interconnected.

② Groundwater level

The phreatic water level and confined water level vary with seasons, climate, lake tides, and other factors. The depth of the phreatic water level is about 0.5–1.0 m from the ground surface, and the depth of the water level of the first confined aquifer is 9 m.

③ Groundwater temperature

The groundwater temperature is affected by the air temperature within a burial depth of 4 m, and the water temperature below 4 m is generally stable at 16–18°C.

④ Hydraulic connection

According to the results of hydraulic connection observation, because there is a flood prevention sheet pile under the flood wall near the project, there is no direct hydraulic connection between the phreatic water and the river water.

⑤ Groundwater parameters

The results of indoor permeability tests for each stratum are shown in Table 3.3.

4. Tunnel design

The inner diameter of the tunnel segment in this section is 5500 mm, the outer diameter is 6200 mm, and the thickness is 350 mm. The entire ring of the segment is

Table 3.2 Table of physical and mechanical properties of strata in cross-river section

Level number	Strata name	Strata thickness (m)	Bottom elevation (m)	Water content (%)	Gravity (kN/m^3)	Void ratio	Force of cohesion (kPa)	Angle of internal friction (°)	Standard penetration test
①	Miscellaneous fill	2–6	−2.14–−2.23						
$②_2$	Grey clayey silt	10.5–13.6	−11.37–−12.64	33.5	18.1	0.96	10	26.0	5
$⑤_1$	Grey clay	3.5–5.1	−16.14–−16.47	42.1	17.4	1.20	14	13.5	
$⑤_2$	Grey clayey silt	3.9–4.5	−20.37–−20. 64	35.5	18.0	1.02	17	18.5	
⑥	Dark green silty clay	4.3–4.4	−27.77–−24.94	24.4	19.4	0.72	38	22.0	
$⑦_1$	Sandy silt	12.30	−37.99	35.1	20.3	1.04	0	33.0	40
$⑦_2$	Silty fine sand	22.86	−60.85	28.2	19.8	0.78	0	37.0	50
⑨	Silty fine sand	Not penetrated	Not penetrated	25.2		0.71	0	35.5	50

Table 3.3 Permeability coefficient table of cross-river section stratum

Level number	Strata name	Permeability coefficient (cm/s)	
		Vertical direction	Horizontal direction
②	Clayey silt	4.53×10^{-4}	3.21×10^{-4}
④	Silt clay	4.05×10^{-8}	6.40×10^{-8}
⑤	Silty clay	5.72×10^{-8}	6.44×10^{-8}
⑥	Silty clay	8.38×10^{-8}	9.51×10^{-8}
⑦$_1$	Sandy silt	5.34×10^{-4}	6.34×10^{-4}
⑦$_2$	Swilty fine sand	8.00×10^{-4}	1.07×10^{-3}
⑨$_1$	Swilty fine sand	1.20×10^{-3}	1.96×10^{-3}
⑨$_2$	Gravel fine sand	2.13×10^{-3}	3.50×10^{-3}

composed of one small top block, two standard blocks, two adjacent blocks, and one large bottom block, with a ring width of 1200 mm, using a through-joint assembly. A large tenon is set in the middle of the segment cross-section to bear the thrust of the jack, an elastic sealing groove is set on the outer arc side, and a groove is set on the inner arc side. The entire ring surface is tightly fitted at the concave and convex tenon grooves, connected by 17 M30 longitudinal bolts between the rings, and connected by 2 M30 circumferential bolts between the blocks. The segment material is C55 high-strength concrete, with a seepage resistance level of 1.0 MPa, and the rebar model is HPB235, HRB335. The segment connection bolts are made of steel with a strength grade of 5.6 and 5.8.

3.6.2 *Accident Process*

Before the accident, the shield advancement had been completed, and the open excavation part of the middle ventilation shaft had also been completed. The bypass tunnel began to freeze on April 26, 2003, the freezing working face was set in the down-line tunnel, and it basically reached the strength on May 30, and began to excavate from the up-line to the down-line. The bypass tunnel was planned to be completed on June 20, and then the vertical ventilation shaft was excavated in sections.

On June 28, due to a malfunction of a small refrigeration unit supplying cold to the downward line freezing pipe, freezing was stopped for about 7.5 h. Due to the reduction in the strength of the freezing body, the construction party temporarily suspended the excavation of the bypass tunnel and sealed the excavation face. On July 1, without detecting the supplementary freezing effect, the construction party resumed the excavation of the bypass tunnel. When the excavation reached a position 80 cm away from the opposite tunnel, a liquid nitrogen freezing hole was found to be leaking. The leakage point gradually expanded, and pressurized water in the sandy loam and fine sand layer quickly poured into the bypass tunnel, followed by a large

a)Ground collapse b)Buildings on the ground are heavily tilted

Fig. 3.36 Situation in the accident area

amount of quicksand. The incoming water brought a large amount of heat, causing the freezing body to melt quickly, causing a collapse in a short time. The large loss of water and soil around the upward line tunnel led to an imbalance in tunnel pressure, eventually leading to complete damage to the tunnel at the bypass tunnel. Subsequently, the incoming water flowed into the downward line tunnel, the stratum loss area continued to expand, the downward line tunnel was gradually destroyed, the completed shaft above quickly subsided, and the underground continuous wall of the shaft cut off the tunnel below. The large loss of water and soil also caused varying degrees of cracks and subsidence on the ground in the surrounding area, causing three buildings to tilt severely. The flood wall evolved from cracks and subsidence to collapse, leading to dangerous situations such as flood wall cofferdam pipe gushing. The situation in the accident area is shown in Fig. 3.36.

3.6.3 Cause of the Accident

1. Long-term stop of freezing

The failure of the refrigeration equipment led to a stop of freezing for 7.5 h. Such a long interruption of freezing would significantly increase the temperature of the freezing body. Plus, the accident occurred in the summer, the freezing body was already in a relatively weak state, and the heat exchange rate with the outside world was also fast. After a long stop of freezing, the freezing body has partially lost its reinforcement and water-stopping effect, creating conditions for groundwater to pour in.

2. Unfavorable hydrogeological conditions

The accident area is located in a confined water stratum. This type of stratum has a huge pressure potential. Once there is a problem with the water stop during the construction process, the confined water is very likely to break through the stratum and cause quicksand, which in turn leads to disturbance and displacement of the stratum, causing tunnel structure damage, forming a vicious cycle, and rapidly expanding the danger. Since the tunnel is located by the river, even if the anti-seepage pile of the flood wall cuts off the hydraulic connection between the confined water and the river water, the water supply in the stratum is still very abundant, and it is very difficult to block once a leak occurs.

3. Resuming excavation too early

According to the requirements of the freezing method construction plan, the freezing time needs to reach 50 days before excavation can be carried out. The upward line began to freeze on May 11, and it only met the requirement of 50 days of freezing time when the accident occurred. However, the failure of the freezing equipment meant that the actual freezing temperature could not meet the excavation requirements. The construction party resumed excavation too early without fully verifying the freezing effect, which ultimately led to the accident.

3.7 Analysis of the Partial Collapse Accident of a Subway Station Pit

3.7.1 Project Background

1. Project introduction

The pit where the accident occurred is the construction pit of a certain cross interchange station. The station is located at the intersection of two main city roads, in a bustling urban area. The station is a two-level side station, with a total length of about 480.6 m, a width of 20.35–44.5 m, the first underground level is the hall level, and the second underground level is the platform level.

2. Surrounding environment

The terrain near the station fluctuates greatly from east to west and is relatively flat from north to south, with dense buildings around. Due to the fact that some urban planning has not been realized during the construction of the station, the current roadside site is relatively wide, and there are dense underground pipelines under the road. About 350 m south of the intersection on the west side is a shopping mall that is still in business, with two underground floors, supported by piles and prestressed anchor cables. Because some of the anchor cables have entered the pit of the station, they were chiseled out before the construction of this pit; on the east side is a car 4S shop, a 2-story steel structure, with manually dug pier foundations, buried 6–8 m deep.

3. Geological conditions

According to the detailed survey report of geotechnical engineering, within the exposed depth range, the strata of the station area can be divided into several unit layers from top to bottom, and each geotechnical layer is divided into several sub-layers according to different lithology and engineering properties. The distribution is as follows.

(1) ⟨1–1⟩Fill (Q_4^{ml}); ⟨6–1⟩Silty clay (Q_4^{al}).
(2) ⟨10–1⟩Silty clay (Q_{2-3}^{al+pl}).
(3) ⟨10–1–1⟩Silty clay (Q_{2-3}^{al+pl}); ⟨10–2) Clay (Q_{2-3}^{al+pl}).
(4) ⟨11–1⟩Sandy soil with clay (Q_2^{al+pl}); ⟨1–1⟩Fill soil (Q_4^{ml}).
(5) ⟨6–1⟩Silty clay (Q_4^{al}); ⟨10–1⟩Silty clay (Q_{2-3}^{al+pl}).
(6) ⟨10–1–1⟩Silty clay (Q_{2-3}^{al+pl}); ⟨10–2⟩clay (Q_{2-3}^{al+pl}).
(7) ⟨11–1⟩sandy soil with clay (Q_2^{al+pl}).

4. Design of retaining structure

The retaining structure uses ϕ1000@1200 drilled pile, the standard section uses 3 supports, the first support is concrete support, the second and third supports are steel supports with a diameter of 800 mm and a wall thickness of 16 mm; at the end position, due to the large depth of the foundation pit, three supports are used and one is replaced.

3.7.2 Accident Process

On December 30, 2012, the south end of the station well had been excavated to a depth of about 17 m, and the first reinforced concrete support and the second and third steel supports had been installed. A worker found a micro-crack on the surface of the first concrete and the smoothing between the retaining piles showed peeling. Soon after, seepage was found on the wall of the south end of the station well, accompanied by subsidence of the ground near the pit and road surface cracking. About 0.5 h later, the south end retaining pile suddenly broke about 10 m below the pile top, the part of the retaining pile near the side wall of the end well was greatly deformed due to traction, the crown beam was damaged, and the foundation pit of the end well partially collapsed. The collapse scene is shown in Fig. 3.37.

a) b)

Fig. 3.37 Partial collapse scene of a subway station foundation pit

3.7.3 Cause of the Accident

1. Leakage of sewage pipe outside the end well

The stratum where the foundation pit is located is mainly silty clay, which has good stability under waterless conditions. However, there is a sewage pipeline near the end well where the accident occurred. The concealed leakage of this sewage pipeline caused the silty clay layer to soften locally, which led to the retaining structure bearing a water and soil load exceeding the design value.

2. Poor reliability of steel waist beams, substandard construction quality

At corner positions such as end wells (especially those that are not expanded), the reliability of steel waist beams is poor. If concrete waist beams are used, similar accidents can be avoided; in addition, the construction unit has some negligence in the construction of the shear stirrups of the steel waist beams, some shear stirrups are omitted, which further weakens the stability of the steel waist beams.

3.8 Collapse Accident of a Subway Section Tunnel (I)

3.8.1 Project Overview

1. Project Introduction

The tunnel section is a parallel single-line horseshoe-shaped tunnel built separately on the left and right, located under two city roads, with a depth of about 14 m, and constructed using the step method. There are multiple pipelines around the tunnel, located about 10 m from the tunnel arch. During the construction period, a temporary vertical shaft was set up in the green space on the southeast side of the road intersection. A cross passage is set up between the vertical shaft and the left and right lines, with a length of about 26.15 m (including the main line segment). The bedrock above the vertical shaft uses ϕ800@1200 drilled pile for retaining, and ϕ1200@1200 rotary spray pile for water stopping between piles. The bedrock below uses straight wall spray anchor support, and the construction is carried out in the open and orderly manner; the cross passage is constructed using spray anchor construction method. After the construction of the tunnel section is completed, the temporary vertical shaft and cross passage are backfilled and the ground is restored.

2. Geological Conditions

(1) Stratum Conditions

As revealed by drilling, the thickness of the Quaternary strata in the field is 2.00–12.70 m, mainly composed of Quaternary Holocene artificial fill (Q_3^{al}), Holocene alluvial layer (Q_4^{al+pl}), and Upper Pleistocene alluvial layer (Q_3^{al+pl}) is composed of. The bedrock in the field area is mainly coarse-grained granite, with porphyry and granitic porphyry interspersed in veins, and mylonite and fractured granite can be seen in different lithological contact zones.

(2) Hydrogeological conditions

There are mainly two types of groundwater in the field area: one is Quaternary pore-confined water, and the other is bedrock fissure water. Quaternary pore-confined water is mainly distributed in the Quaternary flood deposit sand layer, which is the main aquifer. The thickness of the clay layer above the sand layer varies greatly, the water level is slightly undulating, and it has weak confining properties locally. The groundwater level observed by drilling is 2.80–6.50 m deep, with an absolute elevation of 9.20–12.53 m. Bedrock fissure water is mainly stored in the strong and moderately weathered zones of the rock. The water-bearing and permeability of the bedrock are controlled by the structure, tectonics, and degree of fracture development of the rock mass. Due to the anisotropy of the rock mass, coupled with local rock mass fragmentation and joint fracture development, the water richness and permeability of the rock mass are not the same. In the joint fracture development zone of the rock mass and the dense fracture zone of the rock vein, groundwater is relatively enriched, and the permeability is also relatively good. Overall, the development of bedrock fissure water is heterogeneous. The groundwater level observed by drilling is 3.50–6.00 m deep, with an absolute elevation of 14.11–17.46 m. Although the amount of bedrock fissure water is not large, it has a clear hydraulic connection with Quaternary pore-confined water and can be considered under the same head. Overall, the groundwater in the site is moderately rich and has a large amount of water.

3.8.2 Accident Process

Before the accident, the excavation work of the vertical shaft and the horizontal passage had been completed, and part of the support structure had been implemented. The left line of the tunnel main line has been excavated 53 m in the large mileage

Fig. 3.38 Collapse site photo

direction, and the right line has been excavated 38 m in the large mileage; the left line has been excavated 40 m in the small mileage direction, and the right line has been excavated 42 m in the small mileage.

On July 17, 2011, broken rocks and sand fell from the upper part of the arch in front of the steel arch frame near the face of the left line in the large mileage direction of the section. The construction unit immediately organized measures such as shotcrete and sandbag sealing of the face, but due to the large amount of water gushing and strong water pressure, the sealing sandbags were washed away, the gushing water phenomenon at the arch top intensified, and the situation continued to worsen. Soon after, the road surface above the face collapsed, forming a collapse pit with a diameter of about 5 m. The collapse site is shown in Fig. 3.38.

3.8.3 Cause of the Accident

1. Geological reasons

The geological conditions of the accident section are poor, the surrounding rock is a strong weathering zone, its thickness is 0.4–4 m and it is severely altered, and joint fractures are developed, with a clay layer and a sand layer above it. Bedrock fractures are developed, permeability is good, and groundwater is enriched.

2. High groundwater level in the rainy season

The accident occurred during the rainy season. According to the construction unit, there was a large amount of rainfall before the accident, the groundwater level was high, and the water flow was large.

After the loss of the lower sand layer, a sewage pipe near the tunnel broke, causing sewage leakage, and then under the huge water and soil pressure, the stratum of the excavated and unsupported area was broken by mud and sand, leading to this collapse accident.

3. Incomplete grouting reinforcement plan

Under the rich water conditions during the accident, the short-distance drilling and grouting adopted by the construction unit could not effectively reinforce the surrounding rock and the upper sand layer. It should adopt advanced long-distance pre-grouting reinforcement, and the arrangement of the grouting holes, the amount of grouting, and the ratio of grouting should all be able to play a role in water stopping and reinforcement.

3.9 Collapse Accident of a Subway Tunnel Section (II)

3.9.1 Project Overview

1. Project Introduction

This section of the tunnel is located under the city's main road, with a total length of 1234.47 m for the left line and 1213.581 m for the right line. The spacing between the two lines is 15 m at the nearby station of this line, and it gradually changes to 13 m through two curves with a radius of 350 m. The section tunnel sets up a civil air defense section near the station, and has two contact passages, one section pump station, one construction shaft, which is set on the right line. According to ventilation requirements, the construction shaft and construction cross passage of this section are permanent structures, serving as piston wind shafts and piston wind tunnels during operation. The construction shaft is set on the green space north of a gas station, and the construction site is located in this green space.

This section of the tunnel goes under a river culvert and a city road after leaving the nearby station of this line, then goes under a 4-story building of a hospital and 3 nearby 6-story buildings, and after another city road, it goes under 7 6–7 story buildings of a residential area. The accident site is located at the intersection of the two city roads, only 21 m away from the hospital building in horizontal distance. The plan of the accident site is shown in Fig. 3.39, and the section under the hospital is shown in Fig. 3.40.

2. Geological conditions and surrounding environment

The stratum conditions of the accident site: from top to bottom, they are Quaternary soil (miscellaneous fill soil, plain fill soil, ⑦ silty clay, coarse sand, ⑪ silty clay), strongly weathered granite, moderately weathered granite, slightly weathered granite. The areas along the line are mainly residential areas and shops, with busy ground traffic. The existing ground elevation is 26.9–36.9 m. The underlying bedrock is Mesozoic Yanshan late intrusive rock, mainly granite, with porphyritic veins, and locally the thickness of the strong weathering zone is large. In the thicker parts of the granite strong weathering zone, the water content is rich. The engineering geological

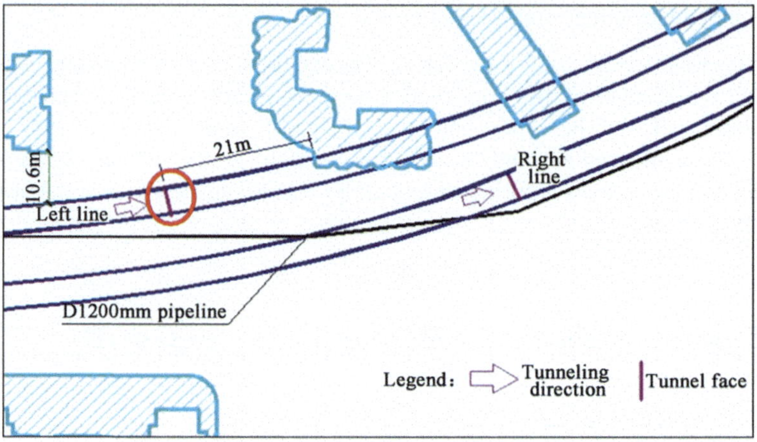

Fig. 3.39 Plan of the accident site

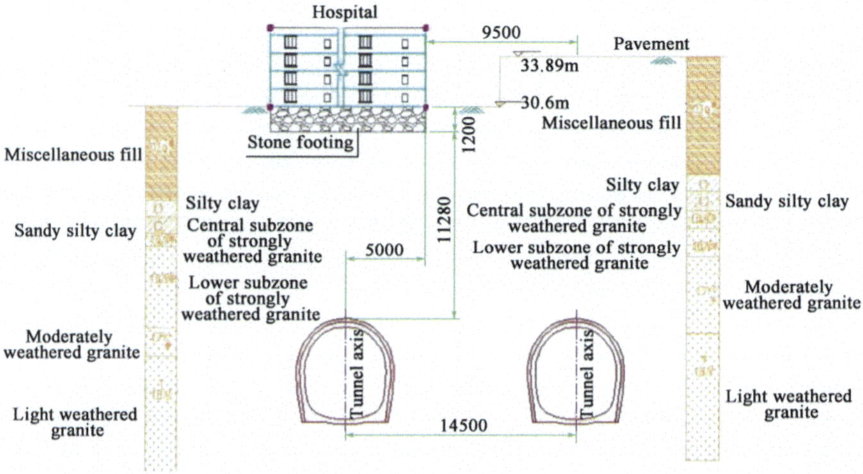

Fig. 3.40 Section under the hospital (unit of measurement: mm)

conditions are relatively simple, the granite strong weathering zone has the characteristics of softening and disintegration when encountering water; the rock in the moderate and slight weathering zone is hard and has high strength. The surrounding rock level is from level II to level IV.

3.9.2 Accident Process

On July 19, 2013, blasting construction was carried out on the left line with a large mileage. After the blasting, a local collapse suddenly occurred on the right side of the face, accompanied by a large amount of yellow mud and water gushing out, along with broken masonry bricks and ceramic pipe fragments.

According to on-site observations, the top of the collapse cavity is about 6 $\times$ 7 m, the depth is about 13 m, the volume of the collapse cavity gradually decreases downwards, and the predicted total volume exceeds 50 m^3. Due to the dense distribution of pipelines under the road at this location, the presence of pipeline wells, and the recent heavy rainfall, the groundwater level is high. This area is often filled with water, resulting in a smaller amount of collapse observed inside the tunnel. Figure 3.41 is a picture of the accident site.

Fig. 3.41 Accident site photo

3.9.3 Cause of the Accident

1. Geological reasons

The geology of the construction site changed abruptly, the surrounding rock changed from level III–IV to level VI, a large area of quicksand appeared in front of the face, and the rock layer above the face was thin. The blasting disturbed the rock layer, causing a sudden water and sand phenomenon.

2. Surrounding environmental reasons

The collapse site is located at the downhill direction of the urban road intersection, the surrounding pipelines are complex, there are unexplored pipeline wells and abandoned pipelines, and the abandoned pipelines are buried deep.

3. Long-term suspension of excavation

Due to the suspension of explosives supply in June, it was not until July 15 that the supply was resumed, and the construction was stopped during this period. During the suspension of excavation, the actual situation on the site changed, leading to unexpected situations.

3.10 Analysis of Other Urban Underground Space Construction Safety Accidents

3.10.1 Water and Sand Gushing Accident at the Shield Machine Outlet of a Subway Station Interval Tunnel

1. Project overview

The interval tunnel where the accident occurred is a double-line tunnel, with the right line being 705.386 m long and the left line being 708.835 m long; the tunnel is buried 8.8–18.7 m deep, the maximum line spacing is 14 m, and there are 4 curves with a radius of 400 m in the interval, with the maximum longitudinal slope being 15.79‰.

2. Accident process

Before the accident, the left line segment was assembled to + 286 rings, and the right line segment was assembled to + 118 rings. On October 29, 2012, when the management personnel conducted a site inspection, they found that sand was gushing from the spiral outlet in the left line tunnel, the side wall of the muck pool collapsed, and multiple damages appeared on the segments of the left and right line tunnels in the curve section. The accident situation is shown in Fig. 3.42.

a)Sand gushing at spiral outlet of shield machine

b)Collapse of side wall of muck pond

c)Curve segment segment damage

Fig. 3.42 Accident site situation

3. Cause of the accident

During the accident, the groundwater head was high, and the improvement effect of the muck was not ideal, which caused the water content in the silo to be too high, causing the soil in the silo to become muddy, resulting in a water and sand gushing accident.

3.10.2 Water Gushing Accident in the Inclined Shaft of a Certain Rail Transit Interval Tunnel

1. Project overview

(1) Project introduction

The accident tunnel is a double-hole single-line tunnel with a total length of 5633 m, using open excavation, drilling and blasting, and a combined TBM construction scheme. Considering safety and schedule requirements, the combined TBM is constructed from the import end to before the coal seam and karst valley area, where

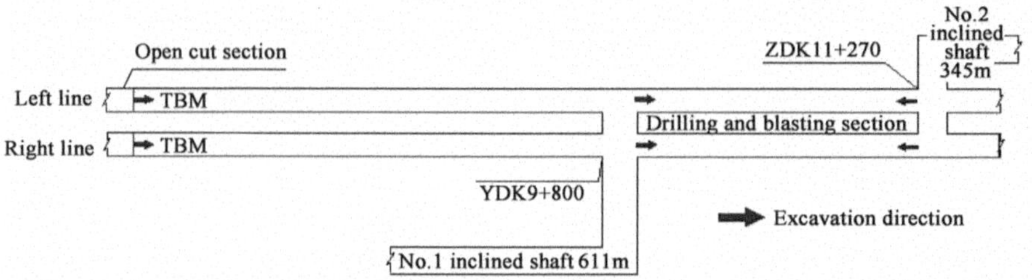

Fig. 3.43 Engineering plan diagram

the No.1 inclined shaft is constructed as a permanent passage for TBM disassembly and transportation, with a total length of 611 m, and from here to the large mileage direction drilling and blasting excavation; at the same time, the No.2 inclined shaft is constructed as a temporary passage 1470 m ahead, and from here to the small mileage direction drilling and blasting excavation, with a total length of 345 m. The engineering plan is shown in Fig. 3.43.

(2) Adverse geology

There are gypsum rock layers, coal mine voids, and karst areas within the tunnel section. The length of the gypsum rock layer is about 25 m, the length of the coal mine void is about 13 m, and the karst areas are intermittently distributed, with a total length of about 470 m. Adverse geological conditions have a significant impact on tunnel construction.

(3) Groundwater

The groundwater in the tunnel section mainly consists of phreatic water, detrital rock pore-fracture water, and carbonate rock fracture-karst water, with some areas containing confined water.

The tunnel area overall belongs to an anticlinal water storage structure, with a carbonate rock aquifer in the core of the anticline, and two thick layers of sandstone aquifers, forming three relatively independent water storage structural units that are not very obviously recharged by each other and receive replenishment from atmospheric precipitation.

2. Accident Progress

On July 1, 2012, geological advance detection construction was carried out on the right line of the main line of the No. 2 inclined shaft. When the drill rod drilled to a depth of 6 m, a small amount of water flowed out of the hole, and the trend gradually increased; when the drill rod drilled to a depth of 8 m, the water flow in the hole continued to increase, and there was water pressure; when the drilling depth reached 10 m, the water volume increased, the water pressure increased, and the drilling machine could not drill.

After the drilling machine could not drill, the drill rod was pulled out. At this time, the water jet distance was about 30–40 cm, the water flow color was turbid yellow, and

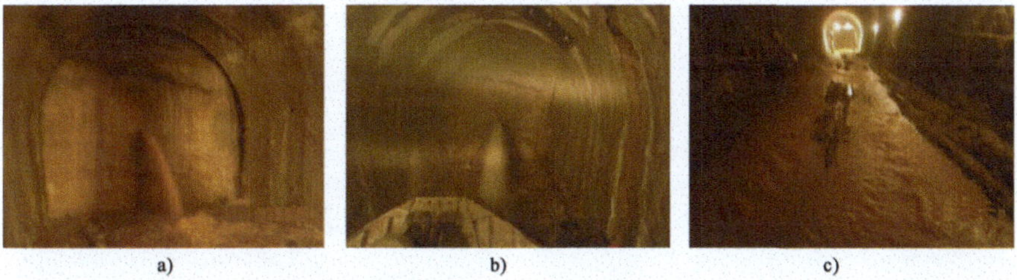

Fig. 3.44 Accident scene photo

there were a lot of mud and sediment; 1 h later, the water jet distance was about 2–2.5 m, the water flow was still turbid yellow, with no significant changes; about 6 h later, the water jet distance reached about 9 m, the water flow color gradually stabilized, the color was deep yellow, the gushing water volume reached about 600 m^3/h, and obvious cracks appeared on the face. The accident scene is shown in Fig. 3.44.

3.10.3 A Subway Station Gantry Crane Tipping Accident

1. Accident process

On July 11, 2013, a gantry crane at a subway station construction site collapsed to the west, scattering equipment debris everywhere. A Sangyong SUV and a Mitsubishi sedan were hit by the main body of the gantry crane, and the vehicles were severely deformed. A Hyundai sedan and a taxi were in the inner and outer spaces of the gantry crane, and the damage was lighter. The accident scene is shown in Fig. 3.45.

2. Cause of the accident

The accident occurred during a heavy rain period, with gusts exceeding 12 levels, and the gantry crane did not take effective windproof measures. Under the influence of inertia, it finally collapsed under the action of the strong wind.

3.10.4 A Residential Building Basement Construction Collapse Accident

At around 13:40 on October 8, 2011, during the concrete pouring construction process of a certain project underground garage, a formwork collapse accident occurred, causing casualties and property losses.

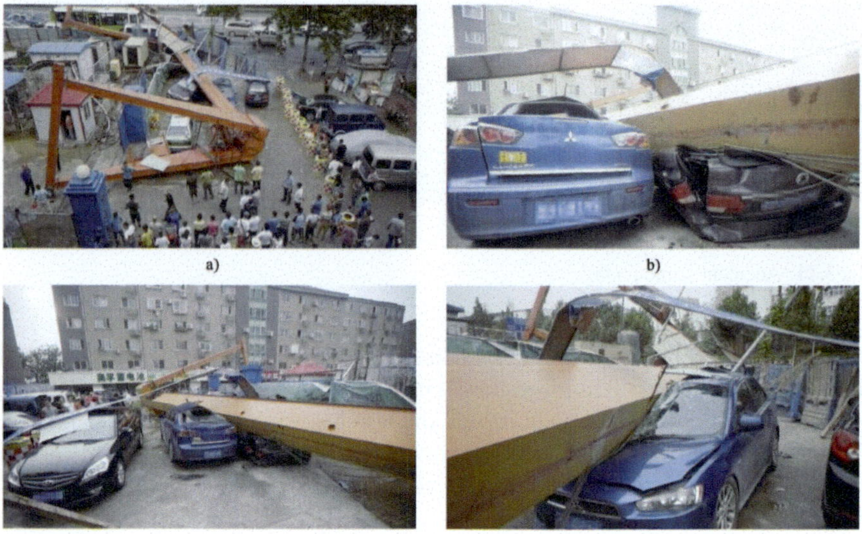

a) b) c) d)

Fig. 3.45 Accident scene photo

1. Project Overview

The accident occurred in the underground garage at the bottom of a residential building. This area is an extension part of the basement, with a floor height of 5.6 m, using cast-in-place concrete construction. The area of concrete to be poured for construction is 600 m^2, and at the time of the accident, 400 m^2 had been completed. The pouring operation of the top plate concrete.

2. Accident process

On October 2, 2011, the formwork support of the collapsed area of the underground garage began to be erected and was completed on October 4. The rebar tying started on October 5 and ended on October 7.

On October 8, five workers checked the formwork and leak plugging work under the formwork; at the same time, 21 workers were pouring concrete on the formwork. The pouring order was shear wall, column cap, and finally the top plate. During this period, someone found that the tie rod bolt at the bottom of the shear wall on the north side of the pouring area was broken, causing bulging of the formwork, concrete outflow, and then 13 workers began to clean up the concrete and repair the bulging.

During the repair work, some workers cleaned up two passages that could be passed by a single-wheel handcart from the bulging place to the east between the formwork supports, removed some rods in the support system, and used a single-wheel handcart to transport the leaked concrete. At the same time, the concrete pouring construction continued on the formwork. When about 400 m^2 of concrete was poured, the workers on the top plate felt a vibration, and the already poured top plate concrete collapsed instantly.

3. Cause of the accident

The onsite workers removed some rods in the support system to clear the leaked concrete, which greatly reduced the overall stability and bearing capacity of the formwork support. At this time, the concrete pouring operation did not stop. Under the load of concrete pouring and vibrating, the support system could not bear the upper load and became unstable, leading to the collapse of the entire newly poured basement top plate.

Chapter 4
Basic Theory of Safety Risk Assessment for Urban Large Underground Space Construction

4.1 Basic Concept of Risk

4.1.1 Definition of Risk

1. General definition of risk

Compared with most certain real things, the concept of risk seems very ambiguous and vague. It is very important to clarify the definition of risk before studying risk. There is still some controversy over the definition of risk. Some people believe that risk is the possibility that a certain activity will lead to some undesired consequences; others believe that risk refers to the possibility that the actual benefits of a project will not reach the minimum acceptable level; others believe that risk refers to the degree of difference between the actual results caused by actions and the expected goal. But regardless of personal understanding, the definition of risk will include a certain "possible" event, and the consequences of this event are two major elements. Therefore, broadly speaking, risk refers to the possibility of a certain event occurring in a specific environment, within a specific time period, and the consequences it brings. Risk arises from the uncertainty of unknown event, that is, the randomness of risk probability. This concept shows that risk is for an action that has not yet been taken. An action has risk if it meets the following two conditions: the result of the action is uncertain; the action will inevitably lead to some kind of gain or loss.

2. Probability definition of risk

(1) General risk—the combination of the consequences and the possibility caused by a certain activity. Suppose the sample space of a certain result that may be caused by a certain activity is Ω, the elements of the sample space are ω. F is a σ algebra defined on Ω, the element in F is f, f is called an event; P is the probability measure on (Ω, F). Therefore, the probability space can be defined as:

S.-X. Lei et al., *Safety Risk Assessment of Urban Underground Large Space Construction*, https://doi.org/10.1007/978-981-97-7437-1_4

$$(\Omega, \mathrm{F}, \mathrm{P}) \tag{4.1}$$

The definition of general risk does not have specific requirements for the consequences caused by activities, so any element in F is the consequence of this activity. Therefore, the probability space (Ω, F, P) can constitute the general definition of general risk. At this point, f can also be called a risk event.

(2) Narrowly defined risks—risks leading to negative consequences

Let C be a random variable defined on the probability space (Ω, F, P):

$$\{\omega : C(\omega) \leq x\} \in, \forall x \in R \tag{4.2}$$

Then C defines a mapping from a risk sample space to a real space. When certain specific mapping rules are adopted, the random variable C will have a definite real meaning. For example, let C represent the equivalent economic loss resulting from the occurrence of some events in Ω, then C is what we know as the loss amount.

On this basis, the narrow definition of risk events is as follows:

$$\mathrm{F} = \left\{\overline{f}\right\} = \{f : \forall\omega \in f, C(\omega) > 0\}, f \in \mathrm{F} \tag{4.3}$$

That is, the narrow definition of risk events specifically refers to events that will cause losses when they occur. Let's denote this type of event (narrowly defined risk event) as $\overline{f}$, and its universal set as $\overline{F}$. $\overline{F}$ is a subset of F. Generally speaking, in the field of engineering, people are concerned about this type of adverse events that can cause disasters and losses. Therefore, unless otherwise specified, the term "risk" in the following text refers to this type of narrowly defined risk.

(3) Basic risk event—the smallest unit defining risk

For a specific activity, there is a wide range of outcomes that can result from it. The multiple results caused by the activity may appear separately, or multiple results may occur at the same time; the subject of the activity may profit, suffer losses, or profit while suffering losses. Therefore, before conducting a risk analysis of an activity, it is necessary to break down all of its possible results into the most basic logical units, that is, basic risk events. Basic risk events are defined as follows:

① A basic risk event is a sample point of the results caused by the activity, that is, it cannot be further divided into sub-events, and there is mutual exclusivity between basic risk events;
② The occurrence of a basic risk event will inevitably cause the activity subject to suffer losses;
③ Multiple basic risk events can be combined into a new event. The new event is still a risk event.

Through the analysis of the previous section, the universal set of risk events $\overline{F}$ is a subset of F, and those single-point sets in $\overline{F}$ are basic risk events, the elements of which is denoted as $\overline{\omega}$. Obviously, $\overline{\omega}$ is an element in the sample space Ω, and its universal set is denoted as Ω. Ω is a subset of the sample space Ω, which is defined as:

$$\overline{\Omega} = \{\omega : C(\omega) > 0\}, \omega \in \Omega \tag{4.4}$$

On the other hand, according to the definition of $\overline{\Omega}$, $\overline{\Omega}$ is also an element in $\overline{F}$, and it satisfies:

$$\forall \overline{f} \in, \overline{f} \subseteq \overline{\Omega} \tag{4.5}$$

Since $\overline{F}$ is a subset of F, the probability measure on (Ω, F) must be defined on $\overline{F}$. Therefore, for any risk event $\overline{f} \in \overline{F}$, its probability measure $P\left(\overline{f}\right)$ and the random variable representing risk loss $C(\overline{\omega}), \overline{\omega} \in \overline{f}$ together constitute the general definition of risk.

4.1.2 *Elements of Risk*

1. Action

Risk is always relative to a specific action. From the perspective of the extension of the concept of risk, due to the universality of risk, without a specific action as a basis, the analysis of risk loses its boundaries, and the selection of risk analysis objectives, risk evaluation reference systems, and risk analysis paths will lose their basis. In this context, risk is mundane and meaningless. From the perspective of the connotation of risk, risk is a measure of the uncertainty of events that have not yet occurred. Without action, there would be no changes in related things and states, and the consequences of uncertainty are naturally unavailable.

2. Risk Events

The concept of "risk" is always linked to a specific event. For example, a business activity may result in a profit, a loss, or a break-even for the business entity. Here, "profit", "loss", or "break-even" are all specific events. According to the definition of risk, since "profit" and "loss" occur with a certain probability and bring certain gains or losses to the action subject, "profit risk" and "loss risk" can be defined, and the corresponding events are called "risk events". The event of "break-even", when limited to the perspective of property gains or losses, does not involve risk, but if the time and energy consumed by the business activity are also considered, this event also involves corresponding risks. Risk events can be defined at different levels, for example, the event of "loss" can be further subdivided into events such as "loss of less than 10,000 yuan" and "loss of less than 100,000 yuan". It can be seen that the definition of risk events is related to the object of study, the perspective of study, and the depth of study. Any event that meets the following conditions can be considered a risk event.

(1) The event is a direct or indirect result of the action under study.
(2) The event is neither impossible nor inevitable.
(3) From the perspective of the study, the occurrence of the event will bring gains or losses to the action subject.

Risk events and risks are two sides of the same coin. Risk events are the concrete manifestation of risks, determining the scale reference for risk losses; on the other hand, the causes of risk events correspond to the formation mechanism of risks. Risks are transformed into losses through risk events, and at the same time, risks can also be seen as one of the inherent attributes of risk events, which is a measure of the degree of harm of risk events under uncertain conditions.

3. Risk factors (hazardous factors)

Risk factors are the necessary conditions for the occurrence of risk events. Essentially, a risk factor is also an event. Depending on the object of analysis, a risk factor of a risk event can also be a risk event itself. For example, "landslide" is a risk event, but for the risk event "debris flow", it is a risk factor. Further generalization, the occurrence of a risk event is actually the result of a chain reaction of a series of events, which forms a "fault tree", as shown in Fig. 4.1. The top event of the fault tree is the risk event, and the other events in the fault tree can be considered as risk factors. Depending on their position in the fault tree, risk factors can be divided into bottom-level risk factors and indirect risk factors.

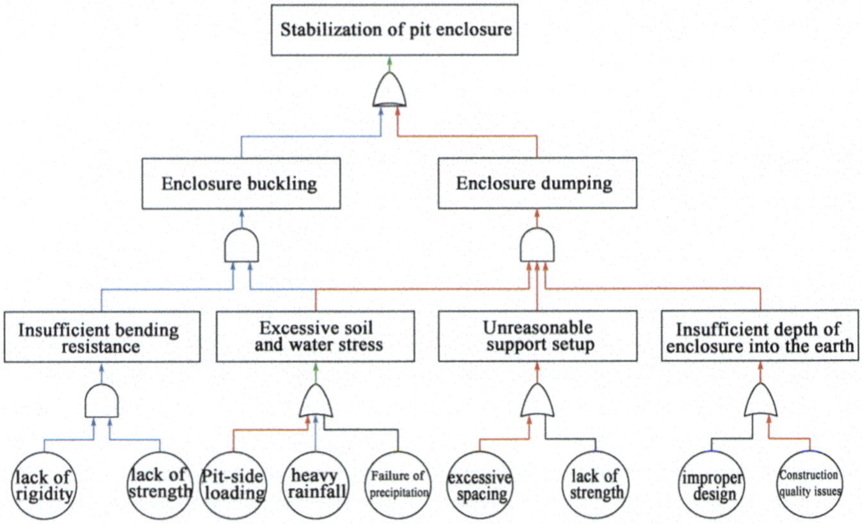

Fig. 4.1 Fault tree

4. Risk pathway

Risk pathway are the bridges connecting risk factors and risk events. In the fault tree, bottom-level risk factors continuously transition upwards along the risk pathway, eventually reaching the top to form a risk event. As shown in Fig. 4.1, the red line represents a risk pathway, and the blue line is another risk pathway. The risk pathway is closely related to the formation path of the risk event. When a risk event occurs, its formation path is unique, but the risk pathway corresponding to its risk may not be unique. In other words, the risk pathway is the set of all possible formation paths of the risk event corresponding to a certain risk.

5. Risk sources

Risk sources can be understood as a specific existing thing, or the state of existence of a thing. The risk source is the root of the existence of risk, and it determines the type of risk event, the existence of risk factors, and the risk pathway. In other words, the risk source determines the structure of the risk system. For example, for the engineering action of building a tunnel, the act of building the tunnel itself is the primary risk source. It determines the risks of this action, such as "roof collapse", "face instability", "support failure", "gas explosion", etc., and the risk factors may include "poor surrounding rock conditions", "high water inflow", "untimely support closure", "substandard grouting quality", "poor management of construction machinery", etc. If there are other risk sources on this basis, such as buildings above the tunnel, the risks of this action will correspondingly increase, such as "building tilt", "building cracking", "construction vibration disturbing residents", etc., and the risk factors will also increase. Therefore, the identification of risk sources is the first step in risk analysis work. Risk events, risk pathway, and risk factors can all be derived from the state of risk sources through logical deduction.

6. Risk system

The risk system is a conceptual whole that includes risk sources, risk events, risk factors, and risk pathway, as well as the physical entities corresponding to this series of abstract concepts and many other elements. The boundary of the risk system is the boundary of the risk problem. Theoretically, the risk system contains all information about risks.

4.1.3 General Properties of Risk

Risks are of the following general nature.

(1) Uncertainty of risk

Uncertainty is the fundamental attribute of risk, and it is a direct inference derived from the definition of risk. The uncertainty of risk does not refer to the uncertainty of the existence of risk or the measurement of risk but refers to the inseparability of the concept of risk and uncertainty. Risk is defined in terms of actions that do not occur and whose results are uncertain. If there is no uncertainty in the action, there is no risk to speak of.

(2) The dialectical unity of objective and subjective aspects of risk

The objectivity of action determines the objectivity of risk. Risk exists objectively and does not shift with human will. Risk analysis is the process of people understanding objective reality through subjective consciousness activities, and it is inevitably affected by individual subjective limitations. Therefore, the cognition of risk can only be a process that constantly approaches objective facts but can never truly reach "objective" risk. This nature of risk makes it difficult for us to conduct rigorous and objective research on risk. The results of risk analysis depend heavily on the model used to describe risk, which gives risk a strong subjectivity. For the same action, the risks obtained from different perspectives and different models are often different and may even be vastly different. Moreover, because the essence of risk is unknown and uncertain, as the action progresses and the known information increases, the information about risk is in a process of constant reduction, which leads to the fact that risk is almost unverifiable in actual situations (unless the same action can be repeated multiple times under the same conditions).

(3) The constancy of risk

As the saying goes, "anything is possible." Although "inevitable events" and "impossible events" exist in people's intuitive experience and pure rational logic, from the perspective of empirical philosophy, there is no action in reality that will absolutely lead to a certain result. This means that actions that have not been taken always have uncertainty, and with uncertainty, there is risk. People can only change the conditions for the formation and Variability of risk within a certain range, reduce the probability of risk accidents, reduce the degree of loss, but cannot completely eliminate risk.

(4) The universality of risk

The universality of risk is a characteristic that all engineering or research objects have. In a broader sense, the history of mankind is a history accompanied by various risks. Since the appearance of humans, they have faced various risks, such as natural disasters, diseases, disabilities, death, wars, etc. With the development of science and technology, the improvement of productivity, the progress of society, and the evolution of humans, new risks have emerged, and the losses caused by risk accidents are getting bigger and bigger. Especially for urban underground engineering due to its special construction environment, once a risk accident occurs, it is often a malignant engineering accident, which will bring huge economic losses and major casualties, and must be given high attention. From the universality of risk, we study the particularity of the construction risk of urban underground engineering, and then control the safety of such special engineering construction.

(5) The randomness of risk events

The occurrence and consequences of risk events are both random. Will a risk event occur? When will it occur? What kind of consequences will it cause after it occurs? These questions all have a strong uncertainty. However, people have found through long-term observation that when external conditions remain unchanged, these factors often follow certain statistical rules. This property of risk events is called randomness.

(6) Relativity of Risk

Risk is always relative to the acting subject. Even the same action that may lead to the same result, for different subjects, the risk is also different. This is because the definition of risk not only contains the objective element of "possibility", but also includes the subjective element of "loss". For the same consequences, due to the different benefits borne by different subjects, the size of the investment, the status of the subject and the resources owned, their cognition, position and ability to bear the risk consequences are different.

(7) Variability of Risk

Everything can be transformed into its opposite under certain conditions. Conditions in this context refer to all risk factors involved in the activity. When these conditions change, it will inevitably cause changes in risk. The variability of risk includes changes in the nature of risk, changes in the consequences of risk, and the emergence of new risks.

(8) Measurability of Risk

The measurability of risk refers to its subjective measurability, which can be a vague concept, a qualitative judgment, or a quantitative value. Strictly speaking, risk cannot be verified, so no matter what kind of risk, we can always establish some kind of measurement method for it.

(9) Variability of risk

The Variability of risk is manifested as the risk of the object changes with the changes in its own conditions and the change of external environment. For example, the risk level of urban underground engineering will increase due to long-term rainfall causing changes in the strength of the strata, and it will also increase or decrease due to changes in the construction team.

4.2 Overview of Construction Safety Risk Assessment

As a natural resource, the development and utilization of underground space has long been the subject of domestic and international concern and attention. With the continuous expansion of urban development scale, increasing development intensity, increasingly tense environmental carrying capacity and strengthening internal actions, while achieving huge benefits from highly concentrated urban development, a series of urban problems such as population expansion, insufficient space capacity, tight land supply, road traffic congestion, deteriorating ecological environment, destruction of historical and cultural features, insufficient infrastructure supply, weak disaster prevention capabilities, etc., have also emerged, bringing serious negative impacts to intensive urban development, forcing people to explore new development spaces—underground spaces.

The development of underground space is a systematic engineering with high technical requirements. The degree of underground space development of a country is closely related to the country's scientific and technological level, degree of industrialization, economic development trend, and the number of technical personnel. In some economically developed countries and regions abroad, due to the early realization of industrialization and modernization, the development of underground space has taken shape as early as the first half of the 20th century. Entering the 21st century, the urbanization process of various countries has further accelerated, and the development of underground space has also achieved substantial development, basically forming a unique yet regular underground space utilization model, which is specifically manifested in the following aspects.

(1) The utilization of underground space is an important approach to sustainable development for a country, increasingly becoming the top choice for the three-dimensional construction of major cities. It interacts and produces synergistic effects with factors of the social development system such as convenient transportation, environmental protection, and economic prosperity.
(2) The trend of focusing on the livelihood functions of underground space utilization is becoming increasingly apparent.
(3) The pursuit of a safe and comfortable living environment is the cornerstone and guarantee for the development of underground space utilization.

The research and development of underground space in our country started relatively late, beginning in the 1960s and 70s of the 20th centuries. After decades of hard exploration, certain experiences and lessons have been accumulated for the development and utilization of underground space resources. Especially during the economic transformation period since the reform and opening up, a large number of large-scale underground space utilization has become the highlight of urban construction. Since the 21st century, the development and utilization of underground space in our country has sprung up like mushrooms after the rain in major cities, and second and third-tier cities and rural areas also consider it in the overall planning of infrastructure. The expansion from the ground to the underground space has become an inevitable trend. At present, there are many large-scale underground complex construction projects in our large cities, with large scale and high level; the construction of underground space has begun to expand in three dimensions, forming a connected network, and loading functions such as business, storage, parking, entertainment, disaster prevention, and municipal functions in it, to expand people's activity space and promote the efficient operation of urban functions.

Although the development and utilization of underground space has many advantages, such as under the guidance of the concept of building a "resource-saving, environmentally friendly" society, the rational development and utilization of underground space resources can expand urban capacity, alleviate urban traffic, optimize the urban environment, etc., due to the large scale of underground space projects, the geology and surrounding environment are extremely complex and there are many uncertain factors, and the construction is difficult, making all links of the design and construction of underground space projects face great safety risks. Improper handling of any link can even lead to catastrophic accidents, causing immeasurable major losses. For example, a survey of accidents in underground spaces in more than twenty cities including Beijing, Shanghai, Shenzhen, and Guangzhou in five types of regions across the country during 2004–2019, including soft soil areas, flood deposit soil layer areas, loess areas, expansive soil areas, and complex geological unit or bedrock areas, found a total of 295 urban large underground space construction accidents, including collapses, falls from heights, water and sand leakage, mechanical injuries, structural deformation and destruction, etc., among which collapse accidents occurred most frequently, and the number of casualties caused by accidents was also high, with an average of about 2.23 casualties per collapse accident, posing a great threat to the safety of construction and operation of underground space projects. Against this background, the importance and urgency of risk management and risk assessment work for underground space projects are increasingly prominent.

The occurrence of numerous underground space engineering accidents has made people more acutely aware of the huge potential risks in underground space engineering, forcing people to conduct a series of related studies to explore the causes and processes of risk accidents, try to find methods to predict the possibility of risk accidents and the preliminary estimation of losses caused after the accident, and formulate a series of risk control measures to reduce the probability of accidents or the loss after risks occur. These studies are inseparable from the research scope of engineering risk assessment. Therefore, systematically carrying out research on underground space engineering risk assessment issues, developing operable risk assessment technologies, formulating construction risk evaluation indicators and quantification standards, and establishing a complete and reliable risk evaluation system can significantly improve the level of risk prevention and control in underground space construction, reduce major risk accidents, reduce construction disaster losses, and have significant and far-reaching significance.

4.2.1 Definition, Purpose, and Principles of Construction Safety Risk Assessment

1. Definition of risk assessment

The so-called "risk assessment" is the process of identifying, analyzing and evaluating risks in order to obtain a qualitative or quantitative perception of the level of risk of an action. The purpose of risk assessment lies in "evaluation" and "estimation". Specifically, risk assessment does not pursue the precise measurement of risk level but uses a series of relatively clear thinking processes and mathematical methods to decompose and concretize the highly abstract concept of risk, guiding the evaluator to establish the logic of risk analysis, and using the evaluator's knowledge, concepts, and some objective data to make a rough judgment of the risk level. Therefore, risk assessment is a work that combines subjectivity and objectivity, and integrates science and art. In the "Tunnel Risk Management Guide" of the International Tunnelling Association (ITA), risk assessment is defined as a collective term including risk identification, risk analysis, risk estimation, and risk control. In the "Tunnel Engineering Risk Assessment Operation Standard Manual" compiled by the International Tunnelling Insurance Group (ITIG), the definition of risk assessment is as follows.

Risk assessment is a systematic process, including the following content.

(1) Distinguish disasters and related risks through risk assessment, as well as the impact of this part of the expenditure on project costs and engineering progress, including the impact on third parties.
(2) Quantify risks, especially those related to engineering progress and costs.
(3) Propose a clear risk management plan, including pre-control measures to eliminate or mitigate risks.
(4) Identify operable risk control measures.
(5) Allocate risks to different responsible parties involved in the project.

2. Purpose of risk assessment

From the above definition, risk assessment should not only be a systematic and procedural work in engineering construction, but also a basic concept of in-depth construction project management. Risk assessment is not only a logical thinking mode, a scientific analysis theory, a series of reasonable and feasible calculation methods, but also a kind of philosophical thought guiding construction management and decision-making. Risk assessment can help construction managers achieve the following goals:

(1) Determine the level of risk issues and take targeted measures. In actual work, the risks encountered by different types of engineering projects are different; the same risk has different impacts on different engineering projects.

    This requires engineering project participants when ranking the risks encountered, a reasonable ranking should be made according to the degree of impact on the project, prioritize them, develop different strategies, and use targeted methods to improve the effectiveness of risk management.
(2) Determine the internal connections of risk issues. Although there are various risk factors in engineering projects, some factors generally appear between internal links. Seemingly unrelated multiple risk factors may originate from the same mechanism. Risk assessment uses scientific calculation methods to gather seemingly unrelated risk factors together, find their internal connections, and formulate safer, more economical, and more efficient disposal plans.

(3) Make a comprehensive judgment and understanding of risks and strive to turn risks into positive factors. Risks and opportunities coexist, where there are risks, there are opportunities, and where there are opportunities, there are risks. Construction safety risk assessment uses reasonable and scientific methods to give people a more correct understanding of risks, and as much as possible turn them into opportunity conditions, in order to achieve the purpose of reducing construction safety risks.
(4) Through the risk assessment work, the possibility of risk occurrence and the loss caused after occurrence can be reduced to a certain extent, the risk can be controlled within a tolerable range, and the best balance of safety and economy can be achieved in conjunction with engineering insurance and other measures.

Construction safety risk assessment, as an important basis for the construction organization plan, safety management measures, and emergency management plan of the project construction party and the construction party, has now become one of the necessary links in the process of underground engineering construction. The level of construction safety risk assessment directly affects the safety and economy of project construction and may even directly determine the success or failure of the project.

3. Principles of underground space construction safety risk assessment

The principles of underground space construction safety risk assessment mainly include:

(1) Goal-oriented principle

The goal-oriented principle of construction safety risk assessment refers to, in order to reduce the possibility of risk occurrence or the degree of loss caused after risk occurrence, formulate and implement a risk assessment plan within the overall framework of the project, control the risk within the acceptable range of the project, and adjust risk control measures through the establishment of a risk dynamic tracking and feedback mechanism. From the perspective of investment, safety, and quality goals, risk assessment serves the project goals. The purpose of risk assessment is to ensure that the project construction safety as far as possible. To minimize the cost of risk in the project on the premise of full and successful completion of the project.

(2) Comprehensive principle

The comprehensive principle of construction safety risk assessment is reflected in the fact that risk assessment is a comprehensive technical activity. Its comprehensiveness stems from the complexity of risk factors, impact range, and occurrence mechanisms in project construction. Because of this complexity, construction safety risk assessment needs to comprehensively apply theories and related technologies from natural sciences, social sciences, engineering technology, systems science, management science, and other disciplines in order to timely discover and effectively identify and assess potential risks in the project.

(3) Dynamic principle

The principle of dynamism in construction safety risk assessment refers to the fact that, in view of underground space engineering has characteristics such as long construction periods, large construction scales, wide involvement, numerous and varied risk factors, and complex coupling between different factors. This means that in the actual construction process, it is necessary to fully consider the impact of existing risks on construction safety, and to constantly respond to the challenges of emerging risks. Therefore, it is necessary to establish real-time tracking mechanisms for existing risks and dynamic feedback mechanisms for new risks, adjust the results of risk assessment dynamically based on risk monitoring information, and ultimately achieve dynamic risk management to ensure the smooth realization of project objectives.

(4) Principle of Rationality

Construction safety risk assessment of the principle of reasonableness means that the risk assessment method needs to be based on the characteristics of the project to make a reasonable choice, so that the risk assessment is more targeted to improve the accuracy of the risk assessment. At the same time, the risk assessment method also needs to consider the knowledge level and technical level of the evaluators in the actual project, conform to the general cognition of the construction site, and overly complex and theoretical risk assessment methods may make it difficult for the on-site evaluators to start, which is not conducive to the realization of the risk assessment objectives.

(5) Principles of Systematization and Informatization

The principle of systematization is reflected in the risk assessment process of underground space engineering, which requires timely summarization of existing engineering theories and experiences, and their systematization and standardization. The reason is that in the implementation of risk assessment, humans are the main body of mastering engineering theoretical knowledge, and human subjective factors occupy a large proportion of the impact on risk assessment. The principle of informatization refers to the need to consider the informatization of engineering theories and existing experiences in the risk assessment process. The premise of risk assessment is the rapid and effective transmission and sharing of information. If the information flow cannot be transmitted quickly and effectively, it will inevitably lead to difficulties in implementing risk assessment. In summary, the systematization and informatization of engineering theories and experiences have a significant impact on risk assessment and need to be given key attention.

4.2.2 Construction Safety Risk Assessment Content

Specific methods of risk assessment need to be matched to the field of their application, the object and the level of knowledge of the assessor. Underground space construction safety risk assessment needs to answer the following questions: ① what is the object of assessment, that is, what type of risk exists in the construction; ② how to assess the risk, that is, what are the risk factors, what assessment methods and assessment methods are used? What is the object of assessment, that is, what type of risk exists in the construction; ③ how big is the risk, that is, what are the risk factors, what are the assessment methods and indicators? How much risk, that is, the probability of occurrence of risky events how high, after the occurrence of loss or gain; ④ how to deal with risk, that is, what attitude should be treated with risk, that is, how to deal with risk? How to deal with the risk, that is, what kind of attitude should be adopted to deal with the risk, whether to choose to avoid, control or accept. In view of the above problems, the risk assessment of underground space construction generally includes the following five basic contents. In view of the above problems, the risk assessment of underground space construction generally includes the following five basic contents.

1. Risk Identification

The task of risk identification is to identify potential sources of risk (such as karst, water sacs, gas, pipelines, above and underground buildings, rivers and lakes, etc.) that may impact the safety of underground space engineering, and to identify and categorize potential risk events (such as collapses, explosions, floods, fires, etc.) that may be triggered by these risk sources. This is done by analyzing the mechanisms of risk occurrence to identify risk factors, i.e., answering the questions "what is the object of assessment" and "what are the risk factors". Generally speaking, the direct causes of risk events include the following categories: ① Unsafe human behavior, such as misoperation or violation of rules due to the subjective negligence of construction personnel, neglect of potential accidents due to insufficient safety awareness of management personnel, design defects due to insufficient technical level or experience of design personnel, etc.; ② Unsafe state of objects, such as poor storage environment of building materials, old construction machinery lacking maintenance, inadequate safety protection of construction personnel, etc.; ③ Accidental events with adverse effects, such as heavy rain, typhoon, earthquake, etc.

Risk identification is the primary task of risk management analysis and has a high degree of difficulty in practical operation. It requires the identifier to have solid professional knowledge and on-site experience. The reason is that the construction of underground space is a systematic project. Any uncertainty in personnel, machinery, environment, and management can potentially become a source of risk. Moreover, the construction of underground projects may involve lifting and hoisting, high-altitude operations, dust and gas, and other high-risk work environments. This makes the types of risk events in underground construction very diverse, with complex and intertwined risk factors. There may also be chain reactions between risk events, which greatly increases the difficulty of implementing risk identification.

2. Risk Analysis

The goal of risk analysis is to conduct a sand table deduction of the project implementation process based on the characteristics of the assessment project, sort out the logical connections between the identified risk events, risk factors, and risk sources, reveal the formation and development mechanism of risks, and find evidence to support risk assessment. To achieve this, the first step of risk analysis can divide the project into several risk assessment units, determine and screen the risk events and risk factors within each assessment unit, as well as the corresponding risk grading standards and acceptance criteria. Risk analysis is the most important link in the risk assessment of underground space engineering. Since the construction of underground space engineering is generally multi-stage, multi-process, and multi-objective, the engineering tasks, engineering characteristics, risk types, and risk management objectives of each stage may differ from each other. Therefore, it is necessary to sort out the assessment objects and assessment logic through risk analysis to improve the pertinence and accuracy of risk assessment.

Another major function of risk analysis is to screen risk events and risk factors. As mentioned earlier, there are various types of risk events in the construction process of underground space engineering, but the importance of these risk events may be completely different. For example, large-scale disasters such as roof collapse and structural damage are obviously more important compared to local accidents such as mechanical injuries and object strikes. For the construction links of underground space, it is not only unnecessary to assess all risk events in detail, but it also greatly increases the complexity and workload of risk assessment, making the risk assessment formalistic and unable to play its due role. Therefore, when conducting risk analysis, it is necessary to determine the granularity of risk assessment based on the risk management objectives of the assessment unit, focus on the main issues, and ensure the operability of risk assessment in actual work.

3. Risk Estimation

Risk estimation is the process of using qualitative or quantitative methods to estimate the likelihood and severity of risk events, provide a judgment of the risk level, and rank the risk events. Risk estimation can be subdivided into the following steps.

(1) Before conducting risk estimation, it is necessary to screen risk factors based on the degree of risk factor impact, the possibility of risk factor occurrence, and the operability of risk assessment. Further, risk assessment indicators and the basis for scoring these indicators are proposed.
(2) Estimate the likelihood of risk events. Risk events are generally low-probability events, and the precision that human subjective concepts can distinguish is often greater than the range of risk event occurrence probabilities, making it very difficult to directly estimate the probability of risk events. Therefore, the estimation of the likelihood of risk events generally uses methods such as the indicator system method, fuzzy comprehensive evaluation method, accident tree analysis method, Bayesian network analysis method, etc.
(3) Estimate the loss caused by the occurrence of risk events. Risk loss refers to the impact of risk occurrence on project objectives, including casualties, economic losses, project delays, environmental impacts, and social public opinion impacts, etc. Risk loss estimation requires specific risk events as hypothetical scenarios,

calculations and quantifications of the above types of losses, and the use of certain principles to integrate various types of losses. Common methods for risk loss estimation include the "highest principle", "weighted average method", "LEC method", etc.

4. Risk decision-making

Risk decision-making is the final step of risk assessment. Regardless of the conclusion of risk estimation, risk is always a "possible but not yet occurred event", not a certain danger or accident. There is no need to adopt an evasive or transfer attitude blindly, and there is no need to give up action because of the existence of risk. On the other hand, risk is not entirely negative, risk often coexists with opportunities, and many times people profit by taking risks. Risk decision-making is based on the conclusion of risk estimation, according to the risk acceptance level and risk response cost of the action subject, the process of adopting different response strategies for different levels of risk.

Common risk response strategies in underground space engineering are divided into the following categories:

(1) Risk avoidance. That is, by changing the construction plan, eliminating or transforming the root cause of the risk, thereby avoiding taking risks. From a safety perspective, risk avoidance is the most effective means, but risk avoidance generally comes with a large cost. Therefore, risk avoidance strategies are generally only adopted for risks that are extremely serious and unbearable when they occur or have a very high probability of occurrence.
(2) Risk disposal. That is, by taking targeted measures to control the possibility of risk occurrence or risk loss, the risk level is reduced to an acceptable level. Generally speaking, most of the risks in underground space construction can be controlled by targeted technical means or enhanced safety management, so risk disposal is the most commonly used risk response strategy in underground space construction. When using risk disposal, it should be ensured that the additional cost brought by the disposal measures is not higher than the expected loss brought by accepting the risk, otherwise other risk response strategies should be considered first.
(3) Risk transfer. This involves transferring some risks that are unwilling to be borne and difficult to effectively cope with to a third party through insurance and other means. Risk transfer is generally used as an alternative strategy for risk disposal or in conjunction with risk disposal and is currently a common strategy for dealing with risks in underground space construction.
(4) Risk acceptance. This means accepting the existence of risk and bearing the risk by the acting entity itself. For risks with low probability and minor consequences, adopting a risk acceptance strategy often achieves the greatest economic benefits.

5. Risk tracking

Risk tracking is carried out after the decision is implemented. Its purpose is to ascertain whether the results of the decision are the same as expected, to evaluate the rationality of the decision, and to find opportunities to refine and improve risk response strategies based on the results of decision evaluation. This information is then fed back to the relevant decision-makers to make future decisions more in line with reality and more operational. Risk tracking in underground space construction needs to run through the entire construction phase, implementing the measures predetermined in the risk response strategy when a risk event occurs; and when the situation changes, feedback is provided in a timely manner, a new risk assessment is carried out, and a new response strategy is formulated. Therefore, construction risk tracking is a real-time, continuous process. It should revolve around the characteristics of construction risk, establish risk monitoring standards, adopt systematic management methods, establish an effective risk warning system, and implement efficient construction risk tracking.

4.2.3 Risk Identification

1. Methods of risk identification

Risk identification is an important basis and premise for risk assessment. Comprehensive and systematic identification of various risks is crucial for improving the accuracy of risk assessment. The main purpose of risk identification is to determine the types of risk events and corresponding risk factors. Risk identification is a comprehensive, qualitative judgment process and is the first step in abstracting complex real-world problems into theoretical models. Therefore, risk identification requires the identifier to have solid basic knowledge and rich practical experience in the relevant field, and also needs to follow certain basic methods and principles. For risk identification, there is currently no fixed, universally applicable method, but in the field of construction risk, risk identification methods can mainly be divided into the following two types.

(1) Based on the identifier's intuitive judgment. On the premise of establishing a detailed understanding of the objective situation of the project, the identifier makes judgments about the possible risks and their causes of the existing project based on past relevant project experience, especially the experience of dealing with accidents and risks, using professional skills.
(2) Based on system theory, safety theory to establish models for analysis. There are many mature models in existing system theory and safety theory. Identifiers can use models to guide their thinking process, more systematically sort out and analyze targets and their interactions, or to identify risks with the help of the conclusions inherent in the models.

The advantage of the former method is that it is simple, easy to operate, and can easily grasp the main issues, but it may result in incomplete identification results, difficulty in considering the special circumstances of the existing project, difficulty in dealing with complex or innovative projects, and certain requirements for the identifier. The latter method requires less of the identifier, is more scientific and promotable, but the process is more cumbersome. In actual risk identification work, a combination of the two methods is often used.

2. The process of risk identification

The general process of risk identification includes data collection, risk source identification, risk event identification, and risk factor identification.

(1) Data Collection

Data collection is the foundation of risk identification and risk assessment. Data collection not only includes the collection of documents in the narrow sense, but also the process of risk assessment participants establishing an intuitive understanding of the actual situation through field investigations. Data collection should be comprehensive, accurate, and objective.

(2) Risk Source Identification

Risk source identification is the prerequisite for determining risk events and risk factors. Risk source identification is based on the data collected earlier, the results of field investigations, and expert consultations, to screen out risk sources that pose safety hazards or may trigger accidents, including sub-item construction, unfavorable geological and hydrological conditions, surrounding environmental facilities, and weak links in construction management. The results of risk source identification can serve as the basis for dividing risk assessment units, and targeted risk assessments can be carried out on this basis.

(3) Risk Event Identification

The purpose of risk event identification is to determine the object of risk assessment. Risk events are directly related to risk sources. Generally speaking, the type of risk source determines the type of risk event. For example, if there are sensitive buildings near the construction site, there naturally exists the risk event of "excessive deformation of the building". Therefore, risk events can often be naturally derived from the conclusions of risk source identification. However, theoretically, everything from serious malignant accidents to minor construction quality issues can be considered as risk events. But different risk events may require different risk analysis methods, risk quantification standards, and risk acceptance criteria. Therefore, during the process of risk event identification, the definition granularity of risk events should be reasonably formulated according to the characteristics of the project to be evaluated, the goals of risk management, and the level of risk analysis technology, to avoid the extreme situations where the identified risk events are too large to be assessed, or the risk events are too small to make the assessment work meaningless.

(4) Risk Factor Identification

The purpose of risk factor identification is to sort out the logic of the assessment and to identify risk assessment indicators. Risk factors are the bridge connecting risk events and risk sources, so risk factor identification can be carried out through "forward push" and "backward push" paths.

① Forward push from the risk source, that is, analyzing the various unsafe states or potential unsafe behaviors or adverse events that may harm construction safety that may exist in the risk source. For example, if there is a risk source of "sewage pipeline", there naturally exists the risk factor of "sewage pipe leakage" that may trigger collapse, environmental pollution, and other risks. However, it should be noted that risk factors and risk events are relative and can be converted to each other. For the project itself, "sewage pipeline" is a risk source, and "sewage pipe leakage" is a risk factor for its own risk; but for the "sewage pipeline", the project itself becomes a risk source, and the disturbance of construction becomes a risk factor.

② Backward push from the risk event, that is, analyzing the causes that may lead to the occurrence of the risk event, thereby determining the risk factors. The backward push process is generally based on past accident experience or professional knowledge and does not need to be based on the premise that the risk source is known. Therefore, not all risk factors identified in this process may be objective and potential risk factors, but relatively new risk sources may also be identified.

Therefore, risk factor identification should be carried out simultaneously in both "forward push" and "backward push" directions.

3. Common Risk Identification Methods

(1) Fault Tree Analysis Method

Fault Tree Analysis (FTA), also known as failure tree analysis, is one of the important analysis methods in safety system engineering and is a deductive safety system analysis method. The Fault Tree Analysis method was proposed by Mearns Watson in 1961 when Bell Labs in the United States conducted a safety analysis of the missile launch system. It is a method of analyzing complex systems. Its definition is "In the system design process, by analyzing various factors (including hardware, software, environment, human factors) that may cause system accidents, drawing a logic diagram (i.e., the accident tree), to determine the various combinations of system accident causes or their occurrence probabilities, to calculate the system accident probability and take corresponding corrective measures, to improve the reliability of the system design method." When establishing the accident tree, the least desirable accident state is usually placed at the top of the tree, defined as the top event, and then the direct factors and reasons that may cause this event (defined as intermediate factors or intermediate events) are found, and the basic reasons that may cause the intermediate accident (defined as the bottom event) are further deduced. The top events, intermediate events, and bottom events are connected into a tree diagram with corresponding representative symbols and logic gates, which is called the accident tree. The methods for establishing the accident tree mainly include the deductive method, the computer-aided synthesis method, and the decision table method. The deductive method is used to establish the accident tree, mainly through human thinking to analyze the process of the occurrence of the top event. The fault tree analysis method starts from the analysis of a specific accident or failure (top event), analyzes its causes layer by layer, until the basic cause of the accident (bottom event)

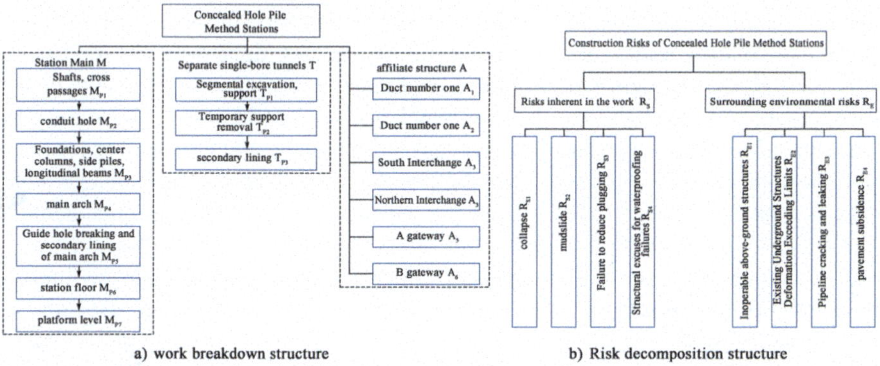

Fig. 4.2 WBS-RBS decomposition structure

is found. These bottom events are also known as basic events, and their data are known or have statistical or experimental results. A typical accident tree (fault tree) is shown in Fig. 4.1.

(2) WBS-RBS Method

The WBS-RBS method is an effective method widely used in risk identification and analysis. The Work Breakdown Structure (WBS) is a mature work breakdown structure in project management, while the Risk Breakdown Structure (RBS) is a risk breakdown structure that is constructed with reference to the WBS principle and has a similar structure to it and is project goal-oriented. By cross-constructing the WBS-RBS matrix, the analysis and judgment of project risks and their transformation conditions can be realized. Because this method not only inherits the advantages of WBS that goes deep into every link of the project life cycle, but also can comprehensively analyze the risks of the internal and external environment through RBS, so it is more suitable for the risk identification work of engineering projects. A typical WBS-RBS decomposition structure is shown in Fig. 4.2.

4.2.4 Risk Analysis

1. The development history of risk analysis methods

The systems established by humans and their environments have been constantly changing in recent decades: science and technology are changing with each passing day, new dangers are emerging, the nature of accidents is constantly changing, the complexity and coupling of systems are constantly increasing, safety concepts are constantly being updated, and safety requirements are increasingly high. These changes pose increasingly new and complex challenges to the construction and operation safety of systems. Although there has been great progress in science and technology over the past few decades, accidents have not been eliminated as a result, and new accidents continue to occur. In order to improve the safety of systems, people have mainly used innovative technical means to achieve their intended goals in the past. However, many studies have found that the root cause of accidents lies in management and organizational factors, which account for 50–90% of the causes of accidents. The construction and development of urban large underground space have gone through a similar process. First, the construction of urban large underground space is a labor-intensive industry, and factors such as organizational management have a significant impact on construction safety; second, the construction of urban large underground space passes through important urban strata, hydrogeological conditions are complex, there are pipelines crisscrossing and coexisting, many uncertain factors make the construction risk increase, leading to an increase in the risk of major safety accidents in the construction of urban large underground space.

It can be seen that from aerospace, chemical, nuclear power and other industries to civil engineering, in order to prevent and reduce the occurrence of accidents, it is necessary to use new and targeted safety construction techniques and management methods. Research generally believes that in order to reduce the probability of accidents, analyzing and studying the impact of organizations on safety is one of the future development directions of the above industries. From the perspective of the organization itself, it is full of many unknown variables, and the variables are interconnected, forming a complex system. What is worth noting is that the feedback loops, non-linear relationships, and emergent phenomena generated by these variables have an unusually complex and variable impact on system safety, ultimately making the general behavioral characteristics of the organization difficult to define or predict. In view of this, in order to better understand the organization, it is necessary to base on multidisciplinary cross and systems thinking, and integrate complex system theory, organizational theory, etc. for systematic research.

In the early 1960s, the Tavistock Institute in the UK proposed the concept of complex socio-technical systems with closely dependent technical and social characteristics, generally referring to industrial organizations with huge energy accumulation, capital and technology intensive. Obviously, the construction of urban large underground space is a typical complex socio-technical system. In this system, the interaction of technical and social characteristics determines system performance. The basic idea of "system safety" is proposed on this basis, that is, to reasonably adopt systematic and predictive management and technical means to identify and control safety risks throughout the life cycle of the project. Over the past few decades, many scholars have conducted related research on safety risks, and the technical level has made significant progress. Some foreign scholars, based on the thinking paradigm of safety risk research, divide risk analysis theories and models into three categories, and divide the models and theories of safety risk analysis research into three generations from a time perspective, as shown in Fig. 4.3.

(1) First generation—Models based on defense-in-depth theory The main idea of the deep defense theory is to improve system security through standardized workflows, establish multi-layer defenses, and increase system redundancy. It should be noted that as the redundancy of the system increases, the complexity of

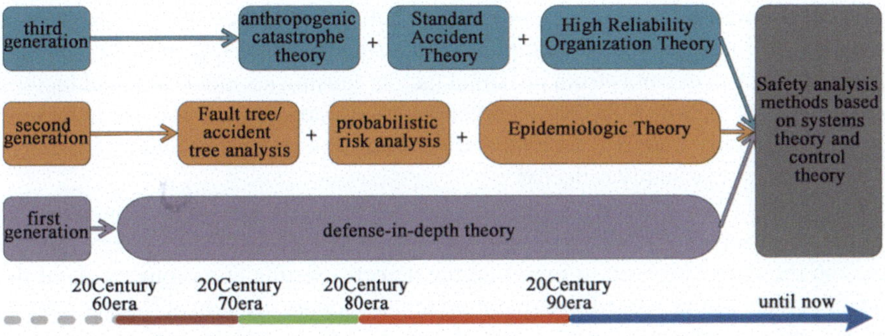

Fig. 4.3 Trend of risk analysis theory and model research

the system also increases accordingly, and the system function will inevitably undergo a certain degree of metamorphosis in its construction or operation process, which enhances the concealment of the process of system development to a dangerous state and weakens the observability of the system state. With the advancement of technology, the increase in system complexity, and the increase in people's requirements for refined risk management, this theory gradually fades out of the historical stage.

(2) Second Generation—Event Analysis Theory Model

The event analysis theory and model believe that the cause of safety accidents is the system state deviation induced by certain specific events, and then analyze and control safety risks by finding the key events or key paths that play a dominant role in the accident process. Event analysis theory captures the essence of safety accidents, conforms to people's logical cognition, can be analyzed with the help of mature mathematical methods, and is still the mainstream theoretical model for safety risk analysis. It can be analyzed with the help of mature mathematical methods and is still the mainstream theoretical model for safety risk analysis. models based on incident analysis theory can be divided into two types as follows.

① Event Sequence Theory and Model

The event sequence theory interprets accidents as the result of a series of events happening in sequence, using an event chain description method. The event sequence model usually uses deduction or induction to analyze the causes of accidents or the consequences of events, such as event trees and fault trees, etc.

② Epidemiological Theory and Model

The epidemiological theory uses disease transmission as a metaphor, believing that accidents are the result of multiple factors coupling, adopting an event network description method. Similar to the event sequence theory, the epidemiological theory is also based on causality to describe the propagation method of the impact produced by events. The difference is that the epidemiological theory has more advantages when describing more complex relationships between factors leading to accidents. In addition to considering more complex relationships between factors, the epidemiological model also expands from proximal factors causing accidents to distal factors, extending pure technical factors to organizational management and other factors, considering a wider range of factors, and the impact of organizational management on accidents is reasonably reflected.

The "Swiss Cheese" model is the most famous representative of the epidemiological model, using cheese slices to represent different aspects of the system, and the holes on the slices are used to represent the defects existing at various levels of the system, which was proposed to explain system accidents. System accidents refer to a state formed by the overlap of potential defects at the organizational management level and external defects at the individual level, which is triggered by an event to cause an accident. As can be seen from the Swiss Cheese model (Fig. 4.4), when the holes on the cheese slices are connected in a straight line, the accident has already occurred. Since the position and size of the holes are constantly changing, it is very difficult for them to form a straight line and cause an accident. However, once a systemic accident occurs, it usually causes serious consequences.

(3) Third Generation—Models based on system theory and Model

The biggest drawback of the second-generation event analysis theory is that the assumptions it introduces will inevitably constrain the analysis process. Due to the limitations of the second-generation theories, some scholars began to try to construct new descriptive theories and models based on the system perspective.

① Descriptive Theory Based on Actual System Behavior

Descriptive theories based on actual system behavior have far-reaching impacts on expanding the research approach to system safety and the causes of accidents, such as the theory of human high-reliability organization, normal accident theory, and disaster theory. The theory of human-made disasters first introduced the organization

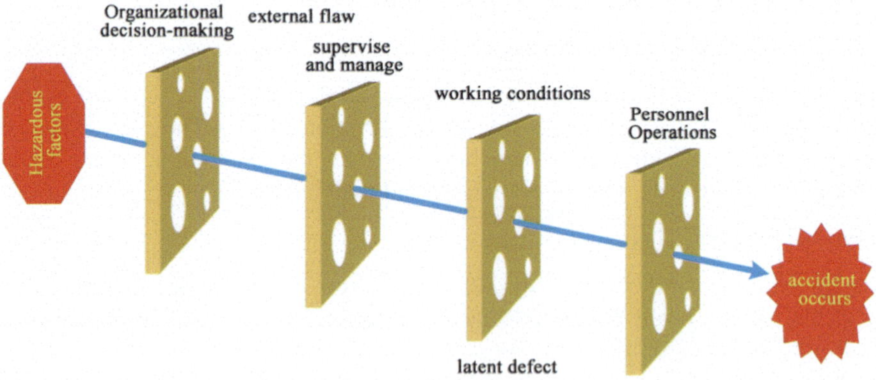

Fig. 4.4 Swiss cheese model

as a factor in the occurrence of accidents, changing people's traditional understanding and interpretation of accidents: (a) The key factor in the occurrence of accidents is organization and management; (b) Accidents require a long incubation period, and the accumulation of a series of abnormal events during the incubation period needs to be considered; (c) Accidents have common patterns of occurrence, which have a positive effect on the analysis and improvement of system safety.

Based on the analysis of the Three Mile Island nuclear accident (the most severe nuclear power plant safety accident in American history), some scholars have established the normal accident theory and given two important characteristics that determine whether a system is safe or not: the degree of coupling and complexity of system interaction. In complex socio-technical systems, the high complexity and tight coupling of the system often lead to events beyond the expected range or beyond the original design during construction or operation. The normal accident theory points out that the occurrence of accidents should be attributed to management problems. Accident analysis should not only explore the causes at the moment of the accident but also trace back to the past.

The basic idea of the high-reliability organization theory is to focus on the ways in which high-reliability organizations improve and ensure the safety of complex systems. To fundamentally limit the occurrence of accidents and achieve as high system performance as possible, an organization usually needs to meet four conditions: (a) Extensive use of system redundancy; (b) Coexistence of decentralized and centralized management modes; (c) Strong organizational learning ability; (d) The organization should unanimously agree that "safety and production should be equally important".

The accident migration model is a theory that analyzes system safety based on control theory and systems theory. Its basic idea is that any work will be constrained by management, function, and safety. The work goals and constraints form individual behavior, and the achievement of established goals requires individuals to adapt to or change constraints. In this process, the individual provides an "efficiency gradient", and management provides a "benefit gradient". Therefore, the system will gradually migrate towards the safety boundary under the joint action of these two gradients until danger is observed or perceived. If danger is not detected or measures are not taken even if danger is detected, the system continues to migrate beyond the safety boundary, leading to an accident. Fig. 4.5 abstractly illustrates the accident migration model with a function.

② Descriptive models based on actual system behavior

The third-generation accident causation and system safety analysis models mainly have three representatives: STAMP (Systems-Theoretic Accident Model and Processes) model, SoTeRiA (Social-Technical Risk Analysis) model, and TOPAZ (Traffic Organization and Perturbation Analyzer) model.

The STAMP model uses system simulation to analyze the dynamic trajectory of related variables and analyzes the impact of certain decisions on related variables by changing model settings, but it does not provide the probability of an event or accident occurring. Therefore, the focus of STAMP is on the social factors in accidents, lacking

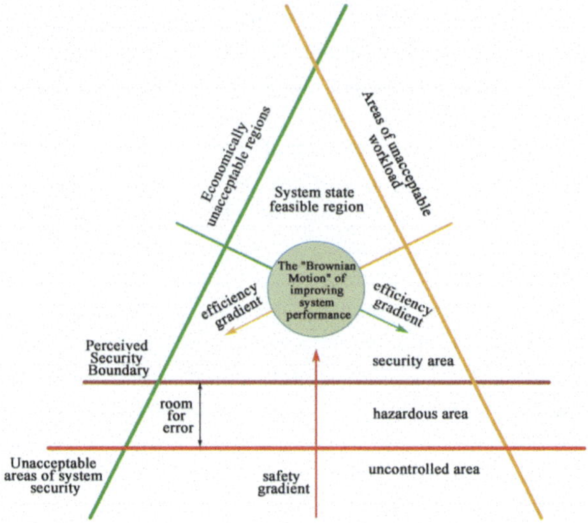

Fig. 4.5 Rasmussen accident migration model

risk analysis at the technical level. The SoTeRiA model, developed based on the STAMP model, systematically establishes the connection between the organization and the technical system through various modeling means, thus being able to analyze the degree of influence of organizational factors on safety risks and estimate the probability of the final accident. However, this model analyzes the impact of the social system on the technical system in complex socio-technical systems, without analyzing their interaction. The TOPAZ model uses "Agent Dynamic Risk Modeling Technology" and Monte Carlo simulation technology to analyze the impact of air traffic controllers, pilots, etc. on the occurrence of accidents. Due to the complexity of the calculation, this model is not suitable for labor-intensive industries such as construction.

2. Classification of risk analysis methods

In theory, any risk event has its mechanism of occurrence, so it can be calculated through a series of state equations using strict probability analysis methods. However, risk systems are often very complex, and in some fields or activities, the use of probability methods to study risks is not feasible. Currently, risk analysis methods are mainly divided into the following two types.

(1) Probability-based analysis methods

Probabilistic Risk Assessment (PRA) is a widely used risk assessment method. It was initially introduced in the nuclear industry and then gradually extended to other complex systems such as chemical production, aviation management, and space missions. Its basic analysis tools are Fault Tree Analysis (FTA) and Event Tree Analysis (ETA). PRA can both estimate the likelihood and consequences of accidents and identify possible accident scenarios.

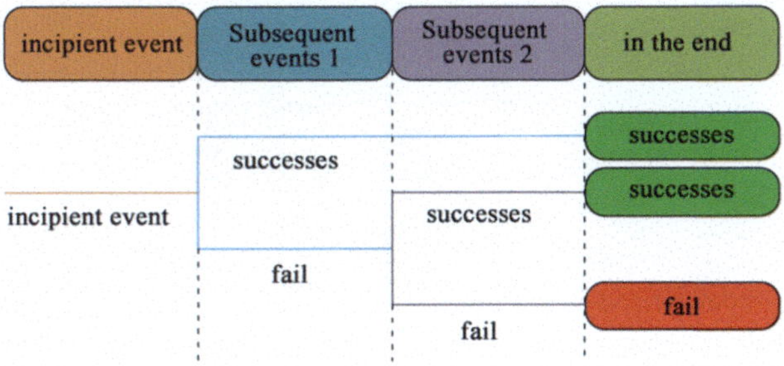

Fig. 4.6 Schematic diagram of the event tree

In fact, FTA is a deductive method that starts with a top event and identifies all direct cause events based on the depth of the analysis. The relationships between different basic events or factors are represented by logical relationships such as AND gates and OR gates, and the probability of the top event occurring can be calculated based on the probability of basic events and logical relationships.

ETA is actually an inductive method that identifies all possible subsequent events from the initial event. These subsequent events are usually the states of safety measures to prevent the initial event from developing into an accident, and the accident scenario is described through the event sequence. If the probabilities of the initial event and subsequent events can be obtained, the size of the accident can be calculated based on each accident sequence (Fig. 4.6).

The advantage of FTA/ETA lies in its ability to describe the three elements of risk and its ease of calculation; hence it is widely used in civil engineering fields such as tunnels and bridges. For instance, FTA is applied to analyze the damage to the surrounding environment caused by highway tunnel construction, while ETA is used to identify adverse events that may be encountered during tunnel excavation and to assist in the selection and design of shield machines. Some scholars have developed a method based on ETA to automatically identify the most likely failure scenarios in the event of a bridge construction collapse. In addition, FTA/ETA is recommended as a risk management tool in risk management guidelines in related fields. There are also models that combine FTA with ETA, using the top event of the fault tree as the initiating event of the event tree, as shown in Fig. 4.7. Because the shape of the model resembles a "tie knot", it is also referred to as the "tie knot" model, the main purpose of which is to identify the causes leading to key events (i.e., top events or initiating events) and to analyze the reliability of safety measures after the occurrence of key events. The "tie knot" model can serve as a qualitative analysis tool for occupational injury risk assessment, providing a basis for port safety design.

Another way to combine FTA and ETA is to connect a fault tree to each event in the event tree (including initiating events and subsequent events). This model has been used to analyze tunnel fire risks, identifying the most necessary faults to eliminate through importance and cost analysis of each basic event. Some scholars

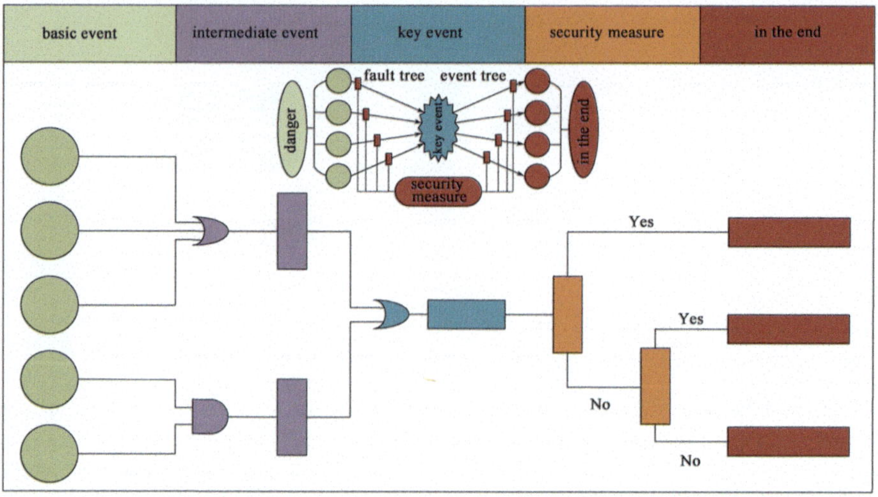

Fig. 4.7 Schematic diagram of the tie knot model

have established a similar model (the "cause-effect diagram" model) for analyzing the safety of traffic diversion systems.

Using FTA/ETA-based risk analysis methods to estimate the probability of each event requires a large amount of statistical data, but such data is often lacking in urban large underground space construction. In contrast, urban large underground space construction relies more on experiential knowledge, so expert judgment often serves as a supplement to objective data.

(2) Indicator-based analysis methods

Analysis methods that mainly establish various indicators are more suitable for assessing social level safety impact factors such as safety culture and safety management, but the drawback of these methods is that it is difficult to derive analysis results such as accident consequences. Conversely, traditional PRA-based risk analysis methods can quantitatively analyze the probability and loss of accidents but cannot incorporate remote factors such as safety culture into the model.

Taking safety culture as an example, foreign experts have summarized three methods for assessing safety culture: anthropological method, accident investigation method, and questionnaire survey method. The anthropological method focuses on qualitative research, usually using interviews or observation methods, which can reflect the actual specific situation in organizational practice, provide more information, but this method requires a lot of time; the accident investigation method more through the analysis of documents and inquiry records obtained during the accident investigation process, can balance the analysis effect and cost well, but in the construction of urban large underground space, such major accidents are relatively rare, and it is difficult to obtain records of ordinary researchers in their investigation process. At present, the questionnaire survey method, as a quantitative research method, is most widely used, mainly suitable for analyzing attitudes, values, etc., but this method cannot reflect the dynamic change process of the organization over a period of time, reflecting only the current state of the organization, and the universality of the survey questionnaire is highly questioned. Survey questionnaires usually need to be specially designed for specific organizations and are difficult to reuse. In addition, the analysis of survey questionnaires results in a correlation model, not a causal model. For example, the sales of sunscreen and ice cream both increase with the rise in temperature. If a survey is conducted on the sales of sunscreen and ice cream, a strong positive correlation will be found between the two, but in fact, there is no causal relationship between the two. In view of this, some scholars now emphasize the combination of questionnaire survey method with other methods. However, no matter which method is adopted, organizational factors such as safety culture are mainly measured and evaluated by constructing indicators. When establishing a model of the impact of organizational social factors on system safety, some scholars also try to construct risk indicators to study, such as evaluating the impact of organizational factors on technical system risk by constructing organizational safety risk indicators. However, this indicator-based analysis method often needs to assume that the indicators are independent of each other, while in fact, there are interactions between organizational factors, and the final system safety is an emergent phenomenon caused by the joint action of these factors. The indicator-based method cannot describe the feedback and delay phenomena in the system. Therefore, some scholars believe that it is very difficult to establish a set of organizational safety indicators, because there is a long path from organizational factors to the final occurrence of accidents, and the organization itself also has complexity and dynamics.

4.2.5 Risk Estimation

1. Measurement of Risk

(1) Nature of Risk Measurement Risk estimation is the process of measuring risk. As can be seen from the above definition, the general definition of risk contains a lot of information, and it is undoubtedly very difficult and inefficient to directly study all the information of risk. Therefore, some kind of measurement method needs to be found for risk. From the perspective of facilitating research, this measure needs to meet the following conditions.

① The result of the measurement is a non-negative real number.
② It is possible to make an intuitive understanding of risk through measurement, which is convenient for comparative analysis.
③ It has certain properties, such as positive homogeneity, weak additivity, translation invariance, etc.

Risk is in terms of the consequences (events) caused by a specific activity, that is, the measurement of risk should be some mapping from F to the set of positive real numbers.

Assume that there are n sample points in the sample space of a certain activity, of which m are elements of basic risk events, which can be arranged in a certain order to be represented as:

$$\Omega = \{\varpi_1, \varpi_2, \ldots, \varpi_m; \eta_1, \eta_2, \ldots, \eta_{n-m}\}, m \leq n \tag{4.6}$$

In the formula, η_i For sample points that are not basic risk event elements, the random variable C, which characterizes risk loss, has the following values:

$$C = \begin{cases} c_i^{\bar{\omega}}, \left\{\bar{\omega}_i\right\}\text{发生}, & i = 1,2,\ldots, m, c_i^{\bar{\omega}} > 0 \\ c_j^{\eta}, \{\eta_j\}\text{发生}, & j = 1,2,\ldots, n-m, c_j^{\eta} \leq 0 \end{cases} \tag{4.7}$$

Assume the distribution law of C is:

$$P\{C = c_i^{\varpi}\} = p_i^{\varpi}, P\left\{C = c_j^{\eta}\right\} = p_j^{\eta} \tag{4.8}$$

Then the risk of the risk event $\overline{f} \in \overline{\mathcal{F}}$ can be measured from the following perspectives.

(2) Absolute measurement

Absolute measurement refers to the use of a numerical value with a specific meaning and unit to measure risk. The advantage of absolute measurement is that the measurement results are clear in meaning, have good operational properties, and the measurement values between different risk events are comparable. The absolute measurement of risk can take the following forms.

① Based on the probability of the occurrence of the risk event

$$R\left(\overline{f}\right) = P\left(\overline{f}\right) = \bigcup_{\varpi_i \in \overline{f}} P\{\varpi_i\} = \sum_{i, \varpi_i \in \overline{f}} p_i^{\varpi} \tag{4.9}$$

In the formula, R is the risk value. This measurement method only considers one aspect of risk, and the information is not comprehensive. It only has practical significance when the risk loss is unacceptable.

② Based on the loss caused by the occurrence of the risk event

$$R\left(\overline{f}\right) = \max\left\{C(\varpi_i), \varpi_i \in \overline{f}\right\} \tag{4.10}$$

This measurement method ignores the most important aspect of risk, so it is rarely used.

③ Based on the expected loss of risk

$$R(\overline{\Omega}) = \sum_{\overline{\omega}_i \in \overline{f}} P\{\overline{\omega}_i\} \times C(\overline{\omega}_i) = \sum_{i, \overline{\omega}_i \in \overline{f}} p_i^{\overline{\omega}} \times c_i^{\overline{\omega}} \tag{4.11}$$

This measurement method takes into account both the probability of risk occurrence and its loss, and is the most commonly used measurement method in actual risk research.

The overall risk of an activity refers to the possibility of all events that may cause the subject of the activity to bear losses and their combined losses. The set of all events that may cause the subject of the activity to bear losses is the universal set of risk events defined by formula (4.3) $\overline{\mathfrak{F}}$, and the maximum element in $\overline{\mathfrak{F}}$ represents the event meaning "there is a risk event". Therefore, the overall risk of the activity can be measured by the following formula.

$$R(\overline{\Omega}) = \sum_{\overline{\omega}_i} P\{\overline{\omega}_i\} \times C(\overline{\omega}_i) = \sum_{i=1}^{m} p_i^{\overline{\omega}} \times c_i^{\overline{\omega}} \tag{4.12}$$

(3) Relative measurement

In actual risk analysis work, there are many difficulties in using the absolute measurement of risk, so we generally prefer to make a relative judgment of risk, such as "high risk" and "low risk", which is actually a relative measurement of risk. Relative measurement is a combination of qualitative conclusions and quantitative conclusions. A relative measurement needs to have two basic elements: a measurement benchmark and a measurement scale. The measurement benchmark determines the specific meaning represented by the relative measurement, and the measurement scale determines the form of the relative measurement and its relationship with the measurement benchmark.

Take the risk matrix commonly used in risk analysis (Table 4.1) as an example, the risk matrix is essentially a relative measurement of risk.The "Level 1", "Level 2", "Level 3", and "Level 4" in Table 4.1 represent the measurement scale, while the measurement benchmark is the probability of a risk event occurring and the equivalent economic loss caused by the occurrence of a risk event. It is not difficult to find that the relative risk measurement defined in Table 4.1 defines a mapping relationship between the relative scale of risk and the expected value of risk loss.

The measurement scale can be a general countable set like the one used in Table 4.1, or it can be a subset of the real number set. As the literal meaning suggests, in the case of an appropriate measurement benchmark, the absolute size of the relative measurement value is not important, so $[0, 1] \in R$ or $[0, 100] \in Z$ can generally be used as the measurement scale. At this time, the lower and upper limits of the scale interval respectively represent the lowest and highest risk levels determined

Table 4.1 Example of risk matrix

Probability of occurrence of a risk event	Amount of equivalent economic loss resulting from the occurrence of a risk event (in millions of dollars)			
	≥10000	5000–10000	1000–5000	<1000
>0.1	Level 1	Level 1	Level 1	Level 2
0.01–0.1	Level 1	Level 1	Level 2	Level 3
0.001–0.01	Level 1	Level 2	Level 3	Level 3
<0.001	Level 2	Level 3	Level 3	Level 4

by the measurement benchmark. Relative measurement has a relatively unified form and is convenient for people to have an intuitive understanding of the risk level, but the meaning of relative measurement is similar to a "score" of the risk level, which cannot be directly calculated, and the relative measurements of different risk events are only comparable when the measurement benchmarks are consistent.

2. Common risk estimation methods

(1) Expert survey method

The expert survey method is also known as the "Delphi method". It is a survey method that solicits the opinions and views of relevant experts or authoritative figures on a certain topic or issue. The survey is limited to experts. The survey is conducted in multiple rounds, generally 3–5 rounds. Each time, the survey respondents are asked to answer a questionnaire with basically the same content, and they are asked to briefly state the reasons for their views. The results of each round of survey are published to all respondents in the next round of survey after being sorted out, so that they can understand the opinions of other experts and the similarities and differences between their views and the majority of expert opinions. The expert survey method is a most commonly used, simplest, and easy-to-operate risk estimation method. This survey method was first used for technology development prediction and is now widely used in the study of many issues in politics, economics, culture, and social development. The specific steps of this method are as follows:

① Based on the results of risk identification, determine the weight of each risk factor to characterize the degree of its impact on risk.
② Determine the grade value of each risk factor. The grade value can be divided into five levels: very large, relatively large, medium, not large, and small. This is a subjective division, and the number and score of grades can be adjusted.
③ Multiply the weight of each risk factor by the corresponding grade score to get the score of this risk factor. The higher the score, the greater the impact on the project.
④ Add up the scores of each risk factor item by item to get the total score of the risk factors of this project, also called the risk degree. The higher the total score, the greater the risk.

To further standardize, a weight value can be determined for the authority of the expert's evaluation. The value of this authority is between 0.5 and 1.0. The risk degree evaluated by each expert is multiplied by their own authority weight, and the product obtained after the calculation is divided by the sum of all expert authority weights, which is the final risk degree of the project.

The expert survey method mainly relies on the experience of experts. The evaluation process is simple, and the reliability of the evaluation results is relatively high, but it is highly subjective and overly dependent on experts. The expert survey method can be simple or complex and is generally used in conjunction with other quantitative estimation methods.

(2) Risk matrix method

The advantage of the risk matrix method is that it comprehensively considers the probability of risk events and the consequences of risk losses to determine the risk level. The theoretical basis of the risk matrix method is to measure risk according to the expected loss of risk, which can be specifically represented as $R = P \times C$, where R represents risk, P represents the probability of risk events, and C represents the equivalent loss caused by risk events. The steps of this method are as follows:

① Calculate the probability P of risk events through probability analysis or semi-quantitative evaluation methods.
② Calculate the equivalent loss value C based on the possible harm caused by the occurrence of risk events. The equivalent loss value C is generally obtained by judging the degree of impact on people, the environment, the project itself and third parties, the construction period and all aspects of society after the occurrence of risk events and converting it according to different conversion standards.
③ Convert P and C into discrete levels according to certain grading standards.
④ Determine the risk level by looking up the corresponding rows and columns in the risk matrix according to the levels of P and C.

(3) Kent Method

The Kent Method (KENT) was applied early in the risk assessment of oil pipeline projects. The United States began to study the Kent hazard index evaluation method in the 1970s, and this method was successfully applied to the risk management of oil and gas pipelines in the 1990s.

The advantage of this method is that it does not need to establish accurate mathematical models and calculation methods, nor does it need to use complex strength theories. Instead, it scores based on the experience of on-site operators and expert opinions, combined with some simple formulas. The accuracy of its evaluation depends on the comprehensiveness of expert experience and the meticulousness and hierarchy of dividing influencing factors.

Taking the crossing of existing tunnel projects as an example, the typical Kent method risk assessment process diagram is shown in Fig. 4.8. The definitions of each index are as follows:

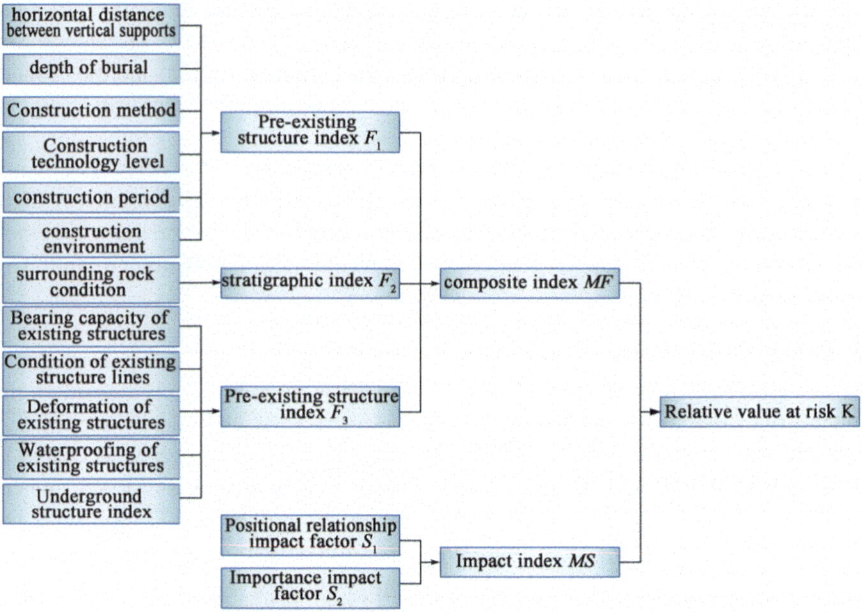

Fig. 4.8 Kent method risk assessment flowchart

① Underground Structure Index F_1

The indices that usually affect the underground structure include cross-sectional geometry, burial depth, construction technology level, construction method, construction period, external environment, etc.

② Stratum Index F_2

The factors affecting the stratum index can be taken as the surrounding rock classification.

③ Existing Structure Index F_3

The main factors affecting the nearby existing structure are the form of the existing structure, the current status of the existing structure, the environment of the existing structure, and the vertical distance between the existing structure and the new structure.

④ Impact Index MS

The impact index MS is jointly determined by the position relationship impact coefficient S_1 of the existing structure and the new structure and the importance impact coefficient S_2 of the existing structure.

Table 4.2 Construction risk level of Kent method

TS	Risk level	TS	Risk level
< 250	Low degree	400–550	High degree
250–400	Moderately	> 550	Extremely high

⑤ Comprehensive Risk Index TS

The comprehensive index refers to considering the underground structure index F_1, stratum index F_2 and existing structure index F_3 to get the basic index, that is:

$$MF = F_1F_2F_3 \tag{4.13}$$

On this basis, the comprehensive risk index TS can be obtained by correcting it with the impact coefficient, that is:

$$TS = \frac{1}{1000}MS \cdot MF = \frac{1}{1000}S_1S_2F_1F_2F_3 \tag{4.14}$$

Based on the calculated TS value, the risk of underground engineering construction can be graded according to Table 4.2.

(4) Analytic Network Process (ANP)

The Analytic Network Process (ANP) is a system analysis method for multi-factor, multi-criteria problems, proposed based on the Analytic Hierarchy Process (AHP). In AHP, the system is divided into several levels, only considering the dominant effect of the upper-level elements on the lower-level elements, and the elements at the same level are considered to be independent of each other. However, the model structure of ANP is more complex, not only having a hierarchical structure similar to AHP, but also having cycles and feedback between the hierarchical structures, and there are internal dependencies and mutual dominance within each hierarchical structure. Therefore, ANP can reflect the complex relationships of mutual interaction and coupling among the elements in the system and has a significant advantage when evaluating complex systems.

The typical ANP network hierarchy is divided into two major parts: the control layer and the network layer, as shown in Fig. 4.9.

① Control layer: Includes a problem goal and several decision criteria. All decision criteria are considered to be independent of each other and are only dominated by the target element. The control layer is actually a typical AHP hierarchical structure.

② Network layer: Composed of all elements dominated by the control layer, the elements are interdependent and mutually dominant, and the elements and levels within are not independent. In this way, the criteria dominated by the control layer are not a simple hierarchical structure, but a network structure where elements are interdependent and feedback to each other.

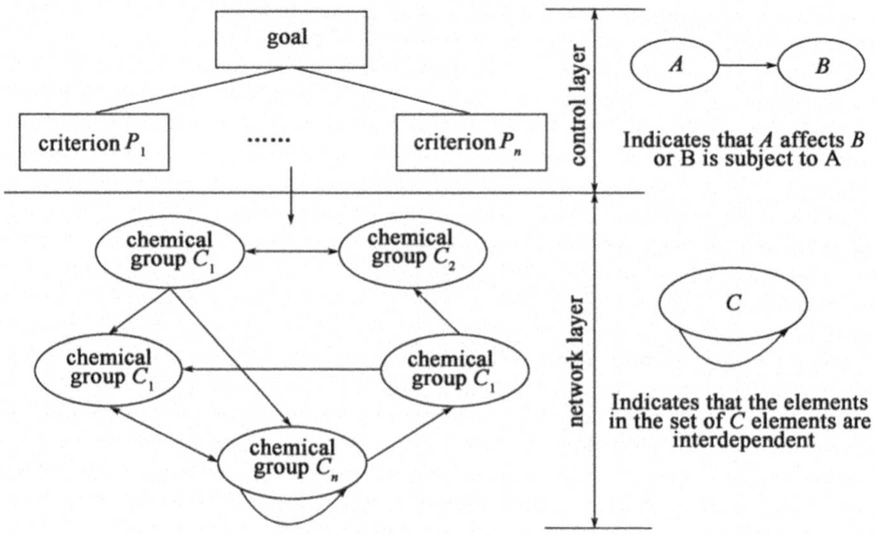

Fig. 4.9 Schematic diagram of ANP network structure

The analysis steps of the Analytic Network Process are as follows.

① Analyze the problem and construct the ANP structure

a. Determine the components of the network: Systematically analyze the problem, determine the evaluation goal, define the evaluation criteria, and classify and summarize the basic factors affecting the problem to form elements and element sets.
b. Construct the control layer: Use the AHP method to construct the hierarchical relationship between the evaluation goal and the evaluation criteria. This hierarchical relationship can be multi-criteria, multi-level, or it can only contain a single evaluation goal.
c. Construct the network layer: Analyze the mutual influence relationship between the element set and the elements and determine the ANP network structure diagram.

The construction of the ANP structure is the most basic and important step in ANP analysis, which generally needs to be carried out through meetings, expert form filling, and other methods.

② Determine the weight of the control layer

The method for determining the weight of the control layer is the same as AHP, and the specific steps are:

a. Take an element Bk of a certain layer as the criterion, and construct a pairwise judgment matrix for the next layer elements C1, C2, …, Cn:

$$B = \begin{bmatrix} b_{11}^k & b_{12}^k & \cdots & b_{1n}^k \\ b_{21}^k & b_{22}^k & \cdots & b_{2n}^k \\ \vdots & \vdots & \ddots & \vdots \\ b_{n1}^k & b_{n2}^k & \cdots & b_{nn}^k \end{bmatrix} \tag{4.15}$$

where, b^k represents the importance of the lower-level element C_i relative to C_j under the criterion B^k. b^k needs to satisfy: $b^k > 0$; $b^k = 1/b^k$; $b^k = 1$.

The value of b^k can adopt the commonly used 1–9 scale method, see Table 4.3.

b. After the comparison matrix is established, a consistency test is required. The consistency conditions are satisfied when CI < 2 and CR < 0.1.
c. When the comparison matrix meets the consistency requirements, the eigenvector corresponding to the maximum eigenvalue of the matrix is the weight vector of each element under criterion Bk.

③ Constructing the ANP Supermatrix

Suppose there are n element groups $D_1, D_2, \ldots, D_n$ in the ANP network layer; each element group has $m_1, m_2, \ldots, m_n$ elements. Then the total number of elements in the network layer is:

$$N = \sum_{i=1}^{n} m_i \tag{4.16}$$

Suppose criterion C_k in the control layer is the evaluation target, a N-order matrix can be constructed, called the ANP supermatrix, expressed in the form of a block matrix as follows:

$$W^{(C_k)} = \begin{bmatrix} W_{11}^{(C_k)} & \cdots & W_{1n}^{(C_k)} \\ \vdots & \ddots & \vdots \\ W_{n1}^{(C_k)} & \cdots & W_{nn}^{(C_k)} \end{bmatrix} \tag{4.17}$$

Table 4.3 Judgment matrix scale table

Hidden meaning	Retrieve a value	Hidden meaning	Retrieve a value
The elements are equally important	1	Element i is extremely more important than element j	9
Element i is slightly more important than element j	3	Medium situation	2, 4, 6, 8
Element i is significantly more important than element j	5	Converse situation	1/3, 1/5…
Element i is strongly more important than element j	7		

In the formula, $W_{ij}^{(C_k)}$ is:

$$W_{ij}^{(C_k)} = \left[\vec{\omega}_{ij}^{(C_k)m1} \; \vec{\omega}_{ij}^{(C_k)mj} \right]_{m_i \times m_j} \tag{4.18}$$

$\omega_{ij}^{(C_k)mj}$ is the weight vector of the influence of elements $e_1^{(i)}, e_2^{(i)}, \ldots e_{mi}^{(i)}$ on $e_{mj}^{(j)}$ under the total criterion C_k, with the sub-criterion being the m_j element $e_{mj}^{(i)}$ in the element group D_j. This vector can be calculated by constructing a comparison matrix for the elements in D_j with $e_{mj}^{(i)}$ as the criterion, according to the aforementioned method.

④ Constructing the Weighted Supermatrix

Under the total criterion C_k, with each element group D_j as the sub-criterion, construct a comparison matrix for each element group $D_1, D_2, \cdots, D_n$, then n n-order weight vectors can be obtained:

$$\vec{\lambda}_i^{(C_k)} = \left\{ \lambda_{1j}^{(C_k)}, \lambda_{2j}^{(C_k)}, \cdots, \lambda_{nj}^{(C_k)} \right\} \tag{4.19}$$

The meaning of $\lambda_{ij}^{(C_k)}$ is under the total criterion C_k Below, the element group D_i affects the weight of the element group D_j. Combining n weight vectors into column vectors, we can obtain the weighted matrix $\Lambda^{(C_k)}$:

$$\Lambda^{(C_k)} = \left[\left(\vec{\lambda}_1^{(C_k)}\right)^T \cdots \left(\vec{\lambda}_n^{(C_k)}\right)^T \right] = \begin{bmatrix} \lambda_{11}^{(C_k)} & \cdots & \lambda_{1n}^{(C_k)} \\ \vdots & \ddots & \vdots \\ \lambda_{n1}^{(C_k)} & \cdots & \lambda_{nn}^{(C_k)} \end{bmatrix} \tag{4.20}$$

The matrix $\Lambda^{(C_k)}$ and the super matrix $W^{(C_k)}$ can be multiplied block by block to obtain the column-normalized weighted super matrix $\overline{W}^{(C_k)}$:

$$\overline{W}_{ij}^{(C_k)} = \lambda_{ij}^{(C_k)} W_{ij}^{(C_k)} \tag{4.21}$$

⑤ Calculate the weight

The element $\overline{\omega}_{ab}^{(C_k)}$ in the weighted supermatrix reflects the one-step dominance of element a over element b under the total criterion C_k. The dominance of a over b can also be obtained by $\sum_{g=1}^{n} \overline{\omega}_{ab}^{(C_k)} \overline{\omega}_{gb}^{(C_k)}$, called two-step dominance, which is an element of $\left[\overline{W}^{(C_k)}\right]^2$, and $\left[\overline{W}^{(C_k)}\right]^2$ is still column-normalized. When $\left[\overline{W}^{(C_k)}\right]^\infty$ exists, any column of $\left[\overline{W}^{(C_k)}\right]^\infty$ is the weight vector of each element's impact on the total criterion Ck.

(5) Other risk estimation methods

Other common risk estimation methods in urban underground engineering construction are shown in Table 4.4.

4.3 Urban Large Underground Space Construction Risk Characteristics and Classification

4.3.1 Urban Large Underground Space Construction Risk Characteristics

According to the definition of urban large underground space engineering mentioned above, compared with general underground engineering, the characteristics of urban large underground space mainly lie in large spatial scale, large plan size, large structural span, ultra-deep underground, complex support and protection, large surrounding environment impact (mutual impact, buildings, active and passive impact, affected objects, becoming load, pipeline structures, destruction causing secondary disasters), and multiple excavation sequences. The above characteristics of urban large underground space determine that its construction is a complex systematic work, which makes its construction risk in addition to the general properties of risk, but also have the following unique characteristics.

1. Diversity

(1) The risk of underground construction increases non-linearly with the increase in excavation depth, excavation width, and structural complexity. Urban underground large spaces are the largest and most complex type of underground engineering, so the safety during construction inevitably faces great challenges. It is only possible to control the safety risks during construction to an acceptable level if there are reasonable construction methods, advanced construction techniques, skilled construction teams, and perfect construction management conditions. The high requirements for construction methods and techniques mean that there is always a risk of serious engineering accidents such as collapses and landslides during the construction process of urban large underground spaces. The high requirements for the construction team and construction management also mean that the frequency of various safety issues caused by human uncertainty is relatively high, such as substandard construction quality due to personnel negligence, mechanical injuries due to misoperation, high-altitude falls due to inadequate protection, or fires and explosions caused by irregular storage operations.

Table 4.4 Common risk assessment methods in urban underground engineering construction

Categorization	Designation	Scope of application
Quantitative estimation methods	Hierarchical analysis	It has a wide range of applications and can be used to analyze problems in the field of social, economic and scientific management. It can be applied to any part of any field, but not to complex systems but not for complex systems
	Monte Carlo method	It is more suitable for application in medium and large projects. Advantages are that it can solve many complex probabilistic arithmetic problems, as well as being suitable for the real test is not allowed to carry out the occasion. It is superior for projects with high costs or long time-consuming experiments It is generally only used for more detailed system analysis and is suitable for situations with complex problems and high precision requirements, especially when selecting and comparing a few feasible solutions, which is more effective
	Reliability analysis method	Analyze the safety probability of the structure to have the intended function in the specified time and under the specified conditions, calculate the reliability index of the structure, and carry out the reliability check of the completed structure. The reliability index of the structure can be calculated, and the reliability check can be carried out on the completed structure. The method is suitable for safety risk analysis of underground structure design. analysis of underground structure design
	Numerical simulation method	Numerical computation software is used to model and simulate the structure, analyze the forces and deformations of the structural design, and carry out the risk assessment of the structure. The method is applicable to the calculation of complex structures and the determination of structural design and construction risk information. This method is applicable to the calculation of complex structures and the determination of structural design and construction risk information
	Fuzzy comprehensive assessment method	The fuzzy comprehensive judgment method is applicable to any part of any system, and its applicability is relatively broad

(continued)

Table 4.4 (continued)

Categorization	Designation	Scope of application
	Equal risk diagram method	This method is suitable for projects that do not require a high degree of precision in the results and only require a rough analysis. At the same time, if only one project and one program are analyzed, the method is relatively cumbersome. At the same time, if only a project and a program analysis, the method is relatively cumbersome, so the method is suitable for a number of similar projects at the same time analysis or a project of multiple options for comparative analysis. Therefore, this method is suitable for the simultaneous analysis of several similar projects or the comparative analysis of several programs of a project
	Control interval memory model	The model is suitable for projects that do not require high precision of results and is only applicable to projects in which the variables are independent of each other or the correlation can be ignored. The model is suitable for projects that do not require high precision of results, and only for projects in which variables are independent of each other or correlations can be ignored
	Neural network approach	Applicable to prediction problems, the cause and effect of the fuzzy relationship between the occasions or pattern recognition and contains fuzzy information occasions. It is not necessary to get the optimal solution, but the main purpose is to quickly find a similar sub-optimal solution; the number of combinations is very large, and the actual solution is almost impossible; the control of a system with high nonlinearity. It is almost impossible to solve the problem in practice, and it is used to control the system with high nonlinearity
	Principal component analysis	The method can be applied in a variety of areas, but the results are only meaningful when comparing relative sizes
Integrated estimation methods	Expert confidence index method	Same as expert survey methodology
	Fuzzy hierarchical integrated assessment method	Its scope of use is consistent with the fuzzy composite judgment method

(continued)

Table 4.4 (continued)

Categorization	Designation	Scope of application
	Engineering analogical analysis	Analyze the project to be evaluated by using the experience of similar projects in the surrounding area or risk incident information, which is suitable for the comprehensive analysis of underground projects. This method is suitable for comprehensive analysis of underground works
	Accident tree method	This method is widely used and is well suited to systems with a high degree of repeatability. In the engineering design stage of the accident query, can use this method to evaluate their safety. The method can be used to evaluate their safety The method is often used for risk identification where there is little direct experience
	Event tree method	This method can be used to analyze system failures, equipment failures, process abnormalities, human errors, etc., and is widely used The method cannot analyze the consequences of parallel production and is not suitable for detailed analysis
	Image mapping	The influence diagram method is similar in applicability to the event tree method, and because it has more advantages than the event tree method, it can also be applied to the analysis of larger systems. It can also be applied to the analysis of larger systems
	Risk assessment matrix approach	The methodology can be modified according to the needs of risk classification, so that it can be applied to different analysis systems, but with a certain amount of engineering experience and data information as a basis. It can be adapted to different analyzed systems on demand, but only with a certain amount of engineering experience and data. It can be applied to the entire system as well as to a specific part of the system
	Fuzzy accident tree analysis	Scope of application is the same as that of the accident tree method, but more applicable than the accident tree method to projects for which basic statistical data are not available

(2) According to the survey results of existing projects, most projects that meet the definition of urban large underground spaces are urban underground complexes, large underground transportation hubs, large underground civil defense projects, etc. Restricted by their functional requirements, these projects have in common that their construction locations are often in the core urban areas with dense populations and numerous above and underground buildings. Due to historical reasons, most of China's urban core areas lacked scientific and systematic long-term planning at the beginning of construction, with buildings of different ages and uneven distribution, underground pipelines tangled and some in disrepair due to age. In addition, the early development of underground engineering had a certain tendency towards blindness and profit-seeking, which means that new large-scale underground projects will face a series of surrounding environmental problems, such as neighboring the base of the existing building or close penetration or side penetration of existing underground structures, the relocation and protection of existing municipal pipelines, and the settlement control of surface roads, bridges, etc.

(3) Many urban large underground space projects are led and funded by the government and built by mobilizing all elite forces, which can play a key leading role in urban development and are important milestone projects. Their construction process is often carried out under the attention of all sectors of society. It can be said that, under the conditions in our country, urban large underground space projects should not only be a high-quality civil engineering project, but also a dazzling image project and an advanced political project, which means that urban large underground space projects are "not allowed to fail". Under such a spotlight, any problem in the construction process of urban large underground spaces will be quickly magnified, any accident will bring public image and social reputation loss to all parties involved in the project, and may even develop into a public relations event, causing extremely adverse social impacts.

2. Complexity of risk formation mechanism

(1) Urban large underground space projects often have large-scale, multi-functional underground structures, and their construction methods and structural forms often have a high degree of complexity and uniqueness. For such underground structures, the bearing mechanical mechanism of the structure itself is already very complex and difficult to analyze through traditional theoretical models. In addition, urban large underground space projects often need to solve the problem of connection and docking with other underground or above-ground spaces, and some urban large underground space projects are expanded from existing structures through methods such as small to large, vertical layering, horizontal expansion, etc. This makes the interaction mechanism between the new structure and the existing structure also one of the problems to be studied. For such a complex multi-structure composite system, designers and builders often find it difficult to obtain sufficient information from past engineering experience, and thus it is difficult to thoroughly and without omission analyze all risk points and corresponding risk formation mechanisms during the construction process.

(2) The construction of large underground spaces in cities requires the collaboration of multiple units and departments, making the impact of human safety factors and uncertainties extremely complex. The conclusions drawn from accident research show that behind every major safety accident, there are common issues in organizational structure, safety management, and personnel awareness. Human safety factors not only have complex internal connections with each other, but under certain objective conditions or events, they can potentially become the fuse that triggers certain accidents. For example, in the major accident of a shield tunnel collapse due to water leakage mentioned earlier, the construction unit did not formulate an emergency plan, and the training and assessment of workers' safety awareness in daily management were not in place. This led to the on-site personnel failing to make a reasonable estimate of the development of the disaster after the signs of the accident appeared. Despite the fact that the accident was essentially unavoidable, they still futilely risked remedial actions, missing the precious opportunity to evacuate. This resulted in a serious accident with casualties under the mudslide triggered by the tunnel collapse. From these cases, it can be seen that human safety factors are not only ubiquitous and difficult to avoid, but also usually lie dormant in a very concealed state, not directly causing serious consequences, making people prone to complacency and overlooking their dangers. However, when the objective conditions are ripe, human safety factors will quickly be exposed, either as the fuse of the accident, becoming the direct cause of the accident; or as a catalyst for the accident, rapidly increasing the severity of the accident consequences. Although such situations are common, they are ultimately a unity of opposites between inevitability and chance: on the one hand, when human safety factors accumulate to a certain extent, they inevitably lead to qualitative changes, making the occurrence of an accident a high probability event; On the other hand, no matter how much human safety factors accumulate, they are not sufficient to independently and definitively cause an accident. There must be some external factors or accidental events that trigger it, which will ultimately induce the accident. So, what kind of role do human safety factors play in an accident? There is often no unified, absolute conclusion. Only by analyzing specific accidents in detail can we clarify the impact mechanism of human safety factors on accidents.

(3) There is a universal correlation and relativity between risks and risk factors during the construction process of large underground spaces in cities. In other words, a certain event is referred to as a risk event in a certain context, but when the context changes, this event may be considered as a risk factor for another event. For example, the "11·15" collapse accident of Hangzhou Metro Line 1 was due to the instability of thc foundation pit base, which led to the "kick" destruction of the underground continuous wall, further developing into the shear of the underground continuous wall in the middle section due to extremely unfavorable stress conditions, the complete instability of the retaining structure, the collapse and surge of the pit edge soil into the foundation pit, and finally the

almost complete collapse of the Fengqing Avenue on one side of the foundation pit. For this event, if we only focus on the foundation pit project itself, then "base instability" or "underground continuous wall shear" may be the final top-level risk event. However, if we focus on a larger range, including the surrounding area, then "surrounding stratum collapse" also becomes a risk event that needs to be discussed, and for this risk event, "base instability" or "underground continuous wall shear" is the cause of its occurrence. In other words, when conducting risk analysis at a larger granularity, "base instability" and "underground continuous wall shear" are risk factors for "surrounding stratum collapse". Although the relativity of risk factors and risk events exists in any risk analysis, in the risk system of urban large underground space construction, this relationship is extremely complex. Under this characteristic, any event or sequence of events in the risk system may not be isolated, but constitute an "event network", which makes risk analysis very complex and difficult to grasp the boundaries of analysis.

3. Complexity of risk factors

In fact, after the urban large underground space construction risk system possesses the aforementioned two characteristics, it can naturally lead to the conclusion that its risk factors are extremely complex. On the one hand, risk factors are linked to risk sources, and the existence of a risk source will naturally lead to a series of risk factors related to it. For example, if there is an existing underground tunnel line, then the ground deformation and construction vibration caused by close-distance construction become risk factors. Obviously, there is a one-to-many relationship between risk sources and risk factors, so when the number of risk sources increases, the number of risk factors will naturally increase geometrically; on the other hand, risk factors are related to the risk formation mechanism, the more complex the risk formation mechanism, the more risk factors. For example, for pipelines, the possible risk events are nothing more than medium leakage, pipeline rupture, medium explosion, etc., and each type of risk has a relatively simple risk path or risk formation mechanism, so we can easily list the risk factors of these risks. However, if it is a complex foundation pit that is excavated in sections and steps, firstly, there are more types of possible risk events, and each type of risk event has many possible risk paths, some of which may even be impossible to simply enumerate or infer through existing experience. Therefore, if you want to list all its risk factors, it is a very difficult thing. In fact, from the perspective of risk analysis, exhausting all risk factors is not only unnecessary, but it may also increase the logical complexity of risk analysis, increase the degree of conclusion dispersion in group decision-making, and have a counterproductive effect in a certain sense. Therefore, the analysis of the urban large underground space construction risk system must grasp the analysis granularity (that is, how to logically break down the problem to what detail), and the analysis boundary (that is, which elements need to be considered, and which elements do not need to be considered).

4. Unpredictability

It has been previously stated that there are certain patterns to the risks of urban large underground spaces. In the case of a well-established database, the risk has measurable characteristics without dispute. However, as the scale of urban large underground space construction continues to grow, especially in large space projects where both the inherent risks of the project and the surrounding environmental risk levels are high, the patterns and characteristics of risk may differ from those summarized in previous databases for general spaces. This leads to the difficulty in predicting risks during the construction process of large spaces. Currently, there are not enough samples of urban large underground space projects, which means that although the risk is measurable, it is still difficult to measure. This is the first layer of meaning for the unpredictability of risk. In addition, as the scale of underground space construction continues to grow, the methods and techniques used in general spaces will show their limitations. Consequently, new techniques, methods, equipment, and devices are developed. Since these are "new" and previously non-existent, even if targeted risk analysis, evaluation, and prediction are carried out, it is still difficult to cover all aspects. This is another layer of meaning for the unpredictability of risk in urban large underground spaces. Perhaps in the future, with the development of social engineering construction levels and the improvement of the urban large underground space project database, the risk of urban large underground spaces will become completely measurable.

5. High Risk

The biggest characteristic of urban large underground spaces is the large scale of the projects, which means that if construction risks occur, the damage is immense. The high risk of urban large underground space construction is determined by three factors: firstly, there are many risk factors. Due to the large scale of the project, there are more factors affecting the construction than in general projects, such as more strata, pipelines, and surrounding buildings (structures) that the construction process will encounter. These factors will affect the entire construction process; secondly, there are many types of risk accidents. In the past, a project would frequently have only one or two types of risk accidents. Large space projects, due to the multitude of factors, have coupled and dynamic characteristics, and the construction process may trigger multiple types of accidents; thirdly, the mutual influence is greater. The impact of the construction process on the surrounding environment is wider and larger in scope compared to general space projects, and there are relatively important buildings and structures around the urban large underground spaces. Therefore, once a risk occurs, it will affect a wider range, have more severe impacts, and more serious consequences.

6. Coupling Amplification

As can be seen from the analysis of "typical urban large underground space construction safety accidents" in the previous section, there are many factors affecting the construction process of urban large underground space. A slight change can affect the whole, and the formation of risk events is inevitably the result of multiple risk factors appearing simultaneously, interacting, and promoting each other. When looking at these risk factors independently, it is not difficult to find that most risk factors are "small events" with extremely limited harm. When risk factors appear independently, they hardly bring about significant changes to the overall risk. However, when several of these "small events" happen to occur at the same time, a certain "chemical reaction" may occur, leading to serious "major accidents". This characteristic indicates that there is a coupling amplification effect between the risk factors of urban large underground space construction, which invisibly increases the probability of risk occurrence or enlarges the risk loss. Therefore, for the construction risk of urban large underground space, the coupling amplification of risk factors cannot be ignored.

7. Dynamic Evolution

Due to the increase in span and scale of urban underground large-space projects, it is difficult for the entire project to be formed at once. As can be seen from the above "Research on Excavation Methods", the column hole method is used most frequently in the construction process of urban underground large-space projects, followed by the frequency of use of the double-side wall guide pit method. Its core idea is to excavate in sections, changing large sections into medium and small sections. However, multiple excavations will cause multiple repeated disturbances to the strata. The process of excavation in the construction of urban underground large-space projects is also the process of unloading the surrounding rock. The more excavation steps, the more complex the evolution of risk, and the more chaotic the development law of risk; the construction period of urban underground large-space projects is often long, and in the risk-bearing environment, the risk-bearing body is under the condition of multiple risk factors for a long time, which also makes the risk under multiple disturbances dynamically evolve throughout the entire construction process.

4.3.2 Urban Underground Large-Space Risk Classification

1. Open excavation single underground large-space construction risk classification

The manifestation and event description of open excavation single underground large-space construction risk events are shown in Table 4.5.

2. Hidden excavation single underground large-space construction risk classification

The manifestation and event description of hidden excavation single underground large-space construction risk events are shown in Table 4.6.

3. Networked expansion underground large-space construction risk classification

The manifestation and event description of networked expansion underground large-space construction risk events are shown in Table 4.7.

Table 4.5 Risk statistics table for open cut single underground large space engineering

Risk events		Incident manifestations	Description of the incident
Engineering own risks	Enclosure instability	Localized structural damage, failure of anchor ropes and anchors, sliding instability of the enclosure structure, overturning instability of the enclosure structure, and so on	The structure itself cracks, leaks, breaks, shears; excessive pressure on the bottom surface of the structure caused footing damage, the formation of sliding surfaces in the soil; the structure together with the soil outside the pit together with the loss of stability, etc.
	Support system destabilization	Localized structural damage, overall instability of supports	Structural fracture, bending, resulting in insufficient structural load-bearing capacity, or support the whole-body deformation failure
	Deformation damage at the bottom of the pit	Excessive deformation of the pit bottom bulge destroys	When the magnitude of soft foundation uplift exceeds a certain range, it will cause the whole pit to be destabilization
Geologic risk	Earth slide	Slip of slope excavation, overall instability, etc.	Slip in pit excavation; soil forms a slip surface, which destabilizes and collapses together with the enclosing structure. Loss of stability and collapse
	Sudden outburst of mud and water	Soil and sand flow on the pit wall, sudden surge at the bottom of the pit, pipe surge at the bottom of the pit, etc.	Failure of the water-stopping curtain leads to a large amount of water, sand and soil gushing into the pit, resulting in soil and water runoff; improper precipitation of pressurized water; failure to set up precipitation wells or the failure of precipitation wells leads to the emergence of water and sand at the bottom of the pit
Surroundings environment risks	Structure destruction	Tilting of buildings above ground, uneven settlement, uplifting of underground structures, cracking, etc.	Uneven settlement of the surrounding soil due to excavation of foundation pits, precipitation, etc., resulting in damage to buildings and structures
	Road and railroad damage	Road surface deformation, cracking, collapse, etc.	Uneven settlement of the surrounding soil due to excavation of foundation pits and precipitation, etc., resulting in Damage to neighboring roads

(continued)

Table 4.5 (continued)

Risk events		Incident manifestations	Description of the incident
	Bridge damage	Bridge tilt, settlement shift, collapse, etc.	Uneven settlement of the surrounding soil due to excavation of foundation pits, precipitation, etc., resulting in Damage to neighboring bridges
	Pipe damage	Pipeline rupture	Structural rupture of the pipeline due to deformation of the soil, rendering it unusable

4.4 Urban Underground Large-Space Construction Safety Risk Analysis Method

4.4.1 Urban Underground Large-Space Construction Risk System Division

1. Urban underground large-space construction risk system division method

The core of urban underground large-space construction risk analysis is inevitably construction mechanics problems. For this issue, there are many mature analysis methods, such as various event sequence models based on probability analysis, or finite element simulation commonly used in the field of safety assessment, or a combination of the two. The reason why these methods can be used is because the system dominated by mechanical mechanisms has the following characteristics.

① All behaviors of the system have a clear physical mechanism. Excluding all external factors, the behavior of the system can theoretically be reproduced.
② Variables in a system have a clear physical meaning and can be calculated or measured.
③ The behavior of the system can be defined as a series of specific events, and probability can be defined on top of the events.
④ Due to the aforementioned characteristics, there is a definite causal relationship between the events in the system, and this causal relationship is mostly monotonic. Therefore, qualitative inferences can be made about the behavior of the system through logical deduction, and the behavior of the system can be quantitatively analyzed in conjunction with probability.

However, describing the urban underground large-space construction risk system from a mechanical perspective alone is undoubtedly incomplete. Through accident case analysis, it is not difficult to find that although the process of an accident is a pure mechanical process, the root cause of the accident and the environment that breeds the accident all have more or less human uncertainty factors. In many cases, human

Table 4.6 Risk statistics table for underground excavation engineering

Risk events		Incident manifestations	Description of the incident
Engineering own risks	Instability of the support system	Excessive structural deformation or encroachment, damage to supporting structures	Sinking of the vault, twisting and deformation of the steel supports, cracking and spalling of the concrete between the supports, cutting of connecting bolts, etc.
Geologic risk	Perimeter rock destabilization	Sheet gangs, falling blocks, sliding slumps, roof rises, collapses	The original equilibrium of geotechnical pressure is destroyed, and the phenomenon of deformation, destruction, and collapse of blocks under pressure
	Sudden outburst of mud and water	Drowning, burying	Water emergencies that result in a large volume of water gushing into the project, resulting in the drowning and death of personnel, or the burial of personnel due to the influx of sediment
Surroundings environment risks	Structure destruction	Tilting of buildings, uneven settlement, upliftment of underground structures, cracking, etc.	Uneven settlement of the surrounding soil due to construction, resulting in building (structural) damage
	Road and railroad damage	Road surface deformation, cracking, collapse, etc.	Uneven settlement of the surrounding soil due to construction, resulting in damage to the surrounding roads damage
	Bridge damage	Bridge tilt, settlement shift, collapse, etc.	Uneven settlement of the surrounding soil caused by the construction, resulting in the destruction of the neighboring bridges
	Pipe damage	Pipeline rupture	Structural rupture of the pipeline due to deformation of the soil, rendering it unusable
		Pipeline leakage	Cracking of sewage pipes, oil and gas pipes, etc., resulting in the overflow of internal fluids and gases, and causing impacts on the project, triggering secondary disasters

Table 4.7 Risk statistics table for expansion and construction projects

Risk events		Incident manifestations	Description of the incident
Engineering own risks	Destruction of existing structures	Structural damage, excessive deformation, instability	Varying degrees of damage to existing structures due to construction disturbances
	Destruction of newly constructed structures	Structural cracking, uneven deformation, etc.	Deformation of existing structures can also lead to some degree of new construction damage to the new structure
Geologic risk	Earth slide	Slip of slope excavation, overall instability, etc.	Landslide occurs during excavation of foundation pit slope; the soil forms a sliding surface, which, together with the retaining structure, loses stability and collapses
	Perimeter rock destabilization	Slip of slope excavation, overall instability, etc.	The phenomenon where the originally balanced rock and soil pressure is disrupted, and the working face deforms and fails under the pressure, resulting in falling blocks and collapsing
	Sudden outburst of mud and water	Runoff of soil and sand from the pit walls, sudden surges at the bottom of the pit, and pipe surges at the bottom of the pit, drowning, burial, etc.	A large amount of water and sand pouring into the pit causes soil erosion; improper precipitation of confined water; failure to install precipitation wells resulted in water seepage and sand turning at the bottom of the pit
Surroundings environment risks	Damage to neighboring buildings	Tilting of buildings above ground, uneven settlement, uplifting of underground structures, cracking, etc.	Uneven settlement of surrounding soil caused by excavation and dewatering of foundation pits, resulting in damage to buildings
	Damage to neighboring roads and railroad	Road surface deformation and cracking, road surface collapse, etc.	Uneven settlement of the surrounding soil due to excavation of foundation pits and precipitation, etc., resulting in damage to neighboring roads
	Damage to neighboring bridges	Bridge tilt, settlement shift, collapse, etc.	Uneven settlement of the surrounding soil due to excavation of foundation pits, precipitation, etc., resulting in damage to neighboring bridges

(continued)

Table 4.7 (continued)

Risk events		Incident manifestations	Description of the incident
	Damage to neighboring pipelines	Pipeline rupture	Structural rupture of the pipeline due to deformation of the soil, rendering it unusable
		Pipeline leakage	Cracking of sewage pipes, oil and gas pipes, etc., resulting in the spillage of internal gases and liquids, leading to secondary effects

uncertainty factors determine the uncertainty of the system. The characteristics of the urban underground large-space construction risk system determine that the human uncertainty factor itself is also a complex subsystem with its own structure, laws, and characteristics. This system composed of human uncertainty factors has the following characteristics that are completely different from the mechanical system:

① The behavior of the system only has logical connections, making it difficult to link with physical mechanisms.
② There are many variables in the system, and the state of each variable has a great deal of uncertainty. Even under relatively consistent external conditions, it is difficult to reproduce the behavior of the system.
③ The behavior of a system can generally be described only in terms of changes in its macroscopic state, which is difficult to define in the form of events, and therefore It is difficult to define the behavior of the system in the form of events, and therefore it is difficult to measure the response of the system by probabilistic methods.
④ The interrelationships between various links within the system are complex, with intertwined cause-and-effect relationships. Even if the system structure is deduced through logical reasoning, it is difficult to intuitively predict the behavior of the system.

In summary, there are two types of components with distinctly different characteristics but are indispensable in the urban large underground space construction risk system: the physical mechanics system dominated by construction mechanics and the organizational management system dominated by human uncertainty factors. The different characteristics of the two types of systems determine that different methods must be used to analyze them, so it is necessary to model them based on the division of each system.

Based on the above characteristics, the urban large underground space construction risk system is divided into the following three subsystems, as shown in Fig. 4.10.

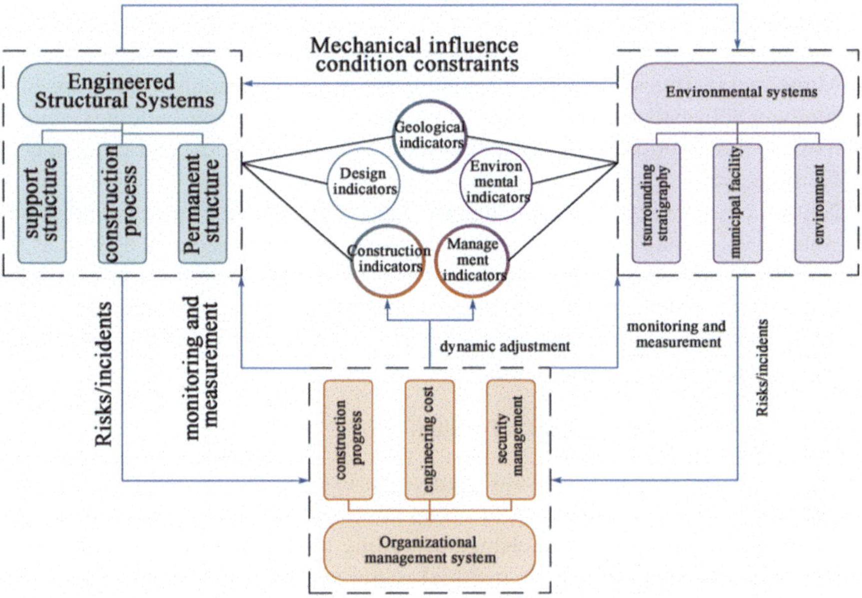

Fig. 4.10 Schematic diagram of the urban large underground space construction risk system

(1) Engineering structure subsystem

The engineering structure subsystem is the core of the urban large underground space system, which includes the following connotations:

① Permanent structures aimed at realizing the function of urban underground large space, such as subway station frame structures, permanent linings of large-span tunnels excavated in secret, and other functional auxiliary structures.

② Permanent or temporary support structures built to stabilize the stratum and provide working space during the construction of urban large underground space, including the retaining structures of foundation pits, internal supports, initial supports of tunnels, temporary inverted arches and partition walls, guide tunnels, etc.

③ Construction process, referring to a series of methods and procedures necessary to complete the permanent structure of the urban underground large space, including pre-construction dewatering and pre-reinforcement, excavation methods of the stratum, construction timing and sequence of support structures, etc.

(2) Environment subsystem

The environment subsystem refers to the system composed of existing entities that existed before the construction of the urban large underground spaceproject within a certain impact range, which includes the following connotations:

① Surrounding strata, including strata and water systems that have mechanical connections with the construction of urban large underground space projects.
② Municipal facilities, including all existing buildings or structures within the environment subsystem, such as surface buildings, underground buildings, municipal pipelines, bridges, power and communication facilities, and other special facilities, etc.
③ Natural environment, including natural vegetation, wildlife, natural landscapes, water sources, and air entities within the environment subsystem, as well as the climate, temperature, humidity, and other natural states within the project area.
④ Social environment, referring to the surrounding community environment, social public opinion environment, and other abstract entities where the urban large underground space project is located.

(3) Organizational management subsystem

The organizational management subsystem refers to the system composed of abstract entities related to the management of urban large underground space construction. The elements of this system can change according to actual needs, including quantifiable material variables such as material requirements and installation progress, as well as soft variables that are difficult to quantify, such as the impact of policy decisions and the rationality of organizational structures. The state of the organizational management subsystem will ultimately affect the construction progress, cost, and safety management level of the project, thereby changing the behavior of the construction mechanics subsystem.

2. Interrelationships between systems

(1) Engineering structure subsystem—Environment subsystem

There is an inseparable relationship in mechanics between the engineering structure subsystem and the Environment subsystem, and it also affects the environment subsystem in some social behaviors. The construction of the engineering structure is based on the transformation of the surrounding environment. Before the construction of the engineering structure, the construction party needs to occupy a part of the surface space as the construction area, and it may also be accompanied by temporary changes to the surrounding environment such as pipeline relocation and traffic diversion. The construction of the project is a process of removing the natural strata in the environment and inserting man-made structures that can balance with it, which is a semi-permanent change to the surrounding environment. During this period, the removal of the natural strata will inevitably cause deformation of the strata outside the retaining structure, thereby causing disturbances to buildings, pipelines, etc. In terms of risk, the risk events in the engineering structure subsystem often exist as superior events or risk factors in the Environment subsystem, such as the collapse of the ground surface caused by the destruction of the tunnel structure.

On the other hand, the Environment subsystem has a restrictive effect on the engineering structure subsystem. The state of the Environment subsystem determines some characteristics in the engineering structure subsystem. For example, the construction method of underground projects in bustling urban areas will be greatly restricted. Due to the limitations of the work site or the consideration of surface traffic, designers will be more inclined to use the tunneling method and minimize the area of open excavation as much as possible. This will significantly change the risk type and risk factors of the entire engineering structure system. In addition, under the restrictions of the surrounding environment, designers may adopt some unconventional engineering measures to avoid affecting the surrounding environment, such as the design of horizontal double holes and upper and lower double hole conversions when passing through existing underground structures at close distances. This situation is an example of increased project risk due to environmental constraints. This restrictive effect is manifested in the risk system, that is, environmental factors will also have an impact on the engineering structure subsystem.

(2) Organizational management subsystem—Engineering structure subsystem and Environment subsystem

The construction of urban large underground space can be seen as a complex process composed of a series of basic control processes. The construction process of urban large underground space starts with the input of certain geometric parameters, geological parameters, and construction parameters. During the construction process, a certain impact on the surrounding soil environment is output according to certain physical laws. This impact is usually observed through the arrangement of monitoring points (such as underground continuous wall inclinometer, building settlement), and construction personnel adjust the next construction parameters by analyzing and predicting the monitoring data. This cyclic process is a dynamic control process. The impact of the construction of urban large underground space on the surrounding environment will be transmitted to the surrounding buildings (structures), pipelines, etc. Under the combined effect of this impact and the parameters of the surrounding buildings (structures), pipelines (such as foundation depth, foundation type), the surrounding buildings (structures), pipelines will produce certain safety risks. After assessing the safety risks, the organization will take corresponding preventive measures. If an accident occurs, it is necessary to adjust the construction plan or carry out remedial activities, both of which are feedback controls. It is worth noting that the execution of these three measures must go through the organization, that is, various monitoring data, risk analysis results, and accidents are input to the organization, and the organization executes corresponding measures based on these inputs.

From this, it can be seen that the organizational management system is the core controller of the urban large underground space construction system, playing an important role in gathering feedback information from all parties of the system and dynamically correcting the behavior of the system.

4.4.2 Urban Large Underground Space Construction Risk System Model

1. System Structure

(1) Engineering Structure Subsystem and Environment subsystem (Construction Mechanics Subsystem)

As mentioned earlier, both the engineering structure subsystem and the Environment subsystem are event-driven systems dominated by physical mechanisms. And because there is a close interaction between the engineering structure and the surrounding environment, the boundary between the two is relatively vague. Therefore, when establishing a risk system model, these two systems are treated as a whole, collectively referred to as the "construction mechanics subsystem". The construction mechanics subsystem of urban large underground spaces the core of the entire construction risk system, and all risk events and factor events are defined in this subsystem. Based on the accident causality diagram obtained from the analysis of typical urban large underground space accident cases, combined with the characteristics of the urban large underground space risk system, the model of the urban large underground space construction mechanics subsystem is formed within the "risk analysis framework" formed earlier, as shown in Fig. 4.11.

The urban large underground space construction mechanics subsystem model is generally divided into event space and state space.

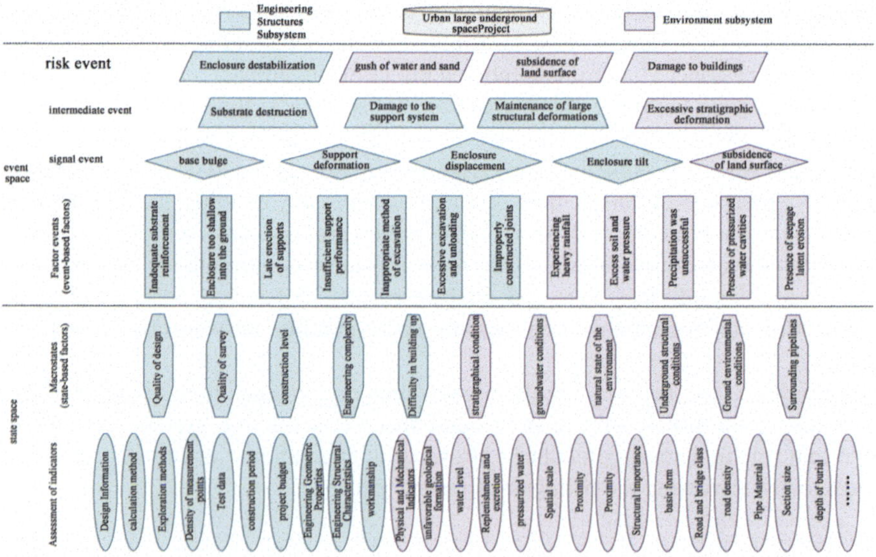

Fig. 4.11 Urban large underground space construction mechanics subsystem model

① Event Space

As the name suggests, the elements in the event space are events (Event) with clear definitions. An event only exists in two states, "occurrence" and "non-occurrence", and its related variable is the probability of the event occurring (p^E). Events are connected by causal relationships. The event layer contains four levels of elements: risk event layer, precursor event layer, intermediate event layer, and factor event layer.

a. Risk Event Layer

The risk event layer is composed of risk events and is located at the top of the event space. Risk events often exist as "results" in the entire event chain, but a risk event can sometimes also be the "cause" of another risk event. Generally speaking, there should not be too many elements in the risk event layer to avoid making risk analysis work too cumbersome.

b. Intermediate Event Layer

The intermediate event layer is located below the risk event layer and plays a bridging role in the event space. Intermediate events refer to those events that can be clearly defined on the path of risk formation and are indispensable. The intermediate event layer can have any number of levels and structures, and each sub-level should follow a "bottom-up" causal relationship. However, too many intermediate event layers not only greatly increase the complexity of the model but may also interfere with the evaluator's logic. Therefore, it is generally recommended that the intermediate event layer be 1–2 layers, so as to help the analyst clarify the event context concisely and succinctly.

Intermediate events can have a one-to-one, one-to-many, or many-to-one relationship with risk events, but there should not be any isolated intermediate events that do not serve as the cause of any risk events. If it is necessary to analyze such events, they should be placed in the risk event layer.

c. Precursor Event Layer

The precursor event layer is a rather special layer in the event space. From a logical relationship perspective, there is a causal relationship between precursor events and risk events (precursor events are the cause of risk events), but the strength of this causal relationship is relatively weak. For example, there is obviously a causal relationship between "ground subsidence" and "ground collapse", but both "ground subsidence" and "ground collapse" have deeper causes, and these deeper causes are more in line with the concept of "cause" in people's minds. The causal relationship between precursor events and risk events is more of a "correlation", that is, the occurrence of a precursor event cannot directly cause a risk event to occur, but the occurrence of a precursor event will increase the credibility of the corresponding risk event. In this way, we can adjust our risk estimates by observing the state of precursor events.

One significant difference between precursor events and other events in the event layer is the ambiguity of their definition. Risk events and intermediate events have clear conceptual boundaries distinguishing their "occurrence" and "non-occurrence". In other words, under any objective circumstances, we can clearly determine whether these events have occurred. However, precursor events, as defined by their literal definition, are almost all inevitable events. For example, the excavation of a pit will inevitably cause the protective structure to tilt and the ground to subside, even if these deformations and subsidence are small enough to be ignored. Therefore, if a definite degree boundary is not assigned to the precursor event, it is meaningless to discuss whether it occurs or not.

The weak causality and ambiguity of precursor events make them not very useful in causal analysis and inference of risks, but these characteristics make them extremely useful in observing and measuring risks. Many times, the risk event chain obtained through causal analysis can help us make a relatively reliable subjective inference about the level of risk before action begins. However, after the action begins, it is difficult to verify whether this inference is accurate through real-time objective information. This is because most of the events in a normally defined risk event chain are low-probability events, and the sample size required to determine low-probability events is almost impossible to meet in reality. In the actual construction process, we can usually only observe changes in some apparent physical indicators, such as deformation of structures and ground, force on important components, cracks in concrete, etc. For a large risk system, the amount of information represented by these indicators is extremely limited, so it is difficult to directly associate them with risk events. But theoretical knowledge and engineering experience tell us that there must be a correlation between these indicator values and construction safety risks. By defining precursor events, a logical association can be established between these measurable physical indicators and risk events, thus enabling real-time observation and measurement of risks during the action process.

d. Factor Event Layer

Factor events, also known as event-type factors, are the smallest elements in the event space. The smallest granularity refers to the fact that factor events have the most detailed definitions and the most minor impacts. Factor events are generally "small events" that have a relatively high probability of occurrence but cause a low degree of adverse impact on system safety when they occur. The impact produced by a single factor event occurring independently is generally negligible, and only when multiple factors occur simultaneously will it cause a significant impact on system safety. Because of these characteristics, the selection of factor events requires careful consideration. On the one hand, the definition of factor events should be as vivid and specific as possible to facilitate decision-makers' judgment; on the other hand, the coverage of factor events should be as comprehensive and complete as possible to reduce the bias in risk assessment. However, this may result in a large number of factor events, making the analysis model overly complex and leading to a significant increase in the workload of analysis. Therefore, when selecting factor events, one should "consider comprehensively, distinguish between primary and secondary, aim accurately, and maintain appropriate granularity."

② State Space

The elements in the state space are indices. The essence of an index is a variable that characterizes the state of the system, so its state (value) can have any number, or even use a continuous scale for measurement. The relationship between the index and the event is similar to the relationship between random variables and random events in probability theory. The state space includes two levels of elements: the macro state layer and the evaluation index layer.

a. Macro State Layer

The macro state layer includes "design quality," "survey quality," "construction level," "project complexity," "construction difficulty," "stratum conditions," "groundwater conditions," "natural environment status," "underground structure conditions," "surface environment conditions," and "surrounding pipeline conditions," a total of 11 indices. These indices basically cover all aspects that can affect the safety risk of urban large underground space construction, so unlike other levels in the model, the number of elements in the macro state layer is finite (it can be simplified according to actual conditions, but generally does not need to be increased). The macro state index is a measure of the degree of impact of the state of a certain aspect of the risk system on safety risk. It is an abstract concept, so its value range can be arbitrarily given and does not have dimensions. For convenience of analysis, a uniform value range, such as a percentage system, is generally given to all macro state indices. The macro state index contains a lot of information, so it is not appropriate to assign values to it directly. It needs to be indirectly assigned through evaluation indices, which are more specific and operable. As the link between the evaluation index and the factor event, the macro state index needs to participate in the probability calculation in the event space, so a mapping relationship between the value of the macro state index and the probability of the factor event needs to be established.

b. Evaluation Index Layer

The evaluation index layer serves the macro state layer. Evaluation indices are variables with clear and specific meanings that can be obtained through observation, calculation, or evaluation. The value of the evaluation index can be a dimensionless number or a specific physical quantity with dimensions, but because it involves conversion to the macro state index, the evaluation index is generally subjected to uniform dimensionless processing. The relationship between the evaluation index and the macro state index is many-to-one, and one evaluation index can only correspond to one macro state index. Sometimes, to make the analysis logic clearer, multiple evaluation indices under one macro state index can have a tree-like hierarchical relationship. Similar to factor events, the selection of evaluation indices should also balance granularity, breadth, and operability. It is generally recommended that the evaluation indices under each macro state index do not exceed two levels, and the number of branches in each level of the tree structure does not exceed five. At the same time, for those evaluation indices that do not have physical meaning or recognized definitions, specific value assignment methods and bases should be given.

(2) Organizational Management Subsystem

Unlike the engineering structure subsystem and the environment subsystem, the significant characteristics of the organizational management subsystem are its abstractness, subjectivity, and the complexity of causal relationships. These characteristics allow the organizational management subsystem to be flexibly extended in terms of conceptual extension and highly refined in terms of conceptual connotation. The so-called flexible extension in conceptual extension means that the boundaries of the system can be freely adjusted to a large extent according to the analyst's cognition and needs. For example, we can consider the organizational management subsystem narrowly as the organization that supervises the construction of large underground urban spaces, or we can consider management culture, management system, engineering contracts and budget estimates, supervision, and execution, etc., all belong to the organizational management subsystem. The high degree of refinement in conceptual connotation means that the granularity of the system has a lot of adjustment space. For example, we can only evaluate the rationality of the setting of an engineering management organization at a coarse granularity, or we can refine it to the skill level, work experience, and even psychological state of each individual involved in the project. Therefore, the richness and meticulousness of the connotation of the organizational management subsystem often far exceed that of the engineering structure subsystem and the environment subsystem. If we use events as the basic unit to model the organizational management subsystem, it will inevitably lead to a highly complex structure, conceptually confusing, difficult to grasp granularity, difficult to quantify, and unoperational model, and the model will inevitably be greatly influenced by the modeler's subjective concept. Therefore, the modeling of the organizational management system must stand from the perspective of the actual decision-maker, grasp the core characteristics of the system, and highly summarize and abstract the related concepts.

For the above reasons, the organizational management subsystem adopts a modeling method similar to the state space in the construction mechanics subsystem, with the abstract "index" as the basic element. The causal relationships in the organizational management subsystem are generalized into hierarchical structures, feedback effects, and constraint relationships between indicators. Since the concept of events is not involved in the organizational management subsystem, the connections between indicators do not have any probabilistic meaning, and therefore will not directly participate in the operation of the event layer in the construction mechanics subsystem.

① Overall structure of the organizational management subsystem

As mentioned earlier, due to the strong subjectivity of the organizational management subsystem, the extension, connotation, and structure of the organizational management subsystem are not unique. However, no matter how the specific implementation is, the modeling of the organizational management subsystem should be carried out within a certain framework and principles. For the risk system of urban large underground space construction, the purpose of modeling the organizational management subsystem is to measure all "safety performance" generated by human management behavior. Figure 4.12 is an example of the structure of an organizational management subsystem model. The changes in the state of the organizational management subsystem are ultimately reflected in the changes in the "safety performance" index, and further reflected in the three sub-indicators of "construction quality", "construction cost", and "construction progress" that have constraint relationships with each other, and through these three sub-indicators to achieve interaction with the construction mechanics subsystem; in order to quantify the safety performance, the organizational management subsystem can be further divided into several state sets such as organizational structure, site management, personnel quality, and engineering contracts, each state set can be further refined to several evaluation indicators, forming a tree-like structure in logic.

② Output end of the organizational management subsystem—Safety performance

"Safety performance" is the core node of the organizational management subsystem and the final measurement object of the organizational management subsystem. Safety performance refers to the impact of various management behaviors during the construction process of the project on the physical process of construction and is an abstract indicator to measure the quality of project management work. Safety performance involves the following three specific contents.

a. Construction quality

Construction quality is a direct reflection of safety performance. The higher the construction quality, the lower the probability that the engineering structure will fail or be unable to meet the requirements of its intended function, which means higher

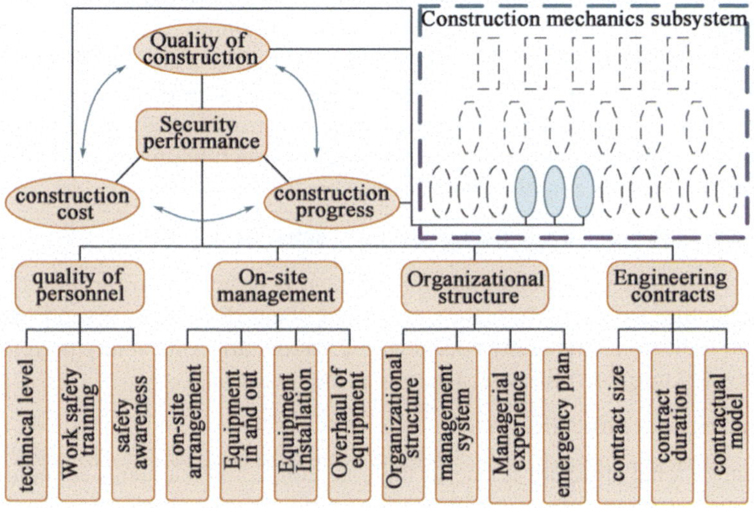

Fig. 4.12 Schematic diagram of the structure of the organizational management subsystem model

safety and lower risk. Generally speaking, the more efficient the management organization, the more complete the management system, and the better the implementation of management measures, the higher the construction quality will be.

b. Construction Cost

Although construction cost is not directly linked to safety in concept, it is still an indispensable part of the safety performance system. Construction cost can be understood as a constraint on safety performance: on the one hand, the construction of management organizations, the employment of management personnel, and the execution of management measures all come with certain costs, which naturally form part of the construction cost; on the other hand, the improvement of management level leads to the improvement of construction quality, and the improvement of construction quality often comes with stricter quality control standards, which may lead to higher rework rates, higher material waste rates, and possibly the purchase of more expensive equipment and materials to improve construction quality, thereby indirectly increasing construction costs.

c. Construction Progress

Like construction cost, construction progress is another constraint on safety performance. Construction progress is determined by the planned construction period and the construction rate. The planned construction period refers to the construction plan set by the construction party under the premise of meeting the contract period, which depends on the rationality of the construction organization scheme and the technical level of the construction team and is a relatively fixed indicator. The construction rate, on the other hand, is subject to dynamic influences from various uncertain factors and is constantly changing. Generally speaking, the planned construction period affects the construction quality, and the construction quality directly relates to the frequency of various construction quality problems, which in turn affects the frequency of management measures such as inspections and rectifications. The implementation of management measures will inevitably slow down the construction rate, thereby reducing the margin of the planned construction period. Therefore, the relationship between construction progress and construction quality is not a simple constraint but involves complex feedback.

③ Input end of the organizational management subsystem

In order to quantify management performance, it is necessary to further refine the factors affecting safety performance and break them down into several operable evaluation indicators. Generally speaking, the factors affecting safety performance can be divided into the following state sets at the first level.

a. Organizational Structure

The organizational structure is one of the core conceptual elements in the organizational management subsystem. When the organizational structure is used as an indicator, its value represents a comprehensive evaluation of the pros and cons of the project management organization. The pros and cons of the project management organization can mainly be evaluated through the following aspects: (a) the rationality of the organizational structure of the project management organization; (b) the qualifications and experience of the management personnel in the project management organization; (c) the completeness of the management system established by the project management organization; (d) whether the project management organization has a comprehensive emergency plan, etc.

b. Personnel Quality

The value of the personnel quality indicator is an evaluation of the comprehensive quality of the workers and technicians who are directly involved in the front-line operations on site.

Personnel quality is an important reflection of the effectiveness of project management work and is also a link that will directly affect safety performance. Generally speaking, personnel quality can be evaluated through the following aspects: (a) the level of professional knowledge and technical skills of the construction personnel; (b) the detail and implementation of the safety production training provided by the project management side to the construction personnel; (c) the level of safety awareness of the construction personnel, and whether the project management side has taken measures to improve the safety awareness of the construction personnel, etc.

c. On-site Management

On-site management is a distinctive and extremely important aspect of civil engineering management. On-site management can be understood as a dynamic implementation of some content in the construction organization plan. In other words, although the work content and process of on-site management are stipulated by the construction organization plan, it needs to face the impact of many unpredictable uncertain factors in the actual implementation process, and its implementation effect also greatly depends on the subjective choices of management personnel. The quality of on-site management is generally evaluated through the following aspects: (a) the reasonableness and compliance of the site layout; (b) the scheduling of equipment entry and exit; (c) the situation of equipment installation and debugging; (d) whether the equipment maintenance is carried out as scheduled, and the actual situation of the equipment and adjustment countermeasures after maintenance feedback.

d. Engineering Contract

The engineering contract indicator reflects the beneficial degree of the engineering contract to project management work, mainly including the following aspects: (a) contract amount, which will affect the engineering budget and its allocation; (b) contract duration, which directly affects the planned duration and thus affects the evaluation of construction progress; (c) contract mode, which affects the project management organization structure and scope of responsibility.

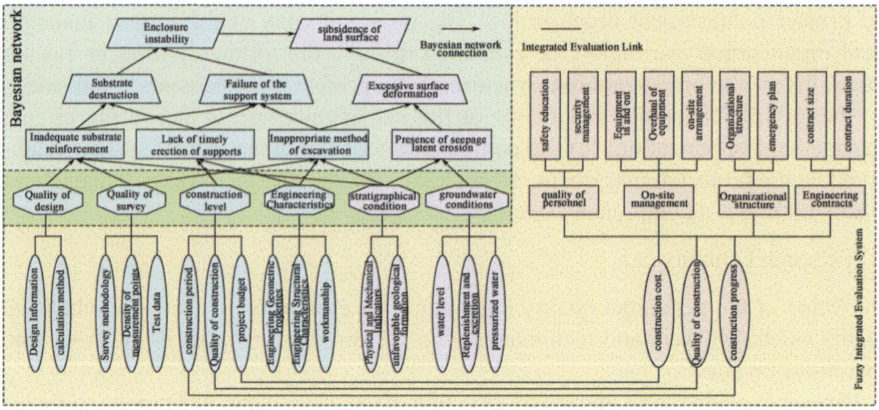

Fig. 4.13 Urban large underground space construction risk inference sub-model

2. Sub-model of Urban large underground space Construction Risk Inference

(1) Model Elements and Structure

The sub-model of urban large underground space construction risk inference is a model formed with the purpose of construction risk inference, based on the urban large underground space construction risk system model, introducing Bayesian network and fuzzy comprehensive evaluation method as analysis methods. As shown in Fig. 4.13, the model mainly consists of two parts: Bayesian network and fuzzy comprehensive evaluation system.

① Bayesian Network

The Bayesian network in the sub-model includes the risk event layer, intermediate event layer, factor event layer, and macro state indicator layer of the urban large underground space construction mechanics subsystem. Among them, the macro state indicators are the leaf nodes of the Bayesian network and the input end of the Bayesian network part; the risk events are the root nodes of the Bayesian network and the target of Bayesian network inference. The connection of the Bayesian network represents the conditional probability relationship between the nodes.

② Fuzzy Comprehensive Evaluation System

The fuzzy comprehensive evaluation system in the sub-model includes the macro state indicator layer, evaluation indicator layer, and the entire organizational management subsystem of the urban large underground space construction mechanics subsystem. Among them, in the construction mechanics subsystem part, the macro state indicators are the targets of fuzzy comprehensive evaluation, and their state space is the domain of fuzzy comprehensive evaluation; the evaluation indicators are the fuzzy evaluation factors, and each macro state indicator and its affiliated evaluation indicators form a tree structure representing a fuzzy evaluation matrix.

The organizational management subsystem in the fuzzy comprehensive evaluation system is equivalent to a sub-model, and the organizational management subsystem itself constitutes an evaluation structure. The three safety performance indicators "construction cost", "construction quality", and "construction progress" are the evaluation targets of the sub-model, and the factors set of the organizational management subsystem under them are the fuzzy evaluation factors of these three evaluation targets. The organizational management subsystem can have its own evaluation domain, which can either follow the fuzzy comprehensive evaluation method of the construction mechanics subsystem or adopt other evaluation methods. There is some conversion relationship between the three evaluation targets of the organizational management subsystem and the corresponding information indicators in the construction mechanics subsystem.

(2) Introduction to Basic Theory

① Bayesian Network

Bayesian networks (BBN) are ideal tools for probabilistic analysis and conditional reasoning of event sequences. A Bayesian network is a directed acyclic graph composed of several nodes. The nodes in the Bayesian network represent a random variable, with the probability distribution of the variable itself attached to the node; the directed arrows between the nodes represent the causal relationship between the variables, characterized by a series of conditional probabilities. Once the structure of a Bayesian network is determined, a conditional probability table (CPT) can be constructed on each non-root node according to the strength of the causal relationship between the nodes, which can fully determine a Bayesian network.

In the past decade or so, modeling analysis using BBN has received great attention. The main advantages of BBN are: (a) It can integrate objective data and expert knowledge into a model; (b) It can graphically represent the causal relationship between variables; (c) It has a solid mathematical theoretical foundation, and uses probability to quantitatively express the propagation of the impact of the causal relationship between variables, thereby reasoning or diagnosing problems; (d) After obtaining new information (evidence), it can calculate the posterior probability of the variable and update the model.

The characteristics of BBN make it an ideal tool for reliability analysis and risk modeling, such as analyzing the impact of organizations on system safety, technical risk analysis, human reliability analysis (HRA) and simulating the entire complex social technical system; BBN is also used in project risk management, such as delay in construction period, cost risk analysis, etc.; RIVAS and others and ZHOU and others use BBN as a data mining technology to analyze the causal relationship between variables or the strength of the causal relationship and use it as the basis for safety management strategies; SOUSA and EINSTEIN use BBN for decision support, they have established a prediction model of the geological conditions in front of the tunnel construction face and a decision model for choosing the excavation method based on BBN; RøED and others combine BBN, FT and Event Sequence Diagram (ESD) to construct a Hybrid Causal Logic (HCL) model to represent the organizational and technical aspects of risk assessment and their impact relationships and developed a software called "Trilith" for analyzing the operational safety of offshore drilling platforms and aviation management safety, while REN and others combined fuzzy set theory with BBN to construct a fuzzy Bayesian network, mainly relying on expert experience for modeling analysis.

② Fuzzy Comprehensive Evaluation Method

Fuzzy mathematics is a new type of mathematical method proposed by American control theory expert ZADEH in 1965. Its research object is the "uncertainty of concept boundaries". It expands traditional set theory with the concept of membership, thereby opening up a new way to study qualitative problems using mathematical methods.

The fuzzy comprehensive evaluation method is a modern comprehensive evaluation method based on fuzzy mathematics. It transforms qualitative evaluation into quantitative evaluation according to the membership theory of fuzzy mathematics, that is, it uses fuzzy mathematics to make an overall evaluation of things or objects constrained by various factors. It has clear results and strong systematicness, can solve fuzzy and difficult to quantify problems well, and is suitable for solving various uncertain problems. The steps of the fuzzy comprehensive evaluation method are as follows:

a. Determine the evaluation factors and evaluation levels

Let $U = \{u_1, u_2, \cdots, u_m\}$ represent the m factors (i.e., evaluation indicators) that characterize the evaluated object, called the factor set;

Let $V = \{v_1, v_2, \cdots, v_n\}$ represent the n evaluation results (i.e., evaluation levels) that characterize the state of each factor, called the comment set.

Here, m is the number of evaluation factors, determined by the specific indicator system; n is the number of comments, generally divided into 3–5 levels.

b. Construct the evaluation matrix

Starting from the factor u_i, judge the degree of membership of this factor to the evaluation level v_i, and obtain the single-factor evaluation vector of the i th factor as: $\bar{d}_i = \{d_{i1}, d_{i2}, \cdots, d_{in}\}$.

Combine the evaluation vectors of m factors to form the overall evaluation matrix D. The overall evaluation matrix determines a fuzzy relationship from U to V:

$$D = \left[d_{ij}\right]_{m \times n} = \begin{bmatrix} d_{11} & d_{12} & \cdots & d_{1n} \\ d_{21} & d_{22} & \cdots & d_{2n} \\ \vdots & \vdots & \ddots & \vdots \\ d_{m1} & d_{m2} & \cdots & d_{mn} \end{bmatrix} \tag{4.22}$$

c. Determine the weights

For a certain evaluation target, the various factors in the evaluation factor set have different impacts on the evaluation target. Therefore, a fuzzy subset A on U needs to be introduced, $A = \{a_1, a_2, \cdots, a_m\}$, known as the fuzzy weight vector. Where $a_i >> 0$ and $\sum a_i = 1$. The fuzzy weight vector reflects the importance of each factor under the given evaluation criteria.

d. Perform fuzzy synthesis

Different rows in D reflect the degree of membership of an evaluated object to the comment set from different single factors. By synthesizing the fuzzy weight vector A with D, the degree of membership of the evaluated object to the comment set from an overall perspective can be obtained, that is, the fuzzy comprehensive evaluation set B. B is a subset on V, $B = \{b_1, b_2, \cdots, b_n\}$.

The general form of fuzzy synthesis is to perform fuzzy transformation B = A $*$ D, where "*" is a fuzzy synthesis operator. The fuzzy synthesis operator should be selected according to the objective nature of the evaluated object and subjective evaluation orientation. Commonly used fuzzy synthesis operators include the following:

(a) Take the minimum—Take the maximum $*(\wedge, \vee)$:

$$b_j = \bigvee_{i=1}^{n} (a_i \wedge d_{ij}), j = 1, 2, \cdots, m \tag{4.23}$$

This operator focuses on considering the main factors, and other factors have little impact on the result, so it is called the main factor determinant operator.

(b) Product—Take the maximum $*(\cdot, \vee)$:

$$b_j = \bigvee_{i=1}^{n} (a_i \cdot d_{ij}), j = 1, 2, \cdots, m \tag{4.24}$$

This operator is close to the take maximum—take minimum operator, considering the correction of ai to dij when considering multiple factors. This operator also highlights the impact of the main factors and is called the main factor prominent operator.

(c) Take the minimum—Bounded sum $*(\wedge, \oplus)$:

$$b_j = \bigoplus_{i=1}^{n} (a_i \wedge d_{ij}) = \sum_{i=1}^{n} (a_i \wedge d_{ij}), j = 1, 2, \cdots, m \tag{4.25}$$

This operator is also a main factor prominent operator.

(d) Weighted average $*(\cdot, \oplus)$:

$$b_j = \bigoplus_{i=1}^{n} (a_i \cdot d_{ij}) = \sum_{i=1}^{n} (a_i \cdot d_{ij}), j = 1, 2, \cdots, m \tag{4.26}$$

THIS operator takes into account all factors in a balanced manner according to the weight, and is suitable for situations where the effects of all factors are considered in a balanced manner.

The characteristics of each operator are shown in Table 4.8.

e. Clarify the evaluation results

Since the fuzzy comprehensive evaluation set B is a fuzzy vector, considering that the actual evaluation result is always clear, it is necessary to clarify the obtained vector to determine the comprehensive evaluation level. The commonly used clarification methods mainly include the following three:

(a) Maximum Membership Method: Suppose $b_p = \overset{n}{\underset{i=1}{\vee}} b_i$, then the evaluation result is at level p;
(b) Median Method: Suppose $i_0 = \min\left\{i \middle| 1 \le i \le n, \sum_{j=1}^{n} b_j = 0.5\right\}$, then the evaluation result is at level i_0 ;
(c) Segment Assignment Method: For $i = 1, 2, \ldots, n$, assign level i with $v_i(0 < v_i < 1)$, and take $b = \sum v_i b_i$, the level corresponding to b is the evaluation level.

(3) Construction of Bayesian Network

① Event-Event (E-E) Connection Conditional Probability Assignment

An event node represents a specific event, and its corresponding random variable is a binary variable, with only "T" (True) and "F" (False) as possible values. Therefore, the independent parameter required to define an event node is only the probability of the event occurring. For the sake of uniformity, for the event node E, the bold symbol "E" represents the random variable representing the event, the occurrence of the event is recorded as "E = T" or "E", and the non-occurrence of the event is recorded as "E = F" or "E", and the corresponding probabilities are recorded as "$p_E = P(E)$" and "$p_E = 1 - p_E = 1 - P(E) = P(E)$".

The strength of the causal relationship represented by the E-E connection in the Bayesian network is characterized by a series of conditional probabilities, which

Table 4.8 The characteristics of fuzzy synthesis operators

Fuzzy synthesis operator	$*(\wedge, \vee)$	$*(\cdot, \vee)$	$*(\wedge, \oplus)$	$*(\cdot, \oplus)$
Reflecting the role of weights	Inconspicuous	Conspicuous	Inconspicuous	Conspicuous
Degree of aggregation	Weak	Weak	Stubborn	Stubborn
Utilizing information from R	Inadequate	Inadequate	Comparatively adequate	Adequately
Typology	Master factor determinant	Primary factor highlighting type	Primary factor highlighting type	Balanced

can be represented in the form of a table, called a Conditional Probability Table (CPT). Generally speaking, the construction of CPT in Bayesian networks requires training with observed samples. However, the construction process of urban large underground space is non-repetitive and original, which is a poor sample problem, and CPT can generally only be constructed by direct assignment method.

The theoretical basis of the direct assignment method is the subjective interpretation of probability. Opposite to the subjective interpretation of probability is the frequency interpretation of probability. In the early study of probability theory, the frequency interpretation of probability dominated. Generally speaking, for an action that can be repeated under the same conditions, if event A occurs M times in all N actions, then its probability P(A) can be approximated by its frequency of occurrence:

$$P(A) \approx \frac{M}{N} \tag{4.27}$$

According to the law of large numbers, when N tends to infinity, the frequency almost everywhere tends to probability, which is the frequency interpretation of probability.

According to the frequency interpretation, probability only makes sense when actions can be repeated infinitely under the same conditions. However, most of the events involved in the study of construction risk are one-time events that cannot be repeated many times or are difficult to ensure that they can be repeated under the same conditions. For these events, the frequency interpretation is not only theoretically untenable, but also makes probability almost inoperable in practical applications.

The subjective interpretation of probability, also known as the Bayesian interpretation, posits that probability represents a reasonable degree of belief in the likelihood of an event occurring, reflecting an individual's state of knowledge and subjective beliefs. In this sense, the probability is called subjective probability. Compared to the frequency interpretation, the strength of the subjective interpretation is that it allows for probability assessments of one-off events. For example, for the question "What is the probability of the Brazilian team winning the next World Cup?", the frequency interpretation considers this question meaningless because "Brazil winning the next World Cup" is not a repeatable event. However, the subjective interpretation can still provide a subjective assessment based on various prior knowledge. The probability wheel is a specific and operable method for subjective probability assessment. Imagine a uniformly textured probability wheel, which only contains black and white continuous areas, and the probability of the pointer stopping at any position after rotation is equal (Fig. 4.14).

When assessing the question "What is the probability of the Brazilian team winning the next World Cup?", the assessor first needs to answer "Is the possibility of Brazil winning greater, or is the possibility of the pointer stopping in the black area greater?" If the possibility of Brazil winning is considered greater, imagine a larger black area, otherwise, imagine a smaller black area, then answer the same question again, repeating this until the assessor cannot distinguish which is greater, the possibility of Brazil winning or the pointer stopping in the black area. At this

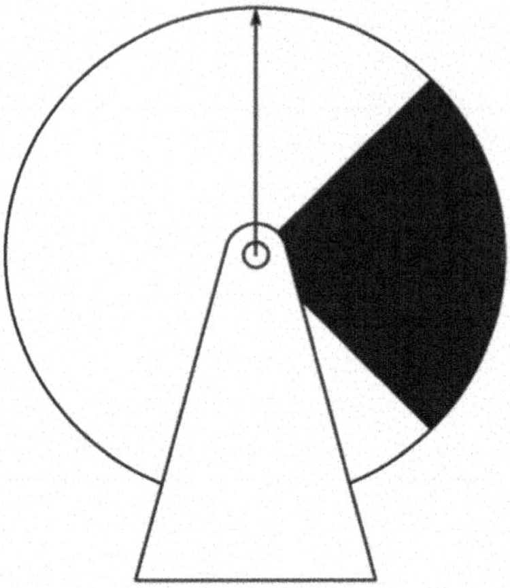

Fig. 4.14 Probability wheel used for subjective probability assessment

point, measure the central angle occupied by the black area and divide by 360° to get the subjective probability of Brazil winning.

The use of the probability wheel for subjective probability assessment is widely used in management science, psychology, and operations research. It should be noted that due to the limited resolution of human subjective concepts, the accuracy of subjective probability assessment is often limited, for example, there is almost no difference in human subjective concepts between an event happening with a probability of 0.4 or 0.401. Fortunately, this is often not a big problem in the application of Bayesian networks, for three reasons: (a) Small differences in probability values generally do not have a large impact on decision-making; (b) In practical applications, it is generally necessary to consider the probabilities of multiple events at the same time, and since probabilities must satisfy the Kolmogorov axioms, there are certain relationships between the probabilities of different events, and these relationships limit the arbitrariness of subjective probabilities; (c) In data analysis, when the amount of data is large enough, the impact of subjective probability is not significant.

For the above reasons, the conditional probability of event-event (E-E) connections in Bayesian networks can be directly assigned based on the concept of subjective probability assessment, using a combination of qualitative and quantitative evaluations. Generally speaking, it is relatively difficult to directly give the quantitative probability value for the conditional probability of one event triggering another event, but it is relatively easy to give a relatively vague qualitative evaluation. Based on psychological principles and the nature of probability, dividing the E-E connection conditional probability into "rarely", "occasionally", "likely", "frequently", "inevitably" five levels is reasonable and easy to assess. The values of the E-E connection conditional probability corresponding to each level of evaluation can refer to Table 4.9.

Table 4.9 Event-event (E-E) connection condition probability evaluation table

Level	Correlation interval	Descriptions
1	(0.9, 1.0]	Inevitably
2	(0.6, 0.9]	Easily
3	(0.3, 0.6]	Ably
4	(0.1, 0.3]	Occasionally
5	[0, 0.1]	Rarely

② Assignment of state-event connection conditional probability

A state node represents the status of a specific existing entity or abstract concept, and defining a state node with n possible states requires n − 1 independent probability parameters. Each state of the state node corresponds to a certain qualitative evaluation, such as "good", "bad", "dangerous", "beneficial", "unfavorable", etc. Although the number of states and qualitative evaluation descriptions of different state nodes may vary, we can always sort these qualitative descriptions according to their contribution to system risk, for example, "poor geological conditions", "large project scale", "high project complexity" are obviously more unfavorable states than "good geological conditions", "small project scale", "low project complexity". After sorting, the qualitative descriptions of each state can be abstracted from the model, and a random variable S can represent its different states. For the sake of uniformity, in the Bayesian network of urban large underground space construction risk, the value of the state node is defined as the natural number S = 1, 2, …, n, and it is stipulated that 1 represents the state node in the state that has the most unfavorable impact on system risk, the larger the value, the more beneficial the impact.

In the risk system model of urban large underground space construction, the occurrence probability of factor events is determined by the level of macro state indicators. Therefore, in the corresponding Bayesian network, it is also necessary to establish a CPT between the factor event node and the macro state indicator node to calculate the occurrence probability of factor events. Since event nodes are all binary, and macro state indicator nodes are generally multi-state, the number of conditional probabilities required to connect the macro state indicator layer and the factor event layer will be extremely large, and it is often unrealistic to determine these conditional probabilities one by one. To simplify decision logic, a method is needed to establish a conditional probability table of factor event nodes with a single indicator.

Since the macro state indicator is an abstract concept that characterizes the state level of a certain aspect of the system, the following assumptions can be made.

a. The impact of macro state indicators on the occurrence probability of factor events is related to the "contribution" of the indicator to the event. The higher the contribution, the greater the corresponding occurrence probability of factor events.
b. The contribution of macro state indicators is determined by the "basic contribution" and "connection strength". The basic contribution is uniform for all macro state indicators and is unrelated to the connection between the indicator and

the event; while the connection strength varies with the strength of the causal relationship between the macro state indicator and the factor event.

Under the above assumptions, suppose that the factor event node E has n related macro state indicator nodes $\vec{S} = S_i$, i = 1, 2,..., n, then when these macro state indicator nodes take values $\vec{x} = x_i$, i = 1, 2,..., n, the conditional probability of the occurrence of factor events can be expressed in the following form.

$$p_{\bar{x}}^{E\bar{S}} = P\left(E \middle| \vec{S} = \vec{x}\right) = f\left[\gamma_{S_i}^{E} \kappa_S(x_i)\right], i = 1, 2, \cdots, n \tag{4.28}$$

In the formula, $\vec{S}$ is a vector composed of random variables corresponding to all macro state indicator nodes related to event E; $\vec{x}$ is a vector composed of the values of $\vec{S}$; $\gamma_{S_i}^{E}$ is the state-event connection strength of the *i*th macro state indicator node related to event E; $\kappa_S(x_i)$ is the basic contribution function of the macro state indicator node; n is the number of macro state nodes related to event E.

State-event connection strength $\gamma_{S_i}^{E}$ It is a parameter that measures the strength of the correlation between macro state indicator nodes and factor event nodes. The connection strength between state and event can be taken from Table 4.10.

The basic contribution function of macro state indicator nodes $\kappa_S(x_i)$ is defined in the range [−1, 1]. According to the definition of the state node value mentioned above, when the ith macro state indicator node takes the value x_i, its basic contribution $\kappa_S(x_i)$ is calculated as follows:

$$\kappa_S(x_i) = \kappa_{S_i} = \frac{m_i - x_i + 1}{m_i} \tag{4.29}$$

In the formula, m_i is the number of states of the ith macro state indicator node.

Based on the basic contribution and connection strength, the contribution of the ith macro state node to the probability of the occurrence of factor events in different states can be calculated, denoted as $\varepsilon_{S_i}^{E} = \gamma_{S_i}^{E} \kappa_{S_i}$. As can be seen from formula (4.29), the contribution of the macro state indicator $\varepsilon_{S_i}^{E}$ and its value x_i are linearly related, and the value of x_i is an arithmetic natural number sequence, which means that the different values of the state indicator correspond to $\varepsilon_{S_i}^{E}$ is an equidistant, linear relationship. In the risk inference model of urban large underground space construction, the probability distribution of the macro state indicator state is obtained through the fuzzy comprehensive evaluation method, which means that the value of the macro state indicator is the embodiment of the analyst's subjective judgment. According to

Table 4.10 State event connection strength evaluation table

Level	Connection strength interval	Descriptions
1	(0.7, 1.0]	Strong link
2	(0.4, 0.7]	Medium term
3	[0.1, 0.4]	Weak link

Weber-Fechner's law, human subjective experience judgment is proportional to the logarithm of the corresponding physical quantity, that is, when the judgment quantity is linearly increasing, the physical quantity is exponentially increasing. Therefore, the contribution of the state indicator to the possibility of event occurrence $\varepsilon_{S_i}^{E}$ should be exponentially related to the increase in the probability of event occurrence. For this reason, the probability conversion function $p^{\Gamma}\left(\varepsilon_{S_i}^{E}\right)$ is introduced to describe this relationship, $p^{\Gamma}\left(\varepsilon_{S_i}^{E}\right)$ should have the following form.

$$p^{\Gamma}\left(\varepsilon_{S_i}^{E}\right) = a^{b\cdot\varepsilon_{S_i}^{E}+c} \tag{4.30}$$

In the formula, a, b, c are parameters.

Since the value range of $\varepsilon_{S_i}^{E}$ is [0, 1], when a factor event node is only related to a macro state node, $p^{\Gamma}(0)$ and $p^{\Gamma}(1)$ they respectively correspond to the lower and upper limits of the probability of the occurrence of the factor event. After giving the upper and lower limits of this probability, there is still a lack of an equation to determine the parameters in formula (4.30), at this time, the method of giving the base a in advance can be adopted. Theoretically, the value of a can be any positive real number greater than 1. According to the common practice in risk assessment, it is generally assumed that the probability between adjacent risk levels is 10 times, so a = 10 can be taken, and then the parameters b and c can be solved.

After determining the connection strength and probability conversion function of all state-event connections on a certain factor event node, similar to event-event connections, it is assumed that the causal mechanisms of all state-event connections pointing to the same factor event node are independent, and the relationship between macro state nodes is logical "or", the specific expression of formula (4.28) can be obtained.

$$p_{\bar{x}}^{E|\bar{S}} = \sum_{i=1}^{n} p^{\Gamma}\left(\varepsilon_{S_i}^{E}\right) - \sum_{i\neq j} p^{\Gamma}\left(\varepsilon_{S_i}^{E}\right)\cdot p^{\Gamma}\left(\varepsilon_{S_j}^{E}\right) + \cdots + (-1)^{n-1}\prod_{i=1}^{n} p^{\Gamma}\left(\varepsilon_{S_i}^{E}\right) \tag{4.31}$$

Through formula (4.31), the CPT of the factor event node can be constructed.

(4) Selection of membership degree of fuzzy comprehensive evaluation

The meaning of the membership degree of the evaluation factor is: when this evaluation factor is the only factor, the membership degree of the evaluation target to the domain. The membership degree of the evaluation factor reflects the strength, level, and level of the evaluation factor itself for the evaluation target. For some qualitative indicators or indicators that are difficult to obtain actual samples, their membership degree can be calculated based on the traditional scoring rating method, introducing the concept of "confidence" in the state of the indicator. When using the confidence method, a rating (extremely low risk, low risk, …) and the confidence level of this judgment [very uncertain (HS), relatively uncertain (NS), certain (S), relatively certain (QS), very certain (VS)] need to be given for each indicator, thus giving the specific form of the membership function. The membership function of the confidence method can be constructed through the following steps.

Let the domain be U, $U = [u_0, u_1]$, F is a fuzzy set on U, then according to the concept of confidence, the conditions that F needs to meet are:

① $Supp(F) = U$
② $\forall\delta > 0, \forall u_i \in [u_0, u_1]$, make $Ker(F) \subset [u_i, u_i + \delta]$

The meaning of condition ② is that the core of F is a point u_i on the domain U. At this point, F's membership degree to U is 1, that is, if F is a classic set rather than a fuzzy set, $F = u_i$. u_i is F's traditional score on U.

Let f(u) be the membership function of F, then the above conditions can be expressed as:

① $0<f(u) \leq 1, u \in [u_0, u_1]$
② There is a unique point ui, so that $f(u_i) = 1$ and

Under the above condition constraints, construct the membership function of the following form:

$$f(u) = \begin{cases} A(u-s)^n, u \leq u_i, A > 0 \\ B(t-u)^n, u > u_i, B > 0 \end{cases}, n > 0, u \in [u_A, u_B] \supset [u_0, u_1] \tag{4.32}$$

In the formula, A, B, s, t are parameters to be determined.

According to the constraint conditions, the boundary conditions of formula (4.32) are:

$$\begin{cases} f(u_A) = f(u_B) = 0 \\ f(u_i) = 1 \end{cases} \tag{4.33}$$

Substitute formula (4.33) into formula (4.32) to get:

$$\begin{cases} s = u_A \\ t = u_B \\ A = (u_i - u_A)^{-n} \\ B = (u_B - u_i)^{-n} \end{cases} \tag{4.34}$$

Let

$$\begin{cases} u_A = u_0 - v \\ u_B = u_1 + v \end{cases}, v > 0 \tag{4.35}$$

And limit the definition domain of f(u) to the above, you can get the specific form of the membership function:

$$f(u) = \begin{cases} \left(\frac{u-u_0+v}{u_i-u_0+v}\right)^n, u \le u_i \\ \left(\frac{u_1+v-u}{u_1+v-u_i}\right)^n, u > u_i \end{cases} n > 0, v > 0, u \in [u_0, u_1] \tag{4.36}$$

In the formula, n and υ are parameters. n reflects the fuzziness of F. When the confidence levels are very unconfident (HS), not very confident (NS), confident (S), quite confident (QS), and very confident (VS), the values of n are 0.25, 0.5, 1, 2, and 4 respectively; the parameter υ determines the shape of the membership function near the boundary of the domain, which can generally be taken as 0.01–0.05.

When the domain U = [0, 1], and ui = 0.5, the graph of formula (4.36) is shown in Fig. 4.15 (υ = 0.01).

From the figure, it can be seen that when the confidence level is low, the values of the membership function of formula (4.36) are relatively large across the entire domain, which violates the general logic rule and can easily lead to an "over-fuzzy" phenomenon when used for evaluation. Therefore, it is necessary to further modify the above membership function and introduce a correction factor α.

$$f'(u) = \alpha f(u) \tag{4.37}$$

The significance of the correction factor α is that when the confidence level is low, it adjusts the membership degree as a whole. Therefore, it can be defined as:

$$\alpha = \begin{pmatrix} 1 - \int_0^{u_i} A(u - u_0 + v)^n du - \int_{u_i}^1 B(u_1 + v - u)^n du + \\ \int_0^{u_i} A(u - u_0 + v)^4 du + \int_{u_i}^1 B(u_1 + v - u)^4 du \end{pmatrix}^{-k} \tag{4.38}$$

In the formula, k is a parameter, which can generally be taken as 2–3. A smaller k value represents a larger adjustment range. α can be obtained through numerical

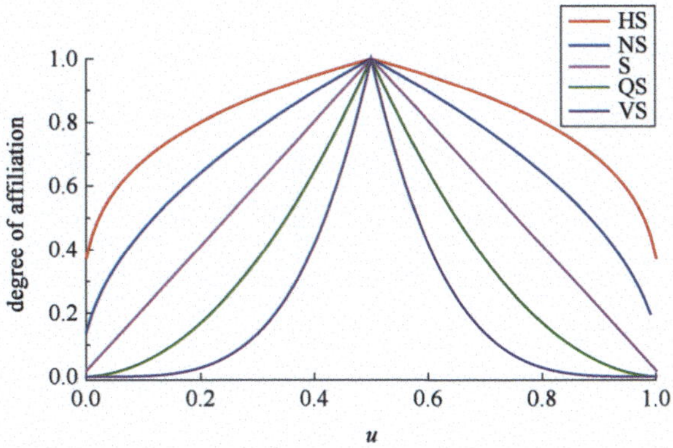

Fig. 4.15 Membership function graph (υ = 0.01)

integration. The values of α when k = 2 and k = 3 are shown in Table 4.11 (υ = 0.01).

The graph of the modified membership function (υ = 0.01, k = 3) is shown in Figures 4.16, 4.17, 4.18, 4.19 and 4.20.

3. Sub-model of risk measurement for urban large underground space construction

Construction risk measurement refers to the process of monitoring and measuring the mechanical responses of engineering structures, surrounding strata, and surrounding environmental facilities after the start of construction, collecting data from these measured indicators, and using the construction risk system model to measure the

Table 4.11 The value of parameter α(υ = 0.01)

Degree of certainty	HS	NS	S	QS	VS
k = 2	0.628	0.725	0.833	0.93	1
k = 3	0.733	0.807	0.885	0.952	1

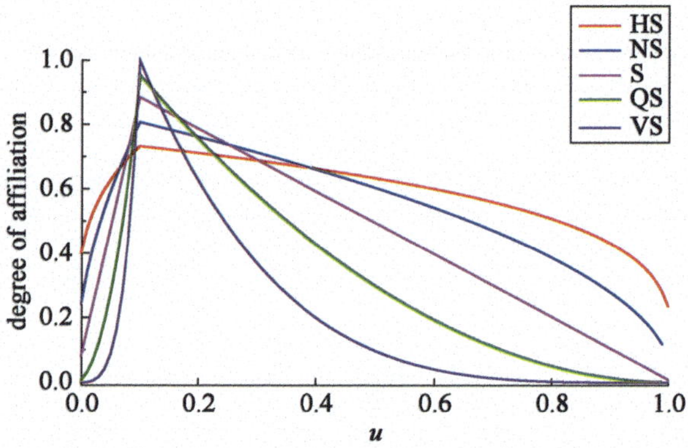

Fig. 4.16 Modified membership function graph ($u_i = 0.1$)

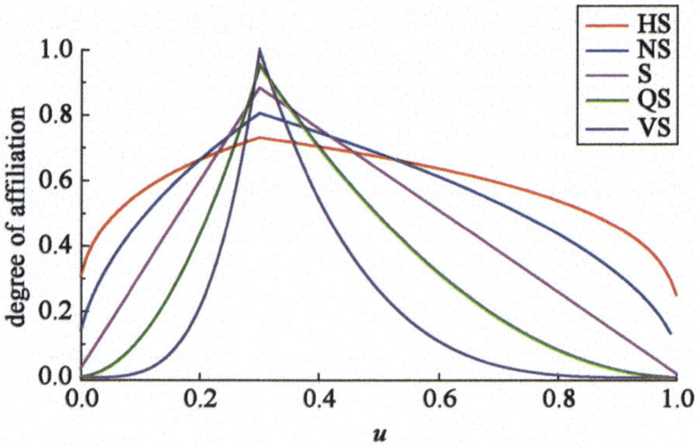

Fig. 4.17 Modified membership function graph ($u_i = 0.3$)

Fig. 4.18 Modified membership function graph ($u_i = 0.5$)

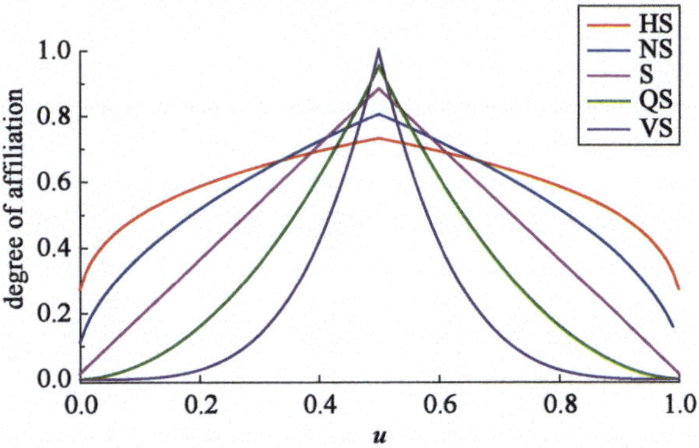

Fig. 4.19 Modified membership function graph ($u_i = 0.7$)

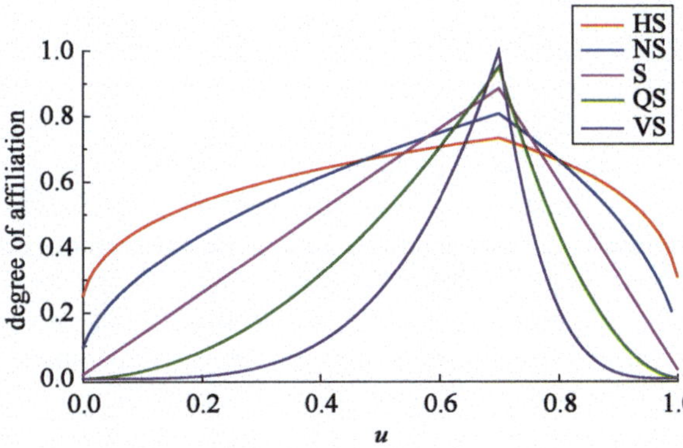

Fig. 4.20 Modified membership function graph ($u_i = 0.9$)

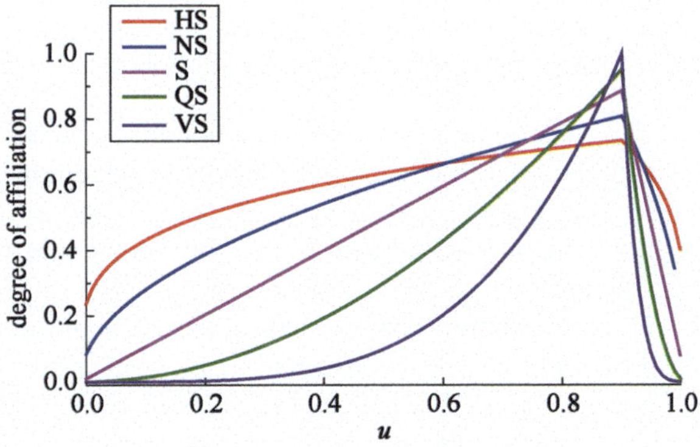

risk level in real time, in order to verify and correct the results of risk assessment. Currently, monitoring and measurement are generally required during the construction process of underground projects, and using the results of monitoring and measurement for risk warning is a most basic risk measurement process. However, current underground engineering monitoring and risk warning schemes often have strong subjectivity and blindness. Underground engineering is a complex whole, sometimes a single measurement point reaching the alarm state does not necessarily mean a surge in overall risk, and sometimes even if all measurement points are within the normal range, the seeds of risk occurrence have been quietly planted. This risk measurement method, which overly focuses on individual indicators and measurement points and lacks a global engineering thinking, undoubtedly has significant drawbacks. Therefore, a more scientific and systematic underground engineering risk measurement method needs to be proposed.

(1) Explanation of "system failure" in construction risk

Before undertaking a certain action, the actor will make a preliminary prediction of the expected purpose, implementation path, and operation method of the activity under the background of existing knowledge, technology, environment, and other factors, to ensure the feasibility of the activity. In other words, in the process of action, if all factors are as expected by the actor, the action will inevitably achieve its expected purpose, and the action will inevitably succeed. However, there is uncertainty in everything. Under the influence of these uncertainties, the results caused by some links in the action will deviate from the expectation, that is, a "misoperation" or "failure" occurs in a certain link. If the adverse effects caused by this misoperation exceed a certain threshold, it will cause the results caused by the action itself to deviate significantly from the expectation. For the action of underground engineering construction, among the many possible results that deviate from expectations, some results will cause loss of life to the construction entity or third party, as well as accompanying huge losses of property, reputation, environment, etc., this type of possible result is a construction risk event. Undoubtedly, once a construction risk event occurs, it means that there has been some degree of failure in the construction process. Based on the above analysis, we can make the following explanations for the construction process and construction risks.

① The construction process is an action that includes a series of construction steps and links to achieve the use function of the engineering structure.

 The construction process can be likened to a "machine" with a specific function in a broad sense, and this broad machine is called a "construction system".

② The construction system needs to ensure safety in the process of realizing structural functions. The construction of a project can be understood as the construction system is in a normal operating state if it is completed smoothly without any accidents; the occurrence of construction risk events can be explained as the failure of the construction system, leading to significantly deviated results from expectations, and these results may not be unique.

③ The occurrence of construction risk events (failure of the construction system) originates from the "misoperation" or "failure" of some links in the system, generally speaking, this is a small probability event.

④ The failure of a single link does not necessarily lead to the occurrence of construction risk events. The failure of the construction system is often the result of the cumulative adverse effects caused by the failure of several links.

Because construction risk has the above characteristics, we can explain construction risk from the perspective of system failure. The prototype of system failure can be seen everywhere in the objective world, such as machine stoppage, computer crash, filament fuse, and even human diseases, can be regarded as an instance of system failure.

(2) Model elements and structure

According to the above, the measurement of construction risk is the process of measuring objective risk through the information actually observed. However, for such an unknown, uncertain, and abstract concept as risk, the information that can be obtained in practice is extremely limited. In the construction risk inference sub-model, the risk value is obtained by evaluating the evaluation indicators and through a series of subjective evaluations and logical deductions. In this process, the status of the evaluation indicators is undoubtedly observable information, but the series of processes that transform into risk values are highly subjective, and the risk values obtained in this way are inevitably subjective. Therefore, the construction risk measurement model must cross this subjective barrier and use more direct and more objective information as model input.

From the analysis of the limitations of construction risk measurement in the first section of this chapter, it can be seen that in the actual construction process, the objective information that people can intuitively obtain is mostly monitoring indicators such as structural deformation and stress. The causal relationship between the size of these monitoring indicators and risk events is direct and objective. Following this concept, the input of the sub-model for measuring the risk of urban large underground space construction should be the intermediate event layer and the sign event layer in the construction system model, and the status of elements below the factor event layer is difficult to obtain through observation. On the other hand, according to the system failure interpretation of construction risk, we can liken the construction process of urban large underground space to a "closed circuit", thereby measuring construction risk by calculating the occurrence probability of intermediate events.

Through the above analysis, the sub-model for measuring the risk of urban large underground space construction can be expanded based on the construction mechanics sub-model, thereby constructing the model shown in Fig. 4.21. The meanings of each part of the model are as follows.

① Construction system

The construction system is an abstraction of the construction actions carried out to achieve structural functions. The lines extending from both sides of the system represent the working circuit of the system. When the circuit is closed, the system is in a normal state, otherwise, it is in a failure state. In this model, the three things or events of "system", "system function", and "system cannot complete function

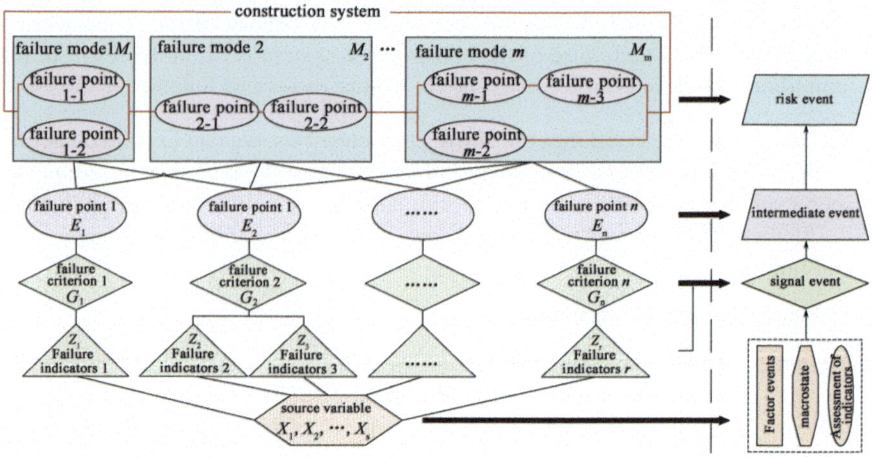

Fig. 4.21 Urban large underground space construction measurement sub-model

(system failure)" are three-in-one. That is to say, the connotation of the system can correspond to a certain engineering structure, a certain construction action, a certain functional goal, or a specific event.

The extension of the system can change with the focus of the modeler. It can build a system for the entire project, or it can regard a specific risk event in the project as a system.

② Failure mode (M)

As the name suggests, the failure mode is a mode in which the system fails. A system can only have a single failure mode, such as the failure mode of a concrete column under axial compression is only shear failure; but more general systems often have multiple failure modes, such as the failure modes of gravity retaining walls are tipping and sliding. When the failure mode is in the "not appeared" state, the circuit connecting its two ends is in a conducting state, and when the failure mode "appears", the connected circuit is disconnected. The failure mode can be understood as a functional module that constitutes the system, and by default, this module is in a normal working state. The condition for the system to work normally is that all functional modules are working normally, so the failure modes are in series. When any failure mode occurs, the system declares failure, and at this time, there is no need to consider whether other failure modes will occur.

In the urban large underground space risk system, the failure mode can be understood as a specific risk event. When a safety accident occurs, the construction party generally needs to suspend construction and carry out self-inspection and rectification. During the rectification process, there will inevitably be human interference with the original risk system, so even if construction is resumed in the future, the risk system at this time is no longer the original risk system. From the perspective of risk, the construction action after resumption of work also becomes not the original construction action due to its new characteristics.

Therefore, the suspension of construction caused by safety accidents meets the definition of system failure, and the corresponding risk events also meet the definition of failure mode.

③ Failure point (E)

A failure point is a sub-event included in each failure mode, equivalent to the bottom risk event in the risk accident tree. Like failure modes, failure points can also be seen as a switch component. A failure mode contains at least one failure point, and any series, parallel, or mixed connections can be formed between the failure points. All the failure points in a failure mode constitute the internal circuit of the failure mode, and the on-off of the internal circuit determines whether the failure mode occurs. Multiple failure modes can share the same failure point, and a failure point can belong to multiple failure modes simultaneously.

In the model of urban large underground space construction risk system, the corresponding element of the failure point is the intermediate event. Intermediate events are generally the inability of a part or some components in the engineering structure to meet functional requirements or damage. In some cases, the damage of a single component can lead to the overall damage of the structure or the structure's inability to meet functional requirements, but generally, the conditions for triggering structural failure will be more complex, such as the collapse of a foundation pit often occurs when multiple conditions such as base damage and internal support failure are met at the same time. By reasonably designing the failure points and their structural relationships, the occurrence conditions and logical relationships of various risk events can be depicted.

④ Failure index (Z)

The failure index is a random variable related to the failure point. With the help of the failure index, the probability of the occurrence of the failure point can be quantitatively calculated p^{fE}. The failure index should meet the following conditions.

a. The failure index should generally be a physical quantity that has a close logical connection with the failure point. For example, for the failure point of "excessive soil pressure", the corresponding failure index should be "soil pressure".
b. The failure index should be a physical quantity that can be obtained by measurement, evaluation, calculation, and other methods.
c. The failure index should not be a probability measure.

If the failure point E_j corresponds to a failure index, then this index is denoted as Z_j, and the specific value of the index is denoted as z_j. Z_j The probability density function is denoted as $f_{Z_j}(z)$; if E_j corresponds to s indices, then each index is denoted as, and the joint probability density function of s indices is denoted as.

⑤ Failure criterion (G)

Similar to the role of the function in reliability theory, the failure criterion is the bridge connecting the failure index and the failure point. The failure criterion defines a subset about the failure index definition space, thereby dividing the value range of the failure index into the "safety domain" and the "failure domain". Through the failure criterion, the occurrence of the failure point can be judged based on the value of the failure index. Generally speaking, a failure point corresponds to a failure index, and the failure criterion at this time can generally be of the following form.

$$G_i = \begin{cases} \left(-\infty, z_i^f\right)或\left(z_i^f, \infty\right) \\ \left(z_i^f - \delta_i^f, z_i^f + \delta_i^f\right) \\ \left(-\infty, z_i^f - \delta_i^f\right) \cup \left(z_i^f + \delta_i^f, \infty\right) \end{cases} \tag{4.39}$$

In the formula, z_i^f and δ_i^f are respectively called the failure index Z_i of failure threshold and failure tolerance. The event of "failure point occurrence" can be expressed as:

$$E_i = z_i \in G_i = \begin{cases} z_i < z_i^f, G_i = \left(-\infty, z_i^f\right), \\ z_i > z_i^f, G_i = \left(z_i^f, \infty\right), \\ z_i^f - \delta_i^f < z_i < z_i^f + \delta_i^f, G_i = \left(z_i^f - \delta_i^f, z_i^f + \delta_i^f\right) \\ \left(z_i < z_i^f - \delta_i^f\right) \cup \left(z_i > z_i^f + \delta_i^f\right), G_i = \left(-\infty, z_i^f - \delta_i^f\right) \cup \left(z_i^f + \delta_i^f, \infty\right) \end{cases} \tag{4.40}$$

When a failure point corresponds to multiple failure indicators, the failure criterion is defined as a region in a multidimensional space of the values of each failure indicator. The failure criterion can be obtained through a finite number of intersection and union operations based on the relationship between the failure indicators.

Failure indicators and failure criteria are the specific implementation of precursor events in the risk system model of urban large underground space construction. The failure indicator defines specific measurement standards and measurement methods for precursor events, and the failure criterion completes the transformation between the quantitative description of precursor events and the qualitative description of intermediate events. Through the multi-layered connection between precursor events (failure indicators and failure criteria), intermediate events (failure points), and risk events (failure modes), the system risk can be measured by measuring precursor events.

⑥ Source variables (X)

Source variables are the fundamental variables that govern the operation of the system and are the root of the system's uncertainty. Source variables are also random variables, but unlike failure indicators, source variables may be "hidden variables" that are difficult to measure, calculate, or even discover.

Failure indicators are essentially functions of source variables, and there is generally a many-to-many relationship between source variables and failure indicators. For example, for the two failure points of "insufficient strength of the retaining structure" and "insufficient stiffness of the retaining structure", their common source variables may include "form of the retaining structure", "thickness of the retaining structure", etc., but obviously, they are governed by more known or unknown other source variables.

Source variables can be seen as the input of the safety risk model, and their status is equivalent to the collection of factor events and state indicators in the risk system model of urban large underground space construction. However, the difference between the two is that factor events and state indicators focus on the logical analysis of the risk system, and their significance is to make the risk analysis logic clear, and their essence is still the subjective concepts and experience of the analyst.

Although evaluation indicators can also be objectively measured, due to the inevitable introduction of a large number of human definitions and judgments in the macro indicator layer and factor event layer, a kind of "blocking effect" on objectivity has been formed, making precursor events the lowest level elements that can be measured in the model when measuring risk objectively. Therefore, in the risk measurement sub-model of urban large underground space construction, these factors that cannot be objectively measured are collectively referred to as source variables. In the actual risk system, source variables are often difficult to exhaust, and the relationship between source variables and model output is extremely complex, so the model mainly emphasizes their conceptual status.

(3) Basic theories and calculation methods

① System failure probability calculation method

Similar to the accident tree, the above safety system failure model is also a reflection of the logical relationship between a series of events. In fact, the conceptual model defined in the above way is a special case of the accident tree (the simple accident tree where there must be an "or gate" under the top event), and it is also applicable to the analysis method of the accident tree. Let the events of "system failure", "the occurrence of the ith failure mode", and "the occurrence of the jth failure point" be "F", "M_i" and "E_j", and the occurrence probabilities of the corresponding events are "p^f" "p_i^{fM}" and "p_i^{fE}", then the following formula can be derived according to the operation rules of events:

$$
\begin{aligned}
p^f &= P\left(\bigcup_{i=1}^{m} F_i^{fM}\right) = \breve{p}^{fM}_{i1,i2,\cdots im} \\
&= \sum_{i=1}^{m} p_i^{fM} - \sum_{i\neq j} \hat{p}_{i,j}^{fM} + \sum_{i\neq j\neq k} \hat{p}_{i,j,k}^{fM} - \cdots + (-1)^{m-1}\hat{p}^{fM}_{i1,i2,\cdots im}
\end{aligned}
\tag{4.41}
$$

where, $\breve{p}_{i_1,i_2,\cdots i_m}$ For $p_{i1}, p_{i2}, \cdots, p_{im}$ the sum of the probabilities of corresponding events and the occurrence of events, is called the m-order total probability, defined as:

$$
\breve{p}_{i_1,i_2,\cdots i_m} = P(F_i \cup F_j \cup \cdots \cup F_m), P(F_k) = p_k \tag{4.42}
$$

For the product of the probabilities of corresponding events, it is called the m-order joint probability, defined as:

$$
\hat{p}_{i_1,i_2,\ldots,i_m} = P(F_i F_j \cdots F_m), P(F_k) = p_k \tag{4.43}
$$

According to the definition of failure mode, once a failure mode occurs, other failure modes will not occur, so the failure modes are mutually exclusive, hence:

$$
\hat{p}_{i,j}^{fM} = \hat{p}_{i,j,k}^{fM} = \cdots \hat{p}^{fM}_{i_1,i_2\cdots i_m} = 0 \tag{4.44}
$$

Therefore:

$$
p^f = \sum_{i=1}^{m} p_i^{fM} \tag{4.45}
$$

That is, the failure probability of the system equals the sum of the probabilities of each failure mode. The above analysis shows that the key to system safety risk analysis is to find out the occurrence probability of each failure mode p_i^{fM}. Similar to the relationship between the system as a whole and the failure modes, the p_j^{fM} occurrence probability of the failure modes can also be calculated from the occurrence probability of the failure points contained in it. However, due to the possible complex mixed connection structure between the failure points and the non-mutual exclusion of the occurrence of each failure point, the calculation of the occurrence probability of the failure mode p_i^{fM} is more complex. For a failure mode n_i that contains M_i serial failure points, we have:

$$
\begin{aligned}
\breve{p}_i^{fM(n_i)} &= P(F_i^M) = P\left(\bigcup_{j=1}^{n_i} F_j^E\right) \\
&= \sum_{j=1}^{n_i} p_j^{fE} - \sum_{j\neq k} \hat{p}_{j,k}^{fE} + \sum_{j\neq k\neq l} \hat{p}_{j,k,l}^{fE} - \cdots + (-1)^{n_i-1}\hat{p}^{fE}_{j_1,j_2,\cdots j_{n_i}}
\end{aligned}
\tag{4.46}
$$

In the formula, $\overset{\smile}{p}_i^{fM\,(n_i)}$ is the probability of the occurrence of the ith failure mode; the subscript "$\cup$)" indicates that this failure mode is a serial system, and the superscript (n_i) indicates that it contains n_i failure points. Assuming that the occurrence of each failure point is mutually independent, formula (4.46) can be further simplified by the independence condition to:

$$\overset{\smile}{p}_i^{fM\,(n_i)} = \sum_{j=1}^{n_i} p_j^{fE} - \sum_{j\neq k} p_j^{fE} p_k^{fE} + \sum_{j\neq k\neq l} p_j^{fE} p_k^{fE} p_l^{fE} - \cdots + (-1)^{n_i-1} p_{j_1}^{fE} p_{j_2}^{fE} \cdots p_{j_{n_i}}^{fE} \tag{4.47}$$

For the failure mode M_i that contains n_i parallel failure points, we have:

$$\overset{\frown}{p}_i^{fM\,(n_i)} = P\left(\bigcap_{j=1}^{n_i} F_j^E\right) = \hat{p}_{j_1,j_2,\cdots j_{n_i}}^{fE} = p_{j_1|j_2,j_3,\cdots,j_{n_i}}^{fE} p_{j_2|j_3,j_4,\cdots,j_{n_i}}^{fE} p_{j_3|j_4,j_5,\cdots,j_{n_i}}^{fE} \cdots p_{j_{n-1}|j_{n_i}}^{fE} p_{j n_i}^{fE} \tag{4.48}$$

In the formula, the subscript "$\dot{\cap}$" indicates that this failure mode is a parallel system; $p_{a|b,c}$ represents the conditional probability of the event corresponding to p_a under the condition that the product events corresponding to p_b and p_c occur, that is:

$$p_{a|b,c} = P(a|bc), P(a) = p_a, P(b) = p_b, P(c) = p_c \tag{4.49}$$

When the independence assumption holds, formula (4.48) becomes:

$$\hat{p}_i^{fM\,(n_i)} = \prod_{j=1}^{n_i} p_j^{fE} \tag{4.50}$$

Formulas (4.46) and (4.47), as well as formulas (4.48) and (4.50), respectively represent the calculation formulas for the occurrence probabilities of serial and parallel failure modes under general conditions and under conditions where failure points are independent. For the mixed connection situation, it can be regarded as a nesting of several serial and parallel subsystems, and the occurrence probabilities of the corresponding failure modes can be calculated by nested application of the above formulas.

Through the above analysis, the calculation of system failure probability is ultimately transformed into the problem of solving the occurrence probabilities of each failure point in the system and their joint probabilities. However, in general, it is difficult to accurately give the occurrence probability of failure points, and can only give a rough range through estimation, evaluation and other methods. Although this practice is very common in the risk assessment work of actual engineering, it is ultimately a compromise under the constraints of limited information and complex operations, and its calculation results are too rough to meet the requirements of real quantitative risk assessment. With the help of the safety system failure model, by establishing the model structure and selecting appropriate failure indicators and failure criteria, the occurrence probability of failure points can be calculated by the following formula:

$$p_i^{fE} = \int\limits_{Z_i \in G_i} f_{Z_i}(z)dz \tag{4.51}$$

In the formula, f_{Z_i} is the probability density function of the failure indicator Z_i. For the failure conditions represented by formulas (4.39) and (4.40) $z_i \in G_i$, a function can always be found to make equivalent to, for example:

$$g_i\left(z_i, z_i^f, \delta_i^f\right) = \begin{cases} z_i \in G_i g_i\left(z_i, z_i^f, \delta_i^f\right) < 0 \\ z_i - z_i^f, G_i = \left(-\infty, z_i^f\right) \\ z_i^f - z_i, G_i = \left(z_i^f, \infty\right) \\ \max\left\{\left(z_i - z_i^f - \delta_i^f\right), \left(z_i^f - \delta_i^f - z_i\right)\right\}, G_i = \left(z_i^f - \delta_i^f, z_i^f + \delta_i^f\right) \\ \min\left\{\left(z_i - z_i^f - \delta_i^f\right), \left(z_i^f - \delta_i^f - z_i\right)\right\}, G_i = \left(-\infty, z_i^f - \delta_i^f\right) \cup \left(z_i^f + \delta_i^f, \infty\right) \end{cases} \tag{4.52}$$

In this way, formula (4.51) can be further transformed into:

$$p_i^{fE} = \int\limits_{z_i \in G_i} f_{Z_i}(z)dz = \int\limits_{g_i(z_i)<0} f_{Z_i}(z)dz = \int\limits_{\Omega_{z_i}} I_i^{fE}(z) f_{Z_i}(z)dz \tag{4.53}$$

In the formula, Ω_{Z_i} is the definition space of Z_i; $I_i^{fE}(z)$ is the failure indicator function of failure point E_i, which means:

$$I_i^{fE}(z) = \begin{cases} 1, g_i(z) < 0 \\ 0, g_i(z) \geq 0 \end{cases} \tag{4.54}$$

On the other hand, since Z_i is a function of the source variable X, so g_i is also a function of the source variable X. In this way, g_i essentially has the form and connotation of the function in reliability theory. If the model input is X (X_1, X_2, ..., X_r), and its joint probability density function is $f_x(X)$, then the failure probability of the ith failure point is:

$$p_i^{fE} = P(E_i) = \int\limits_{g_i<0} f_X(X)dX = \int\limits_{R^n} I_i^{fE} f_X(X)dX \tag{4.55}$$

Formula (4.55) shows that the occurrence probability of failure points can be obtained by integrating the joint probability density function of source variables in the failure domain.

② Calculation method of failure point occurrence probability based on structural reliability theory

From the above analysis, it can be seen that after defining the failure mode and failure point, the probability of the occurrence of risk events can be calculated by solving equation (4.55). Observing equations (4.51) to (4.55), it can be found that although the background and concepts of the above theories are different, the core ideas are very similar to the theory of engineering structure reliability. Therefore, when the probability distribution of the source variables in the model is known, and there is a clear mechanistic relationship between the failure index and the source variables, the probability of the occurrence of failure points can be calculated with the help of structural reliability theory.

a. Introduction to Structural Reliability Theory

Structural reliability theory is the application of probability statistics in structural safety design. In 1947, FREUDENTHAL of the United States discussed structural safety issues and first proposed the mathematical model analysis method of structural reliability. In 1954, Erzhanichev of the Soviet Union proposed the basic theory of first and second moment calculation of reliability and gave the corresponding formula. Later in 1969, CORNELL of the United States established the second moment model of structural reliability and linked the concept of reliability index with the failure probability of the structure. The theory of structural reliability calculation was initially mature at this time. The core idea of structural reliability theory is to introduce randomness into the mechanical calculation of structures. Objectively speaking, there are many uncertainties in the mechanical analysis of structures, such as building materials, loads, and even the geometric dimensions of the structure, which cannot be completely determined. Traditional mechanical analysis ignores randomness, which makes the analysis results theoretically only "valid with a certain probability", but cannot give the specific value of this "certain probability". Structural reliability theory considers these uncertainties and replaces deterministic analysis with uncertainty analysis, thus giving the probability of "safety" or "failure" of the structure under certain conditions.

Structural reliability theory believes that the function of the structure exists in two states, "reliable" and "failure", and the boundary between these two states is the limit state of the structure. The function of the structure can be expressed in the form of an equation:

$$Z = g_X(X_1, X_2, \cdots, X_n) \tag{4.56}$$

In the formula, $g(\cdot)$ is the function of the structure; $X_i(i = 1, 2, \cdots, n)$ is the random variable of structural design. The relationship between the state of the structure and the function is:

$$\begin{cases} Z > 0, \text{The structure is in a reliable state} \\ Z = 0, \text{The structure is in the limit state} \\ Z < 0, \text{The structure is in a failed state} \end{cases} \tag{4.57}$$

The function of the structure defines a hyper-surface in the n-dimensional space about the random variable X_i (I = 1, 2, ..., n), which divides the basic parameter space of the structure into a safe domain (Z > 0) and a failure domain (Z < 0). If the joint probability density function of the random variable X_i (I = 1, 2, ..., n) is $f_x(x_1, x_2, \ldots, x_n)$, and the probability density of the function is $f_z(z)$, then the reliability and failure probability of the structure can be expressed as:

$$\begin{cases} P_s = P(Z > 0) = \int\limits_{Z>0} f_X(x_1, x_2, \cdots, x_n) dx_1 dx_2 \cdots dx_n = \int\limits_{0}^{\infty} f_Z(z) dz \\ P_f = P(Z < 0) = \int\limits_{Z<0} f_X(x_1, x_2, \cdots, x_n) dx_1 dx_2 \cdots dx_n = \int\limits_{-\infty}^{0} f_Z(z) dz \end{cases} \tag{4.58}$$

When the function Z follows a normal distribution with parameters (μ_Z, σ_Z), the failure probability can be expressed as:

$$P_f = \int\limits_{-\infty}^{\frac{-\mu_z}{\sigma_z}} \phi(t) dt = \Phi\left(-\frac{\mu_z}{\sigma_z}\right) \tag{4.59}$$

In the formula, is the probability density function of the standard normal random variable; is the standard normal distribution function. Let the reliability index be, then the relationship between the failure probability and the reliability index is:

$$P_f = \Phi\left(-\frac{\mu_z}{\sigma_z}\right) = \Phi(-\beta) \tag{4.60}$$

From this, it can be seen that there is a one-to-one correspondence between the reliability index and the failure probability.

b. Reliability Calculation Method

Suppose the function is, is the design variable of the structure, and its mean and standard deviation are respectively. The function is Taylor expanded at the mean value, and only the first term is retained for linearization, then the approximate expression of the function is:

$$Z \approx g_X\left(\mu_{X_1}, \mu_{X_2}, \cdots, \mu_{X_n}\right) + \sum_{i=1}^{n} \frac{\partial g_X}{\partial X_i}\Big|_{\mu X_i} \left(X_i - \mu_{X_i}\right) \tag{4.61}$$

When all random variables are independent, the mean and standard deviation of the function are:

$$\begin{cases} \mu_Z \approx g_X\left(\mu_{X_1}, \mu_{X_2}, \cdots, \mu_{X_n}\right) \\ \sigma_Z \approx \sqrt{\sum_{i=1}^{n}\left(\frac{\partial g_X}{\partial X_i}\Big|_{\mu_{X_i}} \sigma_{X_i}\right)^2} \end{cases} \tag{4.62}$$

The definition of the reliability index can be obtained as follows:

$$\beta = \frac{\mu_Z}{\sigma_Z} \approx \frac{g_X\left(\mu_{X_1}, \mu_{X_2}, \cdots, \mu_{X_n}\right)}{\sigma_Z \approx \sqrt{\sum_{i=1}^{n}\left(\frac{\partial g_X}{\partial X_i}\Big|_{\mu_{X_i}} \sigma_{X_i}\right)^2}} \tag{4.63}$$

When the problem to be solved cannot obtain an explicit function due to boundary conditions or material characteristics, the reliability can be calculated using the response surface method. The Response Surface Methodology (RSM) was first proposed by BOX and WILSON in 1951. With the rapid development of computer technology and the application of finite element analysis software, the response surface method plays a pivotal role in the calculation of the reliability of current engineering structures. The response surface method is used to solve the problem of difficult to obtain an analytical form of the function under complex problem boundaries. The response surface method uses a fitted explicit function to replace the real function g(X), making the calculation of reliability possible.

In traditional response surface methods, the fitted explicit function is generally chosen as a second-order polynomial model:

$$\hat{g}(X) = a_0 + \sum_{i=1}^{n} a_i X_i + \sum_{i=1}^{n} a_{ii} X_i^2 + \sum_{1 \le i \le j \le n} a_{ij} X_i X_j \tag{4.64}$$

In the formula $X = (X_1, X_{12}, \ldots, X_n)$, n is the number of random variables; are unknown coefficients, the number of which are n, n, and respectively.

Let the elements of vector A be composed of unknown coefficients, represented as:

$$A = \left(a_0, a_1, \cdots, a_n, a_{11}, a_{22}, \cdots, a_{nn}, a_{12}, a_{13}, \cdots, a_{(n-1)n}\right) \tag{4.65}$$

Obviously, vector A contains elements. To obtain the definite expression of the function, N sample points (N is not less than $C_n^2 + 2n + 1$) need to be selected, calculate the corresponding structural function values of the sample points, thereby constructing N linear equations, and then obtaining the values of each element in vector A. The sample points are generally selected near the center point through the experimental design method, which can generally be taken as the mean point of the base variable. Then the following central composite experimental design method can be used for sampling:

Set the center point as, for each random variable, two level values are taken near, that is, the f value is usually taken as 1. Then, the level values of each random variable are designed for full factorial design, resulting in 2^n sample points, that is. In addition, two points are taken on the axis of each random variable, with corresponding values of, where α is an adjustable parameter, thus forming 2n sample points. Finally, add the center point to form sample points. The values of these sample points and the corresponding function values form a linear equation group:

$$\begin{bmatrix} 1 & x_{11} & \cdots & x_{1n} & x_{11}^2 & \cdots & x_{1n}^2 & x_{11}x_{12} & \cdots & x_{1(n-1)}x_{1n} \\ 1 & x_{21} & \cdots & x_{2n} & x_{21}^2 & \cdots & x_{2n}^2 & x_{21}x_{22} & \cdots & x_{2(n-1)}x_{2n} \\ \vdots & \vdots & \vdots & \vdots & \vdots & \vdots & \vdots & \vdots & \vdots & \vdots \\ 1 & x_{N1} & \cdots & x_{Nn} & x_{N1}^2 & \cdots & x_{Nn}^2 & x_{N1}x_{N2} & \cdots & x_{N(n-1)}x_{Nn} \end{bmatrix} \begin{Bmatrix} a_0 \\ a_1 \\ \vdots \\ a_n \\ a_{11} \\ \vdots \\ a_{nn} \\ a_{12} \\ \vdots \\ a_{(n-1)n} \end{Bmatrix} = \begin{Bmatrix} g(X_1) \\ g(X_2) \\ \vdots \\ g(X_N) \end{Bmatrix} \tag{4.66}$$

Solve this equation group, you can get A, and then get the expression of the approximate function, and then the failure probability can be solved using the first or second order reliability method.

③ Risk approximation calculation method based on safety margin function

The theory of structural reliability provides a scientific and theoretically accurate method for calculating the probability of failure points. However, the downside of reliability theory is that for more complex mechanical problems, the calculation process is very tedious and costly. Especially for urban large underground space projects, due to the complexity of their engineering structures and construction processes, traditional reliability calculation methods based on analytical methods will encounter problems that cannot construct function functions, and can only be solved by numerical simulation combined with Monte Carlo method or response surface method. Even so, due to the more complex uncertainties of the materials (mainly geotechnical materials) and structures involved in underground engineering than above-ground buildings, it is very difficult to obtain the probability distribution characteristics of their design variables through traditional means. In addition, the occurrence of safety risks in the construction of urban large underground space is generally difficult to judge based on a single design indicator (such as the force or deformation of a component), which makes its probability calculation a multi-index, multi-mode structural reliability problem, further increasing the difficulty of calculation. Therefore, although it is theoretically feasible to calculate the probability of safety risks in the construction of urban large underground space using the theory of structural reliability, it lacks practical operability and greatly limits the application of this theory. To address this issue, this section proposes a method for calculating the probability of safety risks in the construction of urban large underground space based on the safety degree function, based on the basic ideas of the theory of structural reliability and the experience summarized in actual engineering.

a. The significance of the safety degree function

The construction risk measurement sub-model mentioned above establishes the relationship between the occurrence probability of risk events and structural mechanical response indicators. However, in actual underground engineering construction, due to the randomness of many factors that affect the mechanical response of the structure, these mechanical response indicators, such as the deformation and force of the structure, are essentially a random variable. Generally speaking, for a random variable X, when its distribution law F(x) is given, we can observe its value x to get the occurrence probability P_E of an event $X \in G_E$ related to this random variable. Generally speaking, when this random variable has repeatable observation conditions, according to mathematical statistics knowledge, we can obtain its probability density function f(x) through the observation sample of random variable X and parameter estimation, and then obtain P_E through the following formula.

$$P_E = \int_{G_E} f(x)dx \tag{4.67}$$

However, if this random variable does not have repeatable observation conditions, we cannot obtain its distribution law, and naturally we cannot perform the calculation of formula (4.67). In the case of extremely limited observation samples, we cannot calculate the probability through strict probability theory and statistical methods. But even in the case of very small sample size, these observation samples will undoubtedly affect our subjective belief in whether the event $X \in G_E$ occurs. Imagine a situation where there is a dice that we don't know whether its mass distribution is uniform, and we only have one chance to roll the dice, then we will naturally think that the probability of this dice roll is greater than other numbers. In other words, in the absence of other information support and without repeatable experimental conditions, we have more reason to believe that the results of a few experiments can represent some situations with the highest probability of occurrence.

In fact, situations like the above are very common in the field of civil engineering. Strictly speaking, the construction of any civil engineering project is a unique action that cannot be repeated under the same conditions, so we cannot know the probability distribution of various mechanical indicators during the construction process. However, common sense and engineering experience tell us that in most cases, the greater the deformation or load of the structure, the higher the possibility of its failure. The essence of this judgment is a subjective belief inference we make based on actual observation results, and its deep logic is consistent with the aforementioned dice-throwing situation. The purpose of proposing the safety degree function is to provide a mathematical description for this process of subjective belief inference.

b. Definition and basic assumptions of the safety degree function

The safety degree function defines the mapping relationship between a physical quantity and the probability of related events occurring when there are no conditions for repeated observations. This mapping relationship can be expressed as:

$$\begin{cases} f_m[x; F(x), G_E] : X \to P_E \\ P_E = P\{X \in G_E\} \end{cases} \tag{4.68}$$

In the formula, f_m is the safety degree function; X is a random variable; x is the value of the random variable X; F(x) is the distribution law of the random variable X; P_E is the probability of event E occurring; G_E is the criterion domain for event E to occur, which is a subset of the value domain of X.

The meaning expressed by formula (4.68) is that the mapping defined by the safety degree function is related to the distribution law of the random variable X. However, this distribution law cannot be obtained under the condition of poor samples. Therefore, in combination with the meaning of the safety degree function, the following basic assumptions need to be made.

(a) The distribution form of the random variable X is known, but the distribution parameters are unknown. Generally speaking, it can be assumed that the random variable X follows a normal distribution;
(b) When only one observation can be made on the random variable X, the observed value x is an unbiased estimate of the expected value μ_X of the random variable X.

Based on the above assumptions, the safety degree function can be defined as follows:

$$f_m(x; \sigma, G_E) = \frac{1}{\sqrt{2\pi}\sigma} \int_{G_E} \exp \sim \left[\frac{-(t-x)^2}{2\sigma^2} \right] dt \tag{4.69}$$

In the formula, σ is the standard deviation of the random variable X.

Similar to formula (4.52), in most cases we can express the event $X \in G_E$ as $Z<0$ by variable substitution $Z = g_X$. But the criterion domain corresponding to the measurable mechanical response indicators in most underground engineering is generally in the following form:

$$G_E = (X_T, \infty) \tag{4.70}$$

In the formula, X_T is the criterion threshold.

The meaning represented by the criterion domain of formula (4.70) is that when the observed value of the random variable X exceeds XT, event E occurs. In underground engineering, the displacement, deformation of the structure, and various loads borne by the structure all apply to this form of criterion domain.

Under this premise, formula (4.69) can be further transformed into the following form.

$$f_m(x;\sigma,X_T)=\frac{1}{\sqrt{2\pi}\sigma}\int_{X_1}\exp\sim\left[\frac{-(t-x)^2}{2\sigma^2}\right]dt \tag{4.71}$$

The concept diagram of the safety degree function is shown in Fig. 4.22.

c. Empirical relationship based on the safety degree function

Compared to the theory of structural reliability, the safety margin function provides a more practical method for calculating the probability of failure points. Given the standard deviation σ of the failure index and the failure criterion, the probability of failure points can be calculated using equation (4.71).

However, because different failure indices have different probability distribution parameters, when applying the safety margin function in practice, the determination of the standard deviation σ and the failure criterion is key to the calculation of the safety margin function, but these two are precisely the more difficult parameters to determine. In addition, according to the definition of the safety margin function, the probability density function on the right side of equation (4.71) is assumed after obtaining the observed value x. This means that the random variable X corresponding to different observed values x is not essentially the same random variable, so its probability distribution may also be different, that is, there is a certain function relationship between the standard deviation σ in equation (4.71) and the observed value x, and this function relationship is very difficult to obtain. Therefore, there are certain difficulties in directly using the safety margin function to calculate the probability of failure points.

It should be pointed out that the essence of the safety margin function is a subjective confidence judgment method rather than a strict probability calculation method, and it allows the inclusion of some subjective judgments and experiences of the analyst in the calculation process. Therefore, from the practical engineering point of view, when we know very little about the randomness characteristics of various factors in construction, the only information we can be sure of is two aspects: one is the actual state of various mechanical response indicators in construction, that is, the real-time values of these indicators obtained through monitoring and measurement;

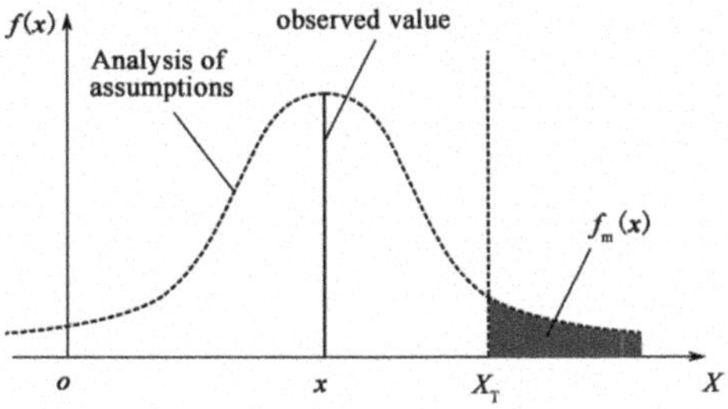

Fig. 4.22 Schematic diagram of the safety degree function

Table 4.12 Empirical relationship between measured indicators and failure probability and safety function

Level	Probability interval	Indicator interval	Descriptive	Safety function
1	(0.03, 0.3]	$(X_T, \infty]$	Extremely possible	$f_m(x) = 10^{-0.53+\frac{x-X_T}{0.5X_T}}$
2	(0.003, 0.03]	$(0.9X_T, X_T]$	Likelihood	$f_m(x) = 10^{-1.53+\frac{x-X_T}{0.5X_T}}$
3	(0.0003, 0.003]	$(0.7X_T, 0.9X_T]$	Infrequent	$f_m(x) = 10^{-2.53+\frac{x-X_T}{0.5X_T}}$
4	[0, 0.0003]	$[0, 0.7X_T]$	Unlikely	$f_m(x) = 10^{-3.53+\frac{x-X_T}{0.5X_T}}$

the other is the maximum allowable values of these indicators without affecting the normal operation of the engineering structure, that is, the control values of the indicators given at the time of structural design. According to the safety warning management principle of the construction site, when the measured index value reaches a certain proportion of the control value, it will trigger the site warning mechanism, and the site will increase the monitoring frequency and focus on the parts where the warning occurs. This safety warning mechanism actually reflects the idea of risk measurement. Therefore, with the basic concept of the safety margin function, combined with the idea of construction site safety warning, an empirical relationship between the measured index and the probability of failure points can be given.

Refer to the "Urban Rail Transit Underground Engineering Construction Risk Management Specification" (GB 50656-2011), the probability of failure points is divided into four levels, namely "very likely", "possible", "occasional" and "unlikely", each level corresponds to a certain probability interval, and the probability intervals of each level are logarithmically equidistant; in addition, refer to the "Building Foundation Pit Engineering Monitoring Technical Specification" (GB50497-2009) and other relevant standards, with the measured index value reaching 70% and 90% of the control value as the boundary for dividing the warning level, and it is believed that the risk reaches the maximum after the measured index value exceeds the control value. The empirical relationship between the measured index and the probability of failure points and the corresponding safety margin function are shown in Table 4.12.

Chapter 5
Coupling Mechanism of Construction Safety Risk in Urban Large Underground Space

5.1 Basic Concepts and Formation Mechanisms of Multi-factor Coupling Risk

5.1.1 Coupling Effect of Risk Factors

Numerous engineering practices show that construction safety risks are influenced by many factors, which are sometimes intertwined and difficult to distinguish. Risk factors, as part of the risk system, will inevitably have connections and interactions with other parts of the system. Risk factors or risk events that move up the risk-causing pathway to higher levels are risk factors acting on the risk system in a logical progression, an evolutionary and developmental process. When there are multiple risk factors in the risk system, the effect of each risk factor on the risk system during its evolutionary development process is often influenced by other risk factors, making the effects of each risk factor "indivisible" and unable to be viewed independently. From the original definition of the term "coupling", the combined effect of risk factors on the risk system during their evolutionary development process is a manifestation of a coupling phenomenon.

From an abstract perspective, risk is generated due to the disturbance caused by actions on an existing natural or social system, thereby bringing uncertainty that the system did not originally possess or was not significant. For the original system, the action is an external effect that stimulates random changes in some links of the system, thereby generating risk factors. There is a causal relationship between risk factors and risk events, and some of the causal mechanisms of risk factors are independent of each other. For example, "falling from height" and "mechanical injury" are risk factors for the risk event of "casualties". Obviously, there is no correlation between the conditional probabilities of "casualties due to falling from height" and "casualties due to mechanical injury", that is, their causal mechanisms are independent. When such risk factors occur simultaneously, the probability of the occurrence of the risk

S.-X. Lei et al., *Safety Risk Assessment of Urban Underground Large Space Construction*, https://doi.org/10.1007/978-981-97-7437-1_5

event is a simple superposition of the conditional probabilities of each factor (i.e., the conditional probability of the occurrence of the risk event when a certain risk factor occurs alone).

However, in most cases, there are more complex connections between the causal relationships of different risk factors. For example, "soft soil layer" and "heavy rainfall" are risk factors for the risk event of "pit collapse". But according to soil mechanics knowledge, the weaker and looser the soil layer, the worse its hydrographic properties. Therefore, when considering the conditional probability of "pit collapse due to heavy rainfall", we also need to consider the characteristics of the soil layer itself, in addition the characteristics of the soil layer will also affect the conditional probability of "pit collapse due to soft soil layer". In other words, there is a certain relationship between the conditional probabilities of "pit collapse due to heavy rainfall" and "pit collapse due to soft soil layer", and their causal mechanisms are not independent. When such risk factors with non-independent causal mechanisms occur simultaneously, these factors will have mutual stimulation and amplification effects, and then develop into larger risk factors. In this process, compared to the situation where these risk factors occur alone, the probability of the occurrence of the risk event will significantly increase. The increased part of the probability can be divided into two parts: ① the superposition increment caused by the superposition of the conditional probabilities of each factor; ② the extra increment caused by the correlation of the causal relationship between each risk factor and the risk event, the risk amplification (or suppression) effect caused by the correlation of the causal mechanisms between risk factors, is called the coupling effect of risk factors (Fig. 5.1).

5.1.2 Coupling Mechanism of Risk Factors

The coupling of risk factors originates from the complexity of the physical mechanisms of risk occurrence. This complexity makes it impossible for us to break down risk factors into the most basic factors with independent causal mechanisms when conducting risk causality analysis. In other words, coupling is caused by "indistinguishability" and "incomplete separation". In the risk system of urban large underground space construction, the most common coupled risk factors are engineering geological factors (such as stratum physical and mechanical parameters) and hydrogeological factors (such as groundwater or rainfall, etc.). For most underground projects, the relationship between strata and water is a core issue. It is well known that the mechanical properties of most strata are strongly related to water. The presence of water reduces the strength of the strata, such as reducing the values of mechanical parameters like c, φ; the presence of water also increases the overall density of the strata, thereby increasing the load acting on the structure; water can also cause changes in the properties of some strata, such as dry natural loess has good performance, but when it encounters water, it collapses, and its performance drops sharply, or sandy soil will liquefy under dynamic action when the water content is high, etc.

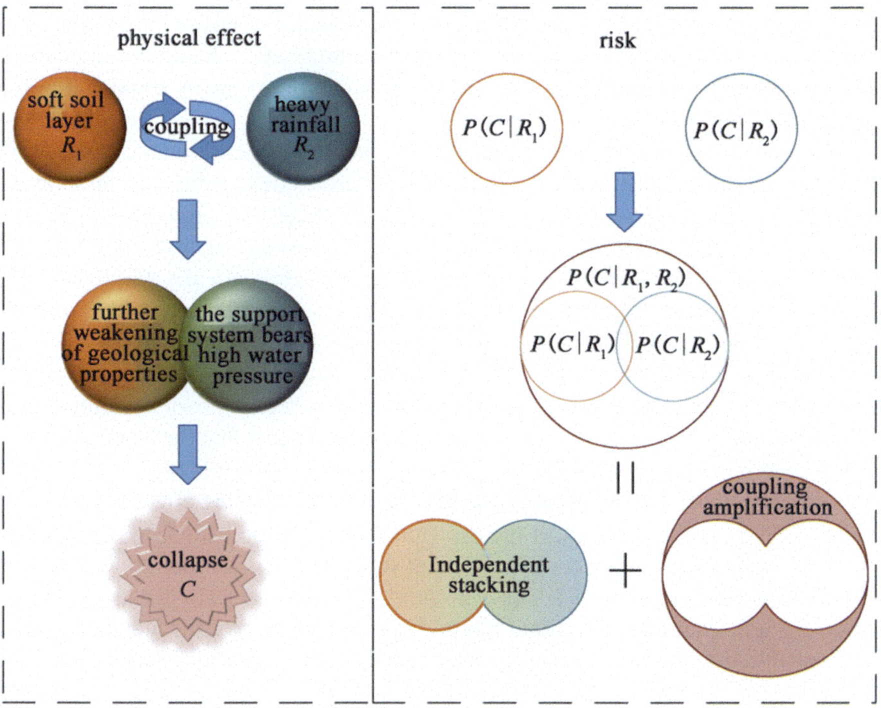

Fig. 5.1 Schematic diagram of risk factor coupling effects

On the other hand, because water exists in the soil particle skeleton or in the pores and cracks of rocks, when the strata deform, the change in porosity will in turn affect the water, causing disturbance to the water pressure field and further adjustment of the groundwater flow field. Due to this strong interrelation, when considering the effect of water, it cannot simply be regarded as an independent load source but must be considered as a whole with the strata.

From a risk perspective, the interaction mechanism between strata and water is reflected in the coupling of related risk factors. For example, in the famous Lan-Yu Railway Humaling Tunnel, due to the encounter with the third series of water-rich fine sand strata, the tunnel's self-stability was extremely poor after excavation, the face instability, prone to collapse, frequent water seepage from the bottom plate, sand and water gushing from the face, and special water sac phenomena, greatly increasing the construction difficulty, severely restricting the construction progress, and causing the tunnel to take 8 years to complete, resulting in huge construction safety risks and schedule delay risks. Looking back at this project, it can be found that the core factor causing a series of problems during the construction process was the fine sand strata. The internal friction angle of fine sand in a wet state is about 38°, and it has slight stickiness, certain erectness, and shear strength. But like most sandy soils, the properties of fine sand are extremely sensitive to changes in water content.

When fine sand changes from a wet state to a dry state, the cohesion of the fine sand will decrease with the disappearance of matrix suction; when the fine sand enters a saturated state, it will quickly soften and mud, which is extremely unfavorable for underground engineering construction.

In fact, a comprehensive view of the project reveals that the ultimate cause of this difficult situation is multifaceted. Firstly, the failure to detect this special stratum during the preliminary survey and route selection stage is a prerequisite for this situation; secondly, the tunnel is constructed using the mining method, which is least adept at dealing with such water-rich weak strata, thus dramatically amplifying the impact of the special strata. In addition, the fine sand stratum has a certain self-stabilizing ability when the water content is low, which is not enough to have such a severe adverse effect on the tunnel excavation. However, once there is confined water, the fine sand stratum becomes an extremely adverse stratum known as "impossible to build a tunnel". It is under the influence of multiple factors that the originally only a few hundred meters long tunnel has become a world-class problem.

From the perspective of risk analysis, there is a clear coupling effect between the main risk factors of the project. If the decision-makers of the Humaling Tunnel had anticipated encountering water-rich fine sand strata before construction, they might have taken various measures in advance to avoid building a tunnel under such extreme conditions. For example, the shield method is more adaptable to water-rich strata. If the construction method is changed in advance, it may turn this disadvantage into an advantage; or consider using the freezing method for pre-reinforcement, which can greatly reduce the construction risk without making major adjustments to the construction plan; or directly avoid this problem fundamentally by changing the route during the planning stage. But as analyzed earlier, in actual risk assessment work, it is often difficult and unnecessary to conduct overly detailed and exhaustive analysis of risk factors, so it is easy to have a "marginal truncation effect" in risk analysis, i.e., analysts often focus on more common, high-probability ordinary events or common combinations of several events, and rarely pay attention to rare occasional events, and generally do not consider rare combinations of special events. This way of thinking may have a certain "luck mentality", but in most cases it is a feasible and economical approach. Under this premise, some risk factors that have a small impact when acting alone may be overlooked or even excluded in the analysis.

However, one problem that must be pointed out is that the selection of risk factors is entirely based on the analyst's subjective judgment, at which point the subjectivity of risk is dominant, but risk also has its objective side. People can make subjective risk as close as possible to objective risk through reasonable analysis methods, i.e., whether a risk analysis method is reasonable depends on whether this method can competently serve as a bridge between subjective and objective. In the case of the Humaling Tunnel, we have reason to believe that the project decision-makers had a fairly comprehensive understanding of the various potential risk factors and possible risks before the start of construction. But even so, the tunnel was excavated according to the original plan and encountered unexpected major problems, indicating that the pre-construction risk assessment deviated from the objective situation. Therefore, if there is no obvious omission in the pre-construction risk analysis, the reason for this deviation is likely to lie in the defects of the risk analysis method, especially the neglect of the coupling effect of risk factors. The essence of the coupling effect of risk factors is mechanical and physical coupling, which is an objective phenomenon that does not depend on human cognition. Therefore, the fundamental purpose of studying the coupling effect of risk factors is to help risk analysts recognize the existence of the coupling effect when facing practical problems, and to measure the size of this coupling effect in a logical and operable way.

5.1.3 Mechanism of Multi-factor Coupling Risk Formation

Through the analysis of accident cases, it is known that the coupling effect of risk factors plays a significant, even decisive role in the formation of many major construction risks. Construction activities inevitably generate risk factors. In most cases, through reasonable planning, surveying, design, and special construction schemes, we can control the adverse effects of risk factors on the construction system within an acceptable range. However, due to the complexity and concealment of the coupling effect, we often find it difficult to accurately predict all possible coupling effects between risk factors. Once construction actions are implemented, the adverse effects of some risk factors may be dramatically amplified under the action of the coupling effect, causing the "combined force" of these adverse effects to exceed the system's expected bearing capacity, breaking through the system's defense barrier, and leading to the occurrence of risk events (Fig. 5.2).

Through analysis, it is not difficult to find that most accidents originate from incorrect risk estimation, and incorrect risk estimation largely stems from the neglect of the coupling effect of risk factors. The occurrence of accidents is generally the result of the accumulation of the system's unsafe state to a certain extent. If we

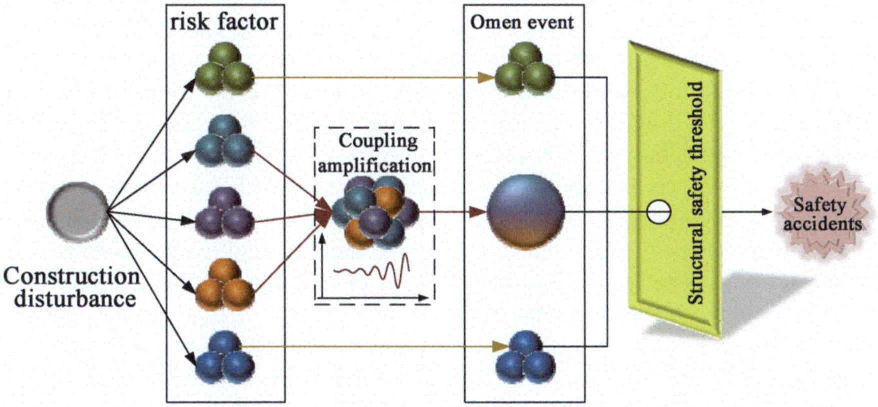

Fig. 5.2 Schematic diagram of the mechanism of multi-factor coupling risk formation

assume that this unsafe state has a certain quantitative benchmark, then we can correspondingly measure the "contribution" of each risk factor to the system's unsafe state. Admittedly, the appearance of risk factors will increase the degree of system insecurity. When there are enough risk factors, it is almost inevitable that the degree of system insecurity will break through the threshold. However, in many cases, the number of risk factors in the system is not so large that the accumulation of each factor's contribution can make the degree of insecurity break through the threshold, but accidents still occur. This situation indicates that considering the contribution of each risk factor independently will inevitably lead to incorrect estimation of system risk. In reality, the joint action of multiple risk factors is the root cause of accidents, that is, the coupling effect of risk factors will have a greater impact on system safety. From a quantitative perspective, when multiple risk factors coexist, in addition to the "self-contribution" of each factor to system insecurity, there is also a "coupling contribution" generated by the coupling effect. Since the coupling effect of risk factors objectively exists, if a risk analysis method does not consider the coupling contribution of risk factors, this analysis method will undoubtedly cause a subjective-objective split, thereby deviating its analysis results from the objective situation.

From the analysis in the previous section, it can be seen that the coupling effect of risk factors is not a brand-new concept proposed from an academic perspective, but an objective phenomenon deeply rooted in everyone's logic, knowledge, and experience. Because of this, the concept of the coupling effect of risk factors is often hidden in our unconscious judgments—most people would think that the combined effect of multiple adverse factors would make an action more dangerous, but very few people would separately analyze how much each adverse factor contributes, nor would they pay attention to whether the effect of multiple adverse factors is simply an addition or there is a coupling effect. Admittedly, risk analysis always needs to consider the limitations of objective reality and subjective cognition. Considering the coupling effect of risk factors will undoubtedly increase the accuracy of risk analysis, but it will also greatly increase the complexity of risk analysis work. The coupling effect of risk factors has a certain degree of concealment, and the deep mechanism behind the coupling effect has a certain complexity, requiring the analyst to have a considerable degree of professional knowledge and practical experience. On the other hand, there is currently no universally accepted definition of the coupling effect of risk factors, and it is difficult to quantify its evaluation. Therefore, the coupling effect of risk factors is generally not considered in current risk analysis methods. From the perspective of analysis cost and benefit, this consideration is reasonable, because introducing the coupling analysis of risk factors will increase the cost of analysis while further increasing the subjectivity of risk analysis, and the improvement in accuracy it brings is undoubtedly a manifestation of "subjective cognition". In some risk analyses in fields such as finance and sociology, it is sometimes unnecessary to exchange a slight increase in accuracy for the increased subjectivity and complexity of the model, but it is different in the field of safety risk. As the saying goes, "life is of paramount importance", safety accidents not only bring property loss, but often accompanied by health or even life loss, which is generally considered unacceptable serious loss in our current social environment. Therefore, the estimation of safety risk often adopts a more conservative strategy, that is, the increased cost brought by overestimating risk can often be accepted, but the safety hazards brought by underestimating risk need to be avoided at all costs. In this way, it is very necessary to consider the coupling effect in the analysis of safety risk, not only because in most cases the coupling effect brings a "magnification" of risk, ignoring the coupling effect will lead to an underestimation of risk, but also because the coupling effect has a decisive impact in the formation process of safety risk. Therefore, the ultimate purpose of the study of the coupling effect of risk factors is to clarify its concept, to clearly separate this effect from people's vague experiential cognition, and to give a quantitative measure of the coupling effect of risk factors, to establish a scientific, reasonable and relatively simple method to consider the coupling effect of risk factors in actual risk assessment work, to improve the accuracy and scientific nature of risk assessment.

5.2 Risk Factor Coupling Coefficient

5.2.1 Attributes of the Coupling Effect of Risk Factors

According to Sect. 5.1, the coupling of risk factors is a generalized interaction, which is a qualitative description. In order to better study this effect, its attributes need to be further analyzed. Generally speaking, when discussing a certain effect, the following types of attributes will naturally be involved.

(1) Acting Subject: The acting subject of risk factor coupling is the risk factor, which is a specific event. Coupling action occurs between at least two objects but can also exist between more than two objects. Depending on the number of coupling subjects, it can be divided into dual factor coupling and multi-factor coupling.
(2) Acting Object: According to the definition of risk factor coupling effect, coupling is manifested as the "indivisibility" of the effect of risk factors. Therefore, the coupling effect must be relative to a certain object. Within the risk system, the object of the risk factor is the "result" that exists in the event chain as a risk factor, including a series of upper-level events including the final risk event. For different result events, the strength of the coupling effect of the same pair or group of risk factors may vary. Therefore, the coupling effect of risk factors can only be clearly defined and measured on the premise of a specific object.
(3) Acting Pathway: Risk factors can be of different types. "Rainfall" is an objective physical phenomenon, "operational error" is a subjective action, and "unstable psychological state of personnel" is a state that is difficult to concretize. Therefore, the pathway of risk factor coupling may sometimes be very clear, but sometimes it may be very vague. Depending on the different coupling pathways,

risk factor coupling can be divided into same-type factor coupling and different-type factor coupling. However, the coupling of risk factors will inevitably reflect changes in some physical states in the system. For construction safety risks, risk factors are coupled through a series of mechanical actions.

(4) Manifestation: A "risk factor" refers to an event that may occur but has not yet occurred, so the impact of risk factor coupling can only be described in a probabilistic sense, that is, the coupling effect of risk factors will affect the occurrence probability of its object, which is the occurrence probability of the upper-level event or the final risk event.

(5) Acting Polarity: The coupling effect of risk factors may have a positive impact on risk. That is, the presence of the coupling effect increases the risk, which is called positive coupling or positive coupling. If the coupling effect of risk factors suppresses the risk, it is called negative coupling or negative coupling. In most cases, the polarity of the coupling effect of risk factors is positive coupling.

(6) Acting Intensity: Any action has a degree of strength, and the coupling of risk factors is no exception. The strength of the coupling action is generally measured by the "coupling degree" or "coupling coefficient". The closer the connection between risk factors and the greater the impact, the greater the degree of risk factor coupling, and vice versa, the smaller the degree of risk factor coupling.

Among the above attributes, the size of the coupling action intensity between risk factors is the only quantifiable attribute and is the most concerned attribute in actual research. However, how to define the coupling coefficient largely depends on the reference system used to measure the intensity of the coupling action. Therefore, before defining the coupling coefficient, it is necessary to first clearly define the basic assumptions and concepts of the coupling coefficient.

5.2.2 *Basic Assumptions of Risk Factor Coupling Coefficient*

1. Assumption of Manifest and Hidden Causes

The concept of risk factor coupling is inseparable from the causal relationship between events, especially the causal relationship in the "multi-cause and one-effect" situation. The causal relationship is objectively existing, while causal analysis is a subjective process in which the analyst observes, analyzes, and infers the relationship between events to extract the causal relationship between events. Therefore, the causal relationships that people can generally perceive are subjective. For example, if the conclusion "R is the cause of C" is obtained through causal analysis, it essentially means that the analyst's observation of R and C. A judgment made about the relationship between them. Objectively speaking, the cause of an event is often difficult to exhaust, and this causal analysis actually only reveals that R is one of the many causes of C. Generally speaking, the causes derived from causal analysis have a clear context, indicating that R has a deterministic impact on C, so such causes are called "apparent causes", and the remaining causes that have not been analyzed are called "hidden causes". Obviously, apparent causes and hidden causes are a pair of subjective relative concepts, which depend on the analyst's subjective experience, analytical ability, and willingness. We can make the following logical assumptions about apparent causes and hidden causes:

(1) Apparent causes and hidden causes are relative to a specific result event and a specific context.
(2) Under the necessary and sufficient premise of causal analysis, the impact of apparent causes on the result event is significant and decisive, that is, there is a strong causal relationship between the apparent cause and its result. Hidden causes, on the other hand, are insignificant and secondary, that is, the causal relationship between the hidden cause and its result is weak.
(3) The mechanism of the apparent cause's impact on the result event must be clear and known, while the mechanism of the hidden cause may be difficult to analyze, difficult to perceive, or even uncertain.
(4) Apparent causes are generally limited, and each apparent cause has a clear conceptual boundary, which can be distinguished subjectively, while hidden causes are often difficult to exhaust and distinguish.
(5) At the time of analysis, the state or probability of the apparent cause is generally more certain, while the state or probability of the hidden cause may be vague.

Due to the above characteristics of hidden causes, they are generally regarded as some kind of fixed "background condition" in causal analysis. For example, the causes of a car accident may include "drunk driving", "fatigue driving", "icy roads", "speeding", "mechanical failure", "operational errors", "poor road conditions", "lack of concentration", etc. Among these causes, "drunk driving", "fatigue driving", "icy roads", "speeding" obviously have a stronger causal relationship than other causes. Once one or more of these events occur, the possibility of a car accident is very significant, so they are apparent causes for "car accidents"; while other causes, although they also have a causal relationship with car accidents, this connection is obviously weaker, even if other causes occur at the same time, it cannot be asserted that a car accident will definitely occur. In other words, the existence or non-existence of these causes has a very insignificant impact on the probability of the result event. Based on the above analysis, it can be considered that the influence of hidden causes on the result event is a relatively constant "background condition", which can be expressed as:

$$P(C) = f[P(R)] + f^0 P[(O)] = f[P(R)] + P_c^0 \tag{5.1}$$

In the formula, f represents the abstract function of the cause event's impact on the result event; O represents the composite event composed of all hidden causes.

Since hidden causes are difficult to distinguish and exhaust, no further distinction is made here. Formula (5.1) indicates that the hidden cause's impact on the result event C can be regarded as a basic probability of C occurring, denoted as P_c^0. The concept of basic probability can be explained as: the probability of the result event C occurring after excluding all apparent causes that may cause the event (i.e., ensuring that these apparent causes will not occur).

The magnitude of the base probability caused by hidden causes should depend on the weight of the hidden cause in the set of causes, and this weight depends on the completeness of the causal analysis. Taking the aforementioned car accident problem as an example, suppose "drunk driving", "fatigue driving", "road icing", "speeding", "mechanical failure", "operational error", "poor road conditions", "lack of concentration" are all the objective causes of the accident (the actual objective causes are infinite, here only a limited number are listed for convenience of narration). If a causal analysis concludes that the causes of the accident are "speeding" and "road icing", then the remaining 6 causes become hidden causes in the context of this analysis. Obviously, "speeding" and "road icing", the two apparent causes, undoubtedly have a significant impact on the probability of an accident, but even if these two apparent causes are excluded, the probability of an accident may still be high under the influence of other hidden causes, indicating that the conclusion of this causal analysis is incomplete; if a causal analysis concludes that 7 of the causes, excluding "poor road conditions", are the causes of the accident, then there is only one hidden cause left. After excluding all apparent causes, the probability of an accident occurring under the influence of a single hidden cause will be greatly reduced compared to the previous situation.

2. Assumption of risk factor coupling effect.

The coupling coefficient of risk factors is introduced to measure the strength of the coupling effect of risk factors, so the coupling coefficient needs to match the definition of the aforementioned risk factor coupling effect. For convenience of description, suppose there is a minimum causal unit as shown in Fig. 5.3, where event C is the result (risk event), and events R_1 and R_2 are the causes (risk factors) of C. Figure 5.3a is the situation where only factor R_1 exists, and Fig. 5.3b is the situation where two factors exist.

Let system of Fig. 5.3a be $S_C^{①}$, and Fig. 5.3b be $S_C^{②}$, based on this, the following assumptions are made:

(1) Basic Assumption 1: When no risk factors occur, the possibility of risk event C occurring is 0.

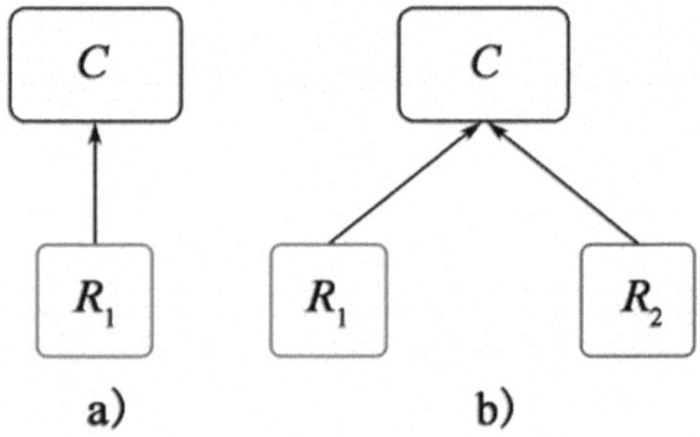

Fig. 5.3 Minimum causal unit

In risk analysis, depending on the amount of information that can be obtained, an event E may be in one of three states: "occurred", "not occurred", and "uncertain", expressed as:

$$\left\{\begin{array}{cc} P(E) = 0, & E \text{ not occurred} \\ 0 < P(E) < 1, & E \text{ uncertain} \\ P(E) = 1, & E \text{ occurred} \end{array}\right\} \tag{5.2}$$

When discussing a set of n events, there may be 3^n possible combinations of states in this set, now referred to as a "situation". "No risk factors occur" refers to a situation where all apparent causes obtained through causal analysis are in a state of non-occurrence, expressed as:

$$\left\{\begin{array}{l} S_C^{①} : P(\overline{R}_1) = 1 \\ S_C^{②} : P(\overline{R}_1\overline{R}_2) = 1 \end{array}\right. \tag{5.3}$$

According to the aforementioned concept of apparent and hidden causes, all factor events R_i in Fig. 5.3 are apparent causes of the result event C, so when all factor events do not occur, the probability of the result event C occurring should be the base probability P_c^0 determined by the hidden causes. However, under the condition of relatively complete causal analysis, the base probability P_c^0 is generally small and has a very limited impact on the possibility of the result event occurring. Therefore, for the sake of simplifying the model, it can be assumed that $P_c^0 = 0$. Thus, Basic Assumption 1 can be expressed as:

$$\left\{\begin{array}{l} S_C^{①} : P(C)_{\overline{R}_1}^{①} = P(C|\overline{R}_1) = 0 \\ S_C^{②} : P(C)_{\overline{R}_1\overline{R}_2}^{②} = P(C|\overline{R}_1\overline{R}_2) = 0 \end{array}\right. \tag{5.4}$$

(2) Basic Assumption 2: Without considering the coupling effect, when a risk factor changes from non-occurrence to occurrence, the possibility of a risk event occurring does not decrease.

In simple terms, the essence of Basic Assumption 2 is that risk factors are elements that have an adverse impact on the security of the system. Without considering the coupling effect, the occurrence of a risk factor cannot make the system safer. In other words, comparing the possibility of a risk event occurring when a risk factor occurs and when it does not occur, the former should be greater than the latter, or at least not less than the latter. Under the condition of satisfying the basic assumption 1, due to the nature of probability, basic assumption 2 naturally holds in the system $S_C^{①}$, and $S_C^{②}$ can be expressed as:

$$P(C)_{R_1R_2}^{2} = P(C|R_1R_2) \geq \max\{P(C|\overline{R}_1R_2), P(C|R_1\overline{R}_2)\} \tag{5.5}$$

(3) Basic Assumption 3: The possibility of a risk event occurring under the same objective conditions does not change, regardless of the results of causal analysis.

Basic Assumption 3 is closely related to the objective existence of risk factors. For the same result event C, different causal analyses may yield different causes, but these different analysis results objectively correspond to the same event, for example, in systems $S_C^{\textcircled{1}}$ and $S_C^{\textcircled{2}}$, there are the following situations:

$S_C^{\textcircled{2}}$ can be seen as two different causal analyses of the same event, but the factor R_1 appears in both. Assuming in $S_C^{\textcircled{1}}$:

$$\begin{cases} S_C^{\textcircled{1}} : P(R_1) = p_1^{\textcircled{1}} \\ S_C^{\textcircled{2}} : P\left(R_1R_2 \cup R_1\overline{R}_2\right) = p_1^{\textcircled{2}} \end{cases} \tag{5.6}$$

Obviously, $p_1^{\textcircled{1}} = p_1^{\textcircled{2}}$ and according to the definition of hidden causes, for any causal analysis instance, hidden causes are in an uncertain state. Since factor R_2 is an explicit cause in one system $S_C^{\textcircled{2}}$ but exists as a hidden cause in the other $S_C^{\textcircled{1}}$, the formula (5.6) corresponds to the same objective situation. Therefore, according to Basic Assumption 3, the following formula holds:

$$\begin{aligned} P(C)_{R_1}^{\textcircled{1}} &= P(C|R_1)p_1 \\ &= P(C)_{R_1R_2 \cup R_1\overline{R}_2}^{\textcircled{2}} \\ &= P\left(C|R_1R_2 \cup R_1\overline{R}_2\right)p_1 \\ &= P(C|R_1R_2)P(R_2)p_1 + P\left(C|R_1\overline{R}_2\right)P\left(\overline{R}_2\right)p_1 \end{aligned} \tag{5.7}$$

Substituting formula (5.5) into formula (5.7), we get:

$$P(C|R_1) \geq P\left(C|R_1\overline{R}_2\right)P(R_2) + P\left(C|R_1\overline{R}_2\right)P\left(\overline{R}_2\right) = P\left(C|R_1\overline{R}_2\right) \tag{5.8}$$

Similarly, considering the single-factor system composed of R_2 and C, we get:

$$P(C|R_2) \geq P\left(C|\overline{R}_1R_2\right)P(R_1) + P\left(C|\overline{R}_1R_2\right)P\left(\overline{R}_1\right) = P\left(C|\overline{R}_1R_2\right) \tag{5.9}$$

Formulas (5.8) and (5.9) are the corollaries derived from Basic Assumption 3.

(4) Basic Assumption 4: When all factor states are unknown, without considering the coupling effect, adding a risk factor to a risk event does not reduce the possibility of the risk event occurring.

Basic Assumption 4 is a conclusion naturally derived from the objectivity of risk. According to the previous analysis, the causes of an event are always objectively present and difficult to exhaust. The work of causal analysis is merely to reveal and select the dominant factors among these causes. Therefore, adding a risk factor to a system is essentially a further causal analysis of it. This behavior merely reveals more information in the system and does not interfere with the system itself. Therefore, the system risk should remain unchanged in this case. The above statement can be expressed as:

$$\begin{aligned} P(C)_{\Omega_R}^{①} &= P(C|R_1)P(R_1) + P(C|\overline{R}_1)P(\overline{R}_1) \\ &= P(C)_{\Omega_R}^{②} \\ &= P(C|R_1R_2)P(R_1)P(R_2) + P(C|R_1\overline{R}_2)P(R_1)P(\overline{R}_2) + \\ & P(C|\overline{R}_1R_2)P(\overline{R}_1)P(R_2) + P(C|\overline{R}_1\overline{R}_2)P(\overline{R}_1)P(\overline{R}_2) \end{aligned} \tag{5.10}$$

In the formula, the subscript Ω_R represents the system in any situation. After substituting formulas (5.4) and (5.7) into formula (5.10), we get:

$$P(C|\overline{R}_1R_2)P(\overline{R}_1)P(R_2) = 0 \tag{5.11}$$

This result is undoubtedly unreasonable. Through analysis, it can be found that the reason for this result is that the basic assumption 1 simplifies the basic probability in a way that does not conform to the objective situation. The term in formula (5.11) $S_C^{①}$ is part of the basic probability discarded by basic assumption 1, but in the system $S_C^{②}$, because $S_C^{②}$ the hidden cause becomes an explicit cause, this part of the probability is explicitly expressed, resulting in this unreasonable result. If the basic probabilities discarded by basic assumption 1 in the two systems $S_C^{①}$ and $S_C^{②}$ are expressed as $P_C^{0①}$ and $P_C^{0②}$ respectively, then formula (5.10) can be rewritten as:

$$\begin{aligned} P(C)_{\Omega_R}^{①} &= P(C|R_1)P(R_1) + P_C^{0①} \\ &= P(C)_{\Omega_R}^{②} \\ &= P(C|R_1R_2)P(R_1)P(R_2) + P(C|R_1\overline{R}_2)P(R_1)P(\overline{R}_2) + \\ &= P(C|\overline{R}_1R_2)P(\overline{R}_1)P(R_2) + P_C^{02} \end{aligned} \tag{5.12}$$

From the relationship between basic probability and the completeness of causal analysis, we know that $P_C^{0①} \geq P_C^{0②}$, by substituting Eqs. (5.7) into (5.12) and eliminating terms, we can obtain:

$$P_C^{0①} = P(C|\overline{R}_1R_2)P(\overline{R}_1)P(R_2) + P_C^{0②} \geq P_C^{0②} \tag{5.13}$$

Thus, we can deduce:

$$P(C|\overline{R}_1R_2)P(\overline{R}_1)P(R_2) \geq 0 \tag{5.14}$$

This result is reasonable. If we want to solve the logical problem in Eq. (5.11) under the condition that the basic assumption 1 is valid, a feasible method is to make Eq. (5.10) an inequality, that is:

$$\begin{aligned} P(C)^{①}_{\Omega_R} &= P(C|R_1)P(R_1) \\ &\leq P(C|R_1R_2)P(R_1)P(R_2) \\ &+ P(C|R_1\overline{R}_2)P(R_1)P(\overline{R}_2) + P(C|\overline{R}_1R_2)P(\overline{R}_1)P(R_2) \\ &= P(C)^{②}_{\Omega_R} \end{aligned} \tag{5.15}$$

Similarly, considering the single-factor system composed of R $_2$ and C, we can get:

$$\begin{aligned} P(C|R_1R_2)P(R_1)P(R_2) &+ P(C|R_1\overline{R}_2)P(R_1)P(\overline{R}_2) \\ &+ P(C|\overline{R}_1R_2)P(\overline{R}_1)P(R_2) \\ &\geq \max\{P(C|R_1)P(R_1), P(C|R_2)P(R_2)\} \end{aligned} \tag{5.16}$$

Equation (5.15) is the mathematical expression of basic assumption 4. Except for basic assumption 1, the rest of the assumptions are based on the basic concepts of causality and probability theory, and are derived through logical deduction. Based on these basic assumptions, we can define the coupling coefficient.

5.2.3 Definition of Risk Factor Coupling Coefficient

1. Principle of Causal Mechanism Independence

As mentioned earlier, the risk factor coupling coefficient is proposed to measure the strength of the coupling effect. To establish this measure, we need to first define the measurement benchmark, that is, to answer the question "how much impact does the risk factor have on the risk when there is no coupling effect".

In the risk system model of urban large underground space construction, risk factors are the underlying events in the event space. The risk factors and risk events are connected by a series of causal relationships composed of event networks. This event network and causal relationship can be represented by a Bayesian network. One of the basic assumptions of the Bayesian network is the conditional independence relationship between network nodes. In other words, there is no significant causal relationship between risk factors as underlying events, and whether each factor event occurs is independent of other risk factors. However, it should be noted that the conditional independence of risk factors and whether there is a coupling effect between risk factors are not concepts in the same category. The conditional independence of risk factors is just the relationship between risk factors, while according to the definition of coupling effect, the coupling effect is a phenomenon that exists when two or more acting subjects (i.e., risk factors) have an impact on the acting object (i.e., the upper-level event as a result of risk factors). Therefore, in the Bayesian network, the strength of the coupling effect of risk factors is related to the conditional probability of its result event. Based on this assertion, the following model is established to study the risk factor coupling coefficient.

As shown in Fig. 5.4a, suppose event node C has n cause event nodes R_i connected by E-E connections pointing to it, i = 1, 2, ..., n. According to the definition of the Bayesian network, describing the causal relationship between this series of events requires 2^n conditional probability parameters. If we can obtain the true values of these 2^n conditional probabilities through some method, then these 2^n conditional probabilities are sufficient to describe all the strengths of the causal relationships between C and R_i and the coupling relationship between any group of factors (R_i, R_j, ..., R_k), i, j, k ∈ {1, ..., n} [Fig. 5.4b]. For example, for the following group of conditional probabilities:

$$\begin{cases} P\left[C|R_1R_2\left(\bigcap_{i\neq1,i\neq2}\overline{R}_i\right)\right] = 0.9 \\ P\left[C|R_1\left(\bigcap_{i\neq1}\overline{R}_i\right)\right] = 0.1 \\ P\left[C|R_2\left(\bigcap_{i\neq2}\overline{R}_i\right)\right] = 0.2 \end{cases} \tag{5.17}$$

Equation (5.17) indicates: when R_1 and R_2 occur separately, the likelihood of C occurring is relatively low, with probabilities of 0.1 and 0.2 respectively. This suggests that the occurrence of R_1 and R_2 has a small impact on whether C occurs. According to the definition of risk factor coupling effect, when there is no coupling effect between R_1 and R_2, the impact of both on the likelihood of C occurring is independent of each other. Therefore, when R_1 and R_2 occur simultaneously, the likelihood of C occurring should not be significantly higher than 0.2. However, the first part of Eq. (5.17) shows that when R_1 and R_2 occur simultaneously, the probability of C occurring increases to 0.9, which is much higher than when R_1 or R_2 occur separately. This suggests that

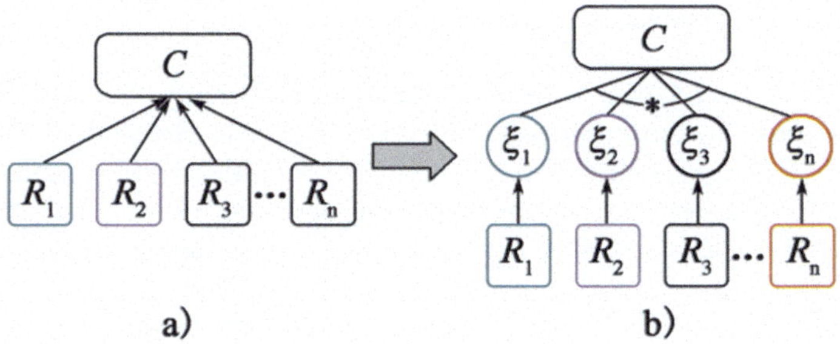

Fig. 5.4 Partial structure of Bayesian network

there is a coupling effect between R_1 and R_2. However, due to the objective existence of the risk factor coupling effect, we cannot obtain the conditional probability value in the first part of Eq. (5.17) under the hypothetical situation of "no coupling effect between R_1 and R_2", so we cannot separate the "probability value increased by the coupling effect of R_1 and R_2" from the total probability value, and naturally cannot measure the strength of the coupling effect. To address this issue, we introduce the concept of "causal mechanism independence".

In the study of Bayesian networks, for the purpose of simplifying network structure and reducing the number of parameters, it is often assumed that the conditional distribution of the Bayesian network has a certain pattern, called "local structure". "Causal mechanism independence" is one of the common local structures. The so-called causal mechanism independence refers to multiple causes independently affecting the same result. For example, if C represents "power outage", R_1 represents "overload", and R_2 represents "typhoon", it is not difficult to find that the causal mechanisms of power outage due to overload and power outage due to typhoon are different, so the conditional probability of C occurring due to R_1 does not affect the conditional probability of C occurring due to R_2. Under the premise of causal mechanism independence, the occurrence of event C can be divided into two sub-events with different causes: ξ_1 represents power outage due to overload, and ξ_2 represents power outage due to typhoon. According to the logical relationship between R_1 and R_2, when one of R_1 and R_2 occurs, C occurs, then it can be noted (the notation still follows the rules of Sect. 3.4.2, the Boolean random variable corresponding to the event node E is denoted by the bold symbol **E**, the occurrence of the event is denoted by "**E** = T" or "E", and the non-occurrence of the event is denoted by "**E** = F" or "$\overline{\text{E}}$"):

$$C = \xi_1 \cup \xi_2 \tag{5.18}$$

The probabilities of ξ_1 and ξ_2 occurring can be expressed as:

$$\begin{cases} P(\xi_1) = P(\xi_1|R_1)P(R_1) + P(\xi_1|\overline{R}_1)[1 - P(R_1)] \\ \triangleq p_{R_1}^{\xi_1} p_{R_1} + p_{\overline{R}_1}^{\xi_1}(1 - p_{R_1}) \\ P(\xi_2) = P(\xi_2|R_2)P(R_2) + P(\xi_2|\overline{R}_2)[1 - P(R_2)] \\ \triangleq p_{R_2}^{\xi_2} p_{R_2} + p_{\overline{R}_2}^{\xi_2}(1 - p_{R_2}) \end{cases} \tag{5.19}$$

In Eq. (5.19), $P(\xi_i|R_i)$ represents the conditional probability of event ξ_i occurring when R_i occurs. Obviously, when the causal mechanism is independent, $P(\xi_1|R_1)$ is unrelated to $P(\xi_2|R_2)$ and these two conditional probabilities can be assigned through simplified models. Since ξ_i essentially represents the same event C, the subscript i is only to distinguish the cause of its occurrence, so when R_1 and R_2 are both the cause of event C, $P(\xi_i|R_i)P(R_i)$ can be understood as a "contribution" of R_i occurrence to the probability of event C occurring. When ξ_i has the same state space as C and has a logical relationship similar to Eq. (5.18), $p_{R_i}^{\xi_i} = P(\xi_i|R_i)$ is called the contribution probability distribution of R_i to C.

From Eq. (5.18), according to the empirical relationship between measured indicators and failure probability and safety function probability operation rules of events, the occurrence probability of C is:

$$P(C) = P(\xi_1) + P(\xi_2) - P(\xi_1)P(\xi_2) \tag{5.20}$$

Substituting Eq. (5.19) into Eq. (5.20) yields:

$$\begin{aligned} P(C) = {} & p_{R_1}p_{R_2}\left(-p_{R_1}^{\xi_1}p_{R_2}^{\xi_2} + p_{R_1}^{\xi_1}p_{R_2}^{\xi_2} + p_{R_1}^{\xi_1}p_{R_2}^{\xi_2} - p_{R_1}^{\xi_1}p_{R_2}^{\xi_2}\right) \\ & + p_{R_1}\left(p_{R_1}^{\xi_1} - p_{R_1}^{\xi_1} - p_{R_1}^{\xi_1}p_{R_2}^{\xi_2} + p_{R_1}^{\xi_1}p_{R_2}^{\xi_2}\right) \\ & + p_{R_2}\left(p_{R_2}^{\xi_2} - p_{R_2}^{\xi_2} - p_{R_1}^{\xi_1}p_{R_2}^{\xi_2} + p_{R_1}^{\xi_1}p_{R_2}^{\xi_2}\right) \\ & + \left(p_{R_1}^{\xi_1} + p_{R_2}^{\xi_2} - p_{R_1}^{\xi_1}p_{R_2}^{\xi_2}\right) \end{aligned} \tag{5.21}$$

On the other hand, according to Bayes' theorem, the occurrence probability of C can also be expressed as:

$$\begin{aligned} P(C) = {} & P(C|R_1, R_2)P(R_1)P(R_2) \\ & + P(C|R_1, \overline{R}_2)P(R_1)[1 - P(R_2)] \\ & + P(C|\overline{R}_1, R_2)[1 - P(R_1)]P(R_2) \\ & + P(C|\overline{R}_1, \overline{R}_2)[1 - P(R_1)][1 - P(R_2)] \\ & \triangleq p_{R_1R_2}^{C}p_{R_1}p_{R_2} + p_{R_1R_2}^{C}p_{R_1}\left(1 - p_{R_2}\right) \\ & + p_{R_1R_2}^{C}\left(1 - p_{R_1}\right)p_{R_2} + p_{R_1R_2}^{C}\left(1 - p_{R_1}\right)\left(1 - p_{R_2}\right) \end{aligned} \tag{5.22}$$

Transforming Eq. (5.22) yields:

$$\begin{aligned} P(C) = {} & p_{R_1}p_{R_2}(p_{R_1R_2}^{C} - p_{R_1R_2}^{C} - p_{R_1R_2}^{C} + p_{R_1R_2}^{C}) \\ & + p_{R_1}(p_{R_1R_2}^{C} - p_{R_1R_2}^{C}) + p_{R_2}(p_{R_1R_2}^{C} - p_{R_1R_2}^{C}) + p_{R_1R_2}^{C} \end{aligned} \tag{5.23}$$

Comparing the same coefficient terms in Eqs. (5.21) and (5.23), a set of four equations can be obtained, and the solution of the equations is:

$$\begin{cases} p_{R_1R_2}^{C} = p_{R_1}^{\xi_1} + p_{R_2}^{\xi_2} - p_{R_1}^{\xi_1}p_{R_2}^{\xi_2} \\ p_{R_1R_2}^{C} = p_{R_1}^{\xi_1} + p_{R_2}^{\xi_2} - p_{R_1}^{\xi_1}p_{R_2}^{\xi_2} \\ p_{R_1R_2}^{C} = p_{R_1}^{\xi_1} + p_{R_2}^{\xi_2} - p_{R_1}^{\xi_1}p_{R_2}^{\xi_2} \\ p_{R_1R_2}^{C} = p_{R_1}^{\xi_1} + p_{R_2}^{\xi_2} - p_{R_1}^{\xi_1}p_{R_2}^{\xi_2} \end{cases} \tag{5.24}$$

The equations in (5.24) can be uniformly expressed as:

$$P(C|\mathbf{R}_1, \mathbf{R}_2) = P(\xi_1|\mathbf{R}_1) + P(\xi_2|\mathbf{R}_2) - P(\xi_1|\mathbf{R}_1)P(\xi_2|\mathbf{R}_2) \tag{5.25}$$

Further transforming Eq. (5.25) yields:

$$\begin{aligned}P(C|\mathbf{R}_1,\mathbf{R}_2) &= P(\xi_1|\mathbf{R}_1) + P(\xi_2|\mathbf{R}_2) - P(\xi_1|\mathbf{R}_1)P(\xi_2|\mathbf{R}_2)\\ &= P(\xi_1|\mathbf{R}_1)P(\xi_2|\mathbf{R}_2) + P(\xi_1|\mathbf{R}_1)[1 - P(\xi_2|\mathbf{R}_2)] + [1 - P(\xi_1|\mathbf{R}_1)]P(\xi_2|\mathbf{R}_2)\\ &= P(\xi_1|\mathbf{R}_1)P(\xi_2|\mathbf{R}_2) + P(\xi_1|\mathbf{R}_1)P(\overline{\xi}_2|\mathbf{R}_2) + P(\overline{\xi}_1|\mathbf{R}_1)P(\xi_2|\mathbf{R}_2)\\ &= P(\xi_1,\xi_2|\mathbf{R}_1,\mathbf{R}_2) + P(\xi_1,\overline{\xi}_2|\mathbf{R}_1,\mathbf{R}_2) + P(\overline{\xi}_1,\xi_2|\mathbf{R}_1,\mathbf{R}_2)\end{aligned} \tag{5.26}$$

On the other hand, according to the properties of probability, we have:

$$\begin{aligned}P(\overline{C}|\mathbf{R}_1,\mathbf{R}_2) &= 1 - P(C|\mathbf{R}_1,\mathbf{R}_2)\\ &= 1 - P(\xi_1|\mathbf{R}_1) - P(\xi_2|\mathbf{R}_2) + P(\xi_1|\mathbf{R}_1)P(\xi_2|\mathbf{R}_2)\\ &= P(\overline{\xi}_1|\mathbf{R}_1)P(\overline{\xi}_2|\mathbf{R}_2)\\ &= P(\overline{\xi}_1,\overline{\xi}_2|\mathbf{R}_1,\mathbf{R}_2)\end{aligned} \tag{5.27}$$

Consider another situation, if R_1 and R_2 occur simultaneously and C occurs only then, repeating the derivation of Eqs. (5.19–5.27), we can obtain:

$$\begin{cases}P(C|\boldsymbol{R}_1,\boldsymbol{R}_2) = P(\xi_1|\boldsymbol{R}_1)P(\xi_2|\boldsymbol{R}_2) = P(\xi_1,\xi_2|\boldsymbol{R}_1,\boldsymbol{R}_2)\\ P(\overline{C}|\boldsymbol{R}_1,\boldsymbol{R}_2) = P(\xi_1|\boldsymbol{R}_1)P(\overline{\xi}_2|\boldsymbol{R}_2) + P(\overline{\xi}_1|\boldsymbol{R}_1)P(\xi_2|\boldsymbol{R}_2) + P(\overline{\xi}_1|\boldsymbol{R}_1)P(\overline{\xi}_2|\boldsymbol{R}_2)\\ \qquad = P(\xi_1,\overline{\xi}_2|\boldsymbol{R}_1,\boldsymbol{R}_2) + P(\overline{\xi}_1,\xi_2|\boldsymbol{R}_1,\boldsymbol{R}_2) + P(\overline{\xi}_1,\overline{\xi}_2|\boldsymbol{R}_1,\boldsymbol{R}_2)\end{cases} \tag{5.28}$$

Observing Eqs. (5.26–5.28), the following general formula can be obtained:

$$P(\boldsymbol{C} = \gamma|\boldsymbol{R}_1,\boldsymbol{R}_2) = \sum_{\gamma_1 * \gamma_2 = \gamma} P(\xi_1 = \gamma_1|\boldsymbol{R}_1)P(\xi_2 = \gamma_2|\boldsymbol{R}_2) \tag{5.29}$$

In the formula, "*" is a basic synthetic operator. When C is a Boolean random variable (at this time ξ_i must also be a Boolean random variable because it has the same state space as C), the basic synthetic operator can be "logical or (∪)", "logical and (∩)", "logical or (⊕)". When C is a binary real random variable, as a natural extension of being a Boolean variable, the basic synthetic operator can be "maximum (max)", "minimum (min)", "addition (+)". Equation (5.29) can be extended to the situation of multiple factor events as shown in Fig. 5.4b:

$$P(\boldsymbol{C} = \gamma|\boldsymbol{R}_1,\cdots,\boldsymbol{R}_n) = \sum_{\gamma_1 * \cdots * \gamma_n = \gamma} P(\xi_1 = \gamma_1|\boldsymbol{R}_1)\cdots P(\xi_n = \gamma_n|\boldsymbol{R}_n) \tag{5.30}$$

Equation (5.30) is the general expression of the independent condition of the causal mechanism, and the related proof can be found in the literature (ZHANG N L; POOLED, 1996). Through this equation, the conditional probability distribution of them can be obtained from the contribution probability distribution of each factor in any group of risk factors that satisfy the independent condition of the causal mechanism.

2. Definition of coupling coefficient

(1) Dual factor coupling

Still taking the Bayesian network shown in Fig. 5.4 as an example, the 2^n conditional probabilities existing in this network can be uniformly expressed as:

$$P(C|\boldsymbol{R}_1, \cdots, \boldsymbol{R}_n) \tag{5.31}$$

Among the 2^n conditional probabilities represented by Eq. (5.31), according to the basic assumption 1 of the risk factor coupling effect, we have:

$$P\left(\boldsymbol{C}|\bigcap_{i=1}^{n}\overline{\boldsymbol{R}}_i\right) = 0 \tag{5.32}$$

Assuming that the relationship between each factor event is "logical OR", and R_j, R_k and $\underset{i\neq j,i\neq k}{\cap} R_i$ are causally independent of each other, then when only R_i, i = 1, …, n only R_j occurs, according to Eq. (5.30), we have:

$$
\begin{aligned}
P\left[C|R_j\bar{R}_k\left(\underset{i\neq j,i\neq k}{\cap}\bar{R}_i\right)\right] &= \sum_{\gamma_j\cup\gamma_k\cup\gamma_i=T} P(\xi_j=\gamma_j|R_j)P(\xi_k=\gamma_k|\bar{R}_k)\,P\left(\underset{i\neq j,i\neq k}{\cup}\xi_i=\gamma_i|\bigcap_{i\neq j,i\neq k}\bar{R}_i\right)\\
&= P(\xi_j|R_j)P(\xi_k|\bar{R}_k)P\left(\underset{i\neq j,i\neq k}{\cup}\xi_i|\bigcap_{i\neq j,i\neq k}\bar{R}_i\right)\\
&+ P(\bar{\xi}_j|R_j)P(\xi_k|\bar{R}_k)P\left(\underset{i\neq j,i\neq k}{\cup}\xi_i|\bigcap_{i\neq j,i\neq k}\bar{R}_i\right)\\
&+ P(\xi_j|R_j)P(\bar{\xi}_k|\bar{R}_k)P\left(\underset{i\neq j,i\neq k}{\cup}\xi_i|\bigcap_{i\neq j,i\neq k}\bar{R}_i\right)\\
&+ P(\xi_j|R_j)P(\xi_k|\bar{R}_k)P\left(\overline{\bigcup_{i\neq j,i\neq k}\xi_i}|\bigcap_{i\neq j,i\neq k}\bar{R}_i\right)\\
&+ P(\bar{\xi}_j|R_j)P(\bar{\xi}_k|\bar{R}_k)P\left(\underset{i\neq j,i\neq k}{\cup}\xi_i|\bigcap_{i\neq j,i\neq k}\bar{R}_i\right)
\end{aligned}
$$

$$
\begin{aligned}
&+ P\left(\bar{\xi}_j|R_j\right)P\left(\xi_k|\bar{R}_k\right)P\left(\overline{\bigcup_{i\neq j,i\neq k}\xi_i}|\bigcap_{i\neq j,i\neq k}\bar{R}_i\right)\\
&+ P\left(\xi_j|R_j\right)P\left(\bar{\xi}_k|\bar{R}_k\right)P\left(\overline{\bigcup_{i\neq j,i\neq k}\xi_i}|\bigcap_{i\neq j,i\neq k}\bar{R}_i\right)
\end{aligned}
\tag{5.33}
$$

From the concept of causal mechanism independence, when a part of the factor events is causally independent of other factor events, this part of the factor events and its result events can be regarded as an independent causal system, which obeys the basic assumption of the risk factor coupling coefficient. Therefore, we have:

$$
\begin{cases}
P\left(\xi_k|\overline{R}_k\right) = P\left(\bigcup_{i\neq j,i\neq k}\xi_i|\bigcap_{i\neq j,i\neq k}\overline{R}_i\right) = 0\\
P\left(\overline{\xi}_k|\overline{R}_k\right) = P\left(\overline{\bigcup_{i\neq j,i\neq k}\xi_i}|\bigcap_{i\neq j,i\neq k}\overline{R}_i\right) = 1
\end{cases}
\tag{5.34}
$$

In Eq. (5.33), all terms on the right side of the equation except the last one are zero, that is:

$$
P\left[C|R_j\overline{R}_k\left(\bigcap_{i\neq j,i\neq k}\overline{R}_i\right)\right] = P\left(\xi_j|R_j\right)
\tag{5.35}
$$

When only R_k occurs, the same can be obtained:

$$
P\left[C|R_j\overline{R}_k\left(\bigcap_{i\neq j,i\neq k}\overline{R}_i\right)\right] = P\left(\xi_j|R_j\right)
\tag{5.36}
$$

When R_j and R_k occur at the same time, we have:

$$
\begin{aligned}
&P\left[C|R_jR_k\left(\bigcap_{i\neq j,i\neq k}\overline{R}_i\right)\right]\\
&= P\left(\xi_j|R_j\right)P(\xi_k|R_k)\\
&+ P\left(\overline{\xi}_j|R_j\right)P(\xi_k|R_k)\\
&+ P\left(\xi_j|R_j\right)P\left(\overline{\xi}_k|R_k\right)\\
&= P\left(\xi_j|R_j\right) + P(\xi_k|R_k)\\
&- P\left(\xi_j|R_j\right)P(\xi_k|R_k)
\end{aligned}
\tag{5.37}
$$

Transforming Eq. (5.37), we get:

$$\begin{aligned}
P\left[C|R_jR_k\left(\underset{i\neq j,i\neq k}{\cap}\overline{R}_i\right)\right] &= P(\xi_j|R_j) + [1-P(\xi_j|R_j)]P(\xi_k|R_k) \\
&= P(\xi_k|R_k) + [1-P(\xi_k|R_k)]P(\xi_j|R_j) \\
&\geq \max\{P(\xi_j|R_j), P(\xi_k|R_k)\} \\
&= \max\left\{P\left[C|R_j\overline{R}_k\left(\underset{i\neq j,i\neq k}{\cap}\overline{R}_i\right)\right], P\left[C|\overline{R}_jR_k\left(\underset{i\neq j,i\neq k}{\cap}\overline{R}_i\right)\right]\right\}
\end{aligned} \tag{5.38}$$

Equation (5.38) shows that under the condition of causal independence of R_j and R_k, their conditional probabilities satisfy the basic assumption 2 of risk factor coupling. On the other hand, we have:

$$\begin{aligned}
P(C|R_j) &= P\left[C|R_j\left(\bigcup_{i\neq j}R_i\right)\right]P\left(\bigcup_{i\neq j}R_i\right) \\
&\quad + P\left[C|R_j\left(\bigcap_{i\neq j}\overline{R}_i\right)\right]P\left(\bigcap_{i\neq j}\overline{R}_i\right) \\
&= \left[\begin{array}{l} P(\xi_j|R_j)P\left(\bigcup_{i\neq j}\xi_i|\bigcup_{i\neq j}R_i\right) + P(\xi_j|R_j)P\left(\overline{\bigcup_{i\neq j}\xi_i}|\bigcup_{i\neq j}R_i\right) \\ +P(\overline{\xi}_j|R_j)P\left(\bigcup_{i\neq j}\xi_i|\bigcup_{i\neq j}R_i\right) \end{array}\right]P\left(\bigcup_{i\neq j}R_i\right) \\
&\quad + \left[\begin{array}{l} P(\xi_j|R_j)P\left(\bigcup_{i\neq j}\xi_i|\bigcap_{i\neq j}\overline{R}_i\right) + P(\xi_j|R_j)P\left(\overline{\bigcup_{i\neq j}\xi_i}|\bigcap_{i\neq j}\overline{R}_i\right) \\ +P(\overline{\xi}_j|R_j)P\left(\underset{i\neq j}{\cup}\xi_i|\bigcap_{i\neq j}\overline{R}_i\right) \end{array}\right]P\left(\bigcap_{i\neq j}\overline{R}_i\right) \\
&= P(\xi_j|R_j) + P(\overline{\xi}_j|R_j)P\left(\bigcup_{i\neq j}\xi_i|\bigcup_{i\neq j}R_i\right)P\left(\bigcup_{i\neq j}R_i\right) \geq P(\xi_j|R_j) \\
&= P\left[C|R_j\overline{R}_k\left(\bigcap_{i\neq j,i\neq k}\overline{R}_i\right)\right]
\end{aligned} \tag{5.39}$$

In addition, assuming that all factor events except R_j and R_k have not occurred, the probability of the occurrence of the result event C is:

$$
\begin{aligned}
P\left(C| \bigcap_{i\neq j, i\neq k} \overline{R}_i\right) &= P\left[C|R_jR_k\left(\bigcap_{i\neq j, i\neq k} \overline{R}_i\right)\right]P(R_j)P(R_k) \\
&+ P\left[C|R_j\overline{R}_k\left(\bigcap_{i\neq f, i\neq k} \overline{R}_i\right)\right]P(R_j)P(\overline{R}_k) \\
&+ \left[C|\overline{R}_jR_k\left(\bigcap_{i\neq j, i\neq k} \overline{R}_i\right)\right]P(\overline{R}_j)P(R_k) \\
&= P\left[C|R_j\left(\bigcap_{i\neq j}\overline{R}_i\right)\right]P(R_j) + P\left[C|R_k\left(\bigcap_{i\neq k}\overline{R}_i\right)\right]P(R_k) \\
&- P\left[C|R_j\left(\bigcap_{i\neq j}\overline{R}_i\right)\right]\left[C|R_k\left(\bigcap_{i\neq k}\overline{R}_i\right)\right]P(R_j)P(R_k) \\
&= P\left[C|R_j\left(\bigcap_{i\neq j}\overline{R}_i\right)\right]P(R_j) \\
&+ \{1 - P[C|R_j(\cap_{i\neq j}\overline{R}_i)]P(R_j)\}P\left[C|R_k\left(\bigcap_{i\neq k}\overline{R}_i\right)\right]P(R_k) \\
&= P\left[C|R_k\left(\bigcap_{i\neq k}\overline{R}_i\right)\right]P(R_k) \\
&+ \left\{1 - P\left[C|R_k\left(\bigcap_{i\neq k}\overline{R}_i\right)\right]P(R_k)\right\}P\left[C|R_j\left(\bigcap_{i\neq j}\overline{R}_i\right)\right]P(R_j) \\
&\geq \max\{P[C|R_j(\cap_{i\neq j}\overline{R}_i)]P(R_j), P[C|R_k(\cap_{i\neq k}\overline{R}_i)]P(R_k)\}
\end{aligned} \tag{5.40}
$$

Equations (5.39) and (5.40) show that the independent relationship of the causal mechanism can also satisfy basic assumptions 3 and 4. From Eq. (5.35) to Eq. (5.40), it can be seen that when the causal mechanisms between two factor events and other factor events under the same result event are independent, the conditional probability related to the occurrence of these two factor events (i.e., when other factor events do not occur) satisfies the basic assumption of risk factor coupling, and is only related to the contribution probability distribution of these two factor events. It can be seen that the meaning represented by the independence of the causal mechanism is essentially that there is no coupling effect between the risk factors that have this relationship. Through the local structure of causal mechanism independence, the conditional probability when there is no coupling effect between risk factors can be obtained, thereby realizing the separation of the coupling effect, and then defining the dual factor coupling coefficient.

Still using the Bayesian network shown in Fig. 5.4 as the background, suppose R_i and R_j are causally independent of other factor events.

① When there is no coupling relationship between R_i and R_j (causal mechanism independent), the conditional probability of the result event C under the condition that only R_i (or R_j) occurs is defined as the "event-event connection strength $\alpha_i|_C$", expressed as:

$$\alpha_i|_C = P\left[C|R_i\left(\bigcap_{k\neq i}\overline{R}_k\right)\right] = P(\xi_i|R_i) \tag{5.41}$$

② When there is no coupling relationship between R_i and R_j, the conditional probability of C under the condition that R_i and R_j occur simultaneously is defined as the "basic conditional probability $\alpha_{ij}^U\Big|_C$", expressed as:

$$\alpha_{ij}^U\Big|_C = P\left[C|R_iR_j\left(\bigcap_{k\neq i,k\neq j}\overline{R}_k\right)\right] = P(\xi_i|R_i) + P(\xi_j|R_j) - P(\xi_i|R_i)P(\xi_j|R_j) \tag{5.42}$$

③ When there is a coupling relationship between R_i and R_j, theoretically, the conditional probability of the result event C under the condition of a single factor occurrence can no longer be obtained based on the principle of causal mechanism independence. However, from a general conceptual perspective, people rarely consider other factors when judging the strength of the causal relationship of a single factor event, and naturally will not consider the coupling effect with other factor events. In addition, the definition of the coupling coefficient also needs a certain benchmark, that is, the question of "what is the situation when there is no coupling effect". Therefore, it is more reasonable to use the conditional probability of the result event when a single factor occurs as the benchmark. Therefore, even when there is a coupling relationship between R_i and R_j, the conditional probability when a single factor R_i or R_j occurs is still represented by Eq. (5.41).

④ When there is a coupling relationship between R_i and R_j, according to the definition of the coupling effect, the conditional probability of the result event C should be different from Eq. (5.42). The conditional probability of C when R_i and R_j occur simultaneously under the existence of a coupling relationship is defined as the "coupling conditional probability $\alpha_{ij}^P\Big|_C$", and the coupling coefficient can be defined as follows:

$$\beta_i^i|_C = \beta_j^i\Big|_C = \begin{cases} \dfrac{\alpha_{ij}^P\big|_C - \alpha_{ij}^U\big|_C}{1-\alpha_{ij}^U\big|_C}, & \alpha_{ij}^P\Big|_C \geq \alpha_{ij}^U\Big|_C \\ \dfrac{\alpha_{ij}^P\big|_C - \alpha_{ij}^U\big|_C}{\alpha_{ij}^P\big|_C}, & \alpha_{ij}^P\Big|_C < \alpha_{ij}^U\Big|_C \end{cases} \tag{5.43}$$

In the formula, $\beta_i^j\big|_C$ is the dual factor coupling coefficient of the factor event denoted by i and the factor event denoted by j to the result event C. From Eq. (5.43), it can be seen that the value range of the dual-factor coupling coefficient $\beta_i^j\big|_C$ is $[-1, 1]$. When $\beta_i^j\big|_C$ is 0, the conditional probability $P\left[C\middle|R_iR_j\left(\bigcap_{k\neq i,k\neq j}\overline{R}_k\right)\right]$ takes the basic conditional probability and is calculated according to Eq. (5.42). In this case, the factor events R_i and R_j have independent effects on C, that is, there is no coupling effect between R_i and R_j; When $\beta_i^j\big|_C$ taking a positive value, the conditional probability $P\left[C\middle|_iR_j\left(\bigcap_{k\neq i,k\neq j}\overline{R}_k\right)R\right]$ is greater than the base conditional probability (when $\beta_i^j\big|_C$ is 1, the simultaneous occurrence of R_i and R_j will make C a certain event), that is, R_i and R_j have a coupling amplification effect, also known as positive coupling; at the same time, when $\beta_i^j\big|_C$ taking a negative value, the conditional probability $P\left[C\middle|R_iR_j\left(\bigcap_{k\neq i,k\neq j}\overline{R}_k\right)\right]$ is less than the base conditional probability (when $\beta_i^j\big|_C$ is -1, the simultaneous occurrence of R_i and R_j will make C impossible), that is, Ri and Rj have a coupling suppression effect, also known as negative coupling. When there is a coupling effect, the object of the coupling coefficient (i.e., the result event C) is called the coupling node.

(2) Multi-factor coupling

As shown in Fig. 5.5a, when there are multiple factors under the coupling node C that have a coupling effect on each other, such as R_1, R_2, …, R_n, there exist C_n^2 two-factor coupling coefficients $\beta_i^j(i, j = 1, 2, \cdots, n; i \neq j)$ between these factor events. According to the definition, the risk factor coupling effect is an objective phenomenon determined by the inherent physical mechanism. Therefore, when there is a coupling effect between two or more risk factors, the strength of the coupling effect when these factors occur together should be a comprehensive reflection of the coupling effects between each pair of factors.

Suppose there are three mutually coupled factors R_1, R_2, and R_3 under C. Assuming that these three factors are all positively coupled, then according to the principle of independent causality, we should have:

$$
\begin{aligned}
\alpha_{123}^{P}|C = P(C|R_1R_2R_3) &> \alpha_{123}^{U}\big|_C \\
&= \sum_{\gamma_1\cup\gamma_2\cup\gamma_3=T} P(\xi_1=\gamma_1|R_1)P(\xi_2=\gamma_2|R_2)P(\xi_3=\gamma_3|R_3) \\
&= P(\xi_1|R_1)P(\xi_2|R_2)P(\xi_3|R_3) + P(\overline{\xi}_1|R_1)P(\xi_2|R_2)P(\xi_3|R_3) \\
&+ P(\xi_1|R_1)P(\overline{\xi}_2|R_2)P(\xi_3|R_3) + P(\xi_1|R_1)P(\xi_2|R_2)P(\overline{\xi}_3|R_3) \\
&+ P(\overline{\xi}_1|R_1)P(\overline{\xi}_2|R_2)P(\xi_3|R_3) + P(\overline{\xi}_1|R_1)P(\xi_2|R_2)P(\overline{\xi}_3|R_3) \\
&+ P(\xi_1|R_1)P(\overline{\xi}_2|R_2)P(\overline{\xi}_3|R_3)
\end{aligned}
$$

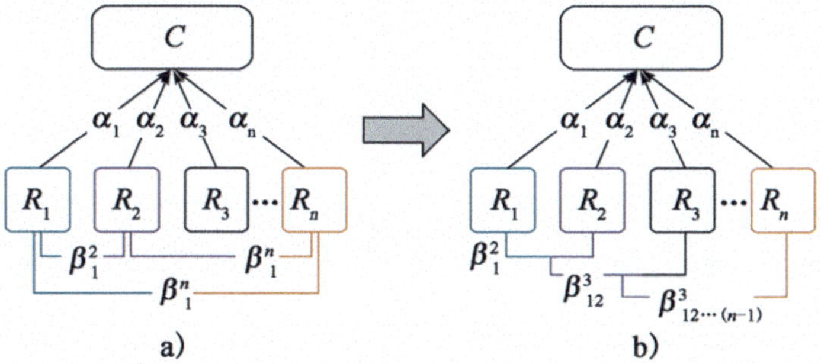

Fig. 5.5 Multi-factor coupling model schematic

$$\begin{aligned}
&= P(\xi_1|R_1) + P(\xi_2|R_2) + P(\xi_3|R_3) - P(\xi_1|R_1)P(\xi_2|R_2) \\
&- P(\xi_1|R_1)P(\xi_3|R_3) - P(\xi_2|R_2)P(\xi_3|R_3) \\
&+ P(\xi_1|R_1)P(\xi_2|R_2)P(\xi_3|R_3)
\end{aligned} \tag{5.44}$$

Theoretically, the coupling effects between multiple factors should exist simultaneously, that is, the coupling effect is independent of the "order" of coupling. Therefore, after simple generalization and transformation of formulas (5.42) and (5.43), it can be found that this order independence can be expressed as:

$$\begin{aligned}
\alpha_{123} &= \left(1 - \beta_{12}^3\right)(\alpha_{12} + \alpha_3 - \alpha_{12}\alpha_3) + \beta_{12}^3 \\
&= \left(1 - \beta_{13}^2\right)(\alpha_{13} + \alpha_2 - \alpha_{13}\alpha_2) + \beta_{13}^2 \\
&= \left(1 - \beta_{23}^1\right)(\alpha_{23} + \alpha_1 - \alpha_{23}\alpha_1) + \beta_{23}^1
\end{aligned} \tag{5.45}$$

In the formula, β_{ij}^k is the three-factor coupling coefficient, defined similarly to (5.43). As mentioned earlier, β_{ij}^k should have some functional relationship with β_i^k and β_j^k. However, if formulas (5.44) and (5.45) are to be satisfied at the same time, it will introduce a very strict constraint condition between β_i^k, β_j^k and β_i^j, which does not conform to the objective situation in most cases. In fact, the coupling effect of more than three risk factors is a "three-body problem", any slight change in the initial conditions will have unpredictable effects on the final result, and there is no general solution. Therefore, we cannot give a multi-factor coupling coefficient under theoretically consistent conditions. To solve this problem, when dealing with multi-factor coupling, it is assumed that the coupling effect is in the form of a "coupling tree", which acts in a certain order layer by layer. As shown in Fig. 5.5b, the coupling effect first occurs between R_1 and R_2, and then acts with R_3 to R_n one by one, expressed as:

$$
\begin{aligned}
\alpha_{12\cdots n} &= f_n\left(\alpha_{12\ldots(n-1)}, \alpha_n, \beta^n_{12\ldots(n-1)}\right) \\
&= f_n\left[f_{n-1}\left(\alpha_{12\ldots(n-2)}, \alpha_{n-1}, \beta^{n-1}_{12\ldots(n-2)},\right), \alpha_n, \beta^n_{12\ldots(n-1)}\right] \\
&= \cdots \\
&= f_n\left\{f_{n-1}\left\{f_{n-2}\left[\cdots, f_2\left(\alpha_1, \alpha_2, \beta^2_1\right), \cdots\right], \cdots\right\}, \cdots\right\}
\end{aligned}
\tag{5.46}
$$

The multi-factor coupling coefficient $\beta^n_{12\cdots(n-1)}$ in formula (5.46) can be taken as:

$$
\beta^n_{12\ldots(n-1)} = \max_{i \neq n}\left\{\frac{\beta^n_i + \left|\beta^n_i\right|}{2}\right\} + \min_{i \neq n}\left\{\frac{\beta^n_i - \left|\beta^n_i\right|}{2}\right\}
\tag{5.47}
$$

It should be noted that the "coupling tree" is not unique in the same problem, the selection of the coupling tree will have a certain impact on the analysis results, and in actual operation, factors with strong coupling effects should be placed at the lower level of the coupling tree as much as possible.

5.2.4 *The Role of Risk Factor Coupling Coefficient*

The introduction of the risk factor coupling coefficient is to quantitatively measure the strength of the risk factor coupling effect. However, due to the objectivity of the risk factor coupling effect, when we actually carry out risk analysis, we can generally consider the coupling effect of risk factors subconsciously based on our own knowledge or experience. However, this unconscious analysis may not only lack accuracy, but also it is very difficult to directly consider the coupling effect of risk factors when encountering complex or unclear coupling relationships. The introduction of the risk factor coupling coefficient can not only improve the accuracy of the coupling effect analysis, but also greatly simplify the logic of risk analysis.

Take the construction of the Bayesian network conditional probability table (CPT) as an example. According to the definition of the Bayesian network, describing a causal relationship of an event node with n factors requires 2^n independent conditional probability parameters, which makes it very difficult to construct a Bayesian network of medium or larger scale using the direct assignment method. The risk factor coupling coefficient is proposed on the basis of the principle of causal mechanism independence. When the causal mechanisms between n factor event nodes are mutually independent, the number of independent parameters required to establish the CPT will be reduced to 2n. However, in actual problems, the causal mechanism independence can generally only be established between a few factor events, and its application conditions are limited. Through the theory of risk factor coupling, we can consider the coupling effect between factors through the risk factor coupling coefficient on the basis of retaining the excellent mathematical properties brought by the principle of causal mechanism independence. In this way, a CPT of an event node with n factors can be constructed through n connection strength parameters and C_n^2 double-factor coupling coefficient, and the required parameters are reduced from 2^n to n(n + 1) /2, significantly reducing the complexity of the assignment parameters while providing a method for quantitatively characterizing the coupling effect of risk factors.

5.2.5 *Calculation Method of Risk Factor Coupling Coefficient*

1. Calculation idea of coupling coefficient

As mentioned earlier, risk has objectivity, but this objectivity generally only exists at the theoretical level. People can only approach objective risk through subjective risk analysis, but they can never fully grasp the full picture of objective risk. The coupling effect of risk factors is also an objectively existing phenomenon, which is not shifted by the cognition of the analyst and the logic of risk analysis. However, the definition of the risk factor coupling coefficient requires us to accurately obtain the risk with the coupling effect and the risk without the coupling effect. Therefore, under the premise that there is an objective coupling effect between the risk factors to be sought, if we use the observation method to measure the risk relatively objectively, then "no coupling effect" is just a situation fabricated due to insufficient subjective cognition, which does not exist in reality and naturally cannot be measured; on the other hand, if we use the inference method to measure the risk subjectively, then the risk "with coupling effect" must be calculated with the help of the risk factor coupling coefficient, thus forming a logical dead zone. This contradiction between subjectivity and objectivity makes the calculation of the risk factor coupling coefficient very difficult. Therefore, the calculation of the risk factor coupling coefficient must adopt a combination of subjective and objective methods, that is, it needs to use both the construction risk inference method and the construction risk measurement method.

2. Method for calculating the coupling coefficient

Figure 5.6 shows a simplified Bayesian network with two factor event nodes R_1, R_2 and a result event node (also the coupling node of R_1 and R_2) C.

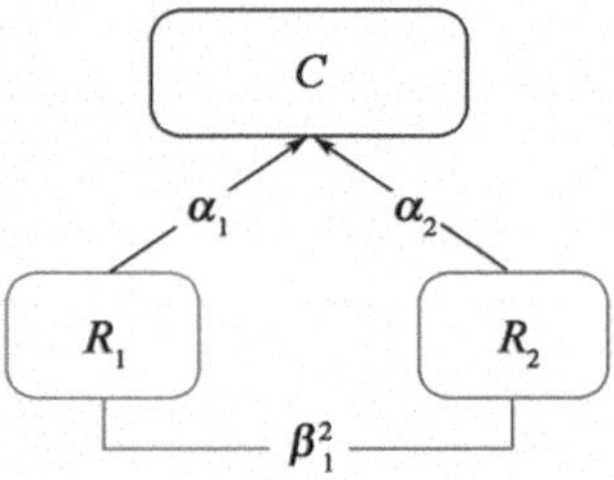

Fig. 5.6 Simplified Bayesian network diagram

Let the coupling coefficient between the two factor events be $\beta_1^2\big|_C$. Consider the following four situations.

(1) R_1, R_2 occur simultaneously, according to the definition of the coupling coefficient, the probability of the occurrence of the result event C is:

$$P(C|R_1R_2) = p^{12} = \begin{cases} \beta_1^2|_C + (1-\beta_1^2|_C) \\ (\alpha_1|_C + \alpha_2|_C - \alpha_1|_C + \alpha_2|_C), & \beta_1^2|_C \in [0,1] \\ (1+\beta_1^2|_C)(\alpha_1|_C + \alpha_2|_C - \alpha_1|_C + \alpha_2|_C), & \beta_1^2 \in [-1,0) \end{cases} \tag{5.48}$$

(2) R_1 occurs, R_2 does not occur, the probability of the occurrence of the result event C is:

$$P\left(C|R_1\overline{R}_2\right) = p_2^1 = \alpha_1|_C \tag{5.49}$$

(3) R_2 occurs, R_1 does not occur, the probability of the occurrence of the result event C is:

$$P\left(C|\overline{R}_1R_2\right) = p_1^2 = \alpha_2|_C \tag{5.50}$$

(4) Neither R_1 nor R_2 occur, according to the basic assumption of the coupling coefficient, the probability of the occurrence of the result event C is:

$$P\left(C|\overline{R}_1\overline{R}_2\right) = p_{12} = 0 \tag{5.51}$$

To obtain the conditional probabilities in formulas (5.48–5.50), the Bayesian network can be expanded using the sub-model of urban large underground space construction risk measurement, and an "observation" node H_i is defined for each factor event R_1 and R_2, and the corresponding observation index η_i and judgment criterion G_i are given for each observation node, and it is defined as:

$$\xi_i = T \Leftrightarrow \eta_i \in G_i \tag{5.52}$$

In the formula, ξ_i is the result event corresponding to the contribution probability distribution of the factor event R_i.

From formula (5.52), it can be seen that the relationship between the observation node and the factor event node is essentially the same as the relationship between the sign event node and the intermediate event node in the urban large underground space construction risk measurement model. Through this definition, the actual values of the conditional probabilities in formulas (5.48–5.51) can be calculated by calculating the values of the observation index η_i under various conditions. Suppose the calculation result is:

$$\begin{cases} \hat{P}(C|R_1R_2) = \hat{p}^{12} \\ \hat{P}(C|R_1\overline{R}_2) = \hat{p}_2^1 \\ \hat{P}(C|\overline{R}_1R_2) = \hat{p}_1^2 \\ \hat{P}(C|\overline{R}_1\overline{R}_2) = \hat{p}_{12} \end{cases} \tag{5.53}$$

To distinguish from the theoretical value, the superscript "^" indicates the actual probability value calculated by the measurement model. Generally speaking, the calculated in formula (5.53) is $\hat{p}_{12} \neq 0$, which contradicts formula (5.51) to some extent. The reason for this phenomenon is that $\hat{p}_{12}$ calculated in formula (5.51) exists as a background probability in this case. According to the analysis of explicit and implicit causes, under the condition of relatively complete causal analysis, this part of the background probability should be very small. However, in many practical problems, a coupling node often has 2 More than one factor event, which leads to when we take one pair of factor events to calculate the coupling coefficient, this pair of factor events is an incomplete set of causes relative to its result event, which may lead to the value $\hat{p}_{12}$ being too large, even larger than $\hat{p}_{12}$ the difference relative to $\hat{p}_{12}$, the risk factor coupling coefficient's basic assumption. To calculate the coupling coefficient, it is necessary to exclude the influence of other factor events (whether they are explicit or implicit causes) on the probability values of R_1 and R_2. For this purpose, suppose there are other factor events under C that are independent of the causal mechanism of R_1 and R_2, represented by R_3 as their sum event, then according to the principle of independent causal mechanism, we have:

$$\begin{cases} \hat{p}^{12} = \hat{P}(C|R_1R_2\overline{R}_3) = P(\xi_1 \cup \xi_2|R_1R_2) + P(\overline{\xi_1 \cup \xi_2}|R_1R_2)P(\xi_3|\overline{R}_3) \\ \hat{p}_2^1 = \hat{P}(C|R_1\overline{R}_2\overline{R}_3) = P(\xi_1 \cup \xi_2|R_1\overline{R}_2) + P(\overline{\xi_1 \cup \xi_2}|R_1\overline{R}_2)P(\xi_3|\overline{R}_3) \\ \hat{p}_1^2 = \hat{P}(C|\overline{R}_1R_2\overline{R}_3) = P(\xi_1 \cup \xi_2|\overline{R}_1R_2) + P(\overline{\xi_1 \cup \xi_2}|\overline{R}_1R_2)P(\xi_3|\overline{R}_3) \\ \hat{p}_{12} = \hat{P}(C|\overline{R}_1\overline{R}_2\overline{R}_3) = P(\xi_1 \cup \xi_2|\overline{R}_1\overline{R}_2) + P(\overline{\xi_1 \cup \xi_2}|\overline{R}_1\overline{R}_2)P(\xi_3|\overline{R}_3) \end{cases} \tag{5.54}$$

The first term on the right side of each equation in formula (5.54) is the four conditional probabilities required to solve the coupling coefficient represented by formulas (5.48–5.51). According to the basic assumption of the risk factor coupling coefficient, for $\xi_1 \cup \xi_2$, R_1 and R_2 undoubtedly constitute its complete explicit cause, so we have $P(\xi_1 \cup \xi_2|\overline{R}_1\overline{R}_2) = p_{12} = 0$, and also:

$$\begin{cases} p^{12} = \hat{p}^{12} - (1 - p^{12})\hat{p}_{12} = \frac{\hat{p}^{12} - \hat{p}_{12}}{1 - \hat{p}_{12}} \\ p_2^1 = \hat{p}_2^1 - (1 - p_2^1)\hat{p}_{12} = \frac{\hat{p}_2^1 - \hat{p}_{12}}{1 - \hat{p}_{12}} \\ p_1^2 = \hat{p}_1^2 - (1 - p_1^2)\hat{p}_{12} = \frac{\hat{p}_1^2 - \hat{p}_{12}}{1 - \hat{p}_{12}} \end{cases} \tag{5.55}$$

Substituting formula (5.55) into formulas (5.48–5.51) yields:

(1) When $p^{12} \geq p_2^1 + p_1^2 - p_2^1 p_1^2$ (i.e. $\hat{p}^{12} \geq \hat{p}_2^1 + \hat{p}_1^2 - \hat{p}_2^1\hat{p}_1^2 - \hat{p}_{12}$,):

$$\beta_1^2\big|_C = \frac{p^{12} - (p_2^1 + p_1^2 - p_2^1p_1^2)}{1 - (p_2^1 + p_1^2 - p_2^1p_1^2)} = 1 - \frac{1 - \hat{p}^{12} - \hat{p}_{12} + \hat{p}^{12}\hat{p}_{12}}{1 - \hat{p}_2^1 - \hat{p}_1^2 + \hat{p}_2^1\hat{p}_1^2} \tag{5.56}$$

(2) When $p^{12} < p_2^1 + p_1^2 - p_2^1 p_1^2$ (i.e.,$\hat{p}^{12} < \hat{p}_2^1 + \hat{p}_1^2 - \hat{p}_2^1 \hat{p}_1^2 - \hat{p}_{12}$):

$$\beta_1^2\big|_C = \frac{p^{12} - \left(p_2^1 + p_1^2 - p_2^1 p_1^2\right)}{p_2^1 + p_1^2 - p_2^1 p_1^2} = \frac{\hat{p}^{12} - \hat{p}^{12}\hat{p}_{12}}{\hat{p}_2^1 + \hat{p}_1^2 - \hat{p}_2^1 \hat{p}_1^2 - \hat{p}_{12}} - 1 \tag{5.57}$$

Note: When $p^{12} = p_2^1 + p_1^2 - p_2^1 p_1^2$, due to the existence of the term $\hat{p}^{12}\hat{p}_{12}$ in the numerator, the coupling coefficient calculated by formulas (5.56) and (5.57) is not 0 (but very close to 0), which is a deviation caused by the basic assumption 1 of the coupling coefficient. Therefore, after correcting formulas (5.56) and (5.57), we get:

$$\beta_1^2\big|_C = \begin{cases} 1 - \frac{1-\hat{p}^{12}-\hat{p}_{12}}{1-\hat{p}_2^1-\hat{p}_1^2+\hat{p}_2^1\hat{p}_1^2}, \hat{p}^{12} \geq \hat{p}_2^1 + \hat{p}_1^2 - \hat{p}_2^1 \hat{p}_1^2 - \hat{p}_{12} \\ \frac{\hat{p}^{12}}{\hat{p}_2^1+\hat{p}_1^2-\hat{p}_2^1\hat{p}_1^2-\hat{p}_{12}} - 1, \hat{p}^{12} < \hat{p}_2^1 + \hat{p}_1^2 - \hat{p}_2^1 \hat{p}_1^2 - \hat{p}_{12} \end{cases} \tag{5.58}$$

5.3 Study on the Coupling Intensity of Risk Factors in Urban Large Underground Space Construction

5.3.1 Case Background

1. Project Overview

The design start and end mileage of Guangzhou South Station is CK0 + 133.900 ~ CK0 + 704.900, and the effective platform center mileage of the station is CK0 + 260.000. The entire station is divided into four underground levels, with the platform being an island structure with a width of 15 m. The station (open-cut section) is 571 m long, and the standard section is 25.54 m wide. The excavation depth of the station pit is 32.12–33.83 m, and the thickness of the top slab cover soil is 0.3–4.2 m. The total building area of the station is 63117 m^2, of which the main building area is 60985 m^2, the auxiliary building area is 2132 m^2, and the plan of Guangzhou South Station is shown in Fig. 5.7.

The construction method of the station is open excavation, and the overall structure is a reinforced concrete frame structure. The first underground level is the underground space transfer area, passenger station part, Guangfo Ring Station Hall, Glory Real Estate Basement and station reserved development space. The second underground level is the subway station hall, Guangfo Ring Station Hall, Glory Real Estate Basement and station reserved development space. The third underground level is the public transfer area and subway equipment management room area, and the fourth underground level is the subway platform level. The cross-section of Guangzhou South Station is shown in Fig. 5.8.

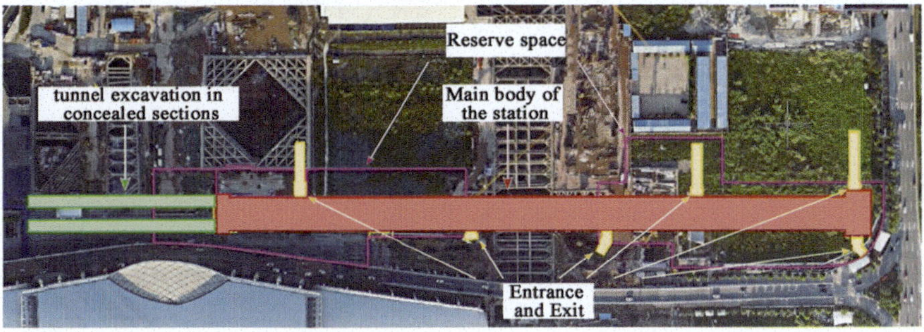

Fig. 5.7 Plan of Guangzhou South Station

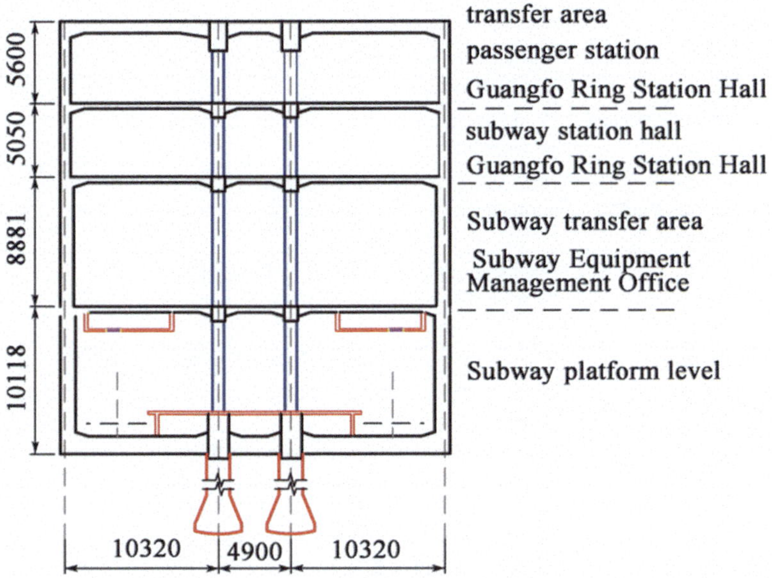

Fig. 5.8 Cross-section of Guangzhou South Station (unit of measurement: mm)

2. Engineering Hydrogeological Overview

(1) Engineering Geology

According to geological data and supplementary surveys, the distribution of rock and soil layers at Guangzhou South Station is obtained, as shown in Table 5.1, and the geological profile is shown in Fig. 5.9.

(2) Hydrogeology

The groundwater in the foundation pit is mainly divided into Quaternary loose pore water and fracture water in bedrock according to the storage method.

Table 5.1 Distribution of rock and soil layers at Guangzhou South Station

Layer number	Name of rock and soil	Main engineering geological features	Classification of geotechnical construction engineering
<1>	Artificial fill	Loose–Slightly compacted, poor water-richness, weak water permeability–Medium degree	I, II
<2-1B>	Mucky soil	Soft plastic, sticky soil, and a slightly permeable layer	II
<2–3>	Muddy medium coarse sand	Loose–slightly dense, with good water richness and moderate permeability	I
<4-2B>	Mucky soil	Soft plastic, poor water yield, micro-permeable layer	II
<5H-1>	Sandy cohesive soil	Plastic, poor water-bearing capacity, weak water permeability	II
<5H-2>	Sandy cohesive soil	Hard plastic, poor water yield, weak water permeability	II
<6H>	Fully weathered granite	Hard soil, poor water-richness, weak water permeability	III
<6H-1>	Fully weathered Huashan rock	Hard soil, poor water yield, weak water permeability	III
<7H>	Strongly weathered granite	Fragmented, poor water yield, weak water permeability	III, IV
<7H-1>	Strongly weathered Huashan rock	Fragmented, poor water yield, weak water permeability	III, IV
<8H>	Moderately weathered granite	The rock mass is relatively fragmented, partially complete, poor in water yield, and weak in water permeability	IV, V
<8H-1>	Moderately weathered huashan rock	The rock mass is relatively fragmented, partially complete, poor in water yield, and weak in water permeability	IV, V
<9H>	Slightly weathered granite	The rock mass is relatively complete, with poor water yield and weak water permeability	V
<9H-1>	Slightly weathered Huashan Rock	The rock mass is relatively complete, with poor water yield and weak water permeability	V

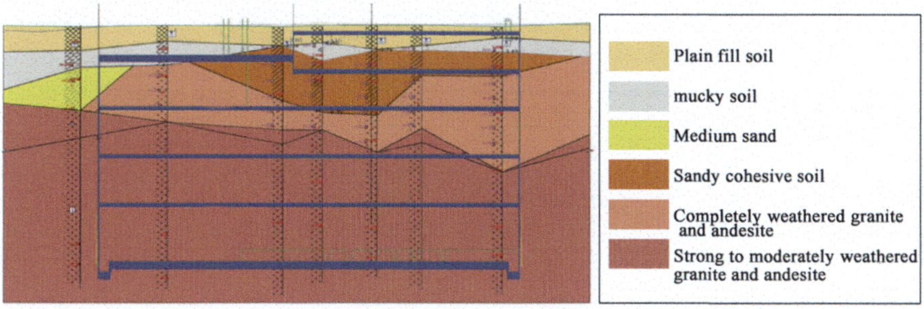

Fig. 5.9 Schematic diagram of the geological section of Guangzhou South Station

Quaternary loose pore water. It is mainly stored in the coarse sand < 2–3 > of the mud in the sea-land interaction phase, and its water content is closely related to the shape, size, particle size distribution, and clay content of the sand. The permeability of <2–3> is generally medium to strong. The artificial fill soil in the other soil layers of the Quaternary has general permeability, while the mud soil, flood deposit, and residual soil layers have the weakest permeability. Generally speaking, the groundwater in the sand layer of the survey area has a unified groundwater surface, and there are slight undulations in places with slightly larger local terrain changes. According to the results of this survey, the main aquifer in this site is covered with soft soil and sticky soil. The groundwater level is usually above the top plate of the main aquifer. In general, except for some areas with confined water, the Quaternary loose pore water in this site is mainly confined water. Some of the artificial fill soil is upper layer stagnant water. The Quaternary loose pore water is partly upper layer stagnant water, stored in the fill layer <1>, and this layer is uneven in composition, has general water richness, and weak to medium permeability.

Fracture water in bedrock. It is mainly stored in the strongly weathered zone and moderately weathered zone of granite, and the storage of groundwater is uneven. In areas where fissures are developed, the water volume is abundant and has a confined nature. According to regional experience, the permeability coefficient is generally 0.4–1.0 m/d. The weathered fissure water in the bedrock is confined water, and the confined water head elevation is 4.67–6.60 m. From the geological profile, the connection between the fracture water in bedrock layer and the Quaternary aquifer is small.

3. Station Construction Method

(1) Vertical stratification, longitudinal segmentation

According to the basic information of the deep foundation pit project of Guangzhou South Station, it is divided into five parts vertically and three areas longitudinally, and each part is constructed in a flow.

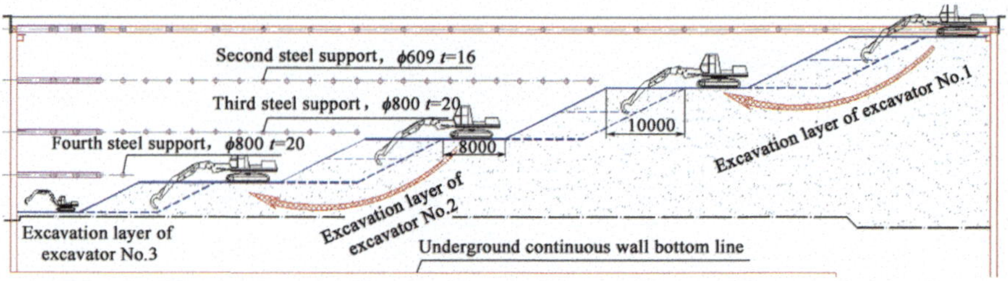

Fig. 5.10 Longitudinal excavation diagram (unit of measurement: mm)

(2) Longitudinal grooving, lateral expansion

The earthwork excavation is carried out in a "relay retreat" manner, as shown in Fig. 5.10, and the excavation is carried out in the order of "first middle groove and then side expansion". The foundation pit earthwork is excavated layer by layer, and each layer is excavated to 0.5 m below the next support, as shown in Fig. 5.11.

4. List of surrounding risk sources

This project is an open excavation station pit project, with a large excavation depth and length, and a complex surrounding environment. The construction safety risk sources of this project mainly include its own risk sources and surrounding environmental risk sources. Since the pit structure itself is relatively simple and has no other ancillary structures, its own risk source is the geology and structure itself. The surrounding environmental risk sources are shown in Table 5.2.

5.3.2 *Establishment of Risk Analysis Model*

1. Risk event identification

Based on the project background conditions and risk source situation, the construction safety risks of this project mainly include the following categories.

(1) Risk of damage, instability, or excessive deformation of the structure itself

The excavation depth of this project is 32.12 to 33.83 m, which is an ultra-deep pit. The structure will inevitably bear a large soil pressure during construction. The pit is a long strip structure with poor support conditions on the long side. If the internal support fails, it will fall into a very unfavorable stress state. The project is located in a dense urban area with existing buildings and pipelines around. If the deformation control is improper, it can easily cause associated deformation of the surrounding strata, thereby threatening the surrounding buildings.

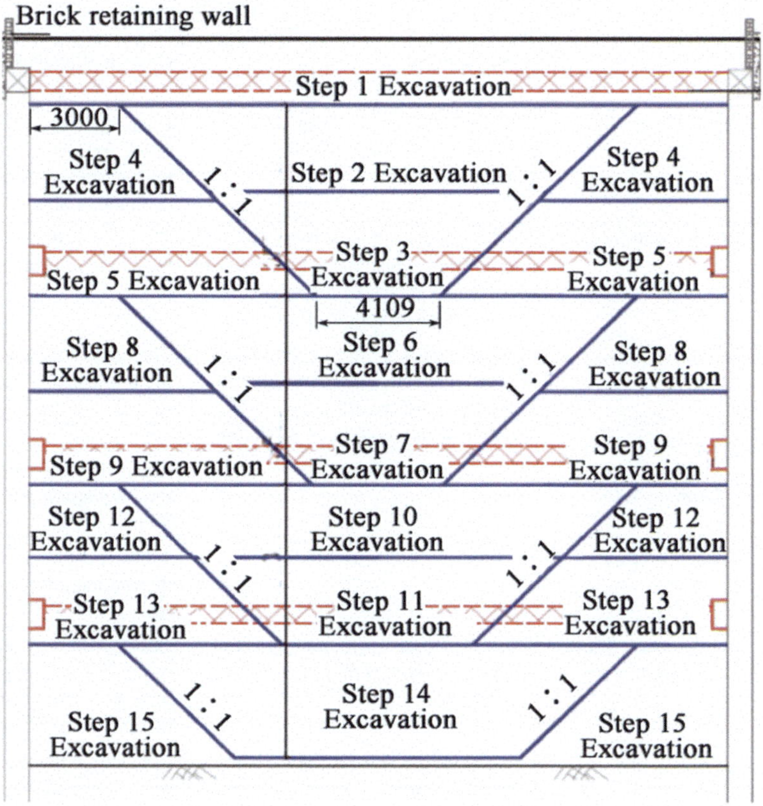

Fig. 5.11 Vertical excavation diagram (unit of measurement: mm)

(2) Damage or loss of use of surrounding buildings and pipelines

There are many ground buildings around this project, including important buildings such as train stations and high-rise buildings. The minimum net distance from the building to the pit is less than 50 m, which is within the influence range of pit excavation. The pipelines near the pit are dense and close, with the smallest net distance only 5 m from the edge of the pit. The pipeline materials are cement, cast iron and other materials with high brittleness. Although they have high rigidity and strength, if they are forced to move due to soil displacement, they are prone to brittle fracture.

2. Establish a construction risk inference Bayesian network model

Refer to the urban large underground space construction risk inference model, and establish the Guangzhou South Station construction risk inference Bayesian network model as shown in Fig. 5.12. The Bayesian network includes 3 risk event nodes, 4 intermediate event nodes, and 5 factor event nodes.

Table 5.2 List of risk sources in the surrounding environment

Serial number	Name of risk source	Risk source situation	Positional relationship
1	Luff Union Square	160 m reinforced concrete, pile foundation 4 m, basically intact appearance	Located on the east side of the foundation pit, with a nearest clear distance of about 130 m
2	Office building	The building is 25 m high and has 10 floors. The beams, slabs, and columns of the building are all made of C30 concrete with a weight of 25 kN/m	The nearest distance to the foundation pit is 50 m
3	Commercial building	The building is 15 m high and has 3 floors. The beams, slabs, and columns of the building are all made of C30 concrete with a weight of 25 kN/m	The nearest distance to the foundation pit is 36 m
4	Shipu Avenue South	Two-way one-lane road	The nearest distance to the foundation pit area is 30 m
5	Hanxi Avenue	Two-way four-lane road	The nearest distance to the foundation pit is 40 m
6	Guangzhou South Railway Station	3-story steel structure, underground diaphragm wall 30 m, basically intact appearance	Located on the west side of the foundation pit, with a nearest clear distance of about 400 m
7	Subway line 2, line 7	Located on the second floor of the Guangzhou South Railway Station, with three exits A, B, and C, the exterior is basically intact	Located on the west side of the foundation pit, with a nearest clear distance of about 370 m
8	Water supply pipe	The pipe diameter is 500 mm, the burial depth is 0.7 m, and the pipeline material is cast iron pipe	In the short side direction of the foundation pit, the nearest clear distance is 5 m
9	Water supply pipe	The pipe diameter is 500 mm, the burial depth is 0.8 m, and the pipeline material is cast iron pipe	In the short side direction of the foundation pit, the nearest clear distance is 6 m
10	Storm sewer	The pipe diameter is 500 mm, the burial depth is 2 m, and the pipeline material is cement pipe	In the long side direction of the foundation pit, the nearest clear distance is 7 m
11	Water supply pipe	The pipe diameter is 200 mm, the burial depth is 0.7 m, and the pipeline material is cast iron pipe	In the long side direction of the foundation pit, the nearest clear distance is 9 m

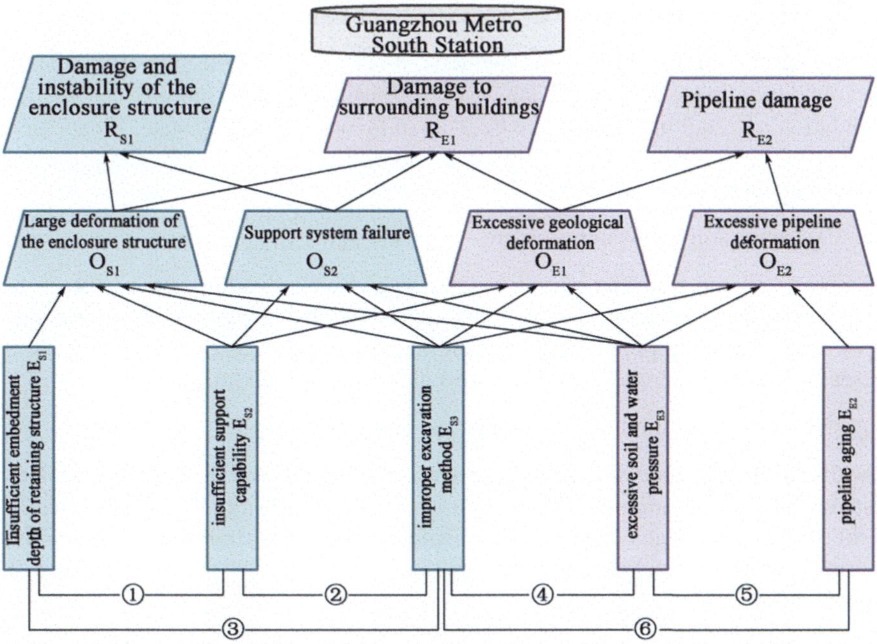

Fig. 5.12 Guangzhou South Station construction risk inference Bayesian network model

3. Coupling factor analysis

There are a total of 6 possible pairs of coupling factors in the model:

(1) Insufficient embedment depth of retaining structure—insufficient support capability (E_{S1}–E_{S2})

Insufficient embedment depth of retaining structure can lead to insufficient ability to resist overturning and "kicking" destruction. If the support performance is insufficient under this condition and cannot provide sufficient restraint, the risk of instability of the retaining structure will significantly increase.

(2) Insufficient support capability—improper excavation method (E_{S2}–E_{S3})

Improper excavation methods generally refer to not excavating in blocks and layers according to the design scheme, which can cause the pre-installed supports to bear extra loads. If insufficient support capability at this time, it is very likely to cause support failure and damage.

(3) Insufficient embedment depth of retaining structure—improper excavation method (E_{S1}–E_{S3})

Insufficient embedment depth of retaining structure combined with improper excavation methods can significantly reduce the anti-overturning safety factor of the retaining structure.

(4) Improper excavation method—excessive soil and water pressure (E_{S3}–E_{E1})

In the case of improper excavation methods, it is difficult for the support system of the foundation pit to form a reliable overall bearing structure in time, causing the foundation pit retaining structure to be in an unfavorable bending state. If excessive soil and water pressure occurs at this time, there is a high risk of bending or shear failure in the retaining structure.

(5) Excessive soil and water pressure—pipeline aging (E_{E1}–E_{E2})

Excessive soil and water pressure is often caused by overloading at the edge of the pit, so this adverse effect is also likely to have a negative impact on the surrounding pipelines. After the pipeline ages, its bending stiffness decreases and its brittleness increases, making it prone to cracking and damage under additional loads.

(6) Improper excavation method—pipeline aging (E_{S3}–E_{E2})

One direct consequence of improper excavation methods is an increase in the lateral displacement of the retaining structure, which in turn causes significant displacement in the strata at the edge of the pit. The displacement of the strata can cause passive deformation of the pipeline, and aged pipelines are likely to be damaged under such passive deformation.

4. Establish a construction risk measurement model corresponding to the Bayesian network model

The risk events in the construction risk inference Bayesian network model of Guangzhou South Station are taken as failure modes, and the intermediate events are taken as failure points to establish the construction risk measurement model of Guangzhou South Station as shown in Fig. 5.13. The failure indicators and failure criteria corresponding to each failure point are shown in Table 5.3.

5.3.3 Numerical Analysis Model

1. Model geometry

Due to the long longitudinal length of the foundation pit of Guangzhou South Station, modeling according to the actual size will lead to a significant increase in modeling workload and calculation cost. Therefore, this model retains the cross-sectional structure form, support system layout, and excavation method of Guangzhou South Station's foundation pit, and takes a longitudinal length of 100 m to establish the model abstractly. According to engineering experience, the depth of influence of foundation pit excavation is about 2 to 4 times the excavation depth, and the width of influence is about 3 to 4 times the excavation depth. In addition, since the foundation pit has symmetry in both structural form and excavation method on the cross-section, a half model is used for calculation. The final model size is 300 m × 112.5 m × 100 m, where the size of the foundation pit is 100 m × 12.5 m × 34 m.

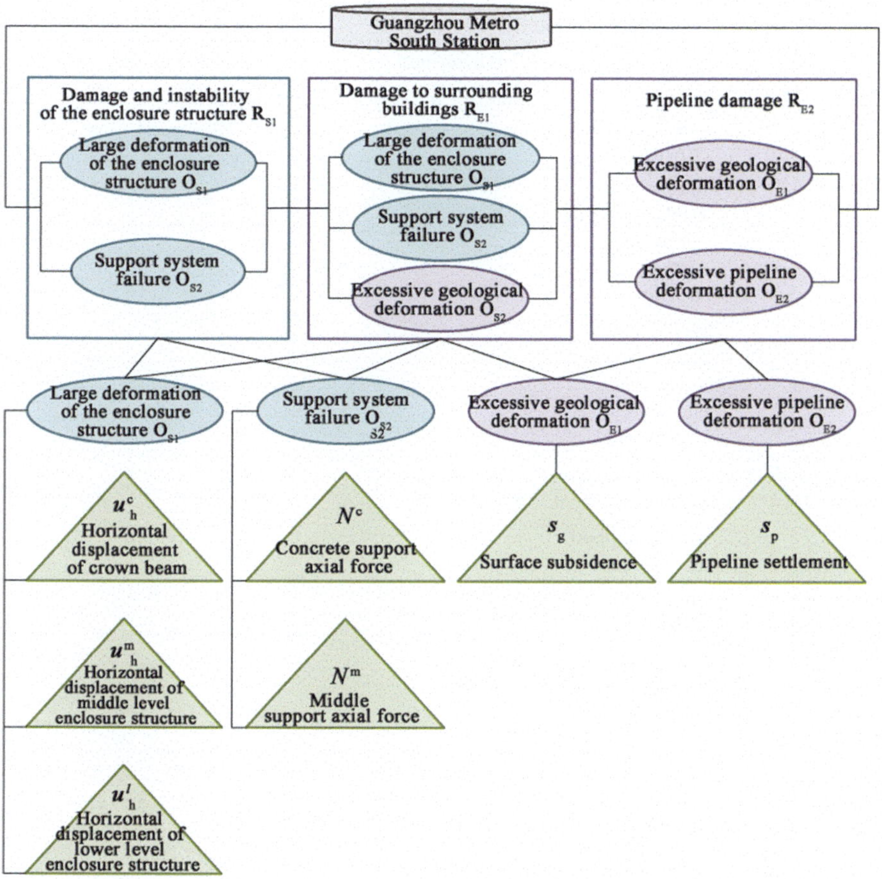

Fig. 5.13 Construction risk measurement model of Guangzhou South Station

The model simplifies the surrounding environment of the pit, including two buildings and two water pipes. The relevant parameters are shown in Table 5.4, and the relative relationship with the pit is shown in Fig. 5.14.

2. Unit and grid division

The Guangzhou South Station pit enclosure and support system consists of underground continuous walls, concrete supports, steel supports, temporary columns, and artificial expanded pile foundations. The stratum, buildings, underground continuous walls, artificial expanded piles, temporary columns, and longitudinal contact beams between the columns are simulated with hexahedral solid units; concrete supports, steel supports, and underground pipelines are simulated with beam units; the guide walls of the underground continuous walls are simulated with shell units. The model

Table 5.3 Table of failure indicators and failure criteria

Point of failure	Failure index	Failure criterion	Functional function
Large deformation of the enclosure structure O_{S1}	Horizontal displacement of crown beam u_h^c	$\max\{u_h^c, u_h^m, u_h^l\} \geq 10(\text{cm})$	$g_{S1} = 10 - \max\{u_h^c, u_h^m, u_h^l\}$
	Horizontal displacement of middle level enclosure structure u_h^m		
	Horizontal displacement of lower-level enclosure structure u_h^l		
Support system failure O_{S2}	Concrete support axial force N^c	$N^c \geq 5(\text{MN})$ 或 $N^m \geq 3(\text{MN})$	$g_{S2} = \min\{5 - N^c, 3 - N^m\}$
	Middle support axial force N^m		
Excessive geological deformation O_{E1}	Surface subsidence s_g	$s_g \geq 10(\text{cm})$	$g_{E1} = 10 - s_g$
Excessive pipeline deformation O_{E2}	Pipeline settlement s_p	$s_p \geq 5(\text{cm})$	$g_{E2} = 5 - s_p$

Table 5.4 Parameters of the surrounding environment of the foundation pit

Serial number	Type	Parameter description
①	Office building	The building is 25 m high and has 10 floors. The beams, slabs, and columns of the building are all made of C30 concrete with a weight of 25 kN/m and a distance of 50 m from the foundation pit
②	Commercial building	The building is 15 m high and has 3 floors. The beams, slabs, and columns of the building are all made of C30 concrete with a weight of 25 kN/m'. It is 36 m away from the foundation pit
③	Water supply pipe	Cast iron pipe, diameter 500 mm, buried depth 0.8 m, 5 m away from the nearest foundation pit
④	Storm sewer	Cement pipe, with a diameter of 500 mm, a burial depth of 2 m, and a distance of 7 m from the nearest foundation pit

has a total of 208,713 solid unit nodes and 2110 structural unit nodes, with 183,922 solid units, 1836 beam units, and 132 shell units. The model grid schematic is shown in Fig. 5.15.

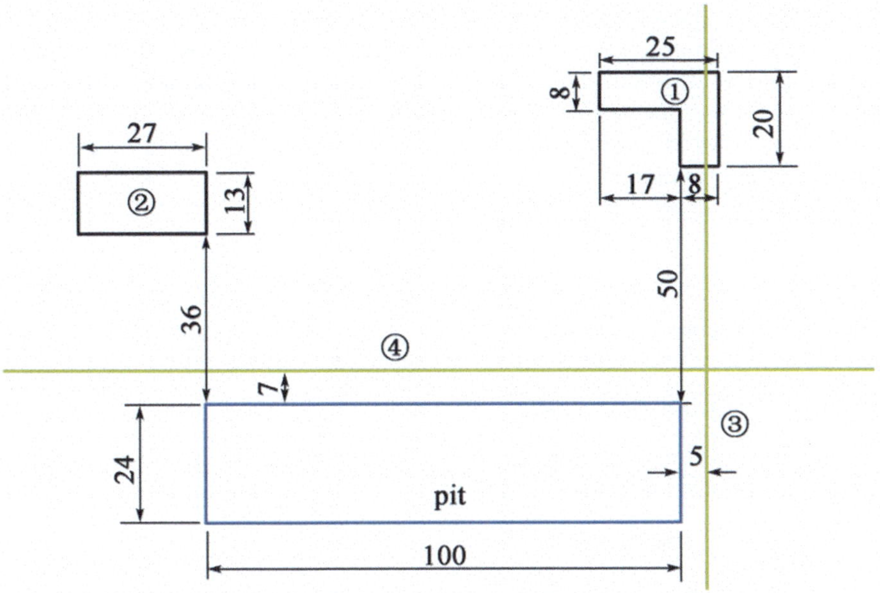

Fig. 5.14 Relative position diagram of the surrounding environment (unit of measurement: m)

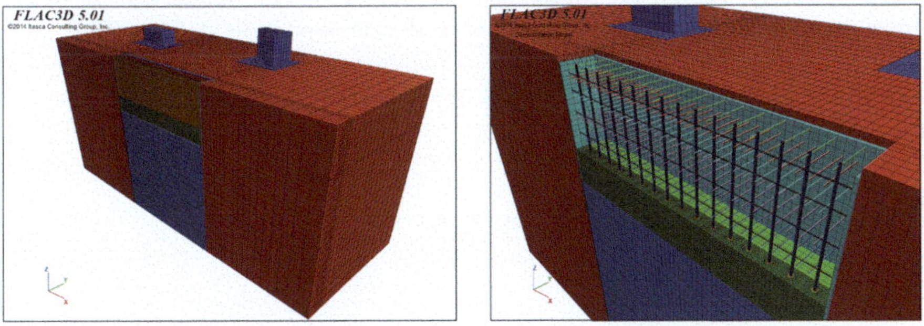

a)The stratum b)Support system

Fig. 5.15 Model grid schematic

3. Constitutive model and material parameters

(1) Stratum parameters

To control the number of design variables, the model simplifies the strata involved in the pit engineering, merging the original site's fill soil, silty clay, muddy soil, and sandy clay layers into a clay layer, 19.4 m thick; fully weathered, strongly weathered, and moderately weathered granite and andesite are merged into a weathered rock layer, 80.6 m thick. According to the site investigation report and literature research of the project, the stratum parameters for numerical simulation are shown in Tables 5.5

and 5.6. Among them, the friction constant M of the clay layer, the internal friction angle φ and cohesion c of the weathered rock layer are model source variables input as random variables, where the probability distribution of the internal friction angle φ and cohesion c of the weathered rock layer follows a normal distribution, and the friction constant of the clay layer is converted from the random variable φ^* according to the following formula:

$$M = \frac{6\sin\varphi^*}{3 - \sin\varphi^*} \tag{5.59}$$

Random variable parameters are shown in the table with an added "*", and parameter values are marked as "mean + standard deviation".

(2) Support structure parameters

The support structure parameters are shown in Table 5.7, where the elastic modulus of concrete and steel are random variables.

4. Simulation of construction sequence

Pit excavation generally needs to be "layered, blocked, symmetrical, balanced", the support system should be constructed as soon as possible, and the bottom plate construction should be completed as soon as possible to form a structure, thereby improving safety. The pit excavation and support steps in the model are consistent with the actual project. For ease of description, the entire pit is divided into Zone 1 (25 m), Zone 2 (50 m), and Zone 3 (25 m). The simulation construction sequence is shown in Table 5.8. The schematic of the earthwork excavation plan on the cross-section of the pit is shown in Fig. 5.16.

Table 5.5 Table of parameters for sandy clay stratum

Name of soil layer	Thickness (m)	Constitutive model	γ (kN/m^3)	ν	λ	e_0	φ^* (°)	Overconsolidation ratio
Sandy clay	19.4	Corrected Cambridge	20	0.25	0.0453	0.6	25.84 + 5.68	1.2

Note 1. Based on experience, the modified Cambridge model parameter κ can be taken as 1/10 of λ
2. * represents random variable parameters, parameter values are marked as "mean + standard deviation", the same applies to the table below

Table 5.6 Parameters of weathered rock formation

Name of soil layer	Thickness (m	Constitutive model	γ (kN/m^3)	E (MPa)	ν	c^* (kPa)	φ^* (°)
Weathered rock	80.6	Mohr–Coulomb	21	125	0.2	65 + 9.42	32 + 4.48

Table 5.7 Mechanical parameters of support structure

Supporting structure	Material	Constitutive model	E^* (GPa)	γ (kN/m^3)	ν
Underground diaphragm wall	C30 concrete	Elasticity	24 + 1.36	25	0.167
Coronal beam	C30 concrete	Elasticity	24 + 1.36	25	0.167
Concrete brace	C30 concrete	Elasticity	24 + 1.36	25	0.167
Steel support	Q235 steel	Elasticity	212 + 10	78.5	0.31
Upright column	Q235 steel	Elasticity	212 + 10	78.5	0.31
connection beam	Q235 steel	Elasticity	212 + 10	78.5	0.31

Table 5.8 Simulation construction sequence table

Serial number	Stage	Serial number	Stage
0	Completion of the equilibrium of geostress	7	The fourth support in Zone 2 and the second support in Zone 3
1	Underground diaphragm wall, ring beam	8	The fifth support in Zone 2 and the third support in Zone 3
2	The second steel support in the first area (the first steel support)	9	Dig to the bottom in Zone 2 and the fourth support in Zone 3
3	The third brace in Zone 1	10	The fifth support in the third district
4	The fourth support in Zone 1	11	Earth excavation completed
5	The fifth support in Zone 1 and the second support in Zone 2	12	Station floor
6	Dig to the bottom in Zone 1 and support the third layer in Zone 2		

5. Layout of measurement points

The layout of the measurement points in the model is shown in Fig. 5.17, with the specific content described below.

(1) Ground settlement measurement line: A, B, C, D, E are set up. Five measurement lines, the vertical distances from the measurement points from near to far are 17.5 m, 22.5 m, 32.5 m, 52.5 m, 72.5 m. Among them, the average value of a total of 9 measurement points at distances of 17.5 m, 22.5 m, and 32.5 m from the foundation pit on measurement lines A, B, and C is used as the failure index "ground settlement s_g".
(2) Deep horizontal displacement measurement line: a total of three measurement lines a, b, and c are set up, with measurement points at 0 m, 2 m, 3.2 m, 5.05

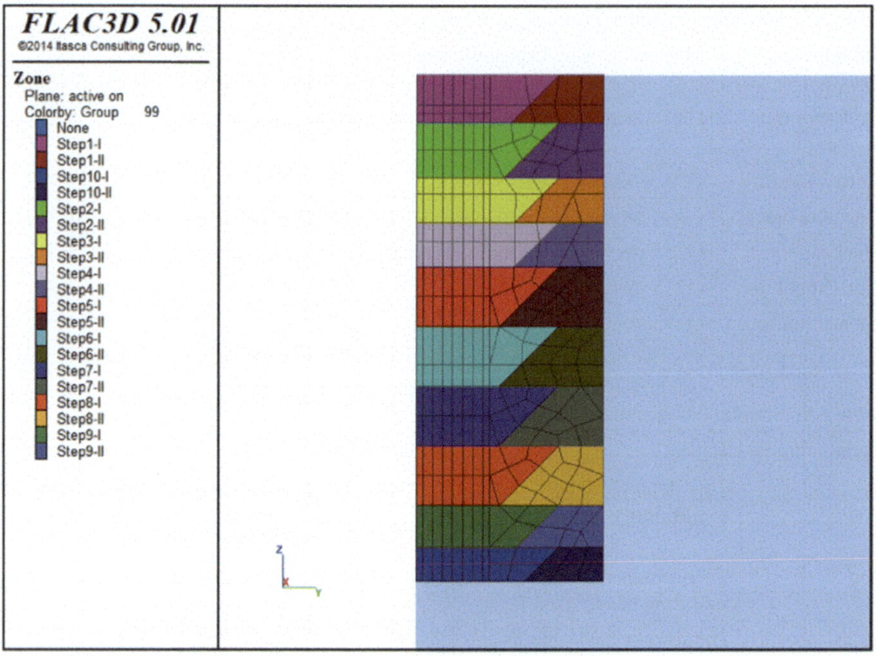

Fig. 5.16 Schematic of the cross-section earthwork excavation plan

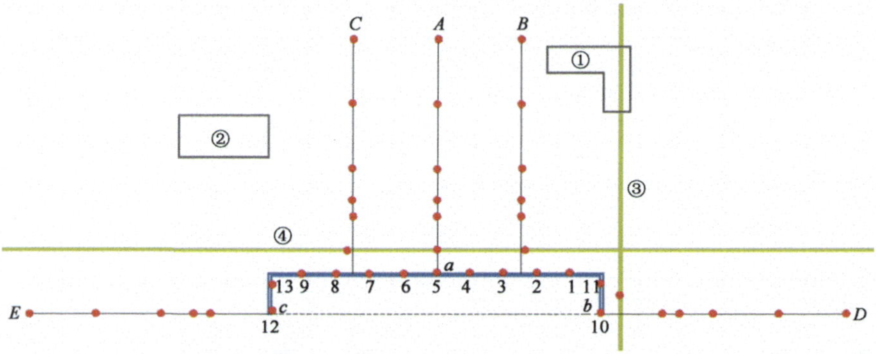

Fig. 5.17 Model measurement point layout diagram1–13-measurement point number; ①-④-foundation pit surrounding environment item number

m, 6.9 m, 8 m, 9.9 m, 11.2 m, 12.9 m, 14.9 m, 16.9 m, 19.4 m, 20.9 m, 22.9 m, 24.9 m, 26.9 m, 28.9 m, 31.7 m, 32.85 m, 34 m. Among them, the average value of 6 measurement points at 9.9 m, 11.2 m, 12.9 m, 14.9 m, 16.9 m, and 19.4 m on line a is used as the failure index "middle layer retaining structure horizontal displacement u_{h}^{m}"; the average value of 6 measurement points at 20.9 m, 22.9 m, 24.9 m, 26.9 m, 28.9 m, and 31.7 m is used as the failure index "lower layer retaining structure horizontal displacement u_{h}^{l}"; the average value of all measurement points on line a is used as the failure index "retaining structure horizontal displacement u_{h}".

(3) Crown beam horizontal displacement measurement point: a total of 13 measurement points are set up around the continuous wall, with a distance of about 10 m between each point. Among them, the average value of the 4th, 5th, and 6th measurement points is used as the failure index "crown beam horizontal displacement u_{h}^{c}".
(4) Peripheral pipeline settlement measurement point: ③ and ④ are pipelines, one measurement point is set up for pipeline ③, and three measurement points are set up for pipeline ④. The average value of the four measurement points is used as the failure index "pipeline settlement s_p".
(5) Support axial force measurement point: a total of five supports are set up, with 16 supports in each row, and one measurement point is set up for each support, 1 m away from the edge of the foundation pit. Among them, the axial force of the nearest to the axis of symmetry in the first support (concrete support) is used as the failure index "concrete support axial force N^{c}"; the axial force of the nearest to the axis of symmetry in the third support is used as the failure index N^{m}"middle layer support axial force".

6. Simulation conditions

To study the coupling effect of 6 groups of coupling factors, a total of 12 sets of simulation conditions are set up in this simulation, see Table 5.9. In the table, "T" represents the occurrence of this factor event, and "F" represents non-occurrence.

The simulation methods of each factor event are described as follows.

(1) Insufficient embedment depth of retaining structure: The depth of the underground continuous wall is reduced by 0.2 times the depth of the foundation pit, from 14 to 7.2 m.
(2) Insufficient support capability: The elastic modulus of the material of the concrete support and steel support is reduced by 30%.
(3) Improper excavation method: On the basis of the original excavation method, one layer of soil is over-excavating each time, and the support is delayed.
(4) Excessive water and soil pressure: An equivalent vertical load of 5 m soil column self-weight is applied within a range of 10 m outside the foundation pit.
(5) Pipeline aging: The elastic modulus of the pipeline material is reduced by 30%.

7. Central composite experimental design

In this calculation, there are a total of 5 design variables, and the meaning, unit, distribution form, mean, and standard deviation of each variable are summarized in Table 5.10.

Selecting an approximate function with cross quadratic terms:

$$\hat{g}_i(X) = a_0 + \sum_{j=1}^{5} a_{ij}X_j + \sum_{j=1}^{5} a_{ij}X_j^2 + \sum_{1\le j<k\le 5} a_{ijk}X_jX_k \tag{5.60}$$

Table 5.9 Simulated working condition table

Condition number	Insufficient embedment depth of retaining structure E_{S1}	Insufficient support capability E_{S1}	Improper excavation method E_{S1}	Excessive soil and water pressure E_{E1}	Pipeline aging E_{E2}
1	F	F	F	F	F
2	T	F	F	F	F
3	F	T	F	F	F
4	F	F	T	F	F
5	F	F	F	T	F
6	F	F	F	F	T
7	T	T	F	F	F
8	F	T	T	F	F
9	T	F	T	F	F
10	F	F	T	T	F
11	F	F	F	T	T
12	F	F	T	F	T

Table 5.10 Summary of design variable information

Design variable	Meaning	Unit	Distribution form	Mean value	Standard deviation
$\varphi^*(X_1)$	Cohesive soil—internal friction angle	°	Normal distribution	25.84	5.68
$c(X_2)$	Weathered rock-cohesion	kPa	Normal distribution	65	9.42
$\varphi(X_3)$	Weathered rock—internal friction angle	°	Normal distribution	32	4.48
$E_c(X_4)$	Elastic modulus of C30 concrete	GPa	Normal distribution	24	1.36
$E_s(X_5)$	Elastic modulus of Q235 steel	GPa	Normal distribution	212	10

The center point of the experiment is selected as the mean point: $X^0 = (25, 84, 65, 32, 24, 212)$. Using the center composite experimental design scheme of parameters $f = 1, \alpha = 0.5$, the sample points are shown in Table 5.11.

Table 5.11 Central composite experimental sample point

Sample point number	X_1	X_2	X_3	X_4	X_5
1	20.16	55.58	27.52	22.64	202
2	20.16	55.58	27.52	22.64	222
3	20.16	55.58	27.52	25.36	202
4	20.16	55.58	36.48	22.64	202
5	20.16	74.42	27.52	25.36	202
6	31.52	55.58	36.48	22.64	202
7	20.16	55.58	27.52	25.36	222
8	20.16	55.58	36.48	25.36	202
9	20.16	74.42	36.48	22.64	202
10	31.52	74.42	27.52	22.64	202
11	20.16	55.58	36.48	25.36	222
12	20.16	74.42	36.48	25.36	202
13	31.52	74.42	36.48	22.64	202
14	20.16	74.42	36.48	25.36	222
15	31.52	74.42	36.48	25.36	202
16	31.52	74.42	36.48	25.36	222
17	31.52	55.58	36.48	22.64	202
18	31.52	55.58	27.52	25.36	202
19	31.52	55.58	27.52	22.64	222
…	…	…	…	…	…
37	25.84	65	34.24	24	212
38	25.84	65	29.76	24	212
39	25.84	65	32	24.68	212
40	25.84	65	32	23.32	212
41	25.84	65	32	24	217
42	25.84	65	32	24	207
43	25.84	65	32	24	212

5.3.4 Calculation Results

Taking each sample point as the model input in turn, taking working condition 1 as an example, the calculation results of each failure index are shown in Figs. 5.18, 5.19, 5.20, 5.21, 5.22, 5.23, 5.24, and 5.25.

The calculated values of each function are shown in Table 5.12.

According to the sample points and the calculated values of the function, the expression of the approximate function is fitted, and the probability p^{fE} of each failure point is calculated by the first and second moment method. The results are shown in Table 5.13.

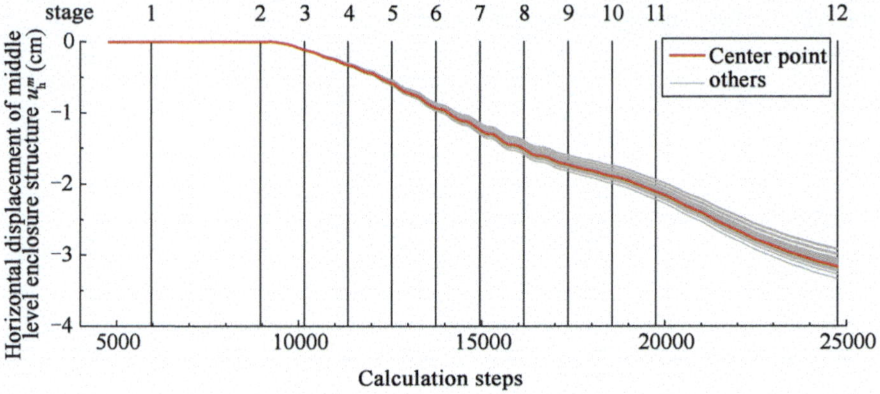

Fig. 5.18 Calculation results of horizontal displacement u_h^m of middle layer retaining structure

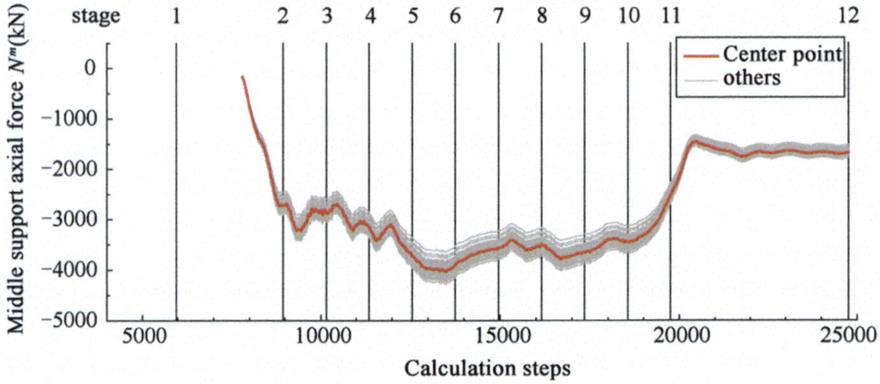

Fig. 5.19 Calculation results of axial force N^m of middle layer support

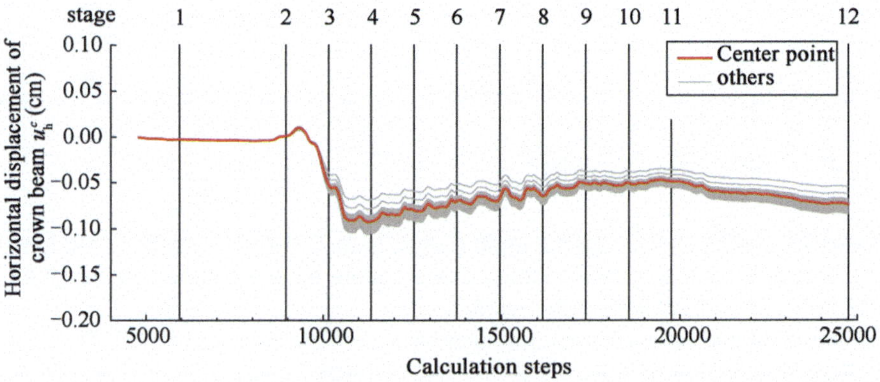

Fig. 5.20 Calculation results of horizontal displacement u_h^c of crown beam

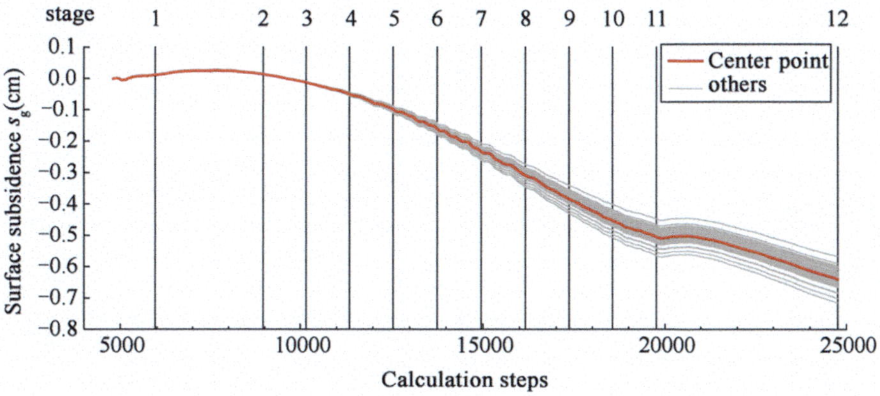

Fig. 5.21 Calculation results of ground settlement s_g

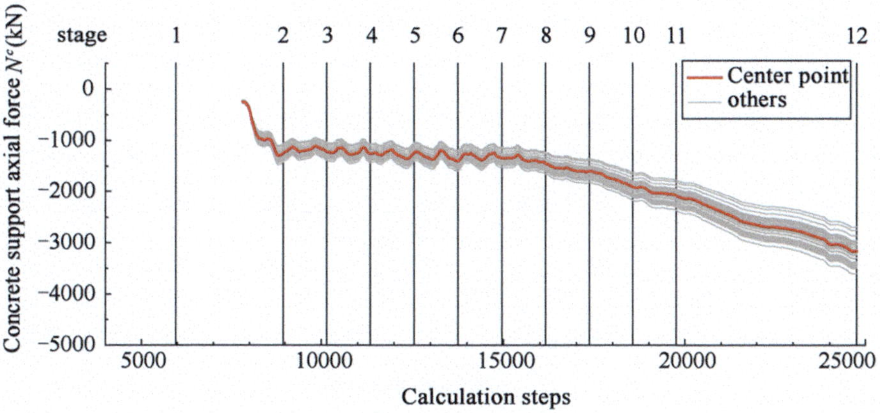

Fig. 5.22 Calculation results of concrete support axial force N^c

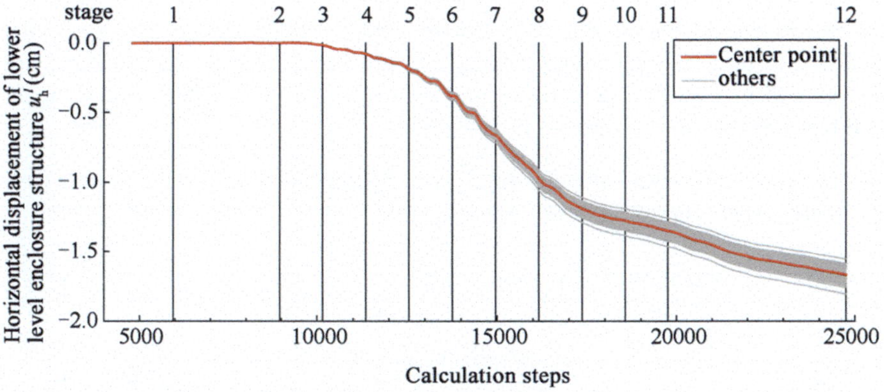

Fig. 5.23 Calculation results of horizontal displacement u_h^l of lower layer retaining structure

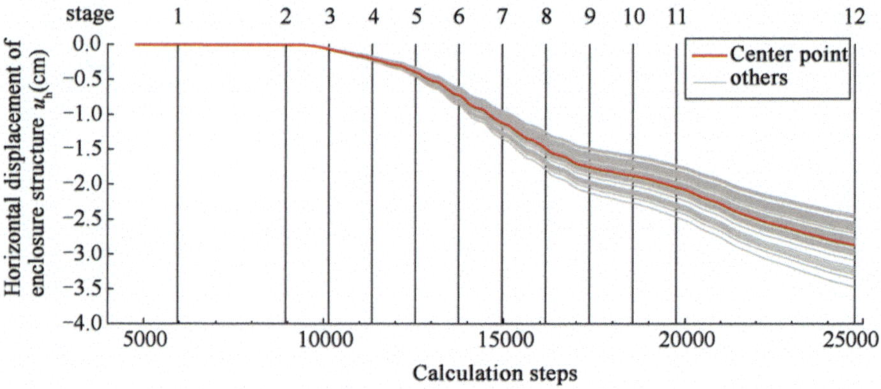

Fig. 5.24 Calculation results of horizontal displacement u_h of retaining structure

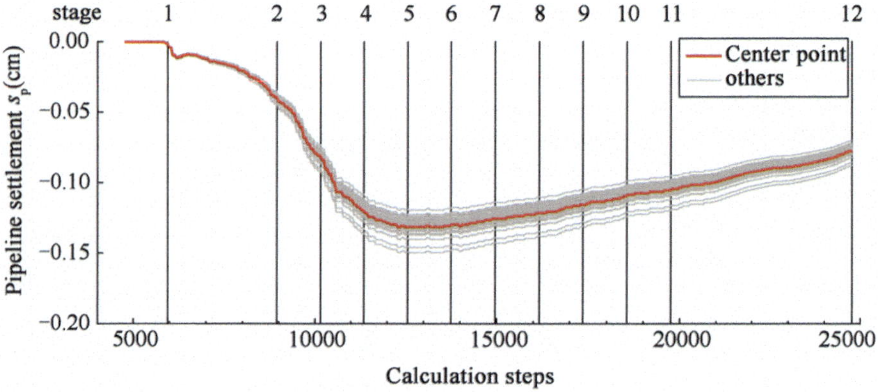

Fig. 5.25 Calculation results of pipeline settlement s_p

Following the above process, calculate the failure probability under each working condition in turn, and substitute it back into the Bayesian network for backward inference to find out the occurrence probability of the intermediate events under each working condition as shown in Table 5.14.

(1) The coupling coefficient of insufficient embedment depth of retaining structure-Insufficient support capability (E_{S1}–E_{S2}) is:

$$p^7_{O_{S1}} = 0.1977 > 0.1386 = p^2_{O_{S1}} + p^3_{O_{S1}} - p^2_{O_{S1}} p^3_{O_{S1}} - p^1_{O_{S1}} \tag{5.61}$$

$$\beta^{E_{S2}}_{E_{S1}}\Big|_{O_{S1}} = 1 - \frac{1 - p^7_{O_{S1}} - p^1_{O_{S1}}}{1 - p^2_{O_{S1}} - p^3_{O_{S1}} + p^2_{O_{S1}} p^3_{O_{S1}}} = 1 - \frac{0.7384}{0.7975} = 0.0741 \tag{5.62}$$

Table 5.12 Table of calculated values of functional functions

Sample point	$g_{S1}(X)$	$g_{S2}(X)$	$g_{E1}(X)$	$g_{E2}(X)$
1	6.8370	1.3363	4.3610	1.8509
2	6.8487	1.3140	4.2988	1.9612
3	6.8276	1.2708	4.3528	1.7102
4	6.9070	1.2575	4.4099	2.3118
5	6.8647	1.3960	4.3502	1.5358
6	6.7816	1.3249	4.3758	1.5305
7	6.8735	1.3891	4.3591	2.2476
8	6.8307	1.3008	4.4021	1.4041
9	6.8133	1.3198	4.3800	1.6153
10	6.8008	1.3575	4.3983	1.8312
11	6.9266	1.2307	4.2843	1.7696
12	6.8507	1.4027	4.3396	1.5729
13	6.8857	1.3396	4.3619	1.8667
14	6.8943	1.4940	4.3368	2.0489
15	6.7881	1.3675	4.4329	1.8864
16	6.8451	1.3170	4.3766	1.5977
17	6.7958	1.2950	4.3905	1.9804
18	6.9030	1.3112	4.3157	1.7979
19	6.8370	1.3363	4.3610	1.8509
…	…	…	…	…
37	6.7667	1.3720	4.3933	2.0295
38	6.8646	1.4132	4.3514	1.7040
39	6.9099	1.4416	4.3667	1.6357
40	6.9235	1.3106	4.3715	1.6999
41	6.6646	1.3980	4.3834	2.1461
42	6.7987	1.2506	4.3791	1.7690
43	6.8583	1.4152	4.3430	2.0142

Table 5.13 Calculation results of the probability of failure point occurrence

Point of Failure	O_{S1}	O_{S2}	O_{E1}	O_{E2}
Failure probability(%)	6.134	13.58	3.469	9.215

In the formula, $p_{O_{S1}}^{i}$ represents the occurrence probability of the intermediate event O_{S1} under the i-th working condition.

(2) The coupling coefficient of insufficient support capability-improper excavation method (E_{S2}–E_{S3}) is:

Table 5.14 Probability of occurrence of intermediate events under various operating conditions (unit: %)

Condition number	Large deformation of the enclosure structure O_{S1}	Support system failure O_{S2}	Excessive geological deformation O_{E1}	Excessive pipeline deformation O_{E2}
1	6.39	14.92	14.3	8.33
2	9.55	15.27	14.53	8.74
3	11.83	31.52	19.38	8.23
4	10.66	26.32	23.87	12.06
5	9.84	19.88	27.44	12.58
6	6.11	14.98	14.23	22.41
7	19.77	33.5	19.79	8.74
8	15.38	56.28	26.41	12.8
9	14.59	26.92	24.33	12.13
10	14.84	35.35	49.01	19.82
11	10.01	20.86	27.78	27.63
12	10.81	26.89	24.65	29.21

$$p^8_{O_{s2}} = 0.5628 > 3462 = p^3_{O_{s2}} + p^4_{O_{s2}} - p^3_{O_{s2}} p^4_{O_{s2}} - p^1_{O_{s2}} \tag{5.63}$$

$$\beta^{E_{s3}}_{E_{S2}}\Big|_{O_{S2}} = 1 - \frac{1 - p^8_{O_{s2}} - p^1_{O_{s2}}}{1 - p^3_{O_{s2}} - p^4_{O_{s2}} + p^3_{O_{02}} p^4_{O_{s2}}} = 1 - \frac{0.288}{0.5046} = 0.4293 \tag{5.64}$$

(3) The coupling coefficient of Insufficient embedment depth of retaining structure -improper excavation method (E_{S1}–E_{S3}) is:

$$p^9_{O_{S1}} = 0.1459 > 0.128 = p^2_{O_{S1}} + p^4_{O_{S1}} - p^2_{O_{s1}} p^4_{O_{s1}} - p^1_{O_{S1}} \tag{5.65}$$

$$\beta^{E_{E1}}_{E_{ES}}\Big|_{O_{E1}} = 1 - \frac{1 - p^{10}_{O_{E1}} - p^1_{O_{E1}}}{1 - p^4_{O_{E1}} - p^5_{O_{E1}} + p^4_{O_{E1}} p^5_{O_{E1}}} = 1 - \frac{0.3669}{0.5524} = 0.3358 \tag{5.66}$$

(4) The coupling coefficient of improper excavation method-excessive water and soil pressure (E_{S3}–E_{E1}) is:

$$p^{10}_{O_{E1}} = 0.4901 > 0.3406 = p^4_{O_{E1}} + p^5_{O_{E1}} - p^4_{O_{E11}} p^5_{O_{E1}} - p^1_{O_{E1}} \tag{5.67}$$

$$\beta^{E_{E1}}_{E_{E|}}\Big|_{O_{E1}} = 1 - \frac{1 - p^{10}_{O_{E1}} - p^1_{O_{E1}}}{1 - p^4_{O_{E1}} - p^5_{O_{E1}} + p^4_{O_{E1}} p^5_{O_{E1}}} = 1 - \frac{0.3669}{0.5524} = 0.3358 \tag{5.68}$$

(5) The coupling coefficient of excessive water and soil pressure-pipeline aging (E_{E1}–E_{E2}) is:

$$p^{11}_{O_{E2}} = 0.2763 > 0.2384 = p^{5}_{O_{E2}} + p^{6}_{O_{E2}} - p^{5}_{O_{E}2}p^{6}_{O_{E2}} - p^{1}_{O_{E2}2} \tag{5.69}$$

$$\beta^{E_1}_{E_{E1}}\Big|_{O_{F2}} = 1 - \frac{1 - p^{11}_{O_{E2}} - p^{1}_{O_{E2}}}{1 - p^{5}_{O_{E2}} - p^{6}_{O_{E2}} + p^{5}_{O_{E2}}p^{6}_{O_{F2}}} = 1 - \frac{0.6404}{0.6783} = 0.0559 \tag{5.70}$$

(6) The coupling coefficient of improper excavation method-pipeline aging (E_{S3}–E_{E2}) is:

$$p^{12}_{O_{E2}} = 0.2921 > 0.2344 = p^{4}_{O_{E2}} + p^{6}_{O_{E2}2} - p^{4}_{O_{E2}}p^{6}_{O_{E2}} - p^{1}_{O_{F2}2} \tag{5.71}$$

$$\beta^{E_2}_{E_{Sz}}\Big|_{O_{E2}} = 1 - \frac{1 - p^{12}_{O_{E2}} - p^{1}_{O_{E2}}}{1 - p^{4}_{O_{E2}} - p^{6}_{O_{E2}} + p^{4}_{O_{E2}}p^{6}_{O_{E2}}} = 1 - \frac{0.6246}{0.6823} = 0.0846 \tag{5.72}$$

The calculation results show that among the six groups of coupling factors in this project, "insufficient support capability—improper excavation method" and "improper excavation method—excessive soil and water pressure" have the strongest coupling effect, with coupling coefficients slightly higher than 0.4 and 0.3 respectively; "Insufficient embedment depth of retaining structure—insufficient support capability" and "improper excavation method—pipeline aging" have a moderate coupling effect, with coupling coefficients close to 0.1; "Insufficient embedment depth of retaining structure—improper excavation method" and "excessive soil and water pressure—pipeline aging" have a weaker coupling effect, with coupling coefficients around 0.02 to 0.05.

5.4 Risk Factor Coupling Effect Grading Standards

Although the coupling coefficient of risk factors can be calculated through numerical simulation combined with Bayesian network inference, due to the limitations of numerical simulation, the coupling coefficient calculated in this way will inevitably have some differences from the actual situation. On the other hand, because the calculation process is cumbersome, it is unrealistic to calculate the coupling coefficient of risk factors one by one in actual analysis work. As pointed out earlier, the concept of the coupling effect of risk factors actually exists in everyone's logical cognition, and the coupling coefficient of risk factors is the quantified form of this concept. Therefore, as long as a connection can be established between people's cognitive understanding of the strength of the coupling effect and the value of the coupling coefficient, the coupling coefficient can be determined through subjective analysis and judgment in actual applications.

According to the definition of the coupling coefficient of risk factors, when the coupling direction is certain, the lower limit (value is 0) and upper limit (value is 1) of the coupling coefficient represent the two extreme situations of "no coupling effect" and "infinite coupling effect". Therefore, when the coupling effect of risk factors is generally present in a risk system, slight changes in the value of the coupling coefficient will have a significant impact on the final risk value. Therefore, based on the analysis of the common construction risk factor coupling effects in urban large underground spaces in the previous section and the calculation results of the coupling coefficient, the strength of the coupling effect is divided into four levels: "extremely strong coupling", "strong coupling", "medium coupling", and "weak coupling". The suggested values of the coupling coefficient under each level are shown in Table 5.15.

Table 5.15 Table of classification standards and value suggestions for coupling coefficients of construction risk factors in urban large underground spaces

Grade	Description	Coupling coefficient value	Remarks
1	Extremely strong coupling	(0.5, 1.0]	Coupling between special extreme adverse factors
2	Strong coupling	(0.25, 0.5]	Coupling of water-land geological factors, coupling of design defects and special external environment, etc.
3	Intermediate coupling	(0.1, 0.25]	Coupling of adverse loads and minor structural defects, coupling of construction quality issues and adverse geological conditions, etc.
4	Weak coupling	(0, 1]	Coupling between management oversight and other factors, coupling between non-major factors, etc.

Note The values listed in the table are for positive coupling situations. When the coupling direction is negative, the coupling coefficient takes the opposite of the corresponding value

Chapter 6
Dynamic Evolution Mechanism of Safety Risks in Urban Underground Space Construction

6.1 Basic Concepts of Risk Evolution in Construction Process

6.1.1 Concept and Connotation of Risk Evolution

The evolution of risk refers to the process of risk changing over time. The causes of risk evolution are diverse. On the one hand, the risk-bearing environment is constantly changing, the changes in risk factors continue to accumulate, and the complex interrelationships between risk factors act as catalysts, causing the risk to possibly undergo magnitude leaps and sudden changes. On the other hand, as the state of the risk-bearing subject continues to advance, the risk pathway will also change, making some originally secondary risk factors become the main factors, leading to a sharp change in risk. Therefore, for activities with a long duration, it is very necessary to deal with risks from an evolutionary perspective.

The evolution of risk contains three connotations (Fig. 6.1):

(1) From the perspective of specific risk events, the evolution of risk is reflected in the increase or decrease of the probability of the event occurring;
(2) From the perspective of the overall action, the evolution of risk is reflected in the replacement of key risk events;
(3) From the perspective of the mechanism of risk formation, the evolution of risk is the process of risk factors gradually leaping to the upper events in the causal network, forming higher-level risk factors and even risk events.

The evolution of risk is objective and universally exists. In some small-scale engineering projects, there is no need to analyze the evolution of risk. However, urban underground space engineering has a magnitude difference from general underground engineering in both spatial scale and time scale, which makes the impact of risk evolution cannot be ignored.

S.-X. Lei et al., *Safety Risk Assessment of Urban Underground Large Space Construction*, https://doi.org/10.1007/978-981-97-7437-1_6

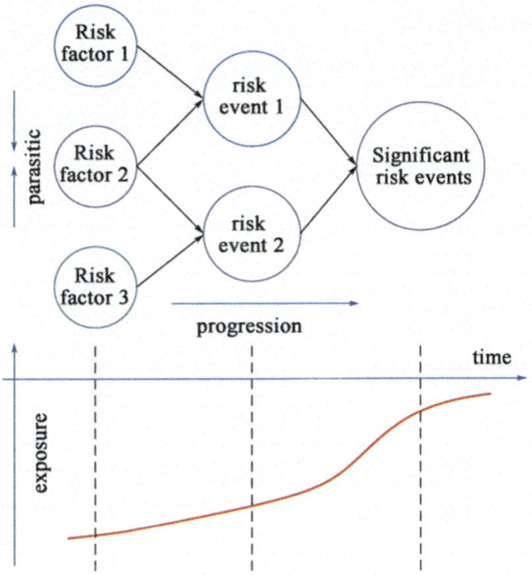

Fig. 6.1 Concept of construction risk evolution

6.1.2 Evolution Mechanism of Construction Risk

According to the previous analysis, the root cause of construction risk is the disturbance brought by construction behavior. Construction disturbances generate risk factors, the existence of risk factors brings uncertainty to construction behavior, which then causes the mechanical response of the structure during the construction process to gradually deviate from expectations, forming signs of accidents, and ultimately causing risk events (accidents) to occur. But one point that needs to be emphasized is that the existence of risk and whether a risk event occurs are two concepts. Risk is ubiquitous, from the moment construction disturbances occur, risk factors are generated, at this time construction behavior has a certain risk, and the risk will accompany the entire construction process until the construction is completed. In this process, affected by changes in the external environment, the advancement of construction sequences, or accidental events caused by various human or natural reasons, the level of construction risk is also changing, and when the risk level exceeds a certain limit, risk events will become very likely or even inevitable. From this perspective, any construction accident is not achieved overnight, but occurs after a series of brewing and development processes due to changes in the state of risk factors under the effect of construction time and space. In other words, the evolution of risk is not only an inherent property and objective law of construction risk, but also a necessary process for risk to transform into an accident.

1. The fundamental driving force of construction risk evolution

From a macro perspective, the evolution of risk can be uniformly interpreted as the process of risk changing over time, but the causes leading to risk evolution are

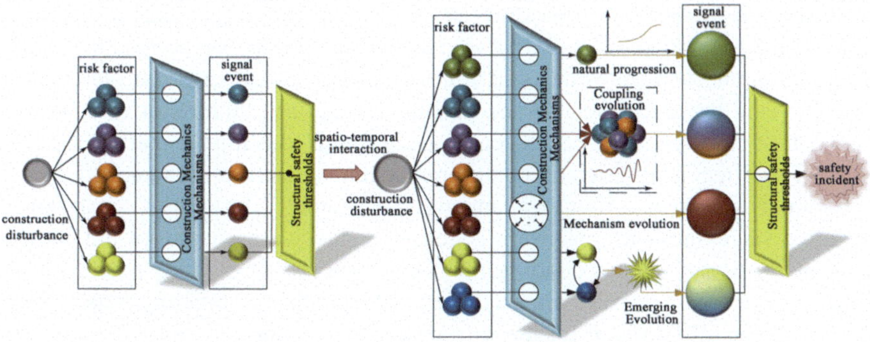

Fig. 6.2 Schematic diagram of construction risk evolution mechanism

diverse. To conduct a detailed study of the risk evolution process, it is first necessary to analyze and summarize the causes leading to risk evolution. According to the definition of risk, the level of risk mainly depends on the state of risk factors and the specific formation mechanism of risk (risk pathway). Therefore, taking this as a starting point, combined with the characteristics of the construction process and the nature of construction risk, the fundamental driving forces of construction risk evolution can be summarized as follows (Fig. 6.2).

(1) Natural evolution of risk factor status

The natural evolution of the status of risk factors is the most direct and common cause of risk evolution. The risk factors that undergo natural evolution during the construction process mainly fall into the following two categories.

① Natural environmental factors, such as groundwater level, rainfall, meteorological conditions, etc. The evolution of these factors is closely related to the construction season, and their driving factor is the natural passage of time.
② Engineering scale factors, such as pit depth, cross-sectional span, etc. The evolution of these factors is mainly due to the multi-stage nature of urban large underground space construction, and their driving factors are the progress of construction and the transition of construction sequence.

(2) Changes in the risk formation mechanism caused by the advancement of construction sequence

When construction moves from one stage to the next, the mechanical mechanism of construction behavior and the degree of disturbance caused by construction to the surrounding environment may undergo significant changes. For example, the pilot tunnel excavation stage of the Pile-Beam-Arch method (PBA method) is a wall disturbance stage mainly based on stratum unloading, but after the completion of the pilot tunnel excavation and entering the tunnel structure construction stage, the disturbance of the construction to the stratum and the surrounding environment will significantly decrease. The direct result of the construction mechanical mechanism is the change in the type of main risk events, and the risk events correspond to the risk formation mechanism, which will affect the size of the effect of risk factors. We can understand the risk formation mechanism as a kind of "filter" for risk factors, which determines which risk factors can "pass through" this filter and affect the risk system. This effect is reflected in the different actual impacts of the same risk factor on risk under different action mechanisms. For example, during the construction excavation stage, the main risk event is generally the collapse and roof fall caused by the instability of the surrounding rock. For such risk events, the nature of the stratum plays a dominant role; when it comes to the permanent structure construction stage, it is generally difficult for the surrounding rock to become unstable, and the main risk event at this stage becomes structural damage caused by the removal of temporary structures. Compared with collapse and roof fall, the influence of stratum properties on such risk events is relatively small, and the performance of the support system itself becomes the dominant factor. Therefore, even if the status of risk factors remains unchanged, the change in the risk formation mechanism will cause the original risk factors to "penetrate" this filter more, making the actual effect of risk factors significantly amplified, thereby leading to the evolution of risk.

(3) Emergent phenomena caused by the feedback control relationship within the construction system

According to the analysis of the risk system in urban large underground space construction, it is a process of interaction and feedback control among the engineering structure subsystem, environmental subsystem, and organizational management subsystem. Systems theory believes that even in the absence of external factors, a system will be in constant change due to the complex interaction of its internal elements, thus triggering many unpredictable emergent phenomena. The emergent phenomena in the urban large underground space construction system are mainly caused by human behavior during the construction process. For example, unreasonable construction periods lead to the construction party blindly chasing the schedule, resulting in a decrease in construction quality, frequent rework due to decreased construction quality, further increasing the pressure of the construction period, forming a negative cycle feedback amplification system. It is not difficult to find that although such emergent phenomena will also be affected by external factors, they fundamentally depend on the structure and boundary conditions of the system itself. Therefore, the risk evolution caused by such reasons is also known as endogenous emergent evolution.

(4) Evolution caused by the coupling effect of risk factors

As mentioned earlier, when risk factors with coupling effects occur simultaneously, their mutual stimulation will develop them into larger risk factors, thereby changing the probability of risk events, a process known as coupling evolution. Coupling evolution can occur independently, or it may be triggered by the above three types of evolutionary factors or occur simultaneously with other types of evolution. Therefore, the existence of coupling evolution makes the evolution of risk more unstable and uncertain, making the probability of risk events have the possibility of sudden changes.

2. The mechanism of construction risk evolution

The construction risk of urban large underground spaces mainly depends on the construction mechanics process. Therefore, any change in a type of factor will ultimately be reflected in the following two aspects, leading to the evolution of risk.

(1) Change the mechanical properties of materials or structures such as rock and soil

There are many factors that can cause changes in the mechanical properties of materials such as rock and soil, such as the rise and fall of groundwater levels, high and low environmental temperatures, creep and fatigue caused by construction loads, and the curing, creep, and corrosion of the material itself. In addition, the construction quality will also be dynamically changed by the management system, causing fluctuations in the mechanical properties of materials or structures.

(2) Change the stress state of the structure

The change in the stress state of the structure is mainly changed by the advancement of the construction sequence. For example, the stress state of the tunnel support structure is obviously different before and after closure, and the stress state of the pit retaining structure will also change with the excavation depth and the erection of the support. In addition, natural factors and management factors will also affect the stress state of the structure, such as increased water pressure caused by rainfall or additional loads caused by illegal stacking.

6.2 Markov-Fuzzy Comprehensive System State Evolution Prediction Model 6.2.1 Model Principles and Basic Theory

1. Further analysis of the mechanism of construction risk evolution

Generally speaking, the occurrence of construction risk events is mostly due to the failure of the engineering structure to meet the expected function. The theory of engineering structure reliability believes that structural failure can be determined by the function function, that is:

$$\chi_f = \frac{\langle g_f(R, S) \rangle}{g_f(R, S)} \tag{6.1}$$

In the formula, χ_f is the indicator function of risk event; $\langle \cdot \rangle$ is the Macaulay bracket function, defined as $\langle x \rangle = (x + |x|)/2$. $g_f(R, S)$ The risk event $\overline{f}$ corresponds to the structural function, which generally has the following form:

$$g_f(R, S) = R(\alpha_1, \alpha_2, \cdots, \alpha_n) - S(\beta_1, \beta_2, \cdots, \beta_m) \tag{6.2}$$

In the formula, R is the resistance effect function of the structure; $\alpha_1, \alpha_2, \cdots, \alpha_n$ is the parameter of R; S is the load effect function of the structure; $\beta_1, \beta_2, \cdots, \beta_m$ is the parameter of S. Thus, the probability of a risk event occurring can be expressed as:

$$P\left(\overline{f}\right) = P\left[g_f(R, S) < 0\right] \tag{6.3}$$

According to the probability definition of risk in Sect. 4.1.1, combined with formulas (6.1), (6.2), and (6.3), it is not difficult to find that construction risk is related to the following factors.

1. The value of $C(\overline{w_i})$

The value of $C(\overline{w_i})$ is determined by the characteristics of the risk event itself. If $\overline{f}$ is a relatively simple event, then the value of can be uniquely determined. Construction safety risk events are often of varying degrees, with significant hazards and wide impacts, so strictly speaking, the value of $C(\overline{w_i})$ for construction safety risk events is often uncertain and not fixed. However, from the perspective of risk analysis, engineers are more concerned about the relative size of the risk rather than its absolute value, so it can generally be considered that $C(\overline{w_i})$ is a constant value that does not change over time.

2. The specific form of functions R and S

Functions R and S reflect the mechanical mechanism of construction, which is jointly determined by the structure of the project itself and the surrounding stratum structure. For urban large underground space projects, the forms of R and S are often extremely complex and cannot be explicitly expressed. However, it is certain that during the construction process of urban large underground spaces, due to the advancement of procedures and changes in construction methods, the specific forms of these two functions will inevitably change, thereby affecting the probability of f' occurrence.

3. The n + m dimensional joint distribution law for the parameters $\alpha_1, \alpha_2, \cdots, \alpha_n, \beta_1, \beta_2, \cdots, \beta_m$, in R and S

Law of g_f, are related to the parameters $\alpha_1, \alpha_2, \cdots, \alpha_n, \beta_1, \beta_2, \cdots, \beta_m$ in R and S. If these parameters are all certain values, then whether this result occurs is already determined, and there is no risk to speak of. Therefore, when studying risk, these parameters must be treated as random variables. In this way, the value of is also a random variable, and its distribution is determined by the distribution of the parameters. If the distributions of the parameters are related to each other, then the distribution of g_f values is at most determined by the n + m dimensional joint distribution law of $\alpha_1, \alpha_2, \cdots, \alpha_n, \beta_1, \beta_2, \cdots, \beta_m$. However, the distribution of some parameters will change over time, such as the distribution of rainfall in different months, the distribution of environmental temperature in different seasons, etc. Even if the specific forms of R and S remain unchanged, the changes in these parameter distributions will also cause the probability of f occurrence to change.

Through the above analysis, it can be found that from a mechanical perspective, the change in the probability of occurrence of construction risk events is driven by two effects: one is the change in the form of the project structure and the way the constructor interacts with the surrounding environment, that is, the mechanistic evolution mentioned earlier, this part of the evolution is driven by the spatiotemporal effect of construction; the second is the change in the probability of occurrence of some risk factors over time, reflected in the quantitative indicators related to them having different probability distributions at different times, that is, the natural evolution mentioned earlier, this part of the evolution is driven by the time-varying effect of the randomness of the risk factors themselves.

As mentioned earlier, the evolution of risk is manifested as the probability of a risk event occurring changes over time, that is:

$$P\left\{\overline{f}\right\} = P\left\{\overline{f}; t\right\}, t \in T \tag{6.4}$$

In the formula, t is time.

The distribution law of the random variable C is related to $P\left\{\overline{f}\right\}$, so for different t, the random variable C has different distributions law, that is:

$$\{C(\omega; t),\ t \in T\} = \{C(t),\ t \in T\} \tag{6.5}$$

$\{C(t),\ t \in T\}$ is a family of random variables defined on (Ω, F, P). According to the definition of a random process, it is known that, $\{C(t),\ t \in T\}$ is a random process defined on the probability space (Ω, F, P), so the measure of risk can be defined by the expectation of the random variable C:

$$R\left(\overline{f}; t\right) = \sum_{\omega_i \in \overline{f}} P\{\overline{\omega}_i; t\} \times C(\overline{\omega}_i) = \sum_{i,\omega_i \in f} p_i^{\overline{\omega}}(t) \times c_i^{\overline{\omega}}(t) \tag{6.6}$$

Equation (6.6) shows that when the event (risk) is fixed, the value of the random variable C remains unchanged, and its probability measure is a function of time t. The risk defined in this way is also a function of time t.

From the above analysis, it can be seen that the evolution of risk can be described by a random process. For one or a finite number of random variables, understanding the distribution function can fully understand the random variables. Similarly, for the random process {X(t), t ∈ T}, in order to describe its characteristics, it is natural to know the distribution function F(t, x) = P{X(t) ≤ x}of X(t) for each t ∈ T, as well as their joint distribution at any n time points. It is no longer a finite number, but a family of joint distributions. The n-dimensional distribution of the random process is defined as:

$$F_{t_1,t_2,\cdots,t_n}(x_1, x_2, \cdots x_n) = P\{X(t_1) \leq x_1,\ X(t_2) \leq x_2, \cdots, X(t_n) \leq x_n\} \quad (6.7)$$

The entirety of all n-dimensional distributions of the random process $\left\{F_{t_1,t_2,\cdots,t_n}(x_1, x_2, \cdots x_n), t_1, t_2, \cdots, t_n \in T, n \geq 1\right\}$ is called the finite-dimensional distribution family of the random process. Knowing the finite-dimensional distribution family of the random process means knowing the joint distribution of any n random variables in{X(t), t ∈ T}, and thus mastering the interdependence between these random variables.

Therefore, studying the evolution of risk is essentially a problem of i finding the finite-dimensional distribution family of the random process $\{C(t),\ t \in T\}$ of risk loss.

2. Modeling ideas

From the definition of the random process of risk, if we want to study risk and its evolution, we need to answer the following two questions.

(1) How to measure risk

According to the definition of risk, if we can give the probability of a certain risk event and establish the random variable C that maps the risk event to the risk loss, we can measure the risk by the numerical characteristics of this random variable. However, with the current level of technology, it is almost impossible to accurately give the probability of a risk event using traditional mathematics and mechanics methods. However, according to the model of urban large underground space construction risk system, we can indirectly measure risk by giving a semi-quantitative description of the state of the construction system and inferring from the Bayesian network.

(2) How to describe the random process of risk evolution

A complete description of a random process requires giving its finite-dimensional distribution family. However, only a few types of random processes can give a clear description of the finite-dimensional distribution family, one of which is the Markov process. The finite distribution family of the Markov process can be completely determined by its transition probability, and the transition probability has its clear logical meaning. This feature makes the Markov process one of the most commonly used random processes, which has been widely used in the study of many practical problems.

In response to the above two problems, this section uses the following methods to establish a risk evolution prediction model.

(1) Use the fuzzy comprehensive evaluation method to measure the macro state of the construction risk system

Fuzzy mathematics is a new type of mathematical method proposed by the American control theory expert ZADEH in 1965. Its research object is the “uncertainty of conceptual boundaries”. It expands the traditional set theory with the concept of membership degree, thus opening up a new way for using mathematical methods to study qualitative problems.

The Fuzzy Comprehensive Evaluation Method is a modern comprehensive evaluation method based on fuzzy mathematics. It transforms qualitative evaluation into quantitative evaluation based on the membership theory of fuzzy mathematics, that is, it uses fuzzy mathematics to make a comprehensive evaluation of things or objects constrained by various factors. It has the characteristics of clear results and strong systematicity, can solve fuzzy and difficult to quantify problems well, and is suitable for solving various uncertain problems. The Fuzzy Comprehensive Evaluation Method fits very well with the characteristics of risk assessment problems, so this model uses this method as a means to express risk specifically.

(2) Using the Markov chain as a tool to describe the stochastic process of risk factor evolution

Because there is a similarity in form between the Markov chain and the Fuzzy Comprehensive Evaluation Method, it is easy to combine the two methods.

Therefore, this model uses the Markov chain as a tool to depict the stochastic process of the evolution of the risk factor's own state.

(3) The combination of the Markov chain and the Fuzzy Comprehensive Evaluation Method

The Fuzzy Comprehensive Evaluation Method is based on fuzzy mathematics, which studies the uncertainty of problem boundaries, and its core concept is membership; the mathematical background of the Markov chain is probability theory, which studies the uncertainty of the possibility of results appearing, and its core concept is probability. Strictly speaking, the two study completely different concepts, but they also have intricate connections.

One of the core issues in fuzzy mathematics is the objective reality of membership. In order to obtain a convincing interpretation of the concept of membership that connects the subjective and objective, the determination of membership is often obtained through fuzzy statistical experiments or the use of objective scales. In fact, fuzzy statistical models and probability statistical models can be converted to each other, and the probability distribution itself can also be used as a membership function.

Therefore, membership has its profound probability connotation, which lays the theoretical foundation for introducing the Markov chain into the Fuzzy Comprehensive Evaluation Method.

3. Markov chain

The Markov chain was proposed by Russian mathematician Andrei Markov (Андрей Андреевич Марков). In the research results published by Markov in 1906–1907, in order to prove that the independence between random variables is not a necessary condition for the Weak Law of Large Numbers and the Central Limit Theorem, he constructed a random process that depends on conditional probabilities, which is called the Markov process. The Markov chain is a special case of the Markov process, that is, the Markov process under discrete state space and discrete time index. The Markov chain model is simple, easy to calculate, and has many excellent mathematical properties.

(1) Definition of the Markov chain

There is a kind of random process, which has the so-called "no aftereffect", that is, to determine the future state of the process, it is enough to know its current situation, and there is no need to understand its past situation. This kind of process is called the Markov process. Specifically, if the state space of this random process is finite or countable (that is, the corresponding random variable is discrete), and it is also discrete in time, then this random process is called the Markov chain. The definition of the Markov chain is as follows:

For a random sequence {X(n), n = 0, 1, ...}, if it only takes a finite or countable number of values (if not otherwise specified, it is represented by the non-negative integer set {0, 1, 2, ...}, and for any n ≥ 0, and any state $i,\ j,\ i_0,\ j_0,\cdots i_{n-1}$, it has:

$$P\{X_{n+1}=j|X_n=i,\ X_{n-1}=i_{n-1},\ldots,\ X_1=i_1,\ X_0=i_0\}=P\{X_{n+1}=j|X_n=i\} \tag{6.8}$$

This random sequence is then referred to as a Markov chain. Equation (6.8) is the mathematical expression of the Markov property under the discrete index set.

In the equation, $X_n = i$ it indicates that the process is in state i at time n, and {0, 1, 2,...} is referred to as the state space of this process, denoted as S.

(2) The transition probability and transition probability matrix of the Markov chain are called conditional probabilities:

$$p_{ij}^{(n)}=P\{X_{m+n}=j|X_m=i\},\quad i,\ j\in S;\ \ m\geq 0;\ \ n\geq 1 \tag{6.9}$$

It is the n-step transition probability of the Markov chain. When n = 1, and it is also stipulated that:

$$p_{ij}^{(0)}=\begin{cases}0, & i\neq j\\ 1, & i=j\end{cases} \tag{6.10}$$

Obviously, the n-step transition probability $p_{ij}^{(n)}$ represents the probability that the process in state i will transition to state j after n steps, and it does not require the states passed by the n–1 step transition. Generally, the transition probability is related to states i, j, and time n.

Specifically, if the transition probability of the Markov chain is only related to states i, j and not related to time n, it is called a "time-homogeneous Markov chain"; otherwise, it is called a "non-time-homogeneous Markov chain".

$p_{ij}^{(n)}$ is called the one-step transition probability of the Markov chain. For a time-homogeneous Markov chain, the relationship between the one-step transition probability p_{ij} and the step transition probability $p_{ij}^{(n)}$ is given by the Chapman-Kolmogorov theorem. That is:

For all, we have:

① $p_{ij}^{(m+n)} = \sum_{k \backslash inS} p_{ik}^{(m)} p_{kj}^{(n)}$

② $P^{(n)} = P \cdot P^{(n-1)} = P \cdot P \cdot P^{(n-2)} \cdots = P^n$.

This theorem shows that all n-step transition probabilities of a time-homogeneous Markov chain are uniquely determined by its one-step transition probability.

When the states of the Markov chain are finite, it is called a finite chain, otherwise it is called an infinite chain. But whether the state is finite or infinite, we can arrange $p_{ij}(i, j \in S)$ into a matrix form, let:

$$\boldsymbol{P} = \left[p_{ij}\right] = \begin{bmatrix} p_{00} & p_{01} & p_{02} & \cdots \\ p_{10} & p_{11} & p_{12} & \cdots \\ \vdots & \vdots & \vdots & \\ p_{i0} & p_{i1} & p_{i2} & \cdots \\ \vdots & \vdots & \vdots & \end{bmatrix} \tag{6.11}$$

P is called the one-step transition probability matrix of the Markov chain, also known as the transition matrix. The Markov chain completely determines the transition matrix, and the transition matrix also completely determines the Markov chain. Since the probability is non-negative and the process must transition to some state, $p_{ij}(i, j \in S)$ has the following properties:

$$\begin{aligned} & p_{ij} \geq 0, \ i, \ j \in S \\ & \sum_{j \in S} p_{ij} = 1, \ \forall i \in S \end{aligned} \tag{6.12}$$

A matrix with the above two properties is called a right stochastic matrix. Similarly, if the second property in Eq. (6.12) is replaced with the sum of the columns being 1, then this type of matrix is called a left stochastic matrix. A matrix that satisfies both row and column sums being 1 is called a doubly stochastic matrix.

$q_i = P\{X_0 = i\}, \ i \in S$ is called the initial distribution of the Markov chain $\{X_n, \ n \geq 0\}$, and the (row) vector $Q^{(0)}$ composed of the initial distribution is the initial distribution vector of the Markov chain. For a time-homogeneous Markov chain $\{X_n, \ n \geq 0\}$, its finite-dimensional distribution can be completely determined by its initial distribution and transition matrix.

6.2.1 Model Algorithm

Let U be the factor set, that is, the set of evaluation indicators; V is the domain, that is, the set of all possible states of the macro state indicator S of the construction system, then the probability distribution vector $p_s(t)$ of the state of this indicator at any time can be expressed as:

$$p_s(t) = A * \begin{bmatrix} d_1^{(t)} \\ d_2^{(t)} \\ \vdots \\ d_m^{(t)} \end{bmatrix} \tag{6.13}$$

In the equation, "*" represents the fuzzy synthesis operator; m is the number of evaluation indicators; $d_i^{(t)}$ is the membership vector of the i-th evaluation indicator at time t to the domain V.

Since the status of each evaluation index is regulated by the Markov chain, Eq. (6.13) can be further written as:

$$\boldsymbol{p}_S(\boldsymbol{t}) = \boldsymbol{A} * \begin{bmatrix} \alpha_1 \overline{\boldsymbol{d}}_1^{(0)} \boldsymbol{P}_1^t \\ \alpha_2 \overline{\boldsymbol{d}}_2^{(0)} \boldsymbol{P}_2^t \\ \vdots \\ \alpha_m \overline{\boldsymbol{d}}_m^{(0)} \boldsymbol{P}_m^t \end{bmatrix} \tag{6.14}$$

In the equation, $d_i^{(t)}$ represents the membership vector of the i-th evaluation index at the t-th step; $\overline{d}_i^{(t)}$ is the normalized membership vector; is the t-th power of the transition matrix of the i-th risk factor; α_i is the normalization coefficient, calculated as follows:

$$\alpha_i = \sum_{j=1}^{n} d_{ij}^{(0)} \tag{6.15}$$

In the equation, $d_{ij}^{(0)}$ represents the membership of the i-th evaluation index to the j-th element in the discourse domain at the initial moment.

Equation (6.14) can be written in an equivalent form:

$$\boldsymbol{p}_S(t) = A * \left(\overline{\boldsymbol{D}}_\Lambda \times \boldsymbol{P}_\Lambda^t \times \boldsymbol{I}_\alpha\right) \tag{6.16}$$

$$\begin{cases} \overline{\boldsymbol{D}}_\Lambda = diag\left(\overline{\boldsymbol{D}}_1^{(0)}, \overline{\boldsymbol{D}}_2^{(0)}, \cdots, \overline{\boldsymbol{D}}_m^{(0)}\right) \\ \boldsymbol{P}_\Lambda = diag(\boldsymbol{P}_1, \boldsymbol{P}_1, \cdots, \boldsymbol{P}_m) \\ \boldsymbol{I}_\alpha = \left[\alpha_1 \boldsymbol{I}^{(n)}, \alpha_2 \boldsymbol{I}^{(n)}, \cdots, \alpha_m \boldsymbol{I}^{(n)}\right]^T \end{cases} \tag{6.17}$$

In the equation, diag (·) represents the diagonal matrix; $\overline{\boldsymbol{D}}_{\Lambda}$ is called the diagonal evaluation matrix, which is a diagonal block matrix composed of single-factor evaluation vectors of m factors; $\boldsymbol{P}_{\Lambda}$ is called the Markov supermatrix, which is a diagonal block matrix composed of Markov transition matrices of m factors; $\boldsymbol{I}^{(n)}$ is the n-order unit matrix; the superscript "(0)" represents the value at the 0th time step, that is, the value at the initial moment of evaluation.

6.2.2 *Construction Method of Markov Transition Matrix*

In risk assessment indicators, according to whether the indicator status will change over time, the indicators can be divided into static indicators and dynamic indicators. For static indicators, their transition matrix is a unit matrix. For dynamic indicators, their evolution over time is controlled by the Markov transition matrix.

In this model, the scoring of risk assessment indicators is given in the form of their membership to the risk level. Therefore, the element in the Markov transition matrix of the assessment indicator p_{ij} means the probability of the state (rating) of the indicator transitioning from state i to state j in one time step. The assumption implied by this concept is that the membership (normalized) of the assessment indicator at a certain point in time represents the probability distribution of the state of the indicator at the current time. The Markov transition matrix of the assessment indicator can be obtained in the following ways.

1. Frequency statistics method

The frequency statistics method is applicable to assessment indicators that can obtain actual data samples through existing similar projects or numerical simulations. The method of calculating the Markov transition matrix using the frequency statistics method is as follows:

(1) Extract the monitoring data corresponding to the assessment indicators and normalize them

Monitoring data is the most representative data reflecting engineering risks. Using monitoring data as samples to generate the Markov transition matrix has the best representativeness. However, due to the different locations of monitoring data points, their values and even orders of magnitude may vary, so the monitoring data needs to be normalized according to their final predicted values (or maximum allowable values).

(2) Determine the time step length and grading threshold, and calculate the transition frequency of monitoring data

The time step length is the actual time length corresponding to one time step of the Markov chain. Its value is determined by the time span of risk assessment, and can generally be taken as hours, days, weeks or months; the grading threshold is the grading standard for risk factors. If the risk level is divided into n levels, there are n−1 grading thresholds.

After determining the time step length and grading threshold, the transition frequency of monitoring data can be calculated. Taking one time step as a unit, the risk grading of monitoring data in the previous time step and the next time step is determined according to the grading threshold. If the monitoring data in the previous time step is at level i level, if it is at level j at the next time step, the frequency of transferring the monitoring data is incremented by 1. After all the data is processed, the transition frequency matrix can be obtained $\left[F_{ij}\right]_{n\times n}$. Subsequently, the Markov matrix can be calculated by the following formula:

$$P_{ij} = \frac{F_{ij}}{\sum_{j=1}^{n} F_{ij}} \tag{6.18}$$

2. Assignment method

The assignment method is suitable for those semi-quantitative evaluation indicators that are difficult to obtain actual samples or have strong subjectivity. The assignment method is a method with strong subjectivity, and its application needs to be based on the evaluator's sufficient understanding of the project to be evaluated, and sufficient professional knowledge and experience. The core idea of the assignment method is to make judgments on the trend of changes in evaluation indicators over a period of time, and then select an appropriate form from some pre-given forms of Markov transition matrices, assign values to the parameters in the matrix, and form the Markov matrix of the evaluation indicators.

In this model, the state space of the evaluation indicators is the risk level, which is an ordered set with quantitative characteristics. The state of the evaluation indicators generally migrates step by step in a short period of time, and the situation of skipping levels is less. Therefore, the assignment method can use the following form of matrix.

(1) Monotonic type

The general form of the monotonic Markov matrix is:

$$P = \begin{bmatrix} 1-\beta & \beta & 0 & \cdots & 0 & 0 \\ \alpha & 1-\alpha-\beta & \beta & \cdots & 0 & 0 \\ 0 & \alpha & 1-\alpha-\beta & \cdots & 0 & 0 \\ \vdots & \vdots & \vdots & \ddots & \vdots & \vdots \\ 0 & 0 & 0 & \cdots & 1-\alpha-\beta & \beta \\ 0 & 0 & 0 & \cdots & \alpha & 1-\alpha \end{bmatrix} \tag{6.19}$$

In the formula, α, β are parameters, which need to satisfy $0 \le \alpha \le 1, 0 \le \beta \le 1$ and $0 \le \alpha + \beta \le 1$.

When $\alpha < \beta$, the matrix indicates that the probability of the indicator transitioning to a higher-level state is greater than the probability of transitioning to a lower level state, so it can describe the situation where the expected value of the indicator monotonically increases and the final value is high. When $\alpha > \beta$, the opposite is true.

The size of parameters α and β is related to the expected change rate and convergence rate of the indicator value, and this rate is relative to the selected time step length. For example, for a 3-order matrix, if $\alpha = 0.1$, $\beta = 0.2$, the matrix power of P can be calculated as:

$$\begin{gathered}
\boldsymbol{P}^2 = \begin{bmatrix} 0.66 & 0.3 & 0.04 \\ 0.15 & 0.53 & 0.32 \\ 0.01 & 0.16 & 0.83 \end{bmatrix}, \quad \boldsymbol{P}^3 = \begin{bmatrix} 0.558 & 0.346 & 0.096 \\ 0.173 & 0.433 & 0.394 \\ 0.024 & 0.197 & 0.779 \end{bmatrix}, \\
\boldsymbol{P}^{39} = \begin{bmatrix} 0.144 & 0.286 & 0.57 \\ 0.143 & 0.286 & 0.571 \\ 0.142 & 0.286 & 0.572 \end{bmatrix}, \quad \boldsymbol{P}^{40} = \begin{bmatrix} 0.143 & 0.286 & 0.571 \\ 0.143 & 0.286 & 0.571 \\ 0.142 & 0.286 & 0.572 \end{bmatrix},
\end{gathered} \tag{6.20}$$

It can be seen that at the 40th time step, this Markov chain has basically converged to its limit distribution [0.143, 0.286, 0.571]. Since the monotonic Markov matrix is irreducible, its limit distribution is its stationary distribution of displacement, that is, the long-term state of the indicator corresponding to the monotonic Markov matrix is independent of its initial state.

The monotonic Markov matrix is suitable for those dynamic indicators that can predict their changing rules, but it is difficult to predict their state after a certain period of time (the duration of an evaluation unit).

(2) Absorbing type

The general form of the absorbing Markov matrix is:

$$\boldsymbol{P} = \begin{bmatrix} 1-\beta & \beta & 0 & 0 & 0 & 0 & 0 \\ \alpha & 1-\alpha-\beta & \beta & 0 & 0 & 0 & 0 \\ 0 & 0 & 1 & 0 & 0 & 0 & 0 \\ 0 & 0 & \beta & 1-\alpha-\beta & \alpha & 0 & 0 \\ 0 & 0 & 0 & \beta & 1-\alpha-\beta & \alpha & 0 \\ \vdots & \vdots & \vdots & \vdots & \vdots & \ddots & \vdots \\ 0 & 0 & 0 & 0 & 0 & \beta & 1-\beta \end{bmatrix} \tag{6.21}$$

In the formula, α, β are parameters, which need to satisfy $0 \le \alpha \le 1$, $0 < \beta \le 1$ and $0 < \alpha + \beta \le 1$.

In this matrix, one or more rows have only diagonal elements as 1 and the rest as 0. If the i-th row has this feature, it means that the i-th state of this Markov chain is an absorbing state. This matrix indicates that the indicator has a trend to change and converge to one or more states. Like the monotonic matrix, parameters α and β determine the rate of change and convergence.

In this type of matrix, due to the presence of absorbing states, the states are not interconnected. In other words, this Markov matrix is a reducible Markov matrix, which can be seen as a combination of several matrix blocks with monotonic matrix forms. Absorbing matrices do not have a unique stationary distribution, and their limit distribution depends on the initial state.

Absorbing Markov matrices are suitable for dynamic indicators that can make relatively clear predictions about their state after a certain period of time.

6.2.3 Model Application Process

The main process of using the Markov-Fuzzy Comprehensive System State Evolution Prediction Model is shown in Fig. 6.3.

1. Establish a weight vector of indicator system

The prediction of the risk system state of urban large underground space construction uses a weight vector of indicator system. The dynamic indicator system can be obtained by adding or replacing some dynamic indicators based on the static indicator system, or it can use a set of indicators different from the static indicator system. Generally speaking, dynamic indicators can be indicators such as excavation depth and excavation length that change with the progress of construction, or they can be indicators such as precipitation, inflow, and temperature that naturally change over time.

2. Indicator system weight vector A

The weight of the dynamic evaluation indicator represents the relative size of the impact of each evaluation indicator on the risk, which is related to the specific formation mechanism of the risk. Therefore, the indicator weight must be given separately for different types of risks.

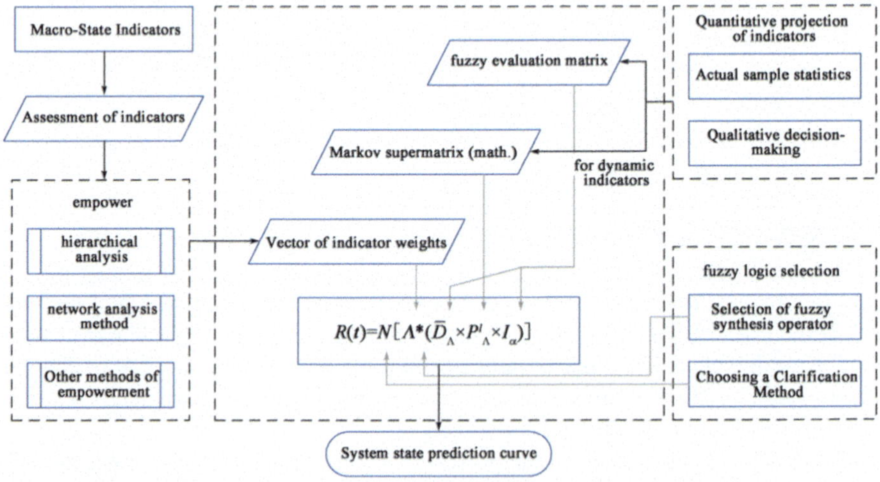

Fig. 6.3 Markov-Fuzzy comprehensive system state evolution prediction flowchart

The method for determining the weight of the evaluation indicator is not unique. The determination of the weight is essentially a decision-making problem, which can be determined by various subjective or objective weighting methods. Theoretically, any set of positive real numbers $\{\alpha_1,\ \alpha_2, \ldots, \alpha_n\}$ that meet the following requirements can be used as weights:

$$\begin{cases} 0 \leq \alpha_i \leq 1 \\ \sum_{i=1}^{n} \alpha_i = 1 \end{cases},\quad i = 1,\ 2,\ \ldots, n \tag{6.22}$$

According to the characteristics of urban large underground space engineering, it is recommended to use the Analytic Hierarchy Process (AHP) or Analytic Network Process (ANP) to obtain the weight of the evaluation indicators.

3. Determine the initial membership degree $d^{(0)}$ of dynamic information indicators

The initial membership degree of dynamic indicators refers to the indicator membership degree at the beginning of the prediction (time step 0), which can be determined according to the method described in Sect. 4.4.2.

4. Assemble the fuzzy evaluation matrix $D_\wedge$

The fuzzy evaluation matrix $D_\wedge$ is a block diagonal matrix composed of membership row vectors of all evaluation indicators in the factor set. The membership degree of the evaluation indicators should be given separately in different areas (each given independently), and the membership degree of dynamic indicators should be updated in real time within different evaluation units (calculated by the Markov transition matrix of dynamic indicators).

After determining the membership degree of all indicators, calculate the normalization coefficient of each indicator in turn and normalize its membership vector. Use α_i to construct the matrix:

$$I_\alpha = \left[\alpha_1 I^{(n)}, \alpha_2 I^{(n)}, \cdots, \alpha_m{}^{(n)}\right]^{\mathrm{T}} \tag{6.23}$$

In the formula, $I^{(n)}$ is an n-order unit matrix.

Then you can assemble the fuzzy evaluation matrix $D_\wedge$. The fuzzy evaluation matrix is an m-row m × n column matrix, where m is the number of evaluation indicators in the factor set, and n is the number of risk division levels in the domain. The elements in $D_\wedge$ are:

$$d_{ij}^{\wedge} = \left(\begin{matrix} d_{ij}, (i-1)n+1 \leqslant j \leqslant in \\ 0, 其他, 1 \leqslant i \leqslant m, 1 \leqslant j \leqslant mn \end{matrix} \right) \tag{6.24}$$

In the formula, $d_{ij}^{\wedge}$ is the element of the i-th row and j-th column of $D_\wedge$; d_{ij} is the j-th component of the membership vector of the i-th evaluation indicator.

5. Assemble the Markov Super Matrix $P_{\wedge}$

The fuzzy evaluation matrix $P_{\wedge}$ is a block diagonal matrix composed of Markov transition matrices of all evaluation indicators in the factor set. The Markov matrix of evaluation indicators is related to the engineering characteristics of different evaluation units. For example, in the evaluation unit with earthwork excavation, indicators such as "plan size" and "height" will have significant changes; while in the evaluation unit that includes the removal of temporary support structures, the "bearing capacity" and "rigidity" of the support structure will decrease to varying degrees. On the other hand, the Markov matrix is also related to the environmental characteristics of the engineering partition itself. Therefore, the Markov matrix should be given separately for each sub-evaluation unit.

When all the Markov matrices of dynamic evaluation indicators are determined, the Markov super matrix $P_{\wedge}$ can be assembled according to the following formula:

$$p_{ij}^{\wedge} = \begin{cases} p_{(i-1) \bmod n+1,\,(j-1) \bmod n+1'}^{\langle(i-1)/\rangle+1} & \langle(i-1)/n\rangle = \langle(j-1)/n\rangle \\ 0, & \text{others} \end{cases} \tag{6.25}$$

In the formula, $p_{ij}^{\wedge}$ is the element of the i-th row and j-th column of $P_{\wedge}$; p_{ij}^{k} is the i-th row and j-th element of the Markov matrix of the k-th evaluation indicator; mod is the modulus operator; $\langle\rangle$ is the integer operator.

6. Carry out system status prediction

When the weight vector A of each evaluation unit, the fuzzy evaluation matrix $D_{\wedge}$, the Markov super matrix $P_{\wedge}$, and the normalization coefficient matrix I_{α} are all given, the evolution prediction of the system macro state indicator can be carried out through formula (6.16).

When using formula (6.16), two aspects of content need to be considered.

(1) Selection of fuzzy synthesis operator

The fuzzy synthesis operator corresponds to a certain decision logic. The fuzzy synthesis operator should be selected based on the characteristics of the evaluation target, the composition of the indicator system, and the decision-making tendency of the evaluator. For the evaluation of urban large underground space construction risk, it is recommended to choose the weighted average operator and the minimum-bounded sum operator.

(2) Selection of clarification method

Strictly speaking, the real output of the fuzzy comprehensive evaluation method is a membership vector. The clarification step is just to provide a perspective for intuitively understanding the evaluation results and will not change the evaluation results. Therefore, the selection of the clarification method can have a certain degree of freedom and subjectivity. In this model, it is recommended to use the segmented assignment method for result clarification.

6.3 Based on System Dynamics Urban Large Underground Space Construction Risk Dynamic Transmission Model

6.3.1 Overview of System Dynamics

1. Introduction to System Dynamics

System Dynamics (SD) is a branch of system science and management science. Based on system theory, control theory, and information theory, it analyzes the system from the internal control mechanism and structure of the system, focusing on the dynamic relationship between the system mechanism and its functional behavior. It is a unity of qualitative analysis and quantitative analysis, and an effective theory, method, and means for analyzing and researching high-order complex nonlinear, high-order, multivariate, multiple feedback complex systems and making scientific decisions.

System dynamics provides a top-down approach to describe project progress, estimate project time, and assess cost risks from a strategic level. It helps project managers understand the impact of project processes on project performance and estimate and grasp the project from a macro perspective. System dynamics establishes a structural model of the system from the microstructure of the system, describes the system structure framework with loop diagrams, describes the logical relationship between system elements with causal relationship diagrams and flow diagrams, describes the quantitative relationship between system elements with equations, and uses specialized simulation software for simulation analysis. The entire analysis process goes from qualitative, semi-quantitative to quantitative, and finally simply converts the quantitative mathematical model into a computer program, using the computer for the final simulation analysis.

System dynamics is widely used in many fields such as economic development, business management, macroeconomic planning, regional economy, energy planning, engineering systems, etc. System dynamics starts from a dynamic perspective, builds system models, displays and controls the behavior patterns and development laws of various physical systems with objective entities or abstract systems based on logical concepts, and then feeds back the system for optimization and control.

2. The relationship between models and systems

The definition given by American system scientist GORDEN G for the system is:

"The so-called system refers to all things that interact and depend on each other, combined according to certain laws". The above definition accurately describes the basic properties, characteristics, and purposes of the system. To further refine, the system should have the following three aspects of meaning: ① a collection of organized objects, such as the solar system; ② a method of organizing and planning objects; ③ a summary of the relationships between different objects.

Simply put, a system is the study of the nature of multiple objects, objective or abstract. and an analysis of their interrelationships. The system has the following four basic characteristics: ① The structure of the system is defined by the objects and processes it describes; ② The system is an induction of reality; ③ Observations of the system can be made through inputs and outputs, inputs are processed and processed internally to form outputs and leave the system; ④ Different parts of the system also interact with each other.

In summary, from a society, a country, to a family, a machine, all have certain characteristics of the system. When examining a system, not only should the static objects inside the system and their dynamic operating rules be examined, but more importantly, the interactions between them need to be analyzed. A large number of facts and experiences tell us that real systems are often large systems with complex structures. For such a system, its overall dynamic behavior characteristics are determined by the interaction between the various parts of the system, not by the operating mechanism of each part itself.

The system dynamics model is an abstract representation of the characteristics and change laws of real systems or objective entities. It is a mapping of the system completed by the researcher according to his own problem needs. Therefore, different models will be established due to different researchers and their research problems. In the modeling process, the principle of simplification should be followed, that is, as much as possible to simplify without affecting the model's imitation of the behavior of the system to be studied. For example, to predict the gross domestic product (GDP) of the national macroeconomic system, an econometric model can be established, and the key factors considered can be total population, total investment, total consumption, etc., while details such as the wage level in Qingdao and national agricultural subsidies can be ignored. Because overall, when considering such a large-scale indicator as GDP, the impact of detail factors can almost be ignored. Similarly, when analyzing the financial operation of a company, other factors such as human resources and production lines can also be appropriately ignored.

Building a model does not mean completely reconstructing the original real system. Researchers need to select appropriate variables based on the research problem and establish a suitably complex model. Since the model is inevitably a simplification and abstraction of the system, it is necessary to carefully evaluate the effect of the model during the modeling process. Generally, the following three aspects are considered: ① Approximation, that is, the similarity between the model and the real system it mimics; ② Reliability, that is, the accuracy of the model's replication of the real system's data; ③ Purpose appropriateness, that is, the degree of conformity between the model and the modeling purpose, which usually reflects the reasonable degree of the model builder's understanding of the model analysis.

The above three aspects emphasize the three key points that need to be focused on when modeling. On the premise of meeting these three requirements, the simpler the model, the better: first, the structure of the simplified model is more concise, covers fewer elements, and the relationship between them is relatively simple, making the model easier to understand and analyze; second, the simplified model reduces unnecessary complexity, facilitates subsequent analysis and understanding, and also shows that the researcher has grasped the key elements in the real system needed to solve the problem; third, the simplified model means that the cost of modeling is lower.

3. System dynamics modeling process

In most cases, the analysis system faced is very large and difficult to analyze directly. For example, the construction safety system is very large, involving many factors and objects, and it is difficult to analyze directly. At the same time, some factors may cause huge losses or accident risks when controlling and adjusting. Therefore, from the perspective of improving the level of analysis or reducing control costs, it is necessary to establish a suitable simplified model for the system under study.

The modeling process is divided into the following steps: first, observe the real system and extract representative data and information, then obtain the model structure framework based on the problem assumption. This step can be regarded as establishing a basic "qualitative" model. Further, it is necessary to refine the problem, that is, to define the constraints and boundary conditions of the system, and collect necessary data to realize the specific model. At this time, the model already has actual data support and can analyze its corresponding model evolution dynamic results. This step can be regarded as establishing a "quantitative" model. In this step, since the evolution dynamic results of the model can be analyzed and observed, it can be contrasted and compared with the real system. And adjust the model according to the comparison results. The entire modeling process is a process of constantly spiraling forward.

6.3.2 Applicability of System Dynamics

1. Characteristics of System Dynamics Application

As a method of system theory research, system dynamics has significant application characteristics, which can be summarized as follows.

(1) Good at handling cyclical problems

Various cyclical economic crises often occur in Western economies, such as short cycles of about 5 years, medium cycles of about 15 years, and long cycles of 40 to 80 years. Now many system dynamics models have explained the causes of various cycles.

(2) Good at handling long-term problems.

System dynamics is a causal mechanistic model, and the internal mechanism of the system is its main emphasis on system behavior, so its simulation time can be relatively long. This is very necessary for studying social and economic systems with large inertia. The famous world model and city dynamics model have simulation times of more than a hundred years. It is particularly worth mentioning that the system dynamics method can introduce potential constraints into the model, and in the process of long-term simulation, it can show the crux of the problem.

(3) Research can still be conducted under conditions of data shortage.

Insufficient data and the difficulty of quantifying certain parameters or relationships is a common problem encountered when studying socio-economic issues. However, under such conditions, system dynamics models can still perform some work. This is because the structure of system dynamics models is based on feedback loops. The theory and practice of dynamic systems show that when multiple feedback loop structures exist, the system behavior pattern is insensitive to most parameters. Thus, although the lack of data brings difficulties to parameter estimation, as long as the estimated parameters fall within their range, the system behavior still shows the same pattern. Under these conditions, system dynamics methods can still be used to study the trends, behavior patterns, fluctuation periods, phase leads and lags, and other issues of system behavior.

(4) Good at handling high-order, non-linear time-varying problems Socio-economic systems are complex, so the equations describing them are often high-order and non-linear, making it difficult to obtain complete information from solving equations using conventional mathematical means. Although methods such as order reduction and linear approximation in mathematical statistics make equation solving easier, the answers obtained may be unreliable and may lose very important parts. Methods like system dynamics that use digital computer simulation as a technical means are quite important for dealing with complex systems characterized by high-order, non-linear time-varying. Although the workload is large and there may not be many innovations in the handling method, it can always provide some information and is a basic means for a considerable length of time.
(5) Emphasizing conditions in the attitude towards prediction

The predictive attitude of system dynamics models is to emphasize the conditions under which results are produced, so conditional predictions can be made in the form of "if...then...".

2. Applicability Analysis System dynamics, as one of the risk modeling theories for complex socio-technical systems, can comprehensively describe various behaviors and characteristics of the system from a macro perspective. For the construction safety system, applying system dynamics theory to the construction safety system can better reflect the changes and recovery of the safety system under different factor disturbances. The variables of the organizational management subsystem, as the distal factors affecting construction safety, are mostly based on a macro perspective to describe the causes of construction safety accidents, such as safety supervision inspections, safety training of construction units, etc., and there is a certain causal relationship and feedback mechanism between the variables, which not only meets the selection criteria of SD model variables, but also is suitable for its simulation mechanism. At the same time, in the safety field, the effect of safety measures often needs to be delayed for a period of time to be observed, and system dynamics modeling can also deal with such problems. This is of great significance for solving the evolution trend of such influence paths in the dynamic evolution system of construction safety

risks throughout the process, such as improving workers' safety values through safety training, and then improving workers' safety operations.

(1) Most factors in the large underground space risk system have non-linear characteristics, and many behaviors show non-linear characteristics, so linear methods are not sufficient to describe such systems. System dynamics systems can generally fully describe the non-linear relationships and complex causal feedback relationships between subsystems within the system.
(2) The large underground space risk system is unstable and unbalanced, so methods for solving stable system problems cannot be used to deal with large underground space risk system problems. System dynamics models can "predict" the development trend and behavioral characteristics of the future safety level of the underground space and can also plan certain subsystems and the entire system according to demand.
(3) The decision-making process in an underground large-space risk system is always associated with a few key factors within the system. Once these key factors are identified, a thorough systematic analysis can help determine the reasons behind the decision-making. The system dynamics model can simulate the degree of change in the system level through key variables.

The underground large-space risk system studies the main factors affecting safety during the underground space production process. It divides these factors into several subsystems, establishes causal relationships between these subsystems, and determines their causal feedback loops. Through system analysis and scientific reasoning, an underground large-space risk system dynamics model is established. The model uses computer simulations to demonstrate the system's performance under different initial conditions and conducts sensitivity analysis on the main factors affecting the system's development. By comparing system analysis and simulation results, suggestions beneficial to system improvement are proposed, providing a reference for formulating scientific safety management decisions.

6.3.3 System Structure Analysis

As mentioned earlier, regardless of the perspective from which a system dynamics model is established, a simplified model suitable for the system under study must be created. Modeling is a feedback process, not a linear arrangement of steps. The model must undergo repeated questioning, testing, and refinement.

1. Determination of System Boundaries

The model of construction management factors based on system dynamics is mainly used to predict and monitor the dynamic changes in cost, quality, and progress during the construction process. It studies the degree of influence of various factors on construction management, identifies key factors for cost control, and inputs change curves in indicators related to technology and environmental systems. This helps form a dynamic positive and negative feedback evaluation for the entire risk analysis.

The risk evolution model route is shown in Fig. 6.4. The system dynamics model is established based on system dynamics principles and construction organization management knowledge. The three major project management goals of cost, quality, and schedule are the main lines for conducting factor analysis. The causal relationships between the main influencing factors are established, and a control model is created. Through simulation, the risk index change curve is obtained to form a dynamic positive and negative feedback evaluation for the entire risk analysis.

The determination of system boundaries is based on the modeling purpose. Elements closely related to the model and whose variable values are important to system characteristics are included in the construction safety system. This allows for the study of structural issues within the construction safety system. System boundary points are the exchange points between this system and the external environment. Such elements are only affected by the external environment and not by the internal system.

2. Core Elements of the System and Their Dynamic Constraints

Safety, cost, and schedule are the three leading organizational performance indicators in a construction project. Among them, safety performance is the focus of the study.

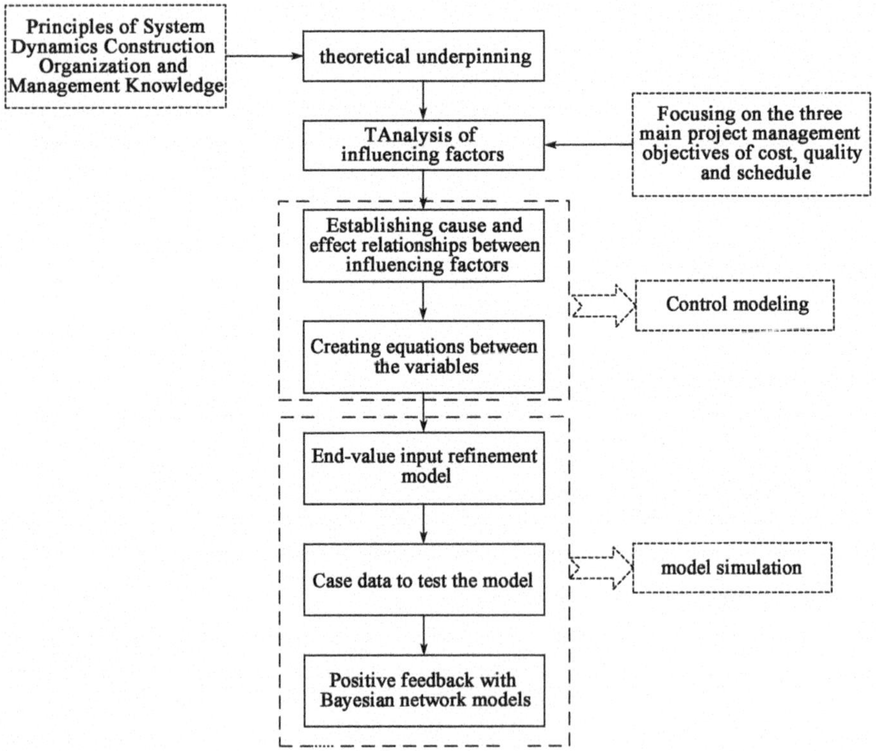

Fig. 6.4 Risk evolution model route map

Organizational safety performance mainly depends on the organization's own structure, but it is also constrained by cost, schedule, and other production performances. There is an interaction and mutual influence between the three, so a qualitative analysis of the three organizational goals (performances) will be conducted based on system dynamics, identifying causal relationships, feedback loops, and system delays between organizational factors.

In actual projects, the project's schedule and cost have a significant impact on safety performance. However, this effect is often overlooked in actual construction, leading to increased costs due to schedule delays or safety hazards. The generation of safety performance is a game process. In the quality control process, low requirements can easily lead to rework, while stringent requirements will lead to more resource and time investment. Both situations will increase costs and cause some schedule delays.

The generation of project duration costs is often due to delays in the project schedule, which require rushing to make up for the progress. Rushing will result in more resource input, increasing costs; at the same time, rushing leads to an increase in workload, on one hand making the supervision of management personnel insufficient, on the other hand, the fatigue caused by long working hours of labor personnel leads to a decline in construction quality, thus causing a large amount of rework and increasing project costs. From the above analysis, it can be seen that in the process of cost control, the control of duration, quality, and cost are inseparable, and there are many internal connections between them. However, in the current cost control methods, these three have not been linked for systematic analysis, and are still in a state of separate control, with each responsible department and management personnel doing their own duties, which will ultimately have a significant impact on the effectiveness of cost control.

In Fig. 6.5, the relationship between the three goals is represented by connecting arrows, each arrow end is marked with a " + " or "−" sign, which is based on the causal relationship of system dynamics for marking and analysis. " + " indicates that the causal relationship between the two is positive, that is, an increase in the cause will lead to an increase in the result. "−" indicates that the causal relationship between the two is negative, that is, an increase in the cause will lead to a decrease in the result. The relationship between duration, quality, and cost can be seen that as long as one of the goals changes, it will inevitably lead to changes in the other two goals. For example, if the quality requirement increases, it will lead to an increase in cost and a delay in duration. A delay in duration will lead to an increase in cost and a decrease in quality, and an increase in cost is equivalent to an increase in cost, that is, the corresponding changes in these three goals will lead to changes in cost.

Based on the above analysis, it can be concluded that changes in quality and duration requirements will cause changes in project costs, and then cause changes in project costs. Therefore, to a certain extent, the control of construction costs of engineering projects is the control of project quality and duration, of course, quality and duration are also constrained by cost (Fig. 6.6).

Therefore, the cost increase caused by quality issues can exist in the form of quality costs; the cost increase caused by duration issues can exist in the form of

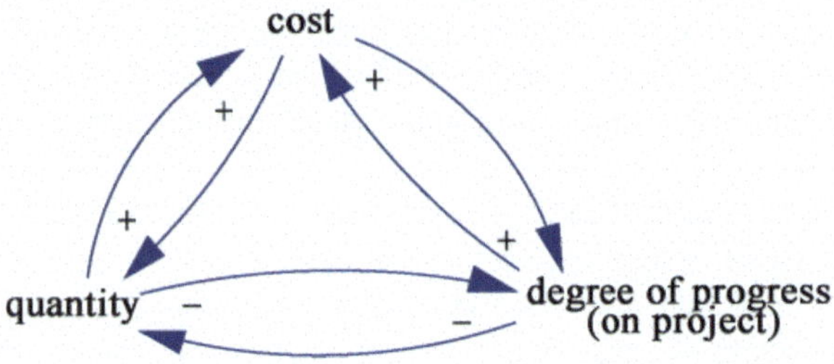

Fig. 6.5 Relationship of the three major goals of project control

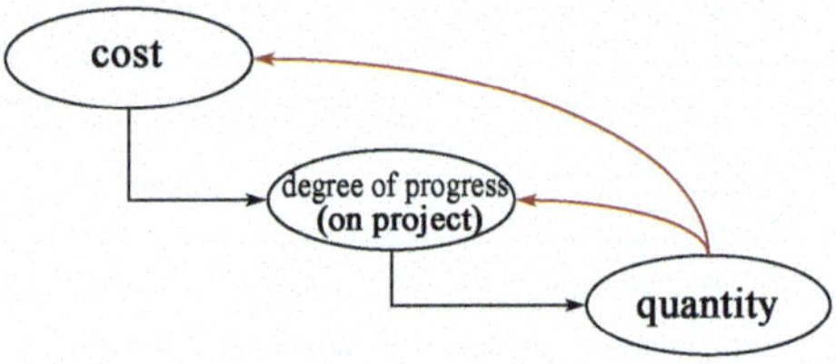

Fig. 6.6 Cost, schedule, quality relationship diagram

duration costs; and the cost increase caused by the actual cost of the project itself, such as the rise in raw material prices or the rise in labor wages, will also lead to an increase in costs.

System dynamics focuses on the macro-systematic study of individual construction project control systems, focusing on studying the project management process at the macro level of the project, with a focus on human-related soft factors and management strategies; system dynamics facilitates the quantification of difficult-to-quantify decision variables, environmental variables, etc., adapts to the requirements of the dynamic system of project management, thus solving the problem that traditional network planning technology alone cannot fully reflect and control all variables.

3. System Evolution Drive Mode

System dynamics suggests that a system, under the dominance of interactions and cyclic feedback among endogenous factors, will exhibit various different apparent behaviors. Although some exogenous factors may also interfere with the system, such interference should be less closely related to the system's own behavior, i.e., there is no complex cyclic feedback, otherwise the system's boundaries should be extended to include these factors within the system. Through previous analysis, in the process of urban large underground space construction, the organizational management subsystem is a relatively independent and complete system. Therefore, without considering the interactions between subsystems at a higher level, the organizational management subsystem should also have its own dynamic behavior. Therefore, the modeling of the organizational management subsystem needs to clarify its evolution driving mode, that is, under unchanged external conditions, which links in the system will spontaneously change, and which links will change first, and which links are affected by it.

As mentioned earlier, the three key elements of the organizational management subsystem are progress, cost, and safety performance. Among these three elements, cost is the element most affected by external factors. The cost of construction mainly depends on the functional goals that the project needs to achieve, and the functional goals are closely related to the type of project, construction environment, construction methods, etc. Therefore, from the perspective of the organizational management subsystem, cost is a relatively static element—it is rarely affected by internal factors of the organizational management subsystem. Therefore, it is hard to imagine cost as a driving element of the organizational management subsystem. Safety performance is the focus and ultimate goal of the organizational management subsystem, that is, although safety performance is a link in many feedback loops in the organizational management subsystem, it is more inclined to the "effect" side in the overall causal network of the system. Therefore, according to the purpose of modeling and basic logical principles, safety performance should be the most dynamic and unpredictable element, so it should not be used as the driving link of the entire subsystem. In this way, the evolution driving mode of the organizational management subsystem should be dominated by progress. Using progress as the driving link of the organizational management subsystem has the following advantages:

(1) The change in progress is a spontaneous and easy-to-understand process. Except in extreme cases (work stoppage, holidays, natural disasters, etc.), the construction progress of a project is bound to naturally increase over time. When external conditions remain ideal, the construction progress can be seen as linearly increasing with time (of course, this also depends on how to measure construction progress), which makes progress very suitable as the driving force for system evolution.
(2) The inflow and outflow rate of progress can be controlled by indicators such as construction speed and rework rate, which are clear in concept and easy to measure. In addition, the factors that can affect the construction speed are also very easy to analyze, and the corresponding causal relationship diagram can be quickly constructed.
(3) Construction progress is the link of the organizational management subsystem. There is a significant correlation between construction progress and construction quality, and there is a clear feedback loop. When construction progress lags, the constructor will take measures to speed up the construction progress, but this will bring about a decrease in construction quality due to rushing, which in turn affects safety performance, reduces construction speed, and increases construction costs; while reasonable or ahead-of-schedule construction progress can easily form a positive incentive, improving construction quality while further accelerating construction speed.

6.3.4 Modeling of Dynamic Transmission Model of Construction Risk in Urban Large Underground Space

1. Archetype of System Dynamics

System Dynamics Archetypes are a fundamental tool for analyzing complex systems, representing a series of minimal system dynamics models with real-world significance. Throughout the development of system dynamics, researchers have established many classic archetypes, covering most of humanity's complex dynamic problems: from individuals and families to organizations, industries, cities, societies, nations, and the world, even activities related to ethnic groups, history, and ecological environments. System dynamics archetypes are built from continuously enhanced feedback ("+" feedback loop), repeatedly adjusted feedback ("−" feedback loop), and time delays. System dynamics archetypes reveal the simple beauty behind managing complex phenomena, allowing us to see more clearly the operation of structures and find leverage points within the structures. The main types of system dynamics archetypes related to safety risks are as follows.

(1) Archetype of safety awareness

As shown in Fig. 6.7, the Archetype of safety awareness describes situations where the effects of various safety measures taken by an organization may have limits. Various safety measures can effectively prevent accidents at the beginning of a project. However, as various safety hazards decrease, the safety awareness of organizational managers will also decline, leading to a decrease in system safety. Construction units emphasize construction safety at the beginning of underground space construction, such as reinforcing the soil body, strengthening monitoring density and frequency, etc., to ensure the safety of the initial stage. But as the shield tunneling gradually enters the normal excavation stage, if no new safety hazards are exposed, the attention or safety awareness of managers may unconsciously decline, reflected in a series of safety measures, thereby increasing the risk of tunnel excavation.

(2) Archetype of safety goal

As shown in Fig. 6.8, the Archetype of safety goal describes how safety goals are eroded. The difference between safety goals and actual system safety will motivate the organization to take measures to improve safety, but the effects of these measures often take some time to manifest; on the other hand, this difference may lower safety goals when facing other pressures. In subway tunnel construction, the construction party will take many safety measures, such as strengthening worker safety training, etc., but the effects of these measures may not be immediate. Therefore, although the organization has made efforts, the difference between safety goals and actual system safety remains large due to the lack of immediate results. At this time, under the influence of production pressure, the organization is likely to lower its safety requirements and shift more resources and attention to areas outside of safety to make compromises.

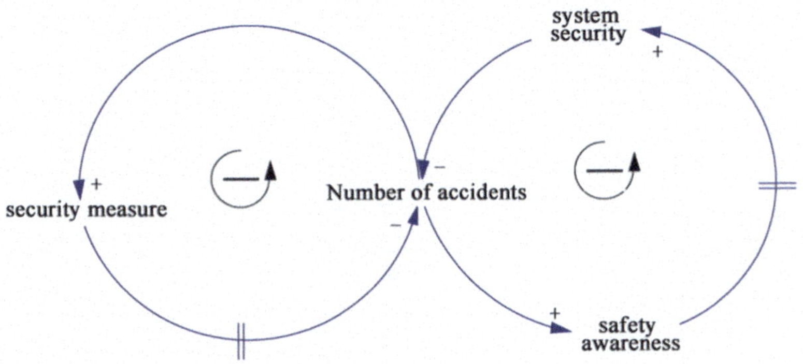

Fig. 6.7 Archetype of safety awareness

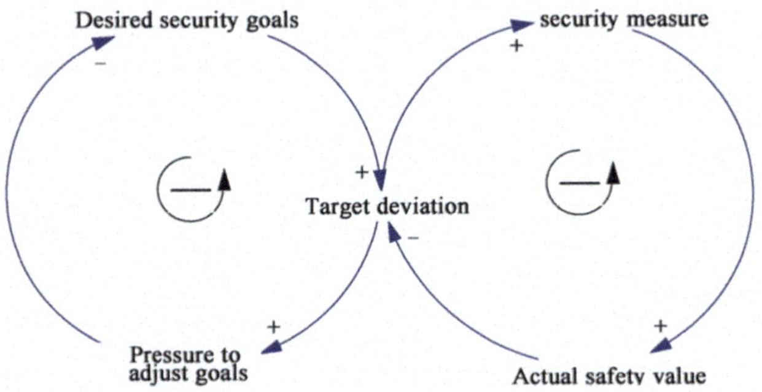

Fig. 6.8 Archetype of safety goal

(3) Archetype of organizational complacency

As shown in Fig. 6.9, the archetype of organizational complacency describes how organizations can become complacent when they have not experienced accidents. With continuous supervision and improvement of the system, the number of accidents will gradually decrease. As accidents no longer occur, these supervisory or improvement measures may be considered too strict. Coupled with possible budget pressures, the investment in supervision or measures will be gradually reduced. This will be reflected in the reduction of safety training or supervisory inspections, thereby continuously increasing the risk of accidents.

2. Causal loop diagram of the organizational management subsystem for urban large underground space construction

Taking the relationship between project progress and cost as an example, based on the foundation of the organizational complacency base model, a system dynamics model is established to analyze the relationship between project progress and cost, explain the mutual influence between the two, and obtain the quantified progress control intensity under cost control (Fig. 6.10).

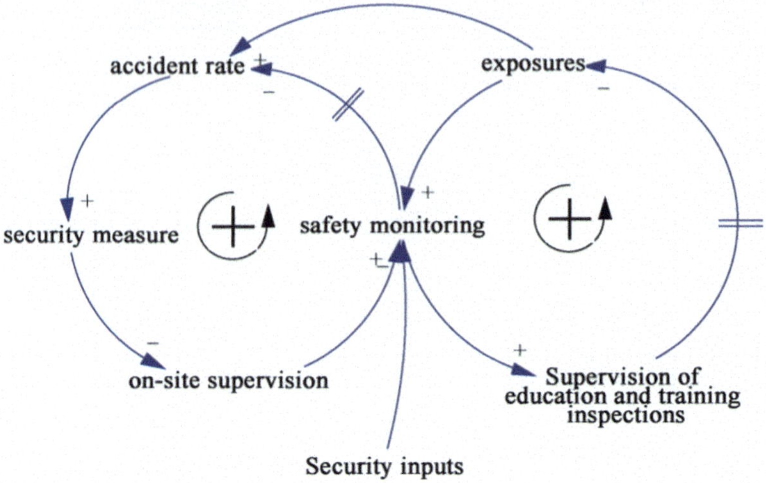

Fig. 6.9 Organizational complacency base model

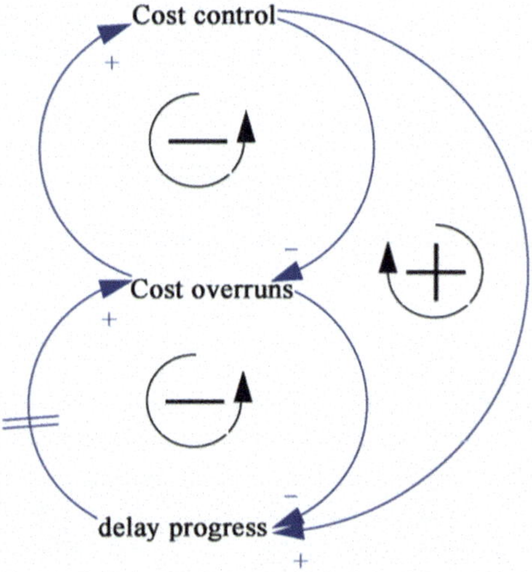

Fig. 6.10 Project progress-cost causal relationship and improvement base model

In the project progress-cost relationship model, stringent cost control measures often harm the interests of the construction party, causing the construction unit to deliberately delay progress in hopes of receiving reasonable remuneration. If this contradiction is not well resolved, it may lead to serious delays in the project. The improved approach to this problem is to adjust cost expenses through progress control measures, forming a regulatory closed loop.

The core of project management is goal control, that is, to achieve or exceed the project's target requirements through various passive or active measures, such as scope, time, cost, quality, and other project goals. Cost control is one of the important means of active measures. For construction projects, due to the irreversibility of the project itself, cost is increasingly valued and recognized by decision-makers and project managers.

Delays are very important in the dynamic generation process. Delays cause system inertia, which may lead to instability, and often make the short-term effects and long-term effects of policies just the opposite. In the model, we need to clarify which factors' lagged responses cannot be ignored based on engineering experience, expert opinions, etc. The delay time is represented when inputting into the model. For example, the response of schedule pressure to construction progress lacks immediacy, a few days of delay will not cause a sharp rise in schedule pressure, and the frequency of accident signs affecting the quality of the project also has a long delay. Poor project quality will not cause or immediately cause safety issues, and there is a delay in feedback to the frequency of accident signs.

Identify project risks from the perspectives of schedule, quality, and cost. According to the feedback principle of system dynamics, use the system dynamics simulation software Vensim PLE to establish a system dynamics causal loop diagram, as shown in Fig. 6.11.

The simplified large underground space construction safety management process has the following three feedback loops.

(1) Frequency of accident signs ↓ → Construction speed ↑ → Construction progress ↑ → Schedule pressure ↓ → Safety margin ↑ → Frequency of accident signs ↓.
(2) Frequency of accident signs ↓ → Budget surplus ↑ → Safety measures ↑ → Safety margin ↑ → Frequency of accident signs ↓.

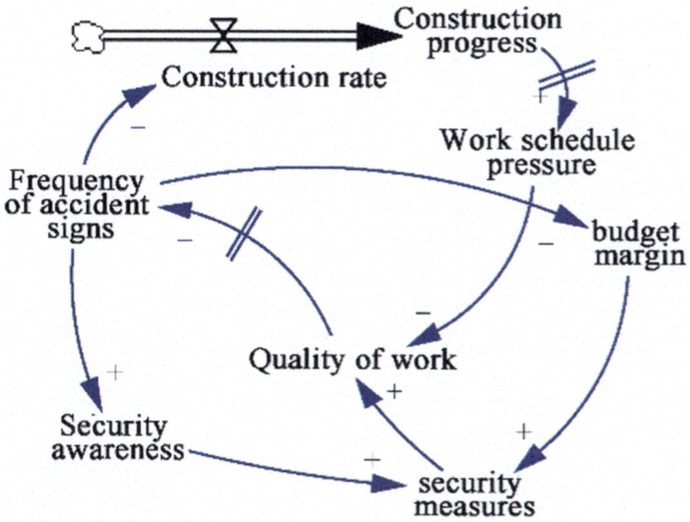

Fig. 6.11 Causal loop diagram of urban large underground space construction organization management subsystem

(3) Frequency of accident signs ↓ → Safety awareness ↓ → Safety measures ↓ → Safety margin ↓ → Frequency of accident signs ↑.

From the causal relationship diagram, it can be seen that risk factors form positive and negative cycles. The frequency of accident signs appearing serves as the initial value. As the frequency of accident signs decreases, the construction speed increases, the construction progress rises accordingly, and the construction progress accelerates, the pressure of the construction period will decrease. When the pressure of the construction period is not tense, the construction quality will increase, and the frequency of accident signs will decrease, forming a positive cycle. As the frequency of accident signs decreases, the budget surplus rises, safety measures increase, safety margin rises, the frequency of accident signs decreases, forming a positive cycle. When the frequency of accident signs decreases, after maintaining a high level of safety awareness for a period of time, there will be a relaxation of vigilance, safety measures will decrease, leading to a decrease in safety quality, and the frequency of accident signs will increase, forming a negative cycle. At the same time, there is a delay reaction time between various factors.

3. Detailed factor analysis of the system

There are many factors involved in construction project management, which can be seen as a system. In the project management system, the manpower required by the task scale and the complex relationship between the project progress and other factors, there are various direct or indirect connections between each factor. From design to construction, debugging includes a large number of interdependent processes. The project management system contains a large number of non-linear, irreversible connections. Therefore, it is often difficult to optimize all activities in actual work. The project management system includes both quantifiable material variables, such as material requirements, installation progress, etc., and also includes a large number of soft variables that are not easy to quantify, such as policy decisions, organizational structures, etc.

(1) Personnel quality

If the on-site management organization is chaotic, there is a lack of safety management personnel, and the distribution of workers is inappropriate, these poor labor organization management factors will directly affect the physiological condition of the construction personnel. If the safety awareness of the construction workers is not high, they are negligent, complacent, self-righteous, or the supervision and management level of the enterprise is low, the supervision is insufficient, the supervisors are derelict, and the supervision system is unreasonable, the safety psychology of the construction personnel will be directly affected. Moreover, when the operator lacks safety knowledge, does not understand safety procedures, and does not have the ability to complete the task, it will directly affect their safety awareness. When the management of the construction project does not organize safety education and training for new entrants, or the intensity of safety education and training is not enough, the safety knowledge level of the construction workers will be at a lower level.

(2) On-site management

The safety subsystem of the equipment is affected by the equipment status, equipment feasibility, equipment applicability, etc. Before construction, the completeness of the design scheme and the rationality of the design calculation will directly affect the applicability of equipment selection, such as choosing equipment that does not meet the requirements due to the lack of stability calculation of the secondary lining formwork trolley and the lack of traction force calculation of the transportation equipment in shield machine. When mechanical equipment has safety hazards, lacks safety acceptance permits, and does not check protection before starting, it will directly affect the feasibility and applicability of equipment use. When the equipment feasibility is poor, the equipment applicability is poor, and the operator has bad habits, the status of the equipment will be directly affected, the equipment wear is severe, the maintenance is not in place, affecting the safety status of the equipment.

(3) Organizational structure

The organizational structure is influenced by emergency plan management, labor organization management, safety management system, and safety supervision management. Construction companies that fail to establish a sound safety production management system as required, or whose establishment and implementation of safety inspection systems do not meet requirements, directly affect the level of emergency plan management, safety education and training, and safety supervision management of construction projects. A well-established safety management system is beneficial for strengthening labor organization management and standardizing safe labor practices. When a company's supervision management is not in place, or when project managers or operators lack safety knowledge, it directly affects the level of machinery and equipment management. If construction techniques are unrealistic or not timely adjusted according to external conditions, the feasibility of safety technical disclosure will be affected and will not meet construction needs. These eight factors work together to affect the safety management status of underground space construction.

4. SD stock flow diagram of urban large underground space construction organization management subsystem

The flow rate system of system dynamics is the basis for determining flow position variables and flow rate variables. Based on the variable set established by analyzing and establishing the causal feedback relationship of large underground space construction management safety, the basic modeling method of system dynamics is applied to draw the system's flow diagram to reflect the influence relationship between various factors within the system.

Based on a full understanding and comprehension of the large underground space construction organization management system, the system is described by constructing a stock flow diagram according to the three major indicators of quality, progress, and cost, i.e., dividing the management subsystem into three subsystems, determining auxiliary variables, exogenous variables, and parameters. Auxiliary variables are used to describe the dependency relationship between flow position and flow rate, and the establishment of exogenous variables and parameters is to describe the impact of the environment on the system. After determining the above flow position, flow rate, auxiliary, exogenous variables, and constants, the entire flow diagram is also determined, as shown in Figs. 6.12, 6.13, and 6.14.

Based on the variable set established by analyzing and establishing the causal feedback relationship of large underground space construction management safety, the SD stock flow diagram of underground space construction system safety can be constructed.

Fig. 6.12 Engineering quality risk flowchart

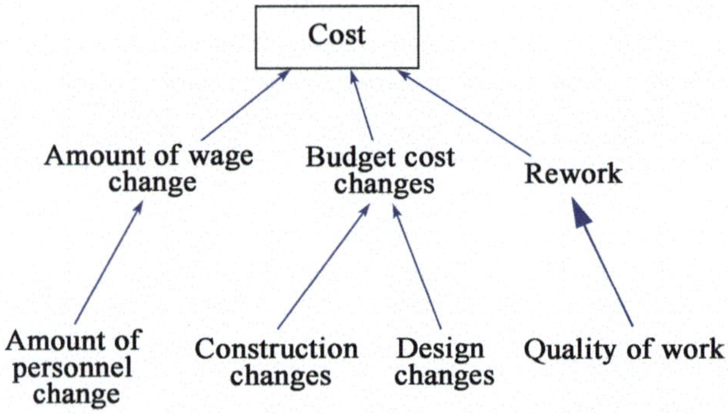

Fig. 6.13 Cost risk flowchart

Fig. 6.14 Construction progress risk flowchart

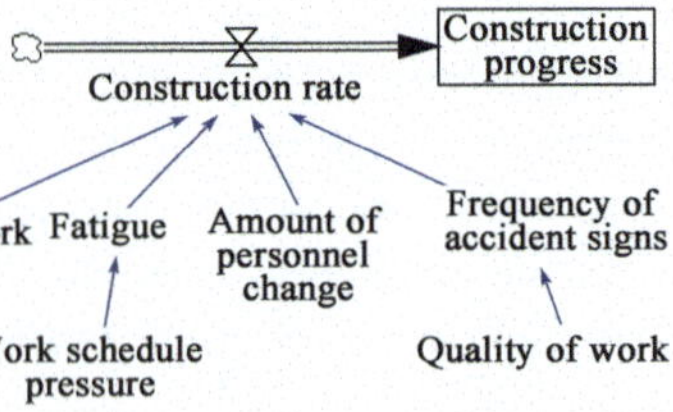

5. Equation and parameter assignment

In addition to constructing a flow diagram at the level of the large underground space construction management system, the model also needs to attach appropriate parameters or equation relationships to the variables in the flow diagram. The assignment of some parameters requires extensive data collection and is determined by methods such as regression analysis, arithmetic mean method, table function method, expert estimation method, etc. Some key parameters need to be accurately estimated through research. In addition, when using the established model for the simulation application of the real system, the initial values of the level variables during the simulation period need to be determined according to the actual situation of the large underground space engineering project. The quantification of factors and equation calculation can be based on the causal relationship diagram and the established system dynamics flow diagram to construct system variable expressions. The determination of variable functions requires repeated debugging and testing before final determination.

Based on the principles of system dynamics and the risk factor variable set, combined with field research, a regression analysis of the quantitative relationship of each variable in the system dynamics model is conducted to obtain the system dynamics equation. Taking Fig. 6.14 as an example, equations are generally established from the upper layer to the lower layer. At the same time, when constructing the model, the principle of simplification is adhered to, and variables that do not have a significant impact on the model operation results are deleted as much as possible, retaining only key factors.

(1) Accumulated variables

The construction progress is the value of the accumulated variable, and the construction efficiency is the rate of change of this accumulated variable, written as a discrete equation, that is:

$$L_{\mathrm{t}}^{2} = L_{\mathrm{t-1}}^{2} + \mathrm{d}t \times \left(I_{\mathrm{t-1}}^{\mathrm{t}} - O_{\mathrm{t-1}}^{\mathrm{t}}\right) \tag{6.26}$$

In the formula, is the current value of the accumulated variable; is the value of the accumulated variable at the previous time step; is the time step length; is the inflow rate of the accumulated variable; is the outflow rate of the accumulated variable.

In construction, we believe that only the construction rate will affect the construction progress, so the calculation of construction progress is:

$$C_{\mathrm{s}}(t) = C_{\mathrm{s}}(t_0) + \int_{t_0}^{t} [C_{\cdot\mathrm{Rate}}]\mathrm{d}t \tag{6.27}$$

where, C_{s} is the construction schedule; C_{Rate} is the construction rate.

(2) Rate equation

The rate variable is the rate of change of the accumulated variable, and the equation is expressed as:

$$\text{rate}\,S(t) = g[L(t), \text{aux}(t), \text{exo}(t), \text{const}] \tag{6.28}$$

where, L(t) is the accumulated variable value at time t; aux(t), exo(t) are the auxiliary variable and exogenous variable values at time t respectively; const is a constant.

During the construction process, the construction rate is also affected by many other factors and is not constant, so it is also a variable related to time, and its equation is as follows:

$$C_{\text{Rate}}\,(t) = QM_1 \times ZP_1(t) + QM_2 \times ZP_2(t) + QM_3 \times ZP_3(t) + QM_4 \times ZP_4(t) \tag{6.29}$$

where, are auxiliary variables characterizing rework, fatigue, personnel changes, and the frequency of accident signs; is the parameter of the related variables, defined as the contribution rate to the accident rate, with a value of 0–1. And:

$$QM_1 + QM_2 + QM_3 + QM_4 = 1 \tag{6.30}$$

(3) Auxiliary equation

The auxiliary equation is the operation equation describing information in the feedback system, and its equation is as follows:

$$\text{aux}(t) = f\left[L(t), \text{aux}^*(t), \text{exo}(t), \text{const}\right] \tag{6.31}$$

where, are other auxiliary variables besides the auxiliary variable to be solved.

In the model, fatigue is an auxiliary variable, which is:

$$ZP_2 = QF_2 \times ZM_2 \tag{6.32}$$

where, is the parameter of the related variables, defined as the coefficient of the effect of the duration pressure on fatigue causing fatigue to increase or decrease.

(4) Table function

In the flowchart, in addition to the relationship between boundary points and linear functions, there are also table function relationships, such as the relationship between progress pressure and work pressure, duration pressure and personnel adjustment coefficient. It is difficult to achieve through simple operations between variables, and it is more convenient to give this non-linear relationship in a graphical way. The following is a detailed introduction to the numerical estimation method of factors with table function relationships, taking the personnel adjustment coefficient as an example.

The personnel adjustment coefficient depends on the duration pressure. Through actual analysis, its dependency is that when there are more than 12 months left to complete the project, the number of personnel is mainly increased or decreased according to the expected number of frontline personnel in the project. When the remaining time is between 2 and 10 months, both the opinions of the project frontline and the existing personnel need to be considered. When only 2 months are left to complete the project, no more personnel will be added or reduced, and the project will be completed based on the existing personnel.

The association between personnel adjustment coefficient and duration pressure is difficult to express with explicit functions, so a corresponding table function can be established based on statistical rules to describe it. The specific content of the table function can be seen in Table 6.1 and Fig. 6.15.

(3) Constant

Parameters that do not change with time are constants, and boundary point parameters are generally constants. Such elements are only affected by the external environment and not by the system itself.

The variable set of the SD model constructed in this study includes level variables, rate variables, auxiliary variables, and constants. The state variable is the accumulated state of the factor, which describes the past or future development trend of the system elements; the rate variable is the change process of the state variable, which determines the trend of the state variable changing with time; the auxiliary variable helps to provide auxiliary information for the complex relationship between the rate variable and the state variable; the constant represents the factors in the system that are approximately unchanged. The defined variable set includes 3 level variables. There

Table 6.1 Scale of personnel functions

Pressure on schedule	0	2	4	8	10	12	14	16
Ratio of personnel	1	0.8	0.6	0.4	0.2	0	0	0

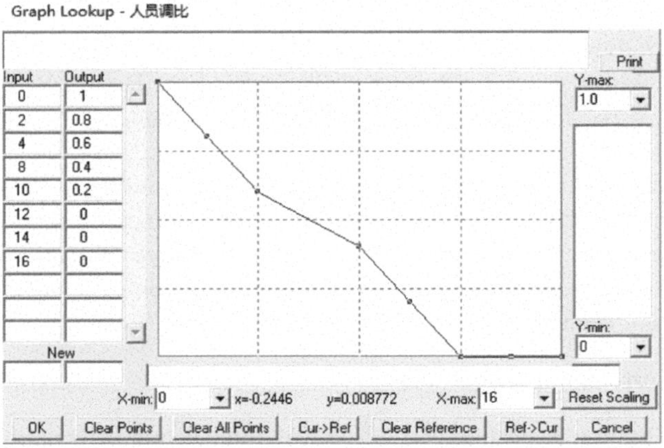

Fig. 6.15 Personnel adjustment coefficient impact factor diagram

is 1 rate variable, 14 auxiliary variables, and 13 constants, totalling 31. The specific meanings of the variables are shown in Table 6.2. All variables are dimensionless, and the larger the value, the better the variable's development direction.

Based on the theory and method of system dynamics, the urban large underground space organization management system can be analyzed. With the help of system dynamics software Vensim, a structure describing the system dynamics model can be established, simulating a continuously changing feedback system, and the simulation results of the model can be analyzed and optimized.

6.4 Networked Large Underground Space Engineering Construction Process Risk Coupling Evolution Analysis

6.4.1 Project Overview

This section takes the newly built Beijing Subway Line 17 Dongda Qiao Station as an example, applies the risk evolution assessment model, assesses the collapse risk during the entire construction process of Dongda Qiao Station, verifies the applicability of the model, and analyzes the evolution law of construction risk for typical urban underground networked extension and construction engineering.

1. Project Background

Dongda Qiao Station is a transfer station of Line 17, which is a "T" shaped passage transfer with the existing Line 6 Dongda Qiao Station, and reserves the conditions for transfer access with the planned Line 28. The plan location and structure of Dongda Qiao Station are detailed in Sect. 2.3.8.

2. Engineering Geology and Hydrogeology

Dongda Qiao Station is located in Chaoyang District, Beijing, which is a plain terrain, relatively flat, with a ground elevation of 39. 2–38. 4 m. The geological section is shown in Fig. 6.16. The physical and mechanical parameters of each stratum are shown in Table 6.3.

The groundwater in the strata of this work point, according to the burial depth, dynamic change characteristics, and the impact on engineering construction, can be divided into upper layer stagnant water (one), Interlayer phreatic water (three), Interlayer phreatic water–confined water (four) and confined water (five). The characteristics of each layer of groundwater are shown in Table 6.4.

3. Site and Surrounding Conditions

Dongda Qiao Station is located at the five-way intersection of Dongda Qiao Road, Workers Stadium East Road, Chaoyangmen Outer Street, and Chaoyang North Road. The station body is set along Dongda Qiao Road and Workers Stadium East Road from south to north. The northwest quadrant is the Workers Stadium East Road

Table 6.2 Meaning of SD model variables

Variability		Variable name	Variable definition	Formula
Horizontal variable	L_1	Level of engineering quality and safety	Dimensionless processing, indicating that the quality and safety indicators of underground space construction works, the larger the score, the higher the level of construction quality	$L_1 = QP_1 \times ZM_1 + QP_2 \times ZM_2 + QP_3 \times ZM_3 + QP_4 \times ZM_4 + QP_5 \times ZM_5 + QP_6 \times ZP_2 + QP_7 \times ZP_3$
	L_2	cost level	Dimensionless treatment, indicating an indicator of the cost of construction of underground space, with larger scores indicating higher levels of cost	$L_2 = QN_1 \times ZN_1 + QN_2 \times ZN_2 + QN_3 \times ZP_1$
	C_S	Level of construction progress	Dimensionless, indicates the level of construction progress in underground space, the higher the score the higher the level of construction progress.	$C_S(t) = C_S(t_0) + \int_{t_0}^{t} [C_{Rate}]dt$
Rate variable	C_{Rate}	Construction rate	Change in construction progress per unit of time	$C_{Rate} = QM_1 \times ZP_1 + QM_2 \times ZP_2 + QM_3 \times ZP_3 + QM_4 \times ZP_4$
Auxiliary variables	ZM_1	Technical level	Overall level of construction technology	$ZM_1 = QS_1 \times CM_1 + QS_2 \times CM_2$
	ZM_2	Pressure on schedule	Level of schedule pressure	$ZM_2 = QT_1 \times CS + QT_2 \times CN_2 + QT_3 \times CM_{13}$
	ZM_4	Managerial level	Level of management of construction sites by managers	$ZM_4 = QS_6 \times CM_5 + QS_7 \times CM_6 + QS_8 \times CM_7 + QS_9 \times CM_8$
	ZM_5	Security measure	Level of completeness of security measures	$ZM_5 = QS_3 \times L_2 + QS_4 \times CM_3 + QS_5 \times ZN_2$
	ZP_1	Rework	Rework level, the higher the score the less likely rework will occur	$ZP_1 = \langle L_1 \text{ TO } ZP_1 \rangle \times L_1$

(continued)

Table 6.2 (continued)

	ZP_3	Volume of personnel changes	Change in construction workforce	$ZP_3 = QS_{10} \times CM_9 + QS_{11} \times CM_{10} + QS_{12} \times CM_{11}$
	ZP_4	Frequency of signs of accidents	Frequency of occurrence of accident signs, the higher the score the lower the frequency of occurrence of accident signs.	$ZP_4 = \langle L_1$ TO $ZP_4\rangle \times L_1$
	ZN_1	Amount of change in wages	Amount of change in personnel wages	$ZN_1 = \langle ZP_3$ TO $ZP_1 > \times ZP_3$
	ZN_2	Changes in budgeted costs	Level of change in construction budget costs	$ZN_2 = QR_1 \times CN_1 + QR_2 \times CN_2$
	CN_2	Design change	Level of design changes	$CN_2 = QR_3 \times CM_{14} + QR_4 \times CM_{15}$
	CM3	Safety awareness	Overall level of safety awareness among construction site personnel	$CM_3 = \langle ZP_4$ TO $CM_3\rangle \times ZP_4$
	CM_{11}	Ratio of personnel	Staffing adjustments	$CM_{11} =< ZM_2$ TO $CM_{11} > \times ZM_2$
	CM_{12}	Frontline expectations staff	First-line expectations of construction staffing levels	$CM_{12} =< ZM_2$ TO $CM_{12} > \times ZM_2$
Constant	ZM_3	Level of training and education	Level of training and education of personnel	
	ZP_2	Weariness	Fatigue level of construction workers, the smaller the score the higher the fatigue of construction workers	
	CM_1	Technical maturity	Level of technological maturity	
	CM_2	Technical applicability	Level of technical suitability and fit	
	CM_5	Management system	Level of completeness of the management system	

(continued)

Table 6.2 (continued)

	CM_6	On-site supervision	Level of on-site supervision	
	CM_7	Emergency plan	Level of emergency preparedness	
	CM_8	Management organizations	Level of institutional integrity of management organizations	
	CM_9	Timing of adjustments	Adjustment time for changes in the number of personnel	
	CM_{10}	Number of personnel	Total number of construction personnel	
	CM_{11}	Construction changes	Level of construction changes	
	CM_{13}	Natural disaster	Level of natural disasters caused by earthquakes, high winds, landslides etc.	
	CM_{14}	Inaccurate surveys	Level of change due to inaccurate geological surveys	
	CM_{15}	Design fault	Level of change due to design errors	
	QP_1	Coefficient of influence of the level of technology on the quality of works	Coefficient of increase or decrease in the quality of the project because of the level of technology on the quality of the project	
	QP_2	Coefficient of influence of schedule pressure on the quality of work	Coefficient of improvement or deterioration of project quality due to schedule pressure	
	QP_3	Coefficient of influence of the level of training and education on the quality of works	Coefficient of increase or decrease in the quality of the works because of the level of training and education	
	QP_4	Coefficient of influence of the level of managers on the quality of works	Coefficient by which the quality of the work is enhanced or diminished by the level of the manager	

(continued)

Table 6.2 (continued)

	QP_5	Coefficient of influence of safety measures on the quality of works	Coefficient of enhancement or diminution of the quality of works because of safety measures on the quality of works	
	QP_6	Coefficient of influence of fatigue on the quality of works	The coefficient by which fatigue acts on the quality of the work to enhance or diminish it	
	QP_7	Coefficient of influence of the level of technology on the quality of works	Coefficient of increase or decrease in the quality of the project because of the level of technology on the quality of the project	
	QS_1	Coefficient of technological maturity on the level of technology	Coefficient by which technological maturity acts on the level of technology to raise or lower it	
	QS_2	Coefficient of technology applicability on the level of technology	Coefficient by which technology suitability acts on the level of technology to increase or decrease it	
	QS_3	Cost impact factor for security measures	Coefficient by which security measures are enhanced or weakened because of the level of technology applied to them	
	QS_4	Coefficient of influence of security awareness on security measures	Coefficient of increase or decrease in security measures because of security awareness on security measures	
	QS_5	Cost impact factor for security measures	Coefficients by which costs are applied to security measures resulting in their enhancement or weakening	

(continued)

Table 6.2 (continued)

	QS_6	Influence of management system on the level of managers coefficient	Coefficient by which the management system acts on the level of managers to raise or lower the level of managers	
	QS_7	Impact of on-site supervision on the level of managers coefficients	Coefficient by which on-site supervision acts to raise or lower the level of managers	
	QS_8	Impact of emergency preparedness on the level of managers coefficient	Factor by which the level of managers is raised or lowered by the effect of contingency planning on their level of competence	
	QS_9	Coefficient of influence of the organizational structure of the management on the level of managers	Coefficient by which the organizational structure of the management organization acts on the level of the manager to raise or lower the level of the manager	
	QS_{10}	Effect of adjustment time on the volume of personnel change coefficient	Coefficient by which the level of technology affects the volume of personnel change, resulting in an increase or decrease in the volume of personnel change	
	QS_{11}	Coefficient of influence of the number of personnel on the volume of personnel changes	Coefficient by which the number of personnel acts on the volume of personnel changes to enhance or diminish the volume of personnel changes	

(continued)

Table 6.2 (continued)

	QS_{12}	Impact of staffing ratios on the volume of staffing changes coefficients	The staffing ratio is a factor applied to the volume of staffing changes to produce an increase or decrease in the volume of staffing changes	
	QN_1	Coefficient of impact of wage changes on costs	Coefficient by which the volume of wage changes acts on costs to raise or lower them	
	QN_2	Cost impact factor for changes in budgeted costs	Factor by which changes in budgeted costs are applied to costs resulting in cost escalation or de-escalation	
	QN_3	Rework impact factor on costs	Coefficients of upliftment or weakening of costs because of the application of rework to costs	
	QR_1	Impact factor of construction changes on changes in budgeted costs	Factor by which changes in budgeted costs are enhanced or diminished by construction changes	
	QN_2	Impact factor of design changes on changes in budgeted costs	Factor by which changes in budgeted costs are enhanced or diminished by the effect of changes in design on changes in budgeted costs	
	QN_3	Impact factor of design errors on design changes	Coefficients by which design errors are applied to design changes resulting in upgrading or downgrading of design changes	

(continued)

Table 6.2 (continued)

	QN_4	Impact factor of inaccurate surveys on design changes	Coefficients by which the frequency of survey inaccuracies is applied to design changes resulting in upgrading or downgrading of design changes	
	QM_1	Contribution of rework to construction rate	Weighting of changes in construction rate due to rework	
	QM_2	Fatigue contribution to construction rate	Weighting of fatigue-induced changes in construction rate	Value 0–1 $QM_1 + QM_2 + QM_3 + QM_4 = 1$
	QM_3	Contribution of the amount of personnel change to the construction rate	Weighting of changes in construction rate due to changes in personnel volume	
	QM_4	Contribution of the frequency of accident signs to the construction rate	Weighting of changes in construction rate due to frequency of occurrence of accident signs	
	QT_1	Coefficient of impact of construction progress on schedule pressure	Coefficient by which construction progress acts on schedule pressure to increase or decrease schedule pressure	
	QT_2	Impact factor of design changes on schedule pressure	Coefficient of increase or decrease in schedule pressure because of construction progress on schedule pressure	
	QT_3	Coefficient of impact of natural disasters on schedule pressure	Coefficient by which construction progress acts on schedule pressure to increase or decrease schedule pressure	
	<QF_2 *TO* ZP_2>	Coefficient of influence of work pressure on fatigue	Coefficient by which fatigue is enhanced or diminished by the effect of duration pressure on fatigue	

(continued)

Table 6.2 (continued)

	<L_1 *TO* ZP_4>	Coefficient of influence of the quality of work on the frequency of accident signs	Coefficient of fatigue enhancement or weakening due to duration pressure acting on fatigue	
	<ZM_2 *TO* CM_{12}>	Coefficient of impact of schedule pressure on frontline expectancies	Coefficient of fatigue enhancement or weakening due to duration pressure acting on fatigue	
	<CM_{12} TO ZP_3>	Coefficient of impact of front-line expectations on the volume of staffing changes	Coefficients by which front-line expectations of personnel act on the volume of personnel change to enhance or diminish the volume of personnel change	
	<ZM_2 TO CM_{11}>	Impact factor of schedule pressure on staffing ratios	Coefficient by which schedule pressure acts on the staffing ratio to increase or decrease it	
	<ZM_2 TO ZP_2>	Coefficient of influence of work pressure on fatigue	Coefficient of fatigue enhancement or weakening due to duration pressure acting on fatigue	
	<ZP_4 TO CM_3>	Coefficient of influence of frequency of accident signs on safety awareness	Coefficient of increase or decrease in safety awareness due to the frequency of occurrence of accident signs.	
	<ZP_3 TO ZN_1>	Coefficient of influence of the volume of personnel changes on the volume of wage changes	The coefficient by which the change in personnel acts on the change in wages to raise or lower the change in wages.	
	<L_1 TO ZP_1>	Coefficient of influence of project quality on rework	The factor by which the quality of the work acts on the rework to enhance or diminish the rework	

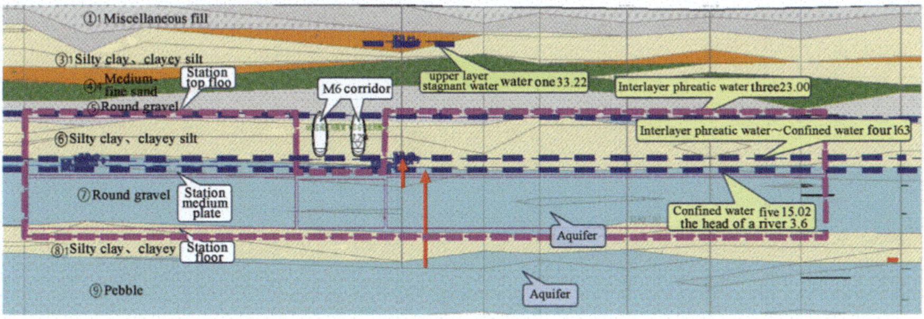

Fig. 6.16 Dongda Qiao station engineering geological section

community, the southwest quadrant is the Blue Island Building, the northeast quadrant has the Baifu International Building and the bus station, and the southeast quadrant is the Dongda Qiao East Community. The surrounding area of the station is a mature community and commercial and business district, and the road has basically achieved planning. Except for the A, B, D passenger entrances and exits occupy the road red line outside the plot, other ventilation pavilion, safety exits, and cooling towers are all located within the planned road red line.

The main road of Dongda Qiao Road at the southern end of the station is currently a two-way five-lane road, with the east and west side roads currently being a one-way motor vehicle lane + non-motor vehicle lane. The Workers Stadium East Road at the northern end of the station is currently a two-way six-lane road, with the east and west side roads currently being a one-way motor vehicle lane + non-motor vehicle lane. The east side of the station is a one-way four-lane Chaoyangmen Outer Street and a one-way six-lane Chaoyang North Road, the west side is a two-way eight-lane Chaoyangmen Outer Street, and the south and north side roads are currently a one-way motor vehicle lane + non-motor vehicle lane. Dongda Qiao Road, Workers Stadium East Road, Chaoyangmen Outer Street, and Chaoyang North Road have large current traffic volumes and busy traffic.

The newly built Dongda Qiao Station is connected to the existing Line 6 Dongda Qiao Station through a transfer passage. Line 6 Dongda Qiao Station is an operating station, which started operation in 2012. The station was completed and opened to traffic in the year. The station is located at the intersection of Chaoyang North Road and Dongda Qiao Road, arranged east–west along Chaoyang North Road. The west end is a two-layered tunnel, the east end is a single-layered tunnel, with an island platform. The platform width is 13 m, and the total length of the station is 255 m.

The newly built Dongda Qiao Station main pilot tunnel is close to the existing Line 6 Chaoyangmen Station–Dongda Qiao Station section, with a horizontal net distance from the tunnel of about 3.14–5.28 m. This section of the line starts from the west side of the intersection of Chaoyangmen Inner Street and East Second Ring Road at Line 6 Chaoyangmen Station, crosses under the East Second Ring Road after leaving the station, then lays along the middle of Chaowai Street towards the east

Table 6.3 Formation physical and mechanical parameter table

Geotechnical layering	Geotechnical name	Gravity γ(kN/m^3)	Cohesion (consolidated quick shear) c(kPa)	Internal friction angle (consolidated quick shear) φ(°)	Compression modulus Es(MPa)	Static lateral pressure coefficient ξ	Coefficient of soil reaction (MPa/m)		Proportionality factor of horizontal resistance coefficient M (MN/m^4)	Characteristic value of foundation bearing capacity f_{ak} (kPa)
							KK_v	KK_h		
①1	Miscellaneous fill	19.5	0	10	–	–	-	-	–	–
③	Clayey silt	20	12	33	5.5	0.36	34	28	25	140
③1	Silty clay	19.8	25	16	5.5	0.39	30	28	10	140
④3	Silty fine sand	20.5	0	25	20	0.39	42	38	20	200
④4	Medium-fine sand	21	0	28	23	0.37	45	40	25	240
⑤	Round gravel	22	0	35	50	0.35	55	60	40	320
⑥	Silty clay	20.2	45	20	12	0.38	36	30	20	200
⑥1	Clayey silt	20.6	16	27	15	0.34	55	32	35	220
⑦	Round gravel	23	0	40	60	0.31	70	65	80	350
⑧	Silty clay	20	40	10	12	0.36	40	32	20	220
⑨	Pebble	24	0	45	70	0.3	75	70	–	500

Table 6.4 List of groundwater characteristics

Groundwater properties	Steady water table (pressure measuring water level of confined water)			Time of observation (year. month)	Aquifer
	Buried depth (m)	Elevation (m)	Water head (m)		
Upper layer stagnant water	5.2	33.22	–	2015.11	Surface fill and silt
Interlayer phreatic water	14.61–15.88	23.03–23.62	–	2008.4–2010.5	④$_3$, ④$_4$, ⑤$_1$, ⑤, ⑥$_3$, ⑥$_2$
	16.2–16.5	22.7–23.0	–	2015.1	
	16.2	22.22	–	2015.11	
Interlayer phreatic water–confined water	19.5–22.78	15.82–19.28	–	2008.4–2010.5	⑥$_2$, ⑦$_1$, ⑦, ⑧$_2$
	23.2–22.5	16.3	–	2015.1	
	22.1	16.32	1.2	2015.11	
Confined water	24.1	15.4	11.6	2015.1	⑧$_2$, ⑨$_1$, ⑨$_2$, ⑨, ⑨$_4$
	23.4	15.02	13.6	2015.11	

and ends at Dongda Qiao Station on the northeast side of the intersection of Chaowai Street, Workers Stadium East Road, and Chaoyang North Road (Table 6.5).

4. Construction Organization Plan

The division of the construction section of Line 17 Dongda Qiao Station is shown in Table 6.6. The construction organization cross-section diagram is shown in Fig. 6.17.

6.4.2 *Risk Analysis Model Establishment*

1. Project Zoning

The total length of Beijing Subway Line 17 Dongda Qiao Station is about 330 m, with a maximum width of about 33 m, and the project volume is large. Therefore, before carrying out the risk assessment work, it is necessary to divide the entire project into regions. According to the structural characteristics of Dongda Qiao Station, the station is divided into five major areas on its horizontal projection: Southeast (SE), Northeast (NE), Southwest (SW), Northwest (NW), and Central (C), as shown in Fig. 6.18.

2. Risk Source List
1. Southeast Area (SE)

The southeast area of Line 17 Dongda Qiao Station is the area near the existing Line 6 Dongda Qiao Station and the planned Line 28 Dongda Qiao Station, as shown in Fig. 6.19. The main risk sources in this area are shown in Tables 6.7 and 6.8.

Table 6.5 The relationship between station and municipal pipeline

Proximal object	Proximity mode	Pipeline size (mm)	Buried depth of pipe bottom (m)
Electric power tunnel	Vertical under-passing	2000 × 2350	11.6
Thermal square ditch	Vertical under-passing	4500 × 2800	10.96
Sewage pipe	Vertical under-passing	D1050	5
Rainwater square ditch	Vertical under-passing	4500 × 3000	4.6
Potable water pipe	Vertical under-passing	D1750	6.15
Thermal square ditch	Vertical under-passing	2600 × 2300	5
Sub-high pressure natural gas pipe	Vertical under-passing	D500	2.3
Rainwater square ditch	Horizontal side wear	4500 × 3000	5
Electric power tunnel	Horizontal side wear	2000 × 2350	11.6
Low pressure gas pipe	Horizontal side wear	D300, D400	约2
Sub-high pressure gas pipeline	Horizontal side wear	D406	约2
Medium pressure gas pipe	Horizontal side wear	D508	约2
Potable water pipe	Horizontal side wear	D600	2.2
Sewage pipe	Horizontal side wear	D1350–1550	5.2

2. Northeast Area (NE)

The northeast area of Line 17 Dongda Qiao Station is the area where the northern transfer passage and Exit B of the station are located, as shown in Fig. 6.20. The main risk sources in this area are shown in Tables 6.9 and 6.10.

(3) Southwest Area (SW)

The southwest area of Line 17 Dongda Qiao Station is the area where Exit D of the station is located, as shown in Fig. 6.21. The main risk sources in this area are shown in Tables 6.11 and 6.12.

(4) Northwest Area (NW)

The northwest area of Line 17 Dongda Qiao Station is the area where Exit A of the station is located, as shown in Fig. 6.22. The main risk sources in this area are shown in Tables 6.13 and 6.14.

Table 6.6 Construction section division of Dongda Qiao station

Station name	Construction range		Length (m)	Mileage
Dongda Qiao station	No. 1 shaft	No. 1 transverse channel	68	
		Northward construction of PBA main body	87	K23 + 353.737–K23 + 440.737
	No. 2 shaft	No. 2 transverse channel	38	
		Southward construction of PBA main body	87	K23 + 527.047–K23 + 440.737
		Northward construction of PBA main body	39	K23 + 533.747–K23 + 572.747
	No. 3 shaft	No. 3 transverse channel	37	
		Southward construction of PBA main body	39	K23 + 572.747–K23 + 611.747
		Northward construction of PBA main body	74	K23 + 616.747–K23 + 690.539
	No. 4 shaft	Construction transfer channel	185	
Ancillary projects	Construction of surplus ventilation pavilion, safety exits channel, barrier-free entrance and entrance/exit			

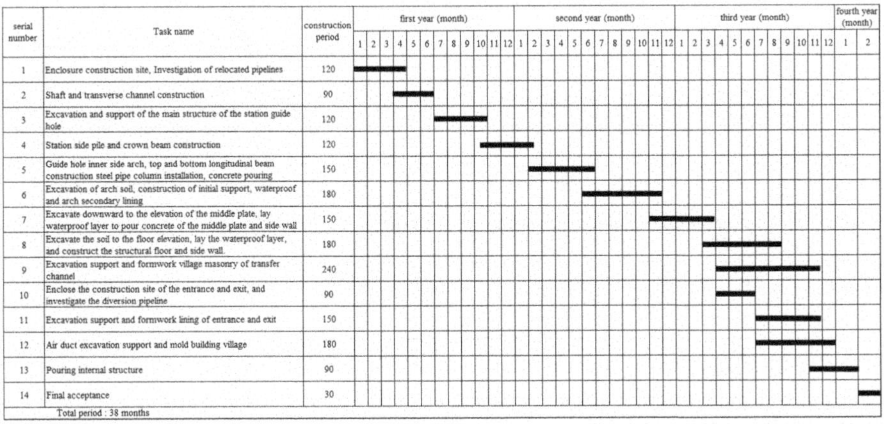

Fig. 6.17 Dongda Qiao station construction organization cross-section diagram

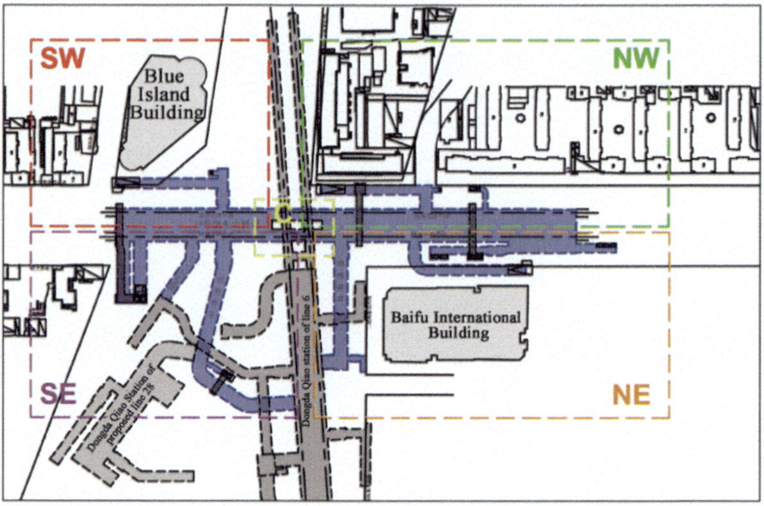

Fig. 6.18 Schematic diagram of project zoning of Line 17 Dongda Qiao station

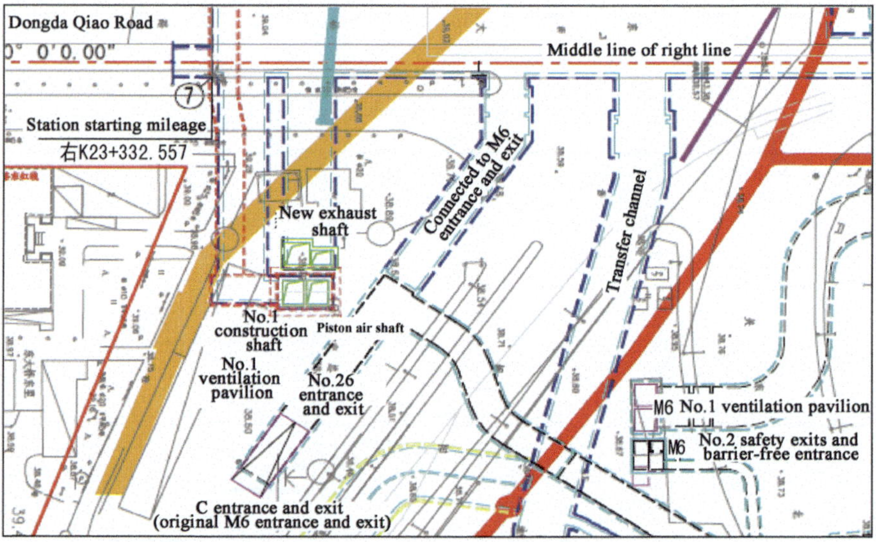

Fig. 6.19 Detailed map of the Southeast area of the station

(5) Central Area (C)

The central area of Line 17 Dongda Qiao Station is the area where the station intersects with the existing Line 6, as shown in Fig. 6.23. The main risk sources in this area are shown in Tables 6.15 and 6.16.

Table 6.7 Southeast project own risk source list

Number	Name of risk source	Information on risk sources
SE-S-1	Main structure of underground excavation of station	Double-layer double-column three-span standard section, 24.5 m wide, 16.87 m high, PBA construction method
SE-S-2	No. 1 piston air duct	The total length is 51 m, and the double-layer single-span structure is constructed by CRD method. The excavation width is 10.2 m, the height is 14.8 m, and the overburden is 17.1 m
SE-S-3	South transfer channel	The total length is 182 m, the excavation width of the standard section is 10.9 m, and the height is 7.2 m. The CRD method is used for construction
SE-S-4	No. 1 construction shaft	The excavation size is 7.9 m × 10.4 m, and the depth is about 34.5 m. The inverted shaft lining method is used for construction
SE-S-5	C entrance and exit	The excavation width is 8.3 m and the height is 6.8–10 m. The overburden is 4.5–19.5 m thick, and the maximum buried depth is about 28.5 m. The CD method of 4 pilot tunnels is used for excavation, and the CRD method of 6 pilot tunnels is used for construction when the section is high

3. Risk Assessment Unit Division

Beijing Subway Line 17 Dongda Qiao Station is a transfer station with the existing Line 6 and the planned Line 28. In addition to the main structure of the station, there are also 2 transfer passages and 4 An entrance and exit, with plans to build new connecting channels to the Dongda Qiao Station of Line 28 on the existing structure in the future. The Dongda Qiao Station of Line 17 is a typical networked extension and construction engineering. According to the characteristics of the project, the project is divided into 13 basic risk assessment units according to the project sub-items and main processes, as shown in Fig. 6.24.

According to the spatial position occupied by each basic risk assessment unit, the basic assessment units can be further subdivided into sub-assessment units, the results are shown in Table 6.17.

4. Risk Identification

Through expert consultation and meeting discussions, the major construction risks involved in the Dongda Qiao Station of Line 17 are summarized according to the Risk Breakdown Structure (RBS), as shown in Fig. 6.25.

5. Bayesian Network Modeling

This project is a networked extension and construction engineering. According to the characteristics of this project, a Bayesian network is established, as shown in

Table 6.8 A list of environmental risk sources around the southeastern region

Number	Name of risk source	Information on risk sources	Relevant projects	Position relation
SE-E-1	Dongda Qiao station of existing Line 6	The double-layer double-column three-span structure has a width of about 22.1 m, a height of about 16 m, and a floor buried depth of 22.9 m	South transfer channel	The connection needs to open a hole with a height of 4.2 m and a width of 10.2 m on the side wall of the existing structure
SE-E-2	No. 3 entrance and exit of existing line 6	Arch crown straight wall structure, 7.1 m wide and 5.4 m high	South transfer channel	Underneath, the minimum vertical distance is about 3.0 m
			C entrance and exit	Connection, connecting with the reserved interface of the existing structure
SE-E-3	No. 1 air duct of existing line 6		South transfer channel	Proximity, minimum horizontal distance 6.1 m
SE-E-4	2000 mm × 2350 mm electric power tunnel	The buried depth is about 11.7 m	South transfer channel	Oblique undercrossing, with a vertical distance of about 8.5 m
			Station main body	Oblique/parallel side under wear, vertical distance of about 7.0 m–3.5 m
SE-E-5	4500 mm × 3000 mm rainwater square ditch	The buried depth is about 5.0 m	South transfer channel	The vertical distance is about 4.0 m
			Station main body	The horizontal distance is about 20 m, and the vertical distance is about 10 m
SE-E-6	D500 mm natural gas pipe	2 roots, high pressure, buried depth about 2.2 m	South transfer channel	Beneath, vertical distance of about 18.2 m
			Station main body	The vertical distance is about 12.8 m
SE-E-7	D300 mm potable water pipe	The buried depth is about 1.73 m	South transfer channel	The vertical distance is about 7.7 m
SE-E-8	D600 mm rainwater pipe	The buried depth is about 3.53 m	South transfer channel	The vertical distance is about 16.7 m

(continued)

Table 6.8 (continued)

Number	Name of risk source	Information on risk sources	Relevant projects	Position relation
SE-E-9	1800 mm × 1600 mm rainwater	Square ditch The buried depth is about 3.5 m	South transfer channel	Crossing obliquely, the minimum vertical distance is about 9.8 m
SE-E-10	4500 mm × 2800 mm thermal square ditch	The buried depth is about 10.96 m	No. 1 shaft cross passage	Beneath, vertical distance of about 1.8 m
			No. 1 piston air duct	Beneath, vertical distance of about 1.8 m
			Main body of station	Oblique undercrossing, vertical distance of about 4.1 m
SE-E-11	D600 mm potable water pipe 1	The buried depth is about 2.5 m	No. 1 piston air duct	Beneath, vertical distance of about 10.3 m
			New exhaust air of No. 1 air duct	Beneath, vertical distance of about 14.6 m
			No. 1 shaft cross passage	Beneath, vertical distance of about 10.3 m
SE-E-12	D600 mm potable water pipe 2	The buried depth is about 1.6 m	C entrance and exit	The vertical distance is about 16 m
SE-E-13	D1750 mm potable water pipe ditch	The buried depth is about 4.4 m	Station main body	The vertical distance is about 10.8 m

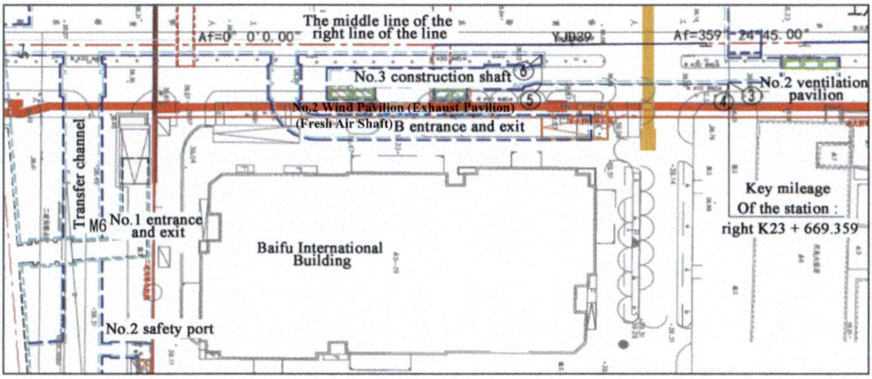

Fig. 6.20 Detailed map of the Northeast area of the station

Table 6.9 List of risk sources of the northeast project itself

Number	Name of risk source	Information on risk sources
NE-S-1	The main structure of station excavation	Double-layer double-column three-span standard section, 24.5 m wide, 16.87 m high, PBA construction method
NE-S-2	The main structure of station excavation	Double-layer three-column four-span standard section, 33.05 m wide, 16.87 m high, PBA construction method
NE-S-3	North transfer channel	The total length is 105 m, the excavation width is 10.9 m, and the height is 7.2 m. The CRD method is used for construction
NE-S-4	No. 3 construction shaft	The excavation size is 5.3 m × 10.7 m, and the depth of 36.5 m is constructed by the inverted shaft wall method
NE-S-5	B entrance and exit	The excavation width is 8.3 m, the height is 6.8–10 m, the overburden thickness is 3.7–19 m, and the maximum buried depth is 28.3 m. The CD method with four pilot tunnels is used for excavation. When the section is high, the CRD method of 6 pilot tunnels is used for construction
NE-S-6	No. 2 air duct	The total length is 51 m. Double-layer single-span structure is constructed by CRD method, the excavation width is 10.2 m, the height is 14.8 m, and the overburden is 17.1 m

Fig. 6.26. The model has a total of 33 nodes, including 5 top-level risk event nodes, 4 intermediate event nodes, 13 factor event nodes, and 11 system status nodes.

(6) Risk Assessment Indicators

The risk assessment indicators of each system status node in Fig. 6.26 are shown in Figs. 6.27, 6.28, 6.29, 6.30, 6.31, 6.32, 6.33, 6.34, 6.35, 6.36, and 6.37. The blue frame in the figure represents that the indicator is regulated by the organizational management subsystem, and the red frame represents that the indicator is a dynamic indicator, controlled by the Markov chain or directly assigned by the progress curve.

7. Coupling Factor Analysis

Through meeting discussions and expert consultations, it is determined that there are the following 4 groups of coupled risk factor pairs in the construction risk system of Dongda Qiao Station.

(1) Inadequate advance reinforcement—presence of weak stratum (E_{S3}–E_{E3})

Advanced reinforcement refers to measures to improve the mechanical properties of the strata to be excavated through methods such as advanced anchor bolts, advanced

Table 6.10 A list of environmental risk sources around the Northeastern region

Number	Name of risk source	Information on risk sources	Relevant projects	Position relation
NE-E-1	Baifu International Building	4 underground floors. 29 floors above ground, frame shear wall structure	Main body of station	Proximity, minimum horizontal distance of about 32.9 m
			B Entrance and exit	Proximity. The minimum horizontal distance is about 9.2 m
NE-E-2	Dongda Qiao station of existing Line 6	The double-layer double-column three-span structure has a width of about 22.1 m, a height of about 16 m, and a buried depth of 22.9 m	North transfer channel	Adjacent/connected, the minimum horizontal distance is 8.5 m, and a hole with a height of 4.2 m and a width of 10.2 m should be opened on the side wall of the existing structure
NE-E-3	The existing No. 1 entrance and exit of Line 6	Arch straight wall structure, 7.1 m wide, about 5.4 m high	North transfer channel	The minimum vertical distance is about 2.0 m
NE-E-4	2000 mm × 2350 mm electric power tunnel	The buried depth is about 8.2 m	No. 3 vertical shaft	Proximity. The minimum horizontal distance is 0.2 m
			Main body of station	Oblique crossing. The vertical distance is 7.0–3.5 m
			B entrance and exit	Parallel undercrossing, the minimum horizontal distance is about 1.2 m
NE-E-5	D600 mm potable water pipe	The buried depth is about 1.6 m	B entrance and exit	Parallel underpass. The minimum horizontal distance is about 5.0 m
NE-E-6	D400 mm rainwater pipe	The buried depth is about 2.2 m	B entrance and exit	Parallel underpass. The minimum horizontal distance is about 3.2 m
NE-E-7	D600 mm rainwater pipe	The buried depth is about 2.0 m	North transfer channel	The minimum vertical distance is 7.1 m, and the minimum horizontal distance is 1.5 m
NE-E-8	D300 mm sewage pipe	The buried depth is about 2.5 m	B Entrance and exit	Parallel underpass. The horizontal distance is 0.4–1.7 m
NE-E-9	D1050 mm sewage pipe	The buried depth is about 5.0 m	Main body of station	The vertical distance is about 9.8 m
NE-E-10	2600 mm × 2300 mm thermal square ditch	The buried depth is about 9.12 m	Main body of station	The vertical distance is about 4.6 m

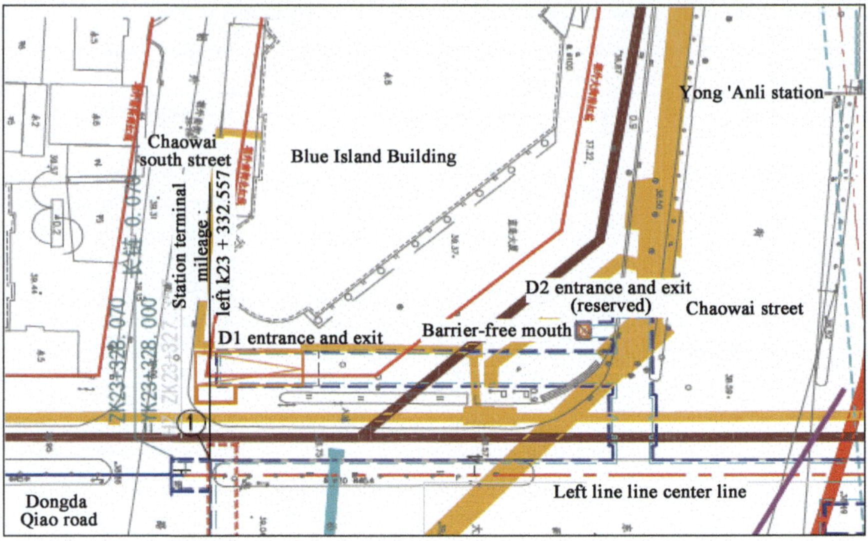

Fig. 6.21 Detailed map of the Southwest area of the station

Table 6.11 List of risk sources of the southwest project itself

Number	Name of risk source	Information on risk sources
SW-S-1	The main structure of station excavation	The basic situation is the same as the southeast area
SW-S-2	D entrance and exit	The excavation width is 8.3 m, the height is 6.8–10 m, the overburden thickness is 3.7–19 m, and the maximum buried depth is 28.3 m. When the section is high, the CRD method of 6 pilot tunnel is used for construction

small conduits, combined with grouting, or measures to pre-support the excavation section using methods such as pipe sheds and horizontal rotary spraying. Advanced reinforcement measures are concealed projects, and their construction quality and effects are difficult to evaluate intuitively, and the reinforcement effect has a high degree of uncertainty. When the stratum properties are poor and the excavation face self-stabilization time is extremely short, unqualified advanced reinforcement can easily lead to instability of the excavation face before the support structure is erected, causing disastrous consequences. Therefore, when carrying out shallow buried tunneling construction in weak strata, qualified advanced reinforcement is especially significant for maintaining the stability of the excavation face and may even become a decisive factor. In summary, it can be considered that there is a significant positive coupling effect between "insufficient advanced reinforcement" and "existence of weak strata".

Table 6.12 A list of environmental risk sources around the southwest region

Number	Name of risk source	Information on risk sources	Relevant projects	Position relation
SW-E-1	Blue Island Building	Frame shear wall structure, 8 floors above ground, 2 floors underground, box Xiao foundation	Main body of station	Proximity, minimum horizontal distance 5.5 m
SW-E-2	4500 mm × 2800 mm thermal square ditch	The buried depth is about 10.96 m	D entrance and exit	Under wear. The vertical distance is about 6.5 m
			Main body of station	Oblique through, vertical distance of about 4.1 m
SW-E-3	D1550 mm sewage pipe 1	The buried depth is about 5.0 m	D entrance and exit	The minimum vertical distance is about 13.0 m
SW-E-4	D1550 mm sewage pipe 2	The buried depth is about 4.8 m	D entrance and exit	Vertical underpass, the minimum vertical distance is about 3.4 m
SW-E-5	D600 mm potable water pipe	The buried depth is about 3.0 m	D entrance and exit	Oblique undercrossing, the minimum vertical distance is about 3.5 m
SW-E-6	1600 mm × 950 mm thermal trench	The buried depth is about 2.4 m	D entrance and exit	Parallel sideways, 4.8 m horizontal distance
			Main body of station	The horizontal distance is about 7.5 m, and the vertical distance is about 12.6 m
SW-E-7	2000 mm × 2350 mm electric power tunnel	The buried depth is 8.2–11.7 m	Main body of station	The vertical distance is about 7.0–3.5 m, and the horizontal distance is about 14.1 m
SW-E-8	D300 mm sewage pipe	The buried depth is about 2.5 m	D entrance and exit	The minimum horizontal distance is about 0.4–1.7 m
SW-E-9	D400 mm gas pipe	The buried depth is about 5.0 m	Main body of station	The vertical distance is about 9.8 m
SW-E-10	D1350–1550 sewage pipe	The buried depth is about 5.2 m	Main body of station	The horizontal distance is about 0.653.8 m, and the vertical distance is about 9.7 m
SW-E-11	D1750 mm potable water pipe	The buried depth is about 9.12 m	Main body of station	The vertical distance is about 4.6 m
SW-E-12	D500 mm natural gas pipe	2 roots, high pressure, buried depth of about 2.2 m	Main body of station	The vertical distance is about 12.8 m

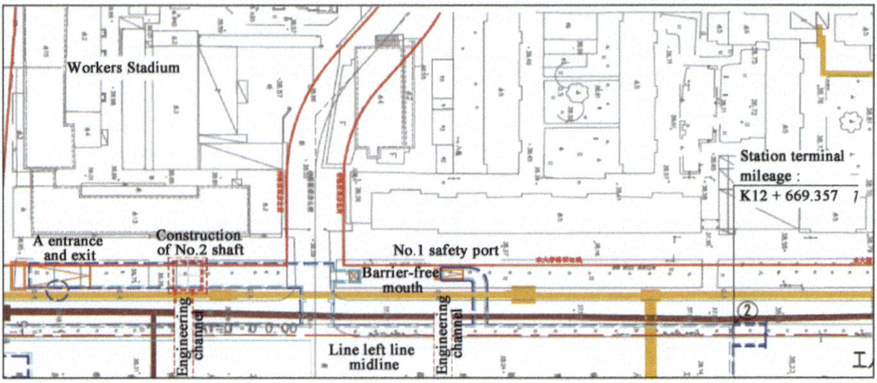

Fig. 6.22 Detailed map of the Northwest area of the station

Table 6.13 List of risk sources of the northwest project itself

Number	Name of risk source	Information on risk sources
NW-S-1	The main structure of station excavation	The basic situation is the same as that of the northeast region
NW-S-2	The main structure of station excavation	The basic situation is the same as that of the northeast region
NW-S-3	No. 2 construction shaft	The excavation size is 9.4 m × 11.7 m, and the depth is about 36.7 m. The inverted shaft lining method is used for construction
NW-S-4	An Entrance and Exit	The excavation width is 8.3 m, the height is 6.8–10 m, the overburden thickness is 4.5–19.5 m, and the maximum buried depth is 28.5 m. The CD method with four pilot tunnels is used for excavation. When the section is high, the CRD method of 6 pilot tunnels is used for construction

(2) Insufficient support design capability—existence of weak strata (E_{S2}–E_{E3})

The theory of support-stratum joint bearing body is one of the basic concepts of modern underground engineering design. Generally speaking, the self-bearing effect of the stratum is the core to ensure the stability of the underground structure, and the role of the support structure is to assist and guide the stratum to form a stable bearing body.

However, in weak strata, due to the low self-bearing capacity of the stratum, the support structure must become the main body of the load, which requires the support structure to have sufficient strength and rigidity. Therefore, in weak strata, the risk brought by insufficient support performance is inevitably significantly greater than in general strata, so it can be considered that there is a positive coupling effect between

Table 6.14 A list of environmental risk sources around the northwest region

Number	Name of risk source	Information on risk sources	Relevant projects	Position relation
NW-E-1	4000 mm × 3020 mm rainwater square ditch	The buried depth is about 5.0 m	No. 2 shaft	Adjacent, horizontal distance 1.0 m
			An entrance and exit	Parallel underpass, the minimum horizontal distance is about 1.8 m
NW-E-2	1600 mm × 950 mm thermal trench	The buried depth is about 2.4 m	No. 2 shaft	Adjacent, horizontal distance is about 0.1 m
			A out of population	Under wear. The minimum horizontal distance is about 0.75 m
			Main body of station	The horizontal distance is about 7.5 m and the vertical distance is about 12.6 m
NW-E-3	Gongti East Road Residential Building	Frame structure, 15 floors	An entrance and exit	Proximity, minimum horizontal distance 7.4 m
NW-E-4	D1550 mm sewage pipe	The buried depth is about 5.0 m	An entrance and exit	Parallel undercrossing, the minimum horizontal distance is about 5.8 m
NW-E-5	D400 mm gas pipe	The buried depth is about 2.0 m	An entrance and exit	Parallel underpass. The minimum horizontal distance is about 4.0 m
NW-E-6	D406 mm gas pipe	The buried depth is about 2.0 m	An entrance and exit	Parallel undercrossing, the minimum horizontal distance is about 8.5 m
NW-E-7	D508 mm gas pipe	The buried depth is about 2.0 m	An entrance and exit	Parallel underpass. The minimum horizontal distance is about 9.8 m
NW-E-8	4500 mm × 3000 mm rainwater square ditch	The buried depth is about 5.0 m	Main body of station	The horizontal distance is about 20 m, and the vertical distance is about 10 m
NW-E-9	2600 mm × 2300 mm thermal trench	The buried depth is about 9.12 m	Main body of station	The vertical distance is about 4.6 m

the risk factors of "insufficient support design capability" and "existence of weak strata".

(3) Encounter heavy rainfall—existence of weak strata (E_{E4}–E_{E3})

The mechanical properties of geotechnical media are significantly affected by their water content, especially in weak strata, where the geotechnical structure is often loose, the porosity is high, and the binding force between solid particles is weak, so

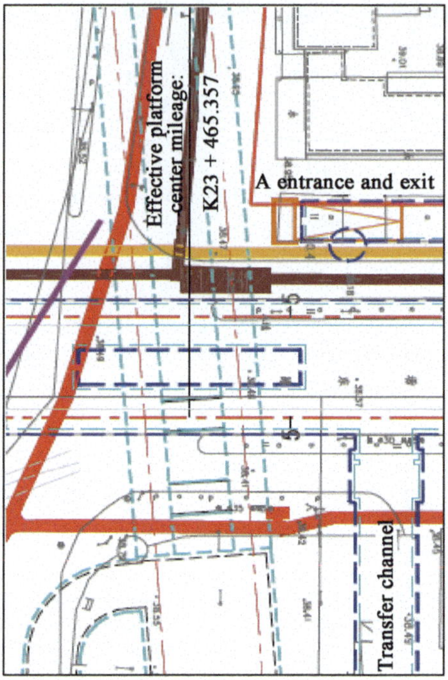

Fig. 6.23 Detailed map of the central area of the station

Table 6.15 List of risk sources of the central district project itself

Number	Name of risk source	Information on risk sources
C-S-1	The main structure of station excavation	Separate single-hole tunnel, 9.9 m wide, 9.52 m high, 5.2 m apart, constructed by CRD method

it often has relatively poor hydrological characteristics. Under normal replenishment and discharge conditions, through various precipitation, water blocking, and drainage measures, the water content of the geotechnical body in the construction area can generally be guaranteed within the allowable range. However, when encountering heavy rainfall, the infiltration and replenishment of groundwater often exceed the maximum amount of water that can be drained by drainage measures, causing the water content of the geotechnical body to increase rapidly. When this situation occurs in weak strata, it often leads to disastrous consequences. Past accident cases show that "heavy rainfall + weak strata" is the number one cause of disaster in urban underground space engineering, so "encountering heavy rainfall" and "existence of weak strata" are also a pair of risk factors with a significant positive coupling effect.

(4) Insufficient support design capability—untimely support closure (E_{S2}–E_{S1})

"Strong support and quick closure" as an important part of the eighteen-character policy of the shallow buried tunneling method are often mentioned together, which

Table 6.16 A list of environmental risk sources around the central t region

Number	Name of risk source	Information on risk sources	Relevant projects	Position relation
C-E-1	Dongda Qiao station of existing Line 6	The double-layer double-column three-span structure has a width of about 22.1 m, a height of about 16 m, and a floor depth of 22.9 m	Separated single-hole tunnel	Proximity. The minimum horizontal distance is about 20.5 m
C-E-2	Interval tunnel of existing Line 6	-	Separated single-hole tunnel	Vertically undercrossing, with a clear distance of about 2.0 m
C-E-3	2000 mm × 2300 mm electric power tunnel	The buried depth is 8.2–11.7 m	Separated single-hole tunnel	The vertical distance is 7.0–3.5 m, and the horizontal distance is about 14.1 m
C-E-4	4500 mm × 3000 mm rainwater square ditch	The buried depth is about 5.0 m	Separated single-hole tunnel	The horizontal distance is about 20 m, and the vertical distance is about 10 m

indicates that when these two requirements are met at the same time, a "1 + 1 > 2" effect can be achieved. Conversely, the untimely closure of the support will inevitably magnify the defects in the performance of the support, and the adverse effects of the two have a superimposed amplification effect. According to the definition of risk factor coupling, "insufficient support design capability" and "untimely support closure" also have a positive coupling effect.

The coupling relationship and coupling tree of risk factors at Dongda Qiao Station are shown in Fig. 6.38. The evaluation of the coupling effect between each pair of factors and the values of the coupling coefficients are shown in Table 6.18.

8. Selection of Bayesian network parameters
(1) Determination of connection strength assessment and probability conversion function

There are a total of 27 connections between event nodes and 20 state-event connections in the model. The connection strength of event-event connections is shown in Table 6.19, and the connection strength of state-event connections is shown in Table 6.20.

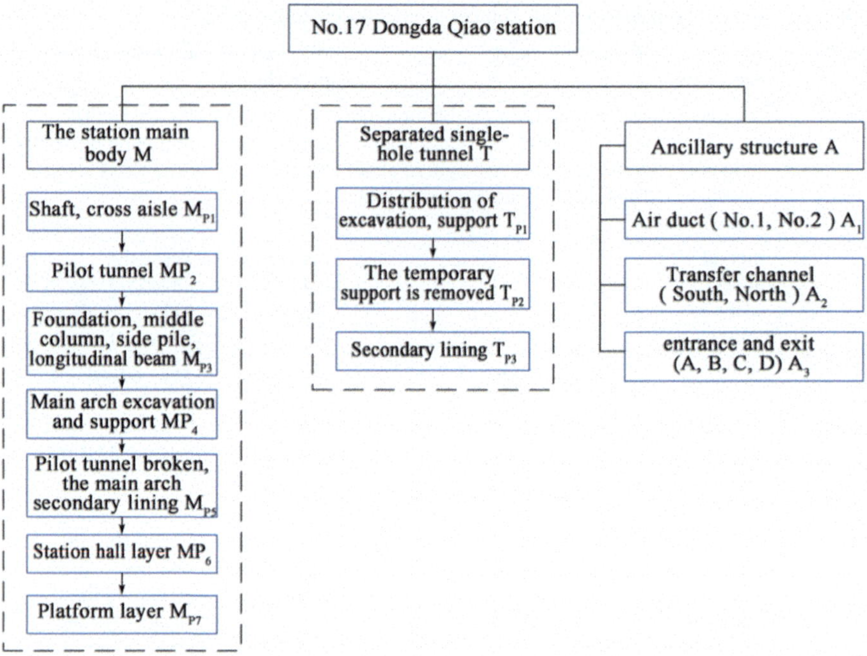

Fig. 6.24 Basic risk assessment units of Dongda Qiao Station of Beijing Subway Line 17

Table 6.17 Risk assessment unit subdivision table

Unit	SE	NE	SW	NW	C
M_{P1}	$M_{P1}{}^{SE}$	$M_{P1}{}^{NE}$		$M_{P1}{}^{NW}$	
M_{P2}	$M_{P2}{}^{SE}$	$M_{P2}{}^{NE}$	$M_{P2}{}^{SW}$	$M_{P2}{}^{NW}$	
M_{P3}	$M_{P3}{}^{SE}$	$M_{P3}{}^{NE}$	$M_{P3}{}^{SW}$	$M_{P3}{}^{NW}$	
M_{P4}	$M_{P4}{}^{SE}$	$M_{P4}{}^{NE}$	$M_{P4}{}^{SW}$	$M_{P4}{}^{NW}$	
M_{P5}	$M_{P5}{}^{SE}$	$M_{P5}{}^{NE}$	$M_{P5}{}^{SW}$	$M_{P5}{}^{NW}$	
M_{P6}	$M_{P6}{}^{SE}$	$M_{P6}{}^{NE}$	$M_{P6}{}^{SW}$	$M_{P6}{}^{NW}$	
M_{P7}	$M_{P7}{}^{SE}$	$M_{P7}{}^{NE}$	$M_{P7}{}^{SW}$	$M_{P7}{}^{NW}$	
T_{P1}					$T_{P1}{}^{C}$
T_{P2}					$T_{P2}{}^{C}$
T_{P3}					$T_{P3}{}^{C}$
A_1	$A_1{}^{SE}$	$A_1{}^{NE}$			
A_2	$A_2{}^{SE}$	$A_2{}^{NE}$			
A_3	$A_3{}^{SE}$	$A_3{}^{NE}$	$A_3{}^{SW}$	$A_3{}^{NW}$	

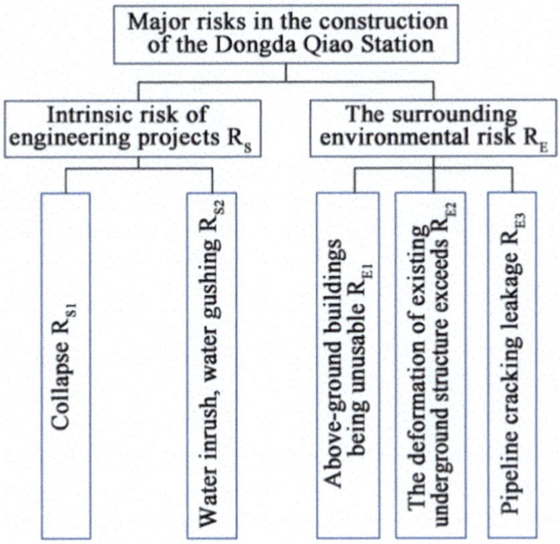

Fig. 6.25 Risk breakdown structure of major construction risks at Dongda Qiao station of Beijing Subway Line 17

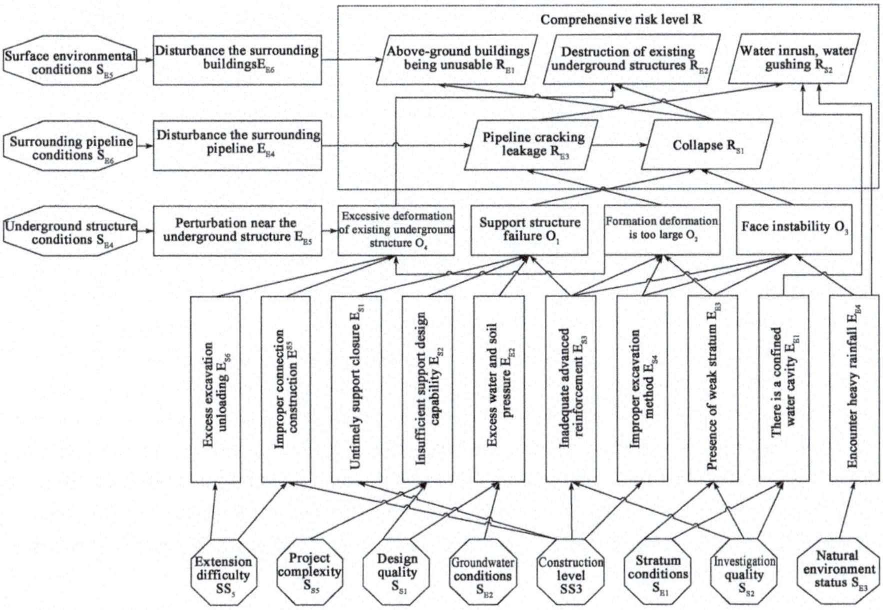

Fig. 6.26 Bayesian network diagram of construction risk system at Dongda Qiao station of Beijing Subway Line 17

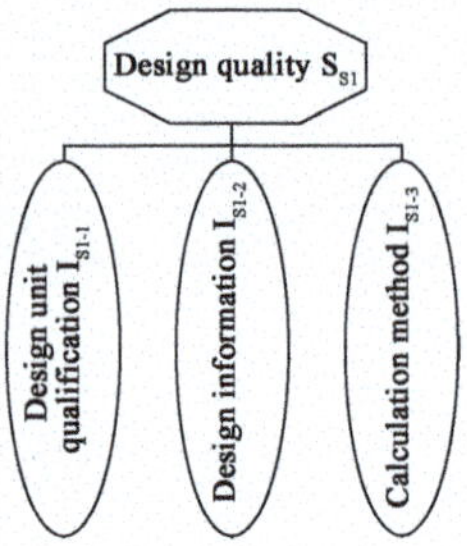

Fig. 6.27 Design quality (S_{S1}) risk assessment indicator

Fig. 6.28 Survey quality (S_{S2}) risk assessment indicator

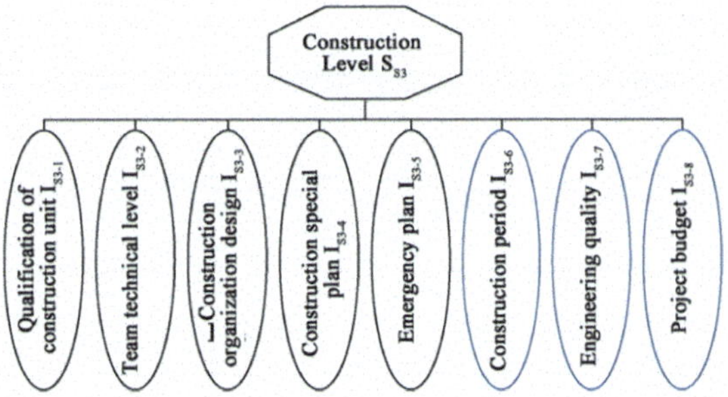

Fig. 6.29 Construction level (S_{S3}) risk assessment indicator

According to the general concept and method of risk assessment, the probability conversion function of the state-event connection uses 10 as the base, and the interval of the occurrence probability of the factor event affected by a single state indicator is [0.003, 0.3]. After substituting into formula (6.4–6.30), the probability conversion function can be obtained:

$$p^{\Gamma}\left(\varepsilon_{S_i}^{E}\right) = 10^{2\varepsilon_{S_i}^{E} - 2.522878} \tag{6.33}$$

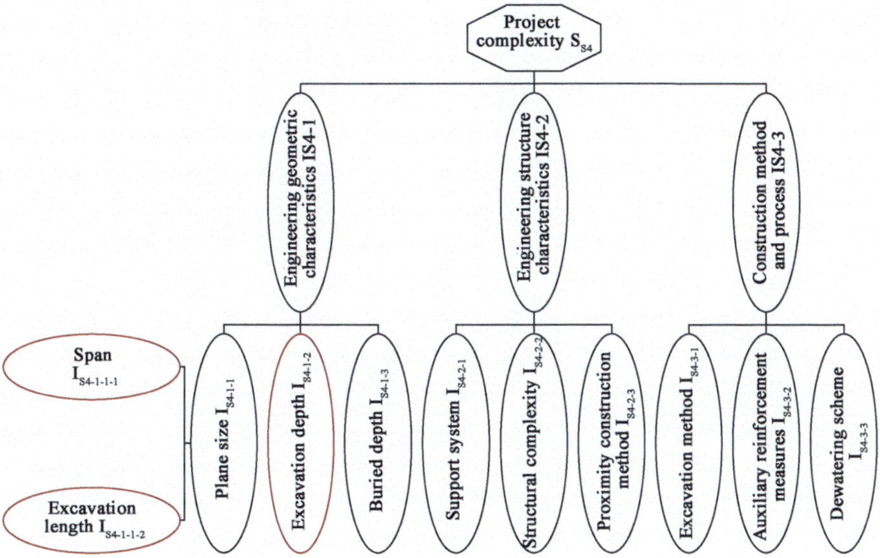

Fig. 6.30 Project complexity (S_{S4}) risk assessment indicator

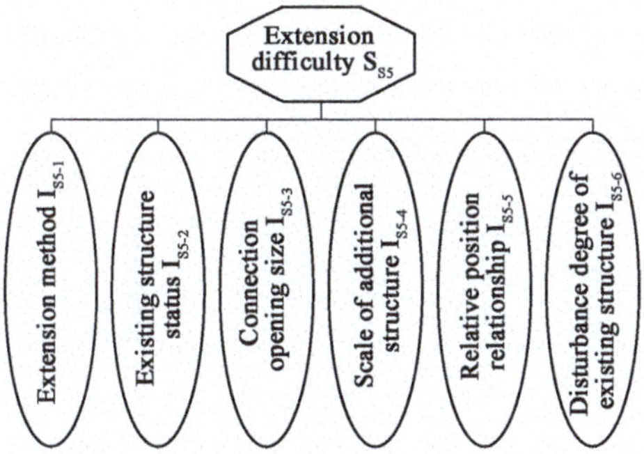

Fig. 6.31 Expansion difficulty (S_{S5}) risk assessment indicator

(2) Establishment of conditional probability table

Based on the results of the connection strength assessment, the CPT of each non-root node in the Bayesian network of the construction mechanics subsystem of Dongda Qiao Station can be established, see Tables 6.21, 6.22, 6.23, 6.24, and 6.25. The number of states of the macro state indicator nodes in the model is uniformly 4, where state 1 represents the most unfavorable situation, and state 4 represents the most favorable situation.

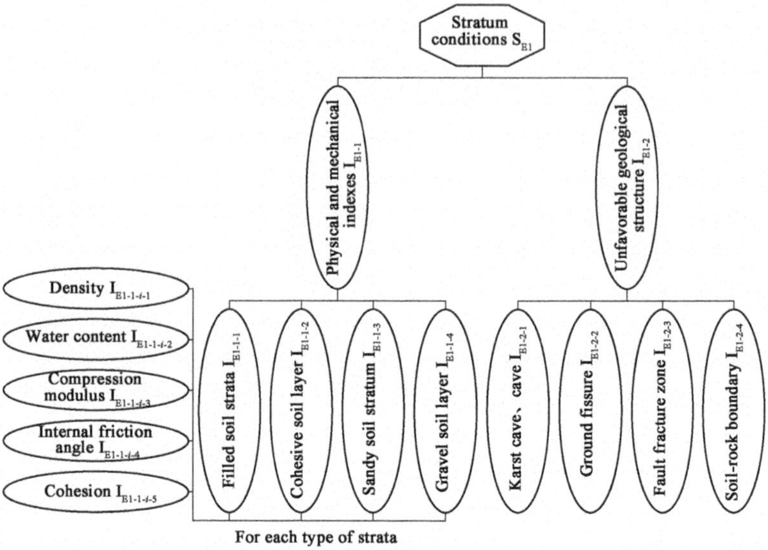

Fig. 6.32 Stratum condition (S_{E1}) risk assessment indicator

Fig. 6.33 Groundwater condition (S_{E2}) risk assessment indicator

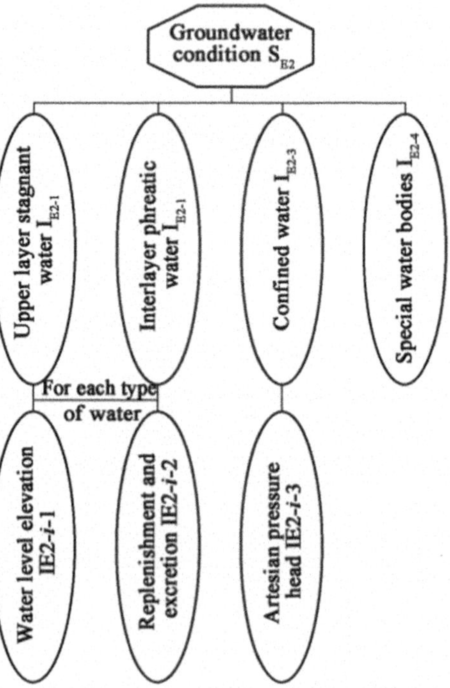

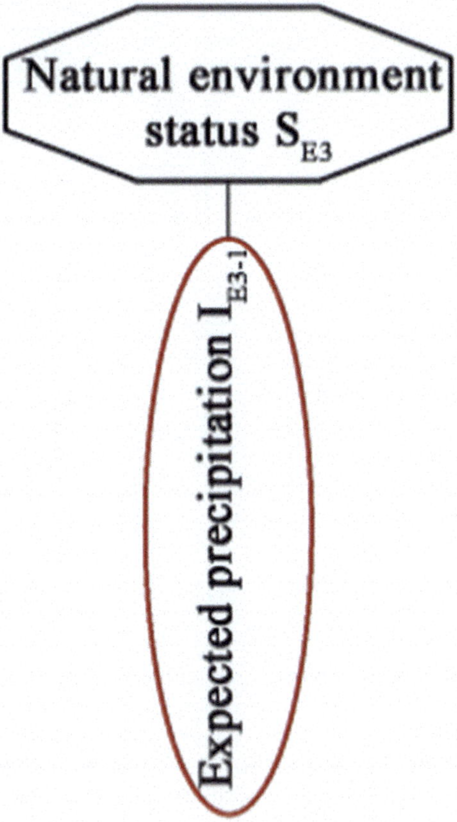

Fig. 6.34 Natural environment status (S_{E3}) risk assessment indicator

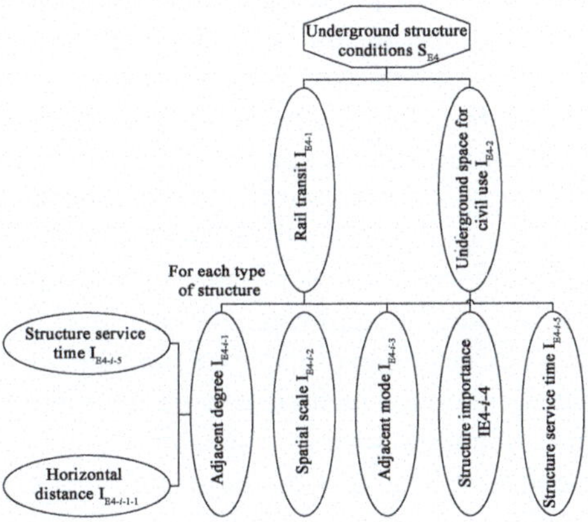

Fig. 6.35 Risk assessment indicators for underground structure conditions (S_{E4})

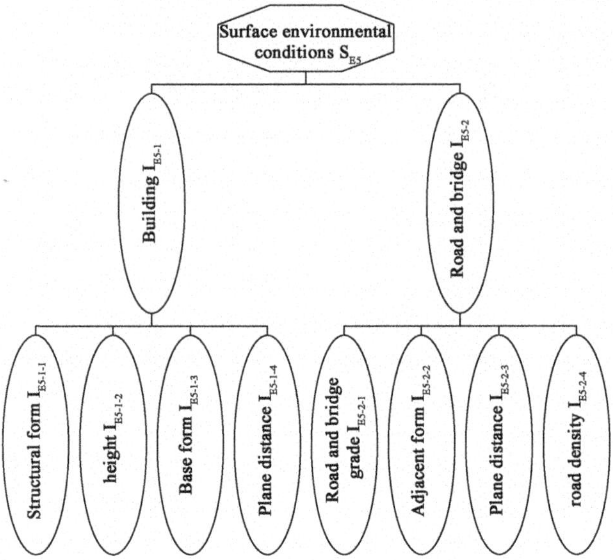

Fig. 6.36 Risk assessment indicators for surface environmental conditions (S_{E5})

9. Membership degree and weight of evaluation indicators

There are a total of 11 categories and 65 underlying evaluation indicators in the model. Among them, the indicators belonging to "Design Quality S_{S1}", "Investigation Quality S_{S2}", "Stratum Condition S_{E1}", "Groundwater Condition S_{E2}", "Natural Environment Status S_{E3}" are global indicators, which are irrelevant to the evaluation unit and project partition; the indicators belonging to "Project Complexity S_{S4}" and "Extension Difficulty S_{S5}" are related to the risk evaluation unit and are assigned values separately for different units; the indicators belonging to "Underground Structure Situation S_{E4}", "Surface Environment Situation S_{E5}", Surrounding Pipeline Situation S_{E6} are related to the project partition and are assigned values separately for different partitions; the indicators belonging to "Construction Level S_{S3}" such as "Construction Period $I_{S3\text{-}6}$", "Project Quality $I_{S3\text{-}7}$", "Project Budget $I_{S3\text{-}8}$" are dynamic indicators controlled by the organizational management subsystem, their membership is automatically calculated by the simulation results of the organizational management subsystem, and the rest of the indicators are all global indicators. The confidence method is used to assign values to each evaluation indicator, and the evaluation domain maintains the same number of level divisions (4 levels) and meanings (level 1 is the most unfavorable) as the macro state indicator nodes, the results are shown in Tables 6.26, 6.27, 6.28, 6.29, 6.30, 6.31, 6.32, and 6.33.

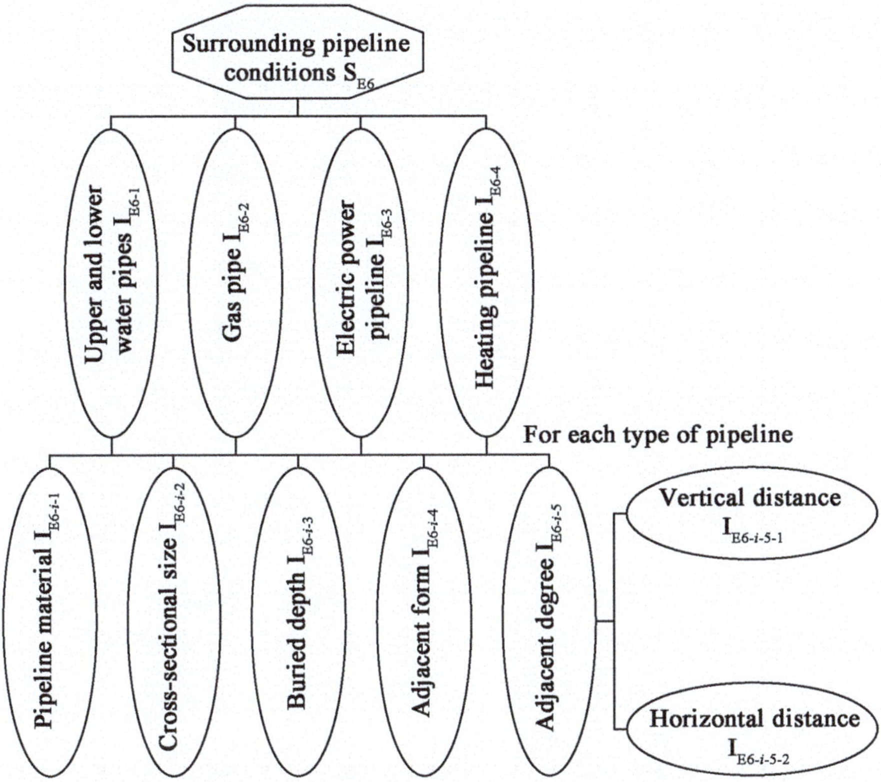

Fig. 6.37 Risk assessment indicators for surrounding pipeline conditions (S_{E6})

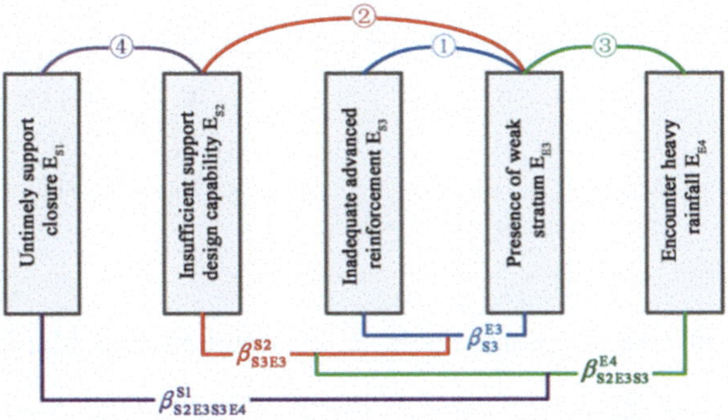

Fig. 6.38 Schematic diagram of the risk factor coupling tree of Dongda Qiao station on Beijing Subway Line 17

Table 6.18 Risk factor coupling effect evaluation table for Dongda Qiao station on Beijing Subway Line 17

Number	Coupling factor pair	Coupled nodes	Polarity of coupling	Strength of coupling effect	Double factor coupling coefficient
①	E_{S3}–E_{E3}	O_1, O_2	Positive coupling	Medium coupling	0.15
②	E_{S2}–E_{E3}	R_{S1}	Positive coupling	Medium coupling–strong coupling	0.25
③	E_{E3}–E_{E4}	O_3	Positive coupling	Strong coupling	0.4
④	E_{S1}–E_{S2}	O_1	Positive coupling	Strong coupling	0.35

Table 6.19 Dongda Qiao station event event-event connection strength table

Serial number	Connection objects	Evaluate	Connection strength	Serial number	Connection objects	Evaluate	Connection strength
1	E_{S1}–O_1	Occasionally	0.2	15	O_2–O_4	Continually	0.8
2	E_{S2}–O_1	Very likely	0.5	16	E_{E6}–R_{E1}	Occasionally	0.2
3	E_{S3}–O_1	Occasionally	0.15	17	E_{E7}–R_{E3}	Very likely	0.5
4	E_{E2}–O_1	Continually	0.7	18	E_{E1}–R_{S2}	Continually	0.8
5	E_{S3}–O_2	Very likely	0.6	19	E_{E4}–R_{S2}	Very likely	0.6
6	E_{S4}–O_1	Continually	0.7	20	O_1–R_{S1}	Necessarily	1.0
7	E_{E3}–O_2	Occasionally	0.2	21	O_3–R_{S1}	Necessarily	1.0
8	E_{S3}–O_3	Very likely	0.5	22	O_2–R_{E3}	Continually	0.8
9	E_{S4}–O_3	Very likely	0.6	23	O_4–R_{E2}	Very likely	0.4
10	E_{E3}–O_3	Occasionally	0.2	24	R_{E3}–R_{S1}	Occasionally	0.3
11	E_{E4}–O_2	Very likely	0.5	25	R_{E3}–R_{S2}	Very likely	0.4
12	E_{S5}–O_4	Occasionally	0.2	26	R_{S1}–R_{E1}	Continually	0.7
13	E_{S6}–O_4	Very likely	0.4	27	R_{S1}–R_{E2}	Continually	0.85
14	E_{E5}–O_4	Very likely	0.35				

10. Evaluation of Time Scale and Dynamic Indicator Markov Matrix

Given that the total construction period for the Dongda Qiao Station on Beijing Subway Line 17 is 3 years and 2 months, the evaluation time step is set to "1 week" after considering the evaluation accuracy and calculation volume. This evaluation uses the assignment method to assign values to the Markov matrix. Since the dynamic indicators used for evaluation have a relatively simple expected change trend within each evaluation unit, the assignment method is used to determine the Markov matrix. The Markov matrix values of each risk evaluation unit are shown in Table 6.34.

Table 6.20 Dongda Qiao station event state–event connection strength table

Serial number	Connection objects	Evaluate	Connection strength	Serial number	Connection objects	Evaluate	Connection strength
1	S_{S1}–E_{S2}	Strong ties	1.0	11	S_{S4}–E_{S2}	Strong ties	0.3
2	S_{S1}–E_{E2}	Medium ties	0.6	12	S_{S5}–E_{S5}	Strong ties	0.7
3	S_{S2}–E_{S3}	Weak ties	0.2	13	S_{S5}–E_{S6}	Strong ties	1.0
4	S_{S2}–E_{E1}	Strong ties	1.0	14	S_{E1}–E_{E1}	Middle ties	0.5
5	S_{S2}–E_{E3}	Strong ties	0.8	15	S_{E1}–E_{E3}	Strong ties	1.0
6	S_{S3}–E_{S1}	Strong ties	1.0	16	S_{E2}–E_{E2}	Strong ties	0.75
7	S_{S3}–E_{S3}	Strong ties	1.0	17	S_{E3}–E_{E4}	Strong ties	1.0
8	S_{S3}–E_{S4}	Strong ties	1.0	18	S_{E4}–E_{E5}	Strong ties	1.0
9	S_{S3}–E_{S5}	Strong ties	0.8	19	S_{E5}–E_{E6}	Strong ties	1.0
10	S_{S4}–E_{S1}	Weak ties	0.3	20	S_{E6}–E_{E7}	Strong ties	1.0

Table 6.21 Bayesian network top-level node CPT with 2 parent nodes

E_{E1}	R_{E3}	R_{S2}	E_{E6}	R_{S1}	R_{E1}	E_{E5}	R_{S1}	R_{E2}	E_{E7}	O_2	R_{E3}
T	T	0.8	T	T	0.7	T	T	0.85	T	T	0.8
	F	0.8		F	0.2		F	0.35		F	0.5
F	T	0.4	F	T	0.7	F	T	0.85	F	T	0.8
	F	0		F	0		F	0		F	0

Table 6.22 Bayesian network top-level node CPT with 3 parent nodes

O_1	O_3	R_{E3}	R_{S3}	E_{S3}	E_{S4}	E_{E3}	O_2
T	T	T	1	T	T	T	0.74
		F	1			F	0.7
	F	T	1		F	T	0.74
		F	1			F	0.6
F	T	T	1	T	T	T	0.7
		F	1			F	0.7
	F	T	0.3		F	T	0.2
		F	0			F	0

Table 6.23 Bayesian network top-level node CPT with 4 parent nodes

E_{S1}	E_{S2}	E_{S3}	E_{E2}	O_1	E_{S3}	E_{S4}	E_{E3}	E_{E4}	O_3
T	T	T	T	0.85	T	T	T	T	0.92
			F	0.85				F	0.74
		F	T	0.85			F	T	0.6
			F	0.85				F	0.6
	F	T	T	0.7		F	T	T	0.92
			F	0.2				F	0.675
		F	T	0.7			F	T	0.6
			F	0.2				F	0.5
F	T	T	T	0.7	F	T	T	T	0.92
			F	0.5				F	0.6
		F	T	0.7			F	T	0.6
			F	0.5				F	0.6
	F	T	T	0.7		F	T	T	0.92
			F	0.15				F	0.2
		F	T	0.7			F	T	0.16
			F	0				F	0

Table 6.24 Bayesian network factor event node CPT with 2 parent nodes

S_{S3}	E_{S1}	S_{S3}	E_{S4}	S_{E3}	E_{E4}	S_{E4}	E_{E5}	S_{E5}	E_{E6}	S_{E6}	E_{E7}
1	0.3	1	0.3	1	0.3	1	0.3	1	0.3	1	0.3
2	0.095	2	0.095	2	0.095	2	0.095	2	0.095	2	0.095
3	0.03	3	0.03	3	0.03	3	0.03	3	0.03	3	0.03
4	0.009	4	0.009	4	0.009	4	0.009	4	0.009	4	0.009

11. Modeling of Organizational Management Subsystem

(1) Model Structure

The organizational management subsystem model of Dongda Qiao Station on Beijing Subway Line 17 is established using the system dynamics software Venisim-PLE, as shown in Fig. 6.39.

(2) System Dynamics Simulation Results

The Vensim-PLE software is used to simulate the SD model of the Dongda Qiao Station organizational management subsystem, and the evolution curves of the main factors are obtained as shown in Fig. 6.40.

Table 6.25 Bayesian network factor event node CPT with 3 parent nodes

S_{E1}	S_{S2}	E_{E1}	S_{S4}	S_{S1}	E_{S2}	S_{S1}	S_{E2}	E_{E2}	S_{S2}	S_{S3}	E_{S3}	S_{S2}	E_{S1}	E_{E3}
1	1	0.321	1	1	0.308	1	1	0.138	1	1	0.305	1	1	0.384
	2	0.122		2	0.106		2	0.086		2	0.102		2	0.203
	3	0.059		3	0.042		3	0.064		3	0.037		3	0.146
	4	0.039		4	0.021		4	0.054		4	0.017		4	0.128
2	1	0.312	2	1	0.306	2	1	0.116	2	1	0.304	2	1	0.333
	2	0.110		2	0.103		2	0.063		2	0.100		2	0.138
	3	0.046		3	0.038		3	0.040		3	0.036		3	0.076
	4	0.026		4	0.018		4	0.031		4	0.015		4	0.057
3	1	0.307	3	1	0.304	3	1	0.106	3	1	0.303	3	1	0.313
	2	0.103		2	0.100		2	0.051		2	0.099		2	0.112
	3	0.039		3	0.036		3	0.029		3	0.035		3	0.048
	4	0.019		4	0.015		4	0.019		4	0.014		4	0.028
4	1	0.304	4	1	0.303	4	1	0.100	4	1	0.303	4	1	0.305
	2	0.100		2	0.099		2	0.046		2	0.098		2	0.102
	3	0.035		3	0.034		3	0.023		3	0.034		3	0.037
	4	0.015		4	0.014		4	0.013		4	0.013		4	0.017

6.4.3 Construction Process Risk Coupling Evolution Analysis Results

According to the construction plan, the planned construction period from the excavation of the construction shaft to the completion of the internal structure pouring at Dongda Qiao Station is 2 years and 10 months.

The analysis time span is considered for 132 weeks (one month is considered as 4 weeks), and the analysis time step is 1 natural week. After obtaining the time course data of the evaluation indicators through the Markov-Fuzzy Comprehensive Evaluation and System Dynamics Simulation, it is imported into the Bayesian network for analysis, and the results are as follows.

(1) Risk Analysis of Each Unit

The construction process risk analysis results of each evaluation unit are shown in Figs. 6.41, 6.42, 6.43, 6.44, 6.45, 6.46, 6.47, 6.48, 6.49, 6.50, 6.51, and 6.52.

2. Overall Construction Process Risk Analysis

According to the start and end times of each unit in Table 6.34, the evaluation results of the overlapping units in time are merged to form the overall construction process risk analysis results as shown in Figs. 6.53, 6.54, 6.55, 6.56, and 6.57.

Table 6.26 Global indicator evaluation results

Evaluation indicators	Evaluation level	Credibility degree	Evaluation indicators	Evaluation level	Credibility degree
$I_{S1\text{-}1}$	4	VS	$I_{S3\text{-}5}$	2	S
$I_{S1\text{-}2}$	4	QS	$I_{E1\text{-}1\text{-}1\text{-}1}$	3	QS
$I_{S1\text{-}3}$	3	QS	$I_{E1\text{-}1\text{-}1\text{-}2}$	4	VS
$I_{S2\text{-}1}$	4	VS	$I_{E1\text{-}1\text{-}1\text{-}2}$	1	S
$I_{S2\text{-}2}$	4	VS	$I_{E1\text{-}1\text{-}1\text{-}4}$	1	VS
$I_{S2\text{-}3}$	4	QS	$I_{E1\text{-}1\text{-}1\text{-}5}$	1	VS
$I_{S2\text{-}4}$	3	QS	$I_{E1\text{-}1\text{-}2\text{-}1}$	2	QS
$I_{S3\text{-}1}$	4	VS	$I_{E1\text{-}1\text{-}2\text{-}2}$	3	QS
$I_{S3\text{-}2}$	4	QS	$I_{E1\text{-}1\text{-}2\text{-}3}$	1	VS
$I_{S3\text{-}3}$	4	VS	$I_{E1\text{-}1\text{-}2\text{-}4}$	1	VS
$I_{S3\text{-}4}$	4	VS	$I_{E1\text{-}1\text{-}2\text{-}5}$	2	QS
$I_{E1\text{-}1\text{-}3\text{-}1}$	2	QS	$I_{E1\text{-}2\text{-}2}$	4	VS
$I_{E1\text{-}1\text{-}3\text{-}2}$	4	S	$I_{E1\text{-}2\text{-}3}$	4	VS
$I_{E1\text{-}1\text{-}3\text{-}3}$	2	QS	$I_{E1\text{-}2\text{-}4}$	4	VS
$I_{E1\text{-}1\text{-}3\text{-}4}$	3	S	$I_{E2\text{-}1\text{-}1}$	1	S
$I_{E1\text{-}1\text{-}3\text{-}5}$	3	HS	$I_{E2\text{-}1\text{-}2}$	2	NS
$I_{E1\text{-}1\text{-}4\text{-}1}$	2	QS	$I_{E2\text{-}2\text{-}1}$	2	QS
$I_{E1\text{-}1\text{-}4\text{-}2}$	1	QS	$I_{E2\text{-}2\text{-}2}$	3	S
$I_{E1\text{-}1\text{-}4\text{-}3}$	3	QS	$I_{E2\text{-}3\text{-}1}$	1	QS
$I_{E1\text{-}1\text{-}4\text{-}4}$	3	QS	$I_{E2\text{-}4}$	4	VS
$I_{E1\text{-}1\text{-}4\text{-}5}$	2	NS	$I_{E3\text{-}1}$	4	VS
$I_{E1\text{-}2\text{-}1}$	4	VS	–	–	–

From the figure, it can be found:

(1) Considering the coupling effect of risk factors, the probability of major risk events occurring during the construction process of Dongda Qiao Station is approximately between 0.006 and 0.2. According to the risk event occurrence likelihood grading standards in the "Code for risk management of underground works in urban rail transit" (GB 50,652—2011), overall, the construction risk level of Dongda Qiao Station is Level II.

Table 6.27 Evaluation unit related indicators evaluation results

Indicator		$I_{S4-1-1-1}$	$I_{S4-1-1-2}$	I_{S4-1-2}	I_{S4-1-3}	I_{S4-2-1}	I_{S4-2-2}	I_{S4-2-3}	I_{S4-3-1}	I_{S4-3-2}	I_{S4-3-3}	I_{S5-1}	I_{S5-2}	I_{S5-3}	I_{S5-4}	I_{S5-5}	I_{S5-6}
M_{P1}	Evaluation level	4	4	4	3	3	4	4	4	4	3	4	4	4	4	4	4
	Credibility degree	VS	VS	VS	S	QS	VS	VS	S	QS	QS	VS	VS	VS	VS	VS	VS
M_{P2}	Evaluation level	–	–	–	3	2	4	3	3	4	3	4	4	4	4	4	4
	Credibility degree	–	–	–	Q S	VS	QS	S	S	QS	QS	VS	VS	VS	VS	VS	VS
M_{P3}	Evaluation level	–	–	–	3	3	2	4	4	4	4	4	4	4	4	4	4
	Credibility degree	–	–	–	QS	QS	S	VS	VS	VS	VS	VS	VS	VS	VS	VS	VS
M_{P4}	Evaluation level	–	–	–	1	1	1	4	2	3	3	4	4	4	4	4	4
	Credibility degree	–	–	–	QS	QS	VS	VS	S	QS	QS	VS	VS	VS	VS	VS	VS
M_{P5}	Evaluation level	–	–	–	1	2	1		4	4	4	4	4	4	4	4	4
	Credibility degree	–	–	–	QS	S	VS	VS	VS	VS	VS	VS	VS	VS	VS	VS	VS
M_{P6}	Evaluation level	–	–	–	1	2	1	4	2	4	3	4	4	4	4	4	4
	Credibility degree	–	–	–	QS	S	VS	VS	SS	VS	QS	VS	VS	VS	VS	VS	VS

(continued)

Table 6.27 (continued)

Indicator		$I_{S4\text{-}1\text{-}1\text{-}1}$	$I_{S4\text{-}1\text{-}1\text{-}2}$	$I_{S4\text{-}1\text{-}2}$	$I_{S4\text{-}1\text{-}3}$	$I_{S4\text{-}2\text{-}1}$	$I_{S4\text{-}2\text{-}2}$	$I_{S4\text{-}2\text{-}3}$	$I_{S4\text{-}3\text{-}1}$	$I_{S4\text{-}3\text{-}2}$	$I_{S4\text{-}3\text{-}3}$	$I_{S5\text{-}1}$	$I_{S5\text{-}2}$	$I_{S5\text{-}3}$	$I_{S5\text{-}4}$	$I_{S5\text{-}5}$	$I_{S5\text{-}6}$
M_{P7}	Evaluation level	–	–	–	1	2	1	4	2	4	2	4	4	4	4	4	4
	Credibility degree	–	–	–	QS	S	VS	VS	SS	VS	QS	VS	VS	VS	VS	VS	VS
T_{P1}	Evaluation level	4	4	4	3	3	4	1	3	2	2						
	Credibility degree	VS	VS	VS	QS	QS	VS	QS	QS	S	QS						
T_{P2}	Evaluation level	–	–	–	3	1	4	4	4	4	4						
	Credibility degree	–	–	–	QS	S	S	QS	VS	VS	VS						
T_{P3}	Evaluation level	–	–	–	3	3	4	4	4	4	4						
	Credibility degree	–	–	–	QS	QS	S	VS	VS	VS	VS						
A_1	Evaluation level	4	4	4	4	2	2	4	2	4	2						
	Credibility degree	VS	VS	VS	VS	S	QS	VS	S	VS	QS						
A_2	Evaluation level	4	4	4	1	2	4	2	3	3	4						
	Credibility degree	VS	VS	VS	QS	QS	QS	QS	QS	S	S						

(continued)

Table 6.27 (continued)

Indicator		$I_{S4-1-1-1}$	$I_{S4-1-1-2}$	I_{S4-1-2}	I_{S4-1-3}	I_{S4-2-1}	I_{S4-2-2}	I_{S4-2-3}	I_{S4-3-1}	I_{S4-3-2}	I_{S4-3-3}	I_{S5-1}	I_{S5-2}	I_{S5-3}	I_{S5-4}	I_{S5-5}	I_{S5-6}
A_3	Evaluation level	4	4	4	1	3	4	3	4	2	4						
	Credibility degree	VS	VS	VS	QS	QS	VS	QS	QS	S	S						

Table 6.28 Evaluation results of relevant indicators for engineering zoning (Southeast region)

Evaluation indicators	Evaluation level	Credibility degree	Evaluation indicators	Evaluation level	Credibility degree
$I_{E4-1-1-1}$	1	QS	I_{E4-2-5}	4	VS
$I_{E4-1-1-2}$	2	QS	I_{E5-1-1}	4	VS
I_{E4-1-2}	4	S	I_{E5-1-2}	4	VS
I_{E4-1-3}	1	QS	I_{E5-1-2}	4	VS
I_{E4-1-4}	3	S	I_{E5-1-4}	4	VS
I_{E4-1-5}	3	QS	I_{E5-2-1}	1	VS
$I_{E4-2-1-1}$	4	VS	I_{E5-2-2}	1	QS
$I_{E4-2-1-2}$	4	VS	I_{E5-2-2}	1	VS
I_{E4-2-2}	4	VS	I_{E5-2-4}	1	VS
I_{E4-2-3}	4	VS	I_{E6-1-1}	2	S
I_{E4-2-4}	4	VS	I_{E6-1-2}	1	VS
I_{E6-1-3}	3	S	I_{E6-3-2}	1	QS
I_{E6-1-4}	2	QS	I_{E6-3-3}	2	S
$I_{E6-1-5-1}$	1	VS	I_{E6-3-4}	1	QS
$I_{E6-1-5-2}$	3	S	$I_{E6-3-5-1}$	1	QS
I_{E6-2-1}	3	QS	$I_{E6-3-5-2}$	1	QS
I_{E6-2-2}	4	QS	I_{E6-4-1}	2	QS
I_{E6-2-2}	1	S	I_{E6-4-2}	1	VS
I_{E6-2-4}	2	QS	I_{E6-4-3}	3	S
$I_{E6-2-5-1}$	1	QS	I_{E6-4-4}	2	QS
$I_{E6-2-5-2}$	3	QS	$I_{E6-4-5-1}$	1	QS
I_{E6-3-1}	2	QS	$I_{E6-4-5-2}$	1	VS

(2) The construction risk of Dongda Qiao Station has obvious stage evolution. During the excavation stage of the pilot tunnel and main arch, the risks of collapse and above-ground buildings being unusable are significantly higher than other stages. Moreover, during the separate double-hole tunnel construction stage, due to the unfavorable relative position of the construction site and the existing Line 6 Dongda Qiao Station interval tunnel (undercrossing), the risk of existing underground structure deformation exceeding limits significantly increases.

Table 6.29 Evaluation results of relevant indicators for engineering zoning (Northeast region)

Evaluation indicators	Evaluation level	Credibility degree	Evaluation indicators	Evaluation level	Credibility degree
$I_{E4-1-1-1}$	2	NS	I_{E4-2-5}	2	NS
$I_{E4-1-1-2}$	1	QS	I_{E5-1-1}	3	QS
I_{E4-1-2}	1	QS	I_{E5-1-2}	2	QS
I_{E4-1-3}	2	S	I_{E5-1-2}	3	S
I_{E4-1-4}	1	S	I_{E5-1-4}	2	QS
I_{E4-1-5}	3	QS	I_{E5-2-1}	1	VS
$I_{E4-2-1-1}$	2	QS	I_{E5-2-2}	1	QS
$I_{E4-2-1-2}$	4	VS	I_{E5-2-2}	2	S
I_{E4-2-2}	2	S	I_{E5-2-4}	2	QS
I_{E4-2-3}	3	S	I_{E6-1-1}	3	S
I_{E4-2-4}	3	S	I_{E6-1-2}	4	S
I_{E6-1-3}	1	S	I_{E6-3-2}	1	QS
I_{E6-1-4}	1	QS	I_{E6-3-3}	2	S
$I_{E6-1-5-1}$	2	S	I_{E6-3-4}	1	VS
$I_{E6-1-5-2}$	3	HS	$I_{E6-3-5-1}$	1	VS
I_{E6-2-1}	4	VS	$I_{E6-3-5-2}$	1	QS
I_{E6-2-2}	4	VS	I_{E6-4-1}	2	QS
I_{E6-2-2}	4	VS	I_{E6-4-2}	1	QS
I_{E6-2-4}	4	VS	I_{E6-4-3}	2	S
$I_{E6-2-5-1}$	4	VS	I_{E6-4-4}	2	QS
$I_{E6-2-5-2}$	4	VS	$I_{E6-4-5-1}$	1	QS
I_{E6-3-1}	2	QS	$I_{E6-4-5-2}$	1	QS

(3) The coupling effect of risk factors has a significant impact on the risk level. Compared to the situation without considering the coupling effect of risk factors, the probability of risk events occurring when considering risk factor coupling has increased by 1–15%. In certain specific time periods, the coupling effect of risk factors can cause a leap in the likelihood of risk events. In addition, it can be found that the collapse risk is significantly affected by the coupling effect of risk factors. Combined with the risk analysis Bayesian network and risk factor coupling relationship analysis, it is known that the more risk factors with direct causal relationships that have coupling effects, the more significant the coupling effect of the risk.

Table 6.30 Evaluation results of relevant indicators for engineering zoning (Southwest region)

Evaluation indicators	Evaluation level	Credibility degree	Evaluation indicators	Evaluation level	Credibility degree
$I_{E4-1-1-1}$	4	VS	I_{E4-2-5}	2	NS
$I_{E4-1-1-2}$	4	VS	I_{E5-1-1}	3	QS
I_{E4-1-2}	4	VS	I_{E5-1-2}	3	Q S
I_{E4-1-3}	4	VS	I_{E5-1-2}	3	S
I_{E4-1-4}	4	VS	I_{E5-1-4}	1	S
I_{E4-1-5}	4	VS	I_{E5-2-1}	1	VS
$I_{E4-2-1-1}$	1	S	I_{E5-2-2}	1	QS
$I_{E4-2-1-2}$	4	VS	I_{E5-2-2}	1	VS
I_{E4-2-2}	3	S	I_{E5-2-4}	1	VS
I_{E4-2-3}	3	S	I_{E6-1-1}	2	QS
I_{E4-2-4}	3	S	I_{E6-1-2}	1	VS
I_{E6-1-3}	2	S	I_{E6-3-2}	1	QS
I_{E6-1-4}	2	QS	I_{E6-3-3}	2	S
$I_{E6-1-5-1}$	1	VS	I_{E6-3-4}	1	QS
$I_{E6-1-5-2}$	1	QS	$I_{E6-3-5-1}$	2	QS
I_{E6-2-1}	3	QS	$I_{E6-3-5-2}$	1	QS
I_{E6-2-2}	4	QS	I_{E6-4-1}	2	QS
I_{E6-2-2}	1	S	I_{E6-4-2}	1	VS
I_{E6-2-4}	2	QS	I_{E6-4-3}	1	S
$I_{E6-2-5-1}$	1	QS	I_{E6-4-4}	1	VS
$I_{E6-2-5-2}$	3	QS	$I_{E6-4-5-1}$	2	QS
I_{E6-3-1}	2	QS	$I_{E6-4-5-2}$	2	QS

(4) The above analysis results show that the multi-factor coupling evolution model of urban large underground space construction risk can achieve quantitative analysis of the multi-factor coupling effect of construction risk and prediction of risk evolution trends. The analysis results are consistent with engineering experience cognition.

Table 6.31 Evaluation results of relevant indicators for engineering zoning (Northwest region)

Evaluation indicators	Evaluation level	Credibility degree	Evaluation indicators	Evaluation level	Credibility degree
$I_{E4\text{-}1\text{-}1\text{-}1}$	4	VS	$I_{E4\text{-}2\text{-}5}$	2	NS
$I_{E4\text{-}1\text{-}1\text{-}2}$	4	VS	$I_{E5\text{-}1\text{-}1}$	2	QS
$I_{E4\text{-}1\text{-}2}$	4	VS	$I_{E5\text{-}1\text{-}2}$	3	QS
$I_{E4\text{-}1\text{-}3}$	4	VS	$I_{E5\text{-}1\text{-}2}$	1	HS
$I_{E4\text{-}1\text{-}4}$	4	VS	$I_{E5\text{-}1\text{-}4}$	2	S
$I_{E4\text{-}1\text{-}5}$	4	VS	$I_{E5\text{-}2\text{-}1}$	1	VS
$I_{E4\text{-}2\text{-}1\text{-}1}$	2	S	$I_{E5\text{-}2\text{-}2}$	1	QS
$I_{E4\text{-}2\text{-}1\text{-}2}$	4	VS	$I_{E5\text{-}2\text{-}2}$	2	S
$I_{E4\text{-}2\text{-}2}$	4	QS	$I_{E5\text{-}2\text{-}4}$	2	QS
$I_{E4\text{-}2\text{-}3}$	3	S	$I_{E6\text{-}1\text{-}1}$	2	QS
$I_{E4\text{-}2\text{-}4}$	3	S	$I_{E6\text{-}1\text{-}2}$	1	VS
$I_{E6\text{-}1\text{-}3}$	2	QS	$I_{E6\text{-}3\text{-}2}$	4	VS
$I_{E6\text{-}1\text{-}4}$	1	VS	$I_{E6\text{-}3\text{-}3}$	4	VS
$I_{E6\text{-}1\text{-}5\text{-}1}$	1	QS	$I_{E6\text{-}3\text{-}4}$	4	VS
$I_{E6\text{-}1\text{-}5\text{-}2}$	3	NS	$I_{E6\text{-}3\text{-}5\text{-}1}$	4	VS
$I_{E6\text{-}2\text{-}1}$	3	QS	$I_{E6\text{-}3\text{-}5\text{-}2}$	4	VS
$I_{E6\text{-}2\text{-}2}$	3	QS	$I_{E6\text{-}4\text{-}1}$	2	QS
$I_{E6\text{-}2\text{-}2}$	1	QS	$I_{E6\text{-}4\text{-}2}$	1	VS
$I_{E6\text{-}2\text{-}4}$	1	VS	$I_{E6\text{-}4\text{-}3}$	1	S
$I_{E6\text{-}2\text{-}5\text{-}1}$	2	S	$I_{E6\text{-}4\text{-}4}$	1	QS
$I_{E6\text{-}2\text{-}5\text{-}2}$	3	NS	$I_{E6\text{-}4\text{-}5\text{-}1}$	1	VS
$I_{E6\text{-}3\text{-}1}$	4	VS	$I_{E6\text{-}4\text{-}5\text{-}2}$	2	QS

Table 6.32 Evaluation results of relevant indicators for engineering zoning (Central region)

Evaluation indicators	Evaluation level	Credibility degree	Evaluation indicators	Evaluation level	Credibility degree
$I_{E4-1-1-1}$	1	VS	I_{E4-2-5}	4	VS
$I_{E4-1-1-2}$	1	VS	I_{E5-1-1}	4	VS
I_{E4-1-2}	3	QS	I_{E5-1-2}	4	VS
I_{E4-1-3}	1	QS	I_{E5-1-2}	4	VS
I_{E4-1-4}	2	S	I_{E5-1-4}	4	VS
I_{E4-1-5}	3	QS	I_{E5-2-1}	1	VS
$I_{E4-2-1-1}$	4	VS	I_{E5-2-2}	1	QS
$I_{E4-2-1-2}$	4	VS	I_{E5-2-2}	1	VS
I_{E4-2-2}	4	VS	I_{E5-2-4}	1	VS
I_{E4-2-3}	4	VS	I_{E6-1-1}	2	QS
I_{E4-2-4}	4	VS	I_{E6-1-2}	1	VS
I_{E6-1-3}	2	S	I_{E6-3-2}	1	QS
I_{E6-1-4}	1	QS	I_{E6-3-3}	2	S
$I_{E6-1-5-1}$	4	QS	I_{E6-3-4}	1	VS
$I_{E6-1-5-2}$	2	QS	$I_{E6-3-5-1}$	3	S
I_{E6-2-1}	4	VS	$I_{E6-3-5-2}$	1	QS
I_{E6-2-2}	4	VS	I_{E6-4-1}	4	VS
I_{E6-2-2}	4	VS	I_{E6-4-2}	4	VS
I_{E6-2-4}	4	VS	I_{E6-4-3}	4	VS
$I_{E6-2-5-1}$	4	VS	I_{E6-4-4}	4	VS
$I_{E6-2-5-2}$	4	VS	$I_{E6-4-5-1}$	4	VS
I_{E6-3-1}	2	QS	$I_{E6-4-5-2}$	4	VS

Table 6.33 Evaluation indicator weight calculation results

Status indicators	Evaluation indicators	Weight	Status indicators	Evaluation indicators	Weight	Status indicators	Evaluation indicators	Weight	Status indicators	Evaluation indicators	Weight
Design quality S_{S1}	I_{S1-1}	0.193	Stratum conditions S_{E1}	$I_{E1-1-1-1}$	0.035	Groundwater conditions S_{E2}	I_{E2-1-1}	0.113	Surrounding pipeline conditions S_{E6}	I_{E6-1-1}	0.047
	I_{S1-2}	0.083		$I_{E1-1-1-2}$	0.054		I_{E2-1-2}	0.371		I_{E6-1-2}	0.049
	I_{S1-3}	0.724		$I_{E1-1-1-3}$	0.061		I_{E2-2-1}	0.074		I_{E6-1-3}	0.038
Investigation quality S_{S2}	I_{S2-1}	0.079		$I_{E1-1-1-4}$	0.046		I_{E2-2-2}	0.039		I_{E6-1-4}	0.006
	I_{S2-2}	0.201		$I_{E1-1-1-5}$	0.077		I_{E2-3-1}	0.345		$I_{E6-1-5-1}$	0.036
	I_{S2-3}	0.519		$I_{E1-1-2-1}$	0.015		I_{E2-4}	0.058		$I_{E6-1-5-2}$	0.054
	I_{S2-4}	0.201		$I_{E1-1-2-2}$	0.037	Underground structure conditions S_{E4}	$I_{E4-1-1-1}$	0.069		I_{E6-2-1}	0.059
Construction level S_{S3}	I_{S3-1}	0.032		$I_{E1-1-2-3}$	0.065		$I_{E4-1-1-2}$	0.004		I_{E6-2-2}	0.010
	I_{S3-2}	0.075		$I_{E1-1-2-4}$	0.040		I_{E4-1-2}	0.095		I_{E6-2-3}	0.013
	I_{S3-3}	0.075		$I_{E1-1-2-5}$	0.050		I_{E4-1-3}	0.116		I_{E6-2-4}	0.008
	I_{S3-4}	0.075		$I_{E1-1-3-1}$	0.022		I_{E4-1-4}	0.079		$I_{E6-2-5-1}$	0.078
	I_{S3-5}	0.032		$I_{E1-1-3-2}$	0.059		I_{E4-1-5}	0.117		$I_{E6-2-5-2}$	0.050
	I_{S3-6}	0.177		$I_{E1-1-3-3}$	0.060		$I_{E4-2-1-1}$	0.028		I_{E6-3-1}	0.079
	I_{S3-7}	0.357		$I_{E1-1-3-4}$	0.026		$I_{E4-2-1-2}$	0.117		I_{E6-3-2}	0.012
	I_{S3-8}	0.177		$I_{E1-1-3-5}$	0.089		I_{E4-2-2}	0.133		I_{E6-3-3}	0.070
Engineering characteristics S_{S4}	$I_{S4-1-1-1}$	0.12		$I_{E1-1-4-1}$	0.055	Surface environmental conditions S_{E5}	I_{E4-2-3}	0.044		I_{E6-3-4}	0.066
	$I_{S4-1-1-2}$	0.06		$I_{E1-1-4-2}$	0.030		I_{E4-2-4}	0.074		$I_{E6-3-5-1}$	0.031
	I_{S4-1-2}	0.099		$I_{E1-1-4-3}$	0.039		I_{E4-2-5}	0.125		$I_{E6-3-5-2}$	0.065
	I_{S4-1-3}	0.055		$I_{E1-1-4-4}$	0.050		I_{E5-1-1}	0.153		I_{E6-4-1}	0.011
	I_{S4-2-1}	0.08		$I_{E1-1-4-5}$	0.000		I_{E5-1-2}	0.184		I_{E6-4-2}	0.060
	I_{S4-2-2}	0.208		I_{E1-2-1}	0.050		I_{E5-1-3}	0.005		I_{E6-4-3}	0.024

(continued)

Table 6.33 (continued)

Status indicators	Evaluation indicators	Weight	Status indicators	Evaluation indicators	Weight	Status indicators	Evaluation indicators	Weight	Status indicators	Evaluation indicators	Weight
	$I_{S4\text{-}2\text{-}3}$	0.046		$I_{E1\text{-}2\text{-}2}$	0.030		$I_{E5\text{-}1\text{-}4}$	0.161		$I_{E6\text{-}4\text{-}4}$	0.042
	$I_{S4\text{-}3\text{-}1}$	0.111		$I_{E1\text{-}2\text{-}3}$	0.005		$I_{E5\text{-}2\text{-}1}$	0.165		$I_{E6\text{-}4\text{-}5\text{-}1}$	0.061
	$I_{S4\text{-}3\text{-}2}$	0.111		$I_{E1\text{-}2\text{-}4}$	0.003		$I_{E5\text{-}2\text{-}2}$	0.099		$I_{E6\text{-}4\text{-}5\text{-}2}$	0.029
	$I_{S4\text{-}3\text{-}3}$	0.111	-				$I_{E5\text{-}2\text{-}3}$	0.173	-		
-							$I_{E5\text{-}2\text{-}4}$	0.062			

Table 6.34 Markov matrix value table

Evaluation unit	Indicator	Type	Parameter			Start-stop time	
			α	β	Absorbing state	Start	End
M_{P1}	Span $I_{S4\text{-}1\text{-}1\text{-}1}$	Static type	–	–	–	0	12
	Excavation length $I_{S4\text{-}1\text{-}1\text{-}2}$	Absorption type	0	0.2	3		
	Excavation depth $I_{S4\text{-}1\text{-}2}$	Monotonic type	0	0.08	–		
	Expected precipitation $I_{E3\text{-}1}$	Static type	–	–	–		
M_{P2}	Span $I_{S4\text{-}1\text{-}1\text{-}1}$	Absorption type	0	0.2	3	12	28
	Excavation length $I_{S4\text{-}1\text{-}1\text{-}2}$	Static type	–	–	–		
	Excavation depth $I_{S4\text{-}1\text{-}2}$	Monotonic type	0	0.14	–		
	Expected precipitation $I_{E3\text{-}1}$	Monotonic type	0	0.08	–		
M_{P3}	Span $I_{S4\text{-}1\text{-}1\text{-}1}$	Static type	–	–	–	28	62
	Excavation length $I_{S4\text{-}1\text{-}1\text{-}2}$	Static type	–	–	–		
	Excavation depth $I_{S4\text{-}1\text{-}2}$	Static type	–	–	–		
	Expected precipitation $I_{E3\text{-}1}$	Monotonic type	0.12	0.1	–		
M_{P4}/M_{P5}	Span $I_{S4\text{-}1\text{-}1\text{-}1}$	Monotonic type	0	0.18	–	62	86
	Excavation length $I_{S4\text{-}1\text{-}1\text{-}2}$	Absorption type	0	0.05	4		
	Excavation depth $I_{S4\text{-}1\text{-}2}$	Monotonic type	0	0.15	–		
	Expected precipitation$I_{E3\text{-}1}$	Monotonic type	0	0.08	–		
M_{P6}	Span $I_{S4\text{-}1\text{-}1\text{-}1}$	Static type	–	–	–	86	106
	Excavation length $I_{S4\text{-}1\text{-}1\text{-}2}$	Monotonic type	0	0.12	–		
	Excavation depth $I_{S4\text{-}1\text{-}2}$	Monotonic type	0	0.03	–		
	Expected precipitation$I_{E3\text{-}1}$	Monotonic type	0.04	0	–		

(continued)

Table 6.34 (continued)

Evaluation unit	Indicator	Type	Parameter			Start-stop time	
			α	β	Absorbing state	Start	End
M_{P7}	Span $I_{S4-1-1-1}$	Static type	–	–	–	106	132
	Excavation length $I_{S4-1-1-2}$	Static type	–	–	–		
	Excavation depth I_{S4-1-2}	Static type	–	–	–		
	Expected precipitation I_{E3-1}	Monotonic type	0	0.05	–		
T_{P1}	Span $I_{S4-1-1-1}$	Monotonic type	–	0.12	–	28	44
	Excavation length $I_{S4-1-1-2}$	Monotonic type	–	0.07	–		
	Excavation depth I_{S4-1-2}	Monotonic type	–	0.06	–		
	Expected precipitationI_{E3-1}	Monotonic type	0.06	0.1	–		
T_{P2}	Span $I_{S4-1-1-1}$	Static type	–	–	–	44	52
	Excavation length $I_{S4-1-1-2}$	Static type	–	–	–		
	Excavation depth I_{S4-1-2}	Monotonic type	0	0.08	–		
	Expected precipitationI_{E3-1}	Monotonic type	0.08	0	–		
T_{P3}	Span $I_{S4-1-1-1}$	Static type	–	–	–	52	62
	Excavation length $I_{S4-1-1-2}$	Static type	–	–	–		
	Excavation depth I_{S4-1-2}	Static type	–	–	–		
	Expected precipitation I_{E3-1}	Static type	–	–	–		
A_1	Span $I_{S4-1-1-1}$	Monotonic type	0	0.11	–	106	132
	Excavation length $I_{S4-1-1-2}$	Monotonic type	0	0.03	–		
	Excavation depth I_{S4-1-2}	Monotonic type	0	0.04	–		
	Expected precipitationI_{E3-1}	Monotonic type	0	0.05	–		

(continued)

Table 6.34 (continued)

Evaluation unit	Indicator	Type	Parameter			Start-stop time	
			α	β	Absorbing state	Start	End
A_2	Span $I_{S4\text{-}1\text{-}1\text{-}1}$	Monotonic type	0	0.14	–	106	132
	Excavation length $I_{S4\text{-}1\text{-}1\text{-}2}$	Static type	–	–	–		
	Excavation depth $I_{S4\text{-}1\text{-}2}$	Monotonic type	0	0.08	–		
	Expected precipitation$I_{E3\text{-}1}$	Monotonic type	0	0.05	–		
A_3	Span $I_{S4\text{-}1\text{-}1\text{-}1}$	Monotonic type	0	0.1	–	112	132
	Excavation length $I_{S4\text{-}1\text{-}1\text{-}2}$	Static type	–	–	–		
	Excavation depth $I_{S4\text{-}1\text{-}2}$	Monotonic type	0	0.06	–		
	Expected precipitation $I_{E3\text{-}1}$	Static type	-	-	-		

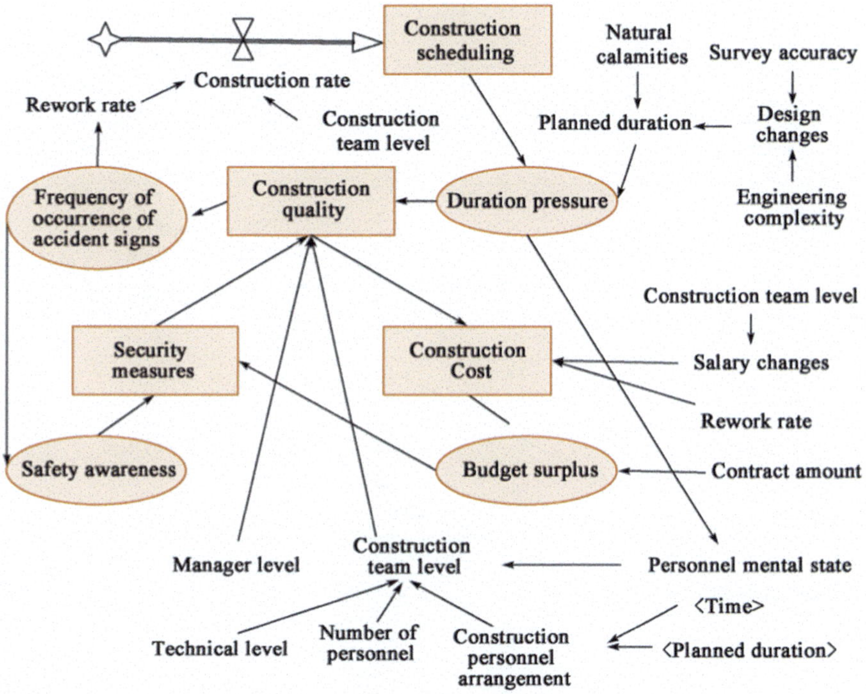

Fig. 6.39 Beijing Subway Line 17 Dongda Qiao station construction organizational management subsystem model

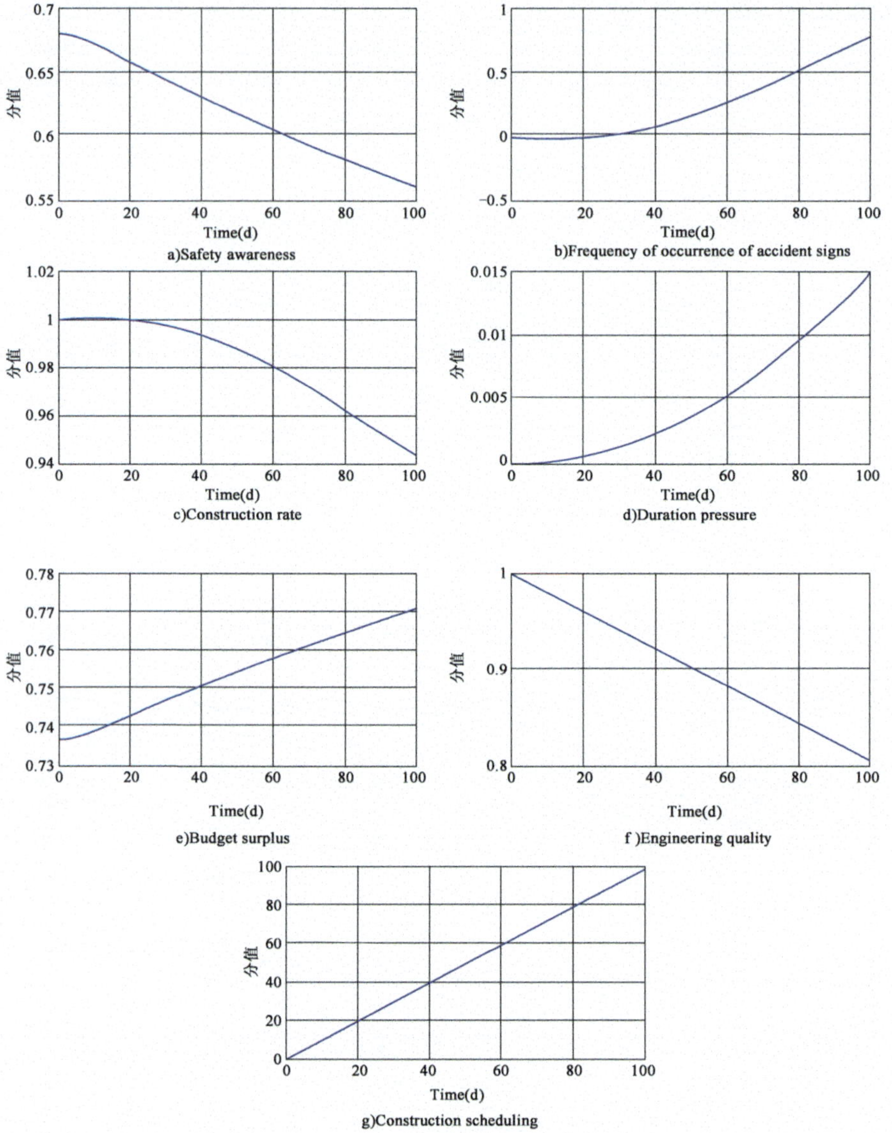

Fig. 6.40 Main factors change curve

Fig. 6.41 MP1 unit risk analysis results

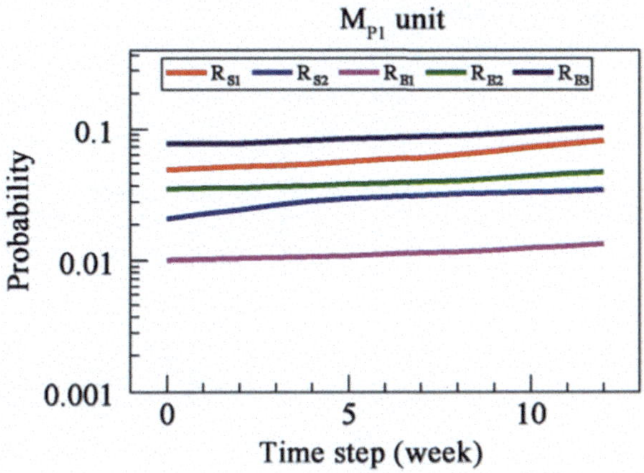

Fig. 6.42 MP2 unit risk analysis results

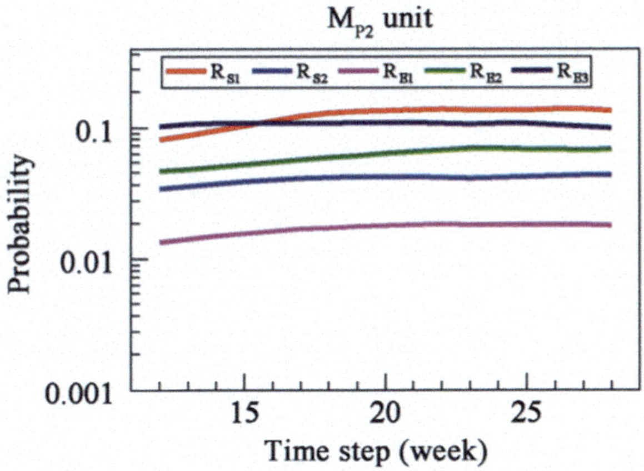

Fig. 6.43 MP3 unit risk analysis results

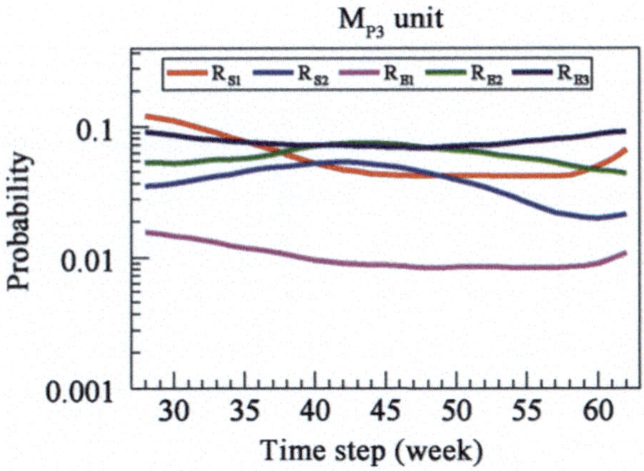

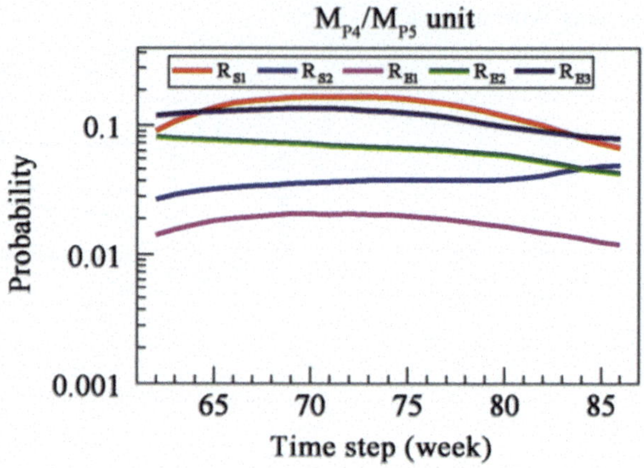

Fig. 6.44 MP4/MP5 unit risk analysis results

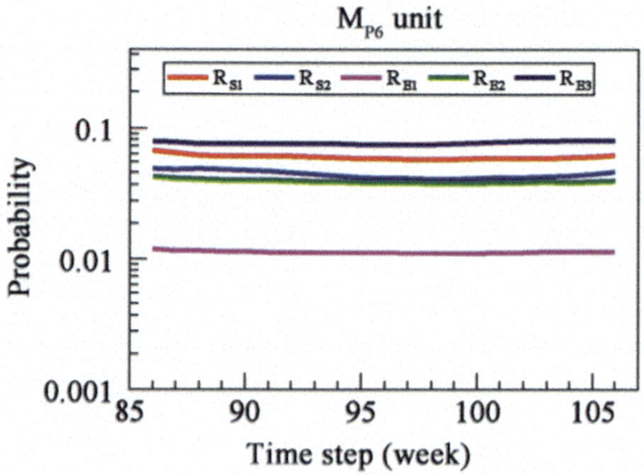

Fig. 6.45 MP6 unit risk analysis results

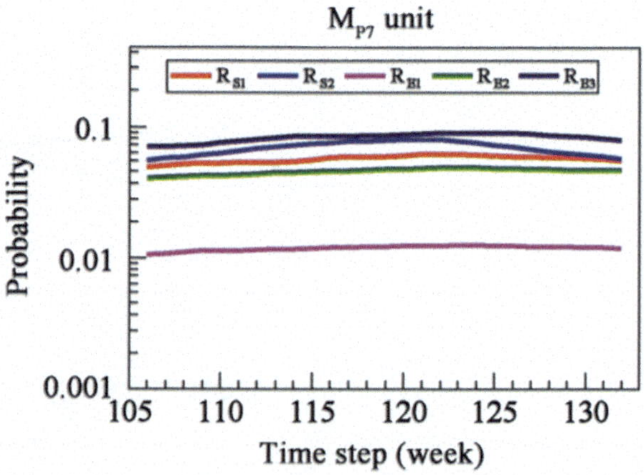

Fig. 6.46 MP7 unit risk analysis results

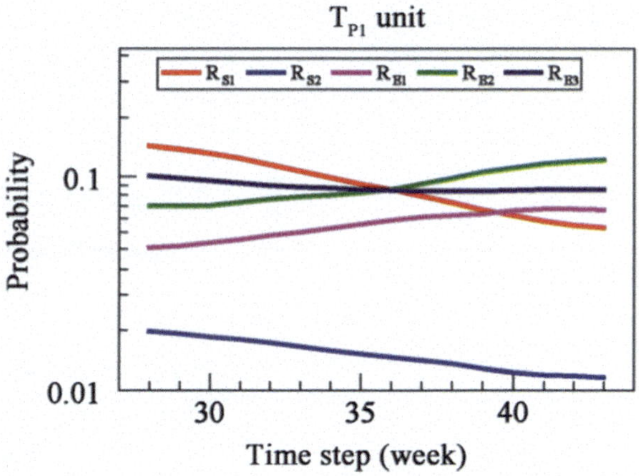

Fig. 6.47 TP1 unit risk analysis results

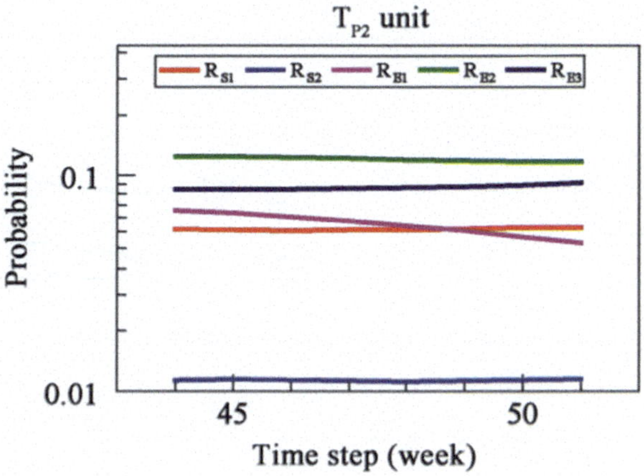

Fig. 6.48 TP2 unit risk analysis results

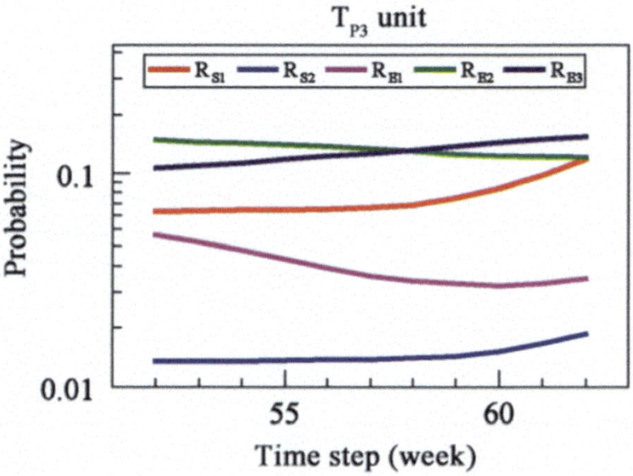

Fig. 6.49 TP3 unit risk analysis results

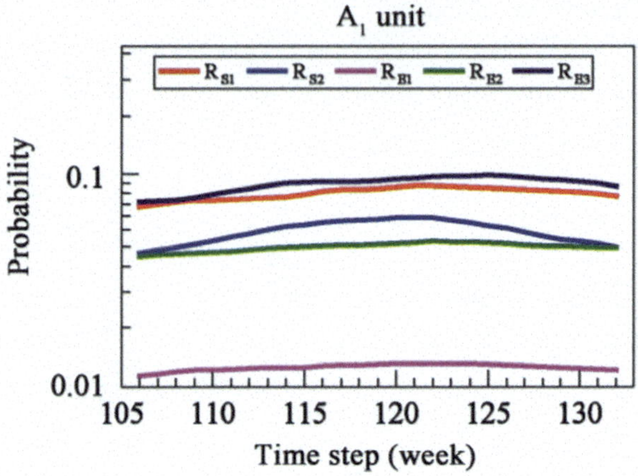

Fig. 6.50 A1 unit risk analysis results

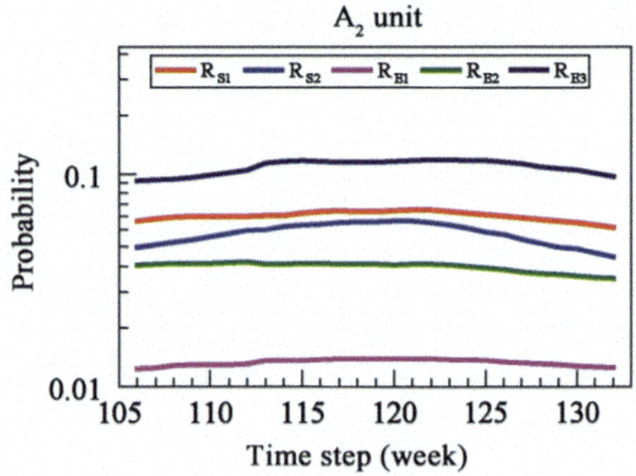

Fig. 6.51 A2 unit risk analysis results

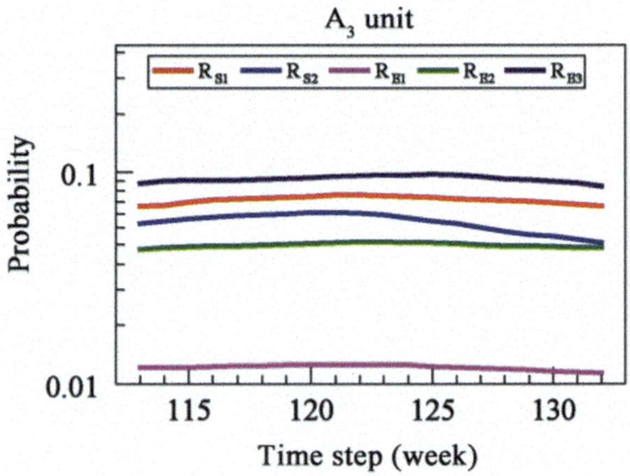

Fig. 6.52 A3 unit risk analysis results

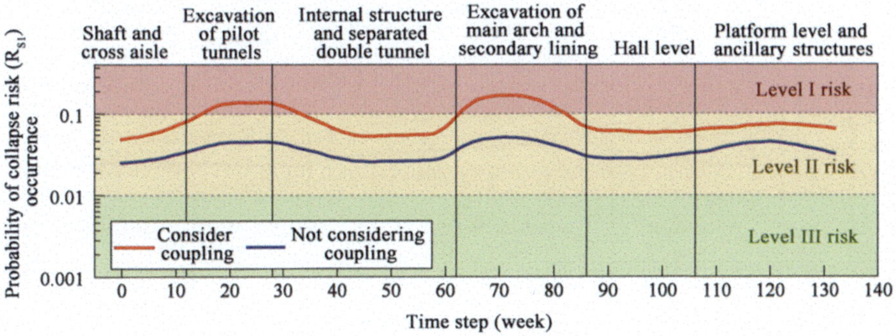

Fig. 6.53 Collapse risk (R_{S1}) analysis results

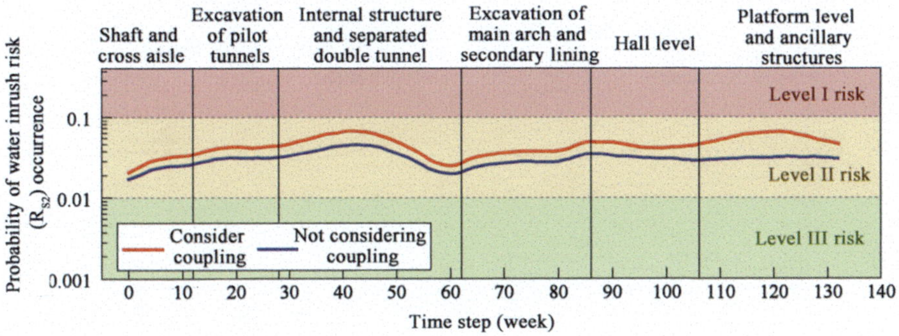

Fig. 6.54 Water Inrush risk (R_{S2}) analysis results

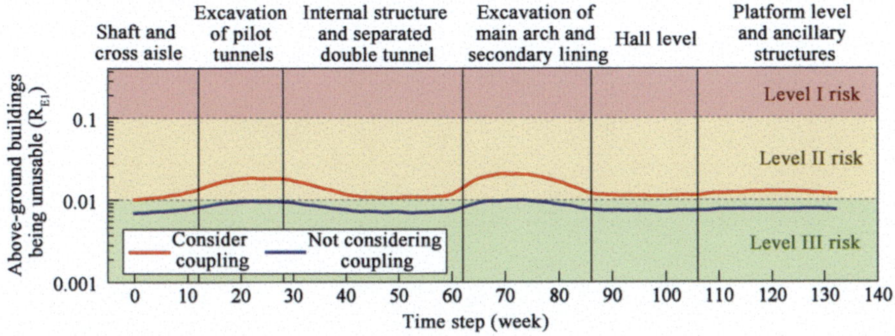

Fig. 6.55 Risk of above-ground buildings being unusable (R_{E1}) analysis results

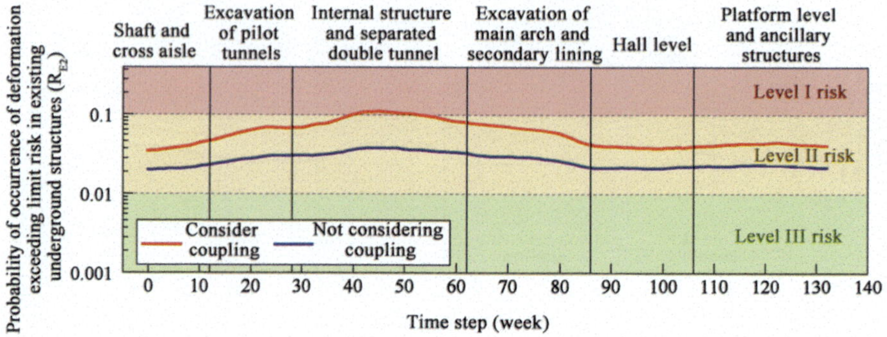

Fig. 6.56 Risk of existing underground structure deformation exceeding limits (R_{E2}) analysis results

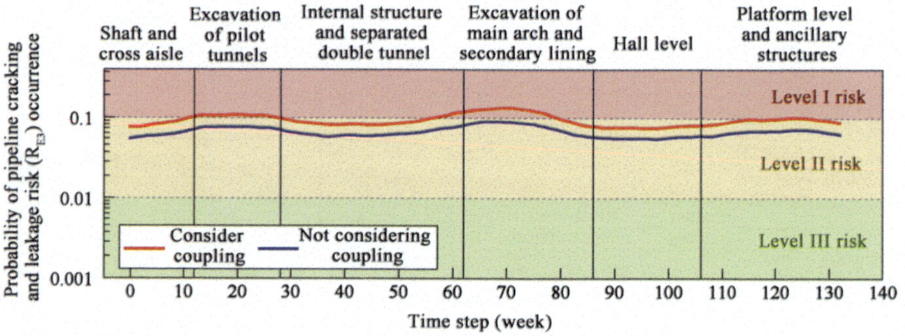

Fig. 6.57 Risk of pipeline cracking and leakage (R_{E3}) analysis results

Chapter 7
Dynamic Assessment System for Safety Risks in Urban Large Underground Space Construction

7.1 Overview

7.1.1 Composition of the Dynamic Assessment System for Safety Risks in Urban Large Underground Space Construction

1. Method of assessing safety risks in urban large underground space construction

The method of assessing safety risks in urban large underground space construction is the core of the dynamic assessment system. The risk assessment method is the unity of risk analysis methods and risk measurement methods, determining the logic of risk assessment, and plays an important role in transforming the knowledge, experience, judgment or other objective information of the evaluators into qualitative or quantitative risk assessments.

The risk assessment method should match the risk assessment objectives. Specifically, the risk assessment method should consider the difficulty of obtaining information required for risk assessment, whether the logic of risk assessment conforms to the general experience in the relevant field, and whether the type of risk assessment results can provide sufficient guidance for actions, etc. For example, financial risk assessment should be based on objective data and provide quantitative assessment results through mathematical calculations; while safety risk assessment generally focuses on risk sources as the main object of concern, providing qualitative or semi-quantitative assessment results for risk events related to risk sources through logical reasoning. Risk assessment methods and risk analysis methods are related but distinct concepts from each other. Risk analysis methods focus on detailed analysis of the mechanisms of risk occurrence, so their calculation process is relatively complex, but the results are relatively accurate; risk assessment methods are derived from risk analysis methods, but risk assessment pays more attention to operability in actual work, its calculation results are relatively rough, and the calculation process is relatively simple.

S.-X. Lei et al., *Safety Risk Assessment of Urban Underground Large Space Construction*, https://doi.org/10.1007/978-981-97-7437-1_7

2. Safety risk assessment indicator system and quantification standards for urban large underground space construction

The safety risk assessment indicator system and quantification standards for urban large underground space construction are the bridge connecting the mechanical process of urban large underground space construction and risk analysis. The risk assessment indicator system is the input end of the risk assessment method, its role is to define the boundary for the information required for risk assessment, and to filter out the most critical and most influential part of the information from the numerous risk-related information; the role of the risk assessment index quantification standard is to transform various types of information into the input form required by the assessment method, providing the basis, method, and standard for transformation. The risk assessment indicator system is closely related to risk factors, and risk factors depend on the type of risk event and the mechanism of risk event occurrence, so the risk assessment indicator system must be closely linked with professional knowledge in the relevant field and have a clear correspondence with the risk event to be evaluated.

3. Safety risk grading standards and acceptance criteria for urban large underground space construction

The role of safety risk grading standards and acceptance criteria for urban large underground space construction is to transform the quantified risk assessment results into semi-quantitative judgments, thereby forming a risk control and response plan that directly guides the construction process.

4. Safety risk assessment process for urban large underground space construction

The process for assessing safety risks in the construction of urban large underground space stipulates the composition of personnel, organizational structure, work steps, and the objectives and outcomes required for each step. The general process of risk assessment mainly includes risk identification, risk analysis, risk assessment, and risk decision-making. The risk assessment of the construction of urban large underground space requires comprehensive analysis from multiple perspectives, collaborative cooperation from multiple units, and dynamic progress over multiple time, making its assessment process more special and complex.

5. dynamic assessment system of construction safety risks of Urban large underground space

The dynamic assessment system of construction safety risks of urban large underground space is a set of application software tailored for actual engineering risk management work. The dynamic risk assessment system is the crystallization and refinement of the results of risk assessment system research and is the implementation tool for risk assessment work.

The relationship between the various components in the dynamic assessment system of construction safety risks of urban large underground space is shown in Fig. 7.1.

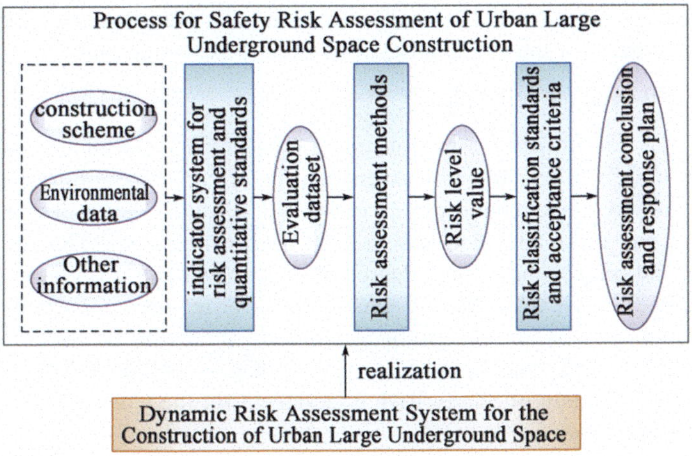

Fig. 7.1 Dynamic risk assessment system for the construction of urban large underground space

7.1.2 Principles and Content of Safety Risk Assessment for the Construction of Urban Large Underground Space

1. Principles of Safety Risk Assessment for the Construction of Urban large underground space

The principles followed in the safety risk assessment of the construction of urban large underground space are as follows.

(1) Classification and zoning principle

The safety risk of the construction of urban large underground space should be assessed for specific risk events within a certain construction area. Due to the large volume of urban underground space projects, the wide range of spaces involved, the numerous construction sequences, and the diverse types of risk events, the use of a classification and zoning assessment strategy can enhance the pertinence of the indicator system and quantitative standards, help to sort out the logic of the assessment and accurately assign indicator scores, and improve the accuracy of risk assessment.

(2) Overall and specific principle

The safety risk assessment of the construction of urban large underground space should be divided into two stages: overall risk assessment and specialized risk assessment. The role of the overall risk assessment is to help assessors and construction personnel establish a macro understanding of construction risks, and at the same time, to screen out projects with lower overall risks that do not need to undergo specialized risk assessment. The role of the specialized risk assessment is to help the construction party identify the main controlling risk events and risk sources and provide a reference for the formulation of specific construction plans.

(3) Static and dynamic principle

The safety risk assessment of the construction of urban large underground space should not only include static risk assessment before construction, but also track the actual situation encountered during construction, adjust the assessment results in real time, and re-identify and reassess risks when necessary. The construction of underground projects has a high degree of uncertainty, and it is common for the geological conditions revealed during construction to be inconsistent with the original geological survey report, and design changes and adjustments to the construction plan often occur, so it is very necessary to carry out dynamic assessment of construction safety risks. (1) Content of Safety Risk Assessment for the Construction of Urban large underground space.

According to the results of construction risk classification research, the main objects of risk assessment for the construction of urban large underground space are:

(1) Overall safety risk

The overall risk refers to the total possibility of various risk events occurring during the entire construction process in the underground large space engineering area of the entire city. In other words, the overall risk is a combination of the possibilities and consequences of a series of uncertain and non-specific risk events. Due to the different formation mechanisms, influencing factors, and degrees of harm of different risk events, the evaluation object of the overall risk evaluation index is no longer the risk event, but the relatively abstract and generalized "overall risk level".

(2) Inherent project risk

The inherent project risk refers to various risk events that may occur in the project structure or strata that directly interact with the project structure, which may cause serious harm, including engineering structure risks and geological risks. The inherent risks of urban large underground space construction projects mainly include the following.

① Pit instability (open excavation). Overturning of the pit retaining structure, shearing of the retaining structure, "kick" damage, and slope instability during slope excavation all belong to pit instability.

② Base heave (open excavation). Sudden water, pipe gushing, sand flow, etc. caused by hydraulic action on the pit bottom plate or retaining structure all belong to base heave.

③ Roof collapse (tunneling). Slip and collapse of the working face, "closing the door", large-scale flake, falling block, etc., all belong to roof collapse.

④ Mud and water inrush (tunneling). Sudden water, mud, and sand gushing on the working face during mining method excavation, or water and mud gushing during shield method excavation, all belong to mud and water inrush.

⑤ Connection opening causes excessive deformation or damage to the existing structure (expansion). Local damage to the opening part of the existing structure caused by the connection opening, or the inability of the entire existing structure to meet normal use requirements, all belong to excessive deformation or damage to the existing structure.

⑥ Excavation unloading causes the existing structure to float (expansion). The floating of the existing structure caused by excavation unloading includes floating caused by groundwater buoyancy and floating caused by stratum elastic reaction.

(3) Surrounding environmental risk

The surrounding environmental risk refers to risk events where surrounding urban facilities and the natural environment are damaged due to construction. The main types of surrounding environmental risks in urban large underground space construction are as follows.

① Damage to ground buildings. The situation where ground buildings cannot meet normal use function requirements due to excessive deformation, uneven settlement, and cracking is considered damage to ground buildings.

② Damage to underground buildings. The situation where underground buildings cannot meet normal use function requirements due to excessive deformation, seepage, and damage to lighting and ventilation conditions is considered damage to underground buildings.

③ Pipeline damage. Pipeline cracking, leakage, gas pipeline explosion, etc., all belong to pipeline damage.

7.1.3 Urban Large Underground Space Construction Safety Risk Assessment Process

Given the dynamic nature of safety risks in urban underground large-space construction, risk assessment should be dynamic and span the entire construction process. Overall, the safety risk assessment of urban underground large-space construction can be divided into pre-construction assessment and construction process assessment.

1. Pre-construction assessment

Pre-construction assessment includes overall risk assessment and special risk assessment during the construction preparation stage, the specific assessment process is as follows.

(1) Establish an assessment team

The assessment team members include industry experts, construction units, design units, regulatory units, owner units, etc., responsible for the implementation, supervision, and final assessment report review of the entire assessment process.

(2) Collect engineering data, clarify the assessment object

The construction unit and the design unit are responsible for the collection and collation of engineering data, summarizing the engineering data involved in the assessment, clarifying the risk engineering and construction plan information, and facilitating the next step of the assessment work. The relevant data to be collected include:

① Engineering geology, hydrogeology, surrounding natural and social environment data, etc.;

② Similar construction experience and typical accident cases;

③ Engineering planning, engineering survey, design and consulting documents, etc.;

④ Data of buildings around the project (including subways, underground pipelines, civil defense facilities, roads, etc.);

⑤ Documents related to construction organization design, special construction plans, etc.;

⑥ Claims of facility ownership units, operation management units, and related parties;

⑦ Other relevant data.

(3) Risk source identification

Based on the collected data, combined with the risk source library of urban large underground space construction, identify the risk sources and compile a list of risk sources. The risk source list should contain detailed information about engineering structure design, geological conditions, construction plans, surrounding pipelines, ground buildings, underground buildings, roads and bridges, rivers and lakes, etc. One of the purposes of compiling the risk source list is to screen and sort out all the information needed for the risk assessment process, provide a reference for the division of assessment units, and enable the assessor to complete the selection of assessment index scores only through the information recorded in the risk source list. Therefore, in the process of compiling the risk source list, the interrelationships between various risk sources can be sorted out together.

(4) Overall risk assessment

For urban large underground spaces that have carried out safety risk assessment at the design stage, the safety risk assessment conclusion at the design stage can be used as the basis for carrying out special risk assessment for construction safety, and there is no need to carry out overall risk assessment for construction safety. However, for urban large underground spaces with complex construction processes, many construction procedures, and many risk sources in the surrounding environment, it is recommended to carry out an overall risk assessment for construction safety as a supplement. The overall risk assessment for urban underground large-space construction mainly includes the following steps.

① Establish an overall indicator system for risk assessment.

The overall risk assessment for construction safety is a pre-set, global, macro, and generalized assessment, so the factors considered in the assessment are also overall, key, and generally do not need to consider factors related to the construction plan. The overall risk assessment indicator system can directly adopt the indicator system recommended in this book or adjust it based on actual conditions, but it should follow the basic principles of establishing an indicator system.

② Questionnaire Scoring.

Design an assessment questionnaire based on the risk assessment indicator system and distribute it to the evaluators. The assessment questionnaire includes basic project information, a list of risk sources, an indicator scores rating table, and an indicator weight rating table.

③ Overall Risk Assessment.

Collect the survey questionnaires, calculate the overall risk score based on the returned results, and determine the overall risk level according to the overall risk level standards. When calculating the overall risk score, the arithmetic mean of the scores of each questionnaire can be used, or the weighted average can be used considering the weight of the evaluators.

④ Preparation of the Overall Risk Assessment Report.

The overall risk assessment report includes the following content.

a. Preparation basis: including relevant laws, regulations, and system documents on risk management in the industry, relevant national and industry standards and norms, project approval documents, project feasibility study reports, engineering geological survey reports, preliminary design documents, field survey data, etc.
b. Project Overview.
c. Assessment process and methods.
d. Assessment conclusions and suggestions: including the overall risk level, objects suggested for specialized risk assessment, risk management plan suggestions, self-evaluation of assessment results, and explanation of remaining issues, etc.
e. Appendices: including the assessment calculation process, evaluator information table, etc.

⑤ Report Review.

Convene experts to review the risk assessment report. If the review is not passed, a re-assessment needs to be conducted based on expert opinions.

(5) Specialized Risk Assessment during the Construction Preparation Stage

For urban large underground space projects with a higher overall risk level (in principle, level II and above), a further specialized risk assessment should be carried out. The specialized risk assessment is the core of construction safety risk assessment, divided into assessment during the construction preparation stage and assessment during the construction process, aiming to estimate and track the possibility of specific risk events and risk losses, and to formulate risk response measures and special construction plans accordingly. The specialized risk assessment during the construction preparation stage mainly includes the following steps.

① Risk Assessment Unit Division.

The purpose of dividing the special risk assessment unit for urban large underground space construction safety is to sort out the assessment logic, clarify risk events, screen risk factors, and reduce the ambiguity of indicator scoring. The specialized risk assessment unit should be divided according to the sub-projects, construction procedures, surrounding environment, and geological conditions. The division of the assessment unit is aimed at facilitating risk assessment, and the following division methods can be specifically adopted.

a. For risk events related to the project itself (instability of open excavation foundation pit, base heave of open excavation, roof collapse of tunnel excavation, mud and water inrush in tunnel excavation), the risk assessment unit can be divided according to the sub-projects, with the key construction links (sub-projects or construction operations steps) within the assessment unit as the risk sources, select the corresponding major risk events as the assessment objects according to the characteristics of the risk sources, and list the environmental risk sources related to the assessment objects (such as rain and sewage pipes in the impact area) as risk factors.
b. For risk events related to environmental facilities (pipeline damage, ground building damage, underground building damage, connection openings causing excessive deformation or damage to existing structures, excavation unloading causing existing structures to float), risk assessment units can be divided according to the type of environmental facilities. The specific important environmental facilities within the assessment unit are taken as the source of risk, and the construction operations that will cause significant disturbances to the source of risk (such as excavation construction, connection openings, etc.) are listed as risk factors.
c. The spatial and temporal effects of urban large underground space construction are mainly reflected in the sequentiality of the construction process and the finiteness of the construction disturbance range. When dividing risk assessment units, this spatiotemporal effect can be considered. For different construction links that have a sequential relationship in time, or different construction parts that have a significant boundary in space, they can be treated as different risk assessment units or sources of risk.
d. For assessment units with higher importance, larger spatial scope, or higher construction danger (such as the main structure of the project), risk assessment units can be further subdivided according to project zoning, etc.

② Establish a special indicator system for risk assessment.

The object of construction safety special risk assessment is specific risk events, so the specialized risk assessment indicator system should be established according to the occurrence mechanism of different risk events. The specialized risk assessment indicator system is divided into static and dynamic indicator systems, the former is used for specialized risk assessment during the construction preparation stage, and the latter is used for specialized risk assessment during the construction process. The specialized risk assessment indicator system can directly adopt the indicator system recommended in this book or adjust it based on the actual situation, but it should follow the basic principles of establishing the indicator system.

③ Questionnaire scoring.

The specialized risk assessment questionnaire is similar to the overall risk assessment questionnaire, but the indicator system scoring table should give different risk assessment units separately.

④ Risk occurrence possibility assessment.

Calculate the risk occurrence possibility score, and determine the possibility level according to the standard of the possibility level of urban large underground space construction risk events.

⑤ Risk loss assessment.

Calculate the risk loss score according to the risk loss calculation formula, and determine the risk loss level according to the standard of the risk loss level of urban large underground space construction.

⑥ Determine the specialized risk level.

According to the risk event occurrence possibility level and risk loss level, determine the specialized risk level through the urban large underground space specialized risk grading standard (risk level matrix).

⑦ Formulate risk response measures.

After determining the specialized risk level, it is necessary to formulate corresponding risk response measures according to the risk acceptance criteria. Common risk response measures include risk rejection, risk disposal, risk transfer, and risk retention. According to the risk acceptance criteria commonly used in the field of civil engineering, when there is a risk event with a specialized risk level of I, construction cannot be carried out. It is necessary to adjust the construction plan and re-conduct the construction specialized risk assessment until the specialized risk level is reduced to level II or below.

⑧ Compile a specialized risk assessment report for the construction preparation stage.

The specialized risk assessment report for the construction preparation stage includes:

a. Basis for compilation. This includes legal regulations and institutional documents related to industry-specific risk management, relevant national and industry standards and norms, geological survey reports, preliminary design documents, construction design documents and review opinions, overall risk assessment results and pre-project risk assessment results, field investigation data, and field testing and monitoring data.

b. Project overview.
c. Assessment process and methods.
d. Assessment content. This includes a list of risk sources, division of risk assessment units, list of risk events, evaluation indicator system, detailed scoring table for the assessment, etc.
e. Assessment conclusions. This includes a summary of risk levels of the assessment objects, suggestions for risk pre-control measures, self-evaluation of assessment results, and explanation of remaining issues, etc.
f. Appendices. This includes the assessment calculation process, assessment personnel information table, etc.

When the overall risk assessment and the specialized risk assessment during the construction preparation phase are carried out simultaneously, their risk assessment reports can be compiled together.

⑨ Report review.

The requirements for the review of the specialized risk assessment report are the same as those for the overall risk assessment.

⑩ Establish a risk tracking table.

The function of the risk tracking table is to track risk events, risk factors, evaluation indicators, changes in risk sources, and the implementation time of risk pre-control measures taken during the process. It mainly includes the following content.

a. Initial status record: Mainly records the status during the risk assessment of the construction preparation phase, including information such as risk level, main risk control factors, risk evaluation indicator values, etc.
b. Current status record: Mainly records the stage status during the risk event tracking process, including information such as risk level, major risk source change records, changes in risk factors, changes in risk evaluation indicator values, etc.

2. Construction process assessment

(1) Conditions for dynamic risk assessment during the construction process

The construction process assessment is only for specialized risk assessment, also known as dynamic risk assessment during the construction process. It is a process of verifying and reassessing the existing risk assessment results based on the latest information collected during the construction period. If any of the following situations occur during the construction process that affect construction safety, a dynamic risk assessment during the construction process should be carried out.

① The emergence of new major risk sources or significant changes in existing risk sources. Possible new major risk sources include unexplored underground obstacles, poor geology, groundwater systems, etc. Significant changes in existing risk sources include the disappearance of risk sources, significant changes in risk source characteristics, or transformations in the nature of risk sources (such as general risk sources transforming into major risk sources under specific circumstances).

② The revealed geological and hydrological conditions are seriously inconsistent with the survey and design documents.

③ Major design changes and changes in construction techniques occur, including changes in excavation methods, changes in support parameters, changes in auxiliary methods, etc.

④ Disaster signs appear during the construction process or accident risks occur.

(2) Dynamic risk assessment process during the construction process

① Update project data and risk source information

Due to the great uncertainty in underground engineering construction, design changes are often made during the construction process due to differences between the actual revealed geological conditions and the preliminary survey data, and the difficulty of construction methods in meeting the predetermined quality control standards. Therefore, for major node projects, it is necessary to update the project data and risk source information before conducting dynamic assessments.

② Establish a dynamic evaluation indicator system

The specialized risk assessment dynamic indicator system can be expanded based on the static indicator system, or it can replace the corresponding static index with more timely real-time measurement indicators. Dynamic evaluation indicators are generally indicators related to construction monitoring projects, reflecting the real-time status of construction risks.

③ Update assessment index information

For dynamic indicators, such as support deformation, pipeline deformation, ground settlement, deformation of surrounding buildings, etc., it is necessary to obtain real-time status through monitoring instruments; the status of some static indicators may also change as the construction progress advances, so it is also necessary to update the relevant information.

④ Conduct dynamic risk assessment

Generate assessment questionnaires with the latest information, repeat questionnaire scoring, risk event occurrence probability evaluation, risk loss evaluation, and special risk grading steps to obtain the dynamic specialized risk level.

⑤ Dynamic risk warning

According to the dynamic specialized risk level, decide on countermeasures based on the specialized risk acceptance criteria. If a risk event occurs with a specialized risk level of I or a risk level that has risen by two levels or more compared to the previous assessment, an emergency plan procedure needs to be initiated, and the risk must be reduced to an acceptable level before construction can continue.

⑥ Compile a dynamic risk assessment report

In addition to the content in the same format as the specialized risk assessment report during the construction preparation stage, the construction process specialized risk dynamic assessment report should also include the following content:

a. Changes in major risk sources.
b. The reassessed risk level and calculation process.
c. Evaluation of the effect of existing risk control measures.
d. Suggestions for adjusted risk control measures.

3. Assessment flowchart

The safety risk assessment flowchart for urban large underground space construction, as shown in Fig. 7.2.

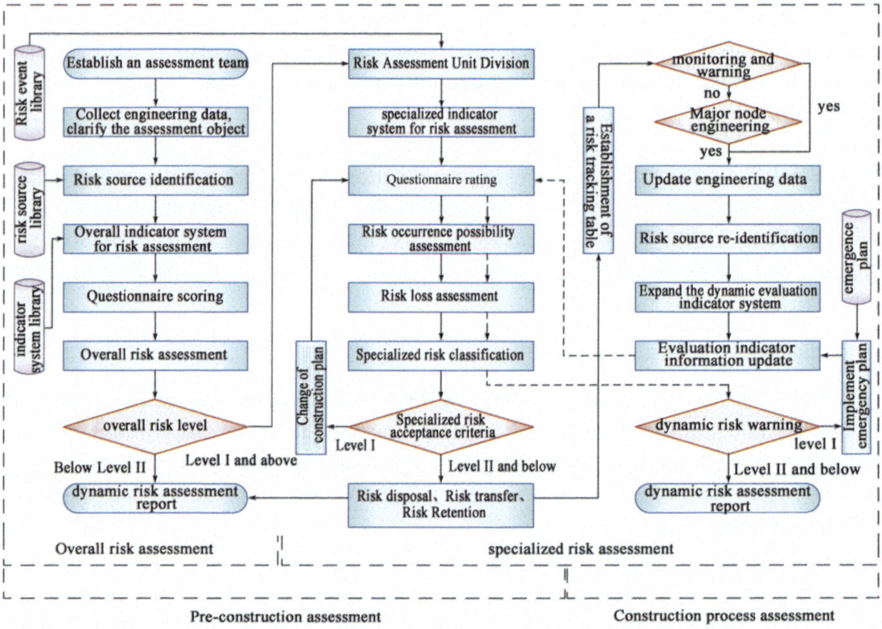

Fig. 7.2 Urban large underground space construction risk assessment flowchart

7.2 Introduction to the Indicator System Method

7.2.1 Indicator System Method

The indicator system method is a method that establishes an evaluation indicator system that reflects the characteristics of the object to be evaluated based on its main influencing factors, quantifies and grades the numerical range of each evaluation indicator, and considers the weight coefficients of each evaluation indicator to make a quantitative score judgment on the object to be evaluated.

1. Indicator and indicator system

An indicator is a numerical measure of a certain aspect of the state or characteristics of an object. A complete indicator generally consists of the indicator name and the indicator value, which reflect the characteristics of the material and quantity of the object. The indicator system (Indicator System) is a comprehensive measurement system composed of complex indicators targeting the same measurement object according to a certain logical structure. It is generally a hierarchical tree structure, but it can also be chain-like, network-like, or flat. The indicator system can carry out comprehensive measurement and quantitative evaluation of the multi-angle, multi-faceted, and multi-level features of the object, avoiding the one-sidedness and limitations of single indicator measurement, but it will also face problems such as conceptual contradictions between indicators, unclear boundaries, and inconsistent granularity. Therefore, the construction of the indicator system requires the designer to have rich and comprehensive relevant professional knowledge, clear logical analysis ability, and a certain art of simplifying problems. When using the indicator system, a "weight coefficient" is generally set for each indicator to represent the relative degree of impact or relative importance of each indicator on the measurement object.

2. The relationship between risk assessment indicators and risk factors

(1) Definition and characteristics of risk factors

Risk assessment indicators are concepts that are related to but different from risk factors. The definition and characteristics of risk factors are:

① Risk factors refer to factors that directly or indirectly affect the probability of risk occurrence or the factors that affect the consequences of the risk after it occurs. Its essence is still an event.

② Risk factors and risks can be converted into each other. For a certain event, if it is the direct cause of the loss, then it is a risk; under other conditions, if it is the indirect cause of the loss, then it is a risk factor.

③ The mechanism of risk formation is closely related to risk factors. Different types of risks have different risk factors.

④ Risk factors are an event, so they have their own probability of occurrence. After different risk factors occur, their contributions to risk vary.

(2) Characteristics of risk assessment indicators

① Risk assessment indicator is a measurable or evaluable quantity with a clear objective meaning.

② Risk assessment indicator is not an event, so they cannot be directly linked to the occurrence of risk. However, by determining the size of the risk assessment indicators, an evaluation and judgment of the nature and probability of risk factors can be formed, thereby determining the size of the risk.

③ There is a corresponding relationship between risk assessment indicators and risk factors, but this corresponding relationship is not necessarily one-to-one. One-to-many, many-to-one, or many-to-many are all possible situations.

④ Risk assessment indicator is established for risk assessment decisions. It should have clear value principles, value standards, and value methods.

Generally speaking, the determination of risk assessment indicators should be based on the analysis results of risk factors. Risk assessment indicators can be formulated by expanding, dividing, or merging and sorting the connotations of risk factors. To facilitate risk assessors to make decisions such as indicator weighting, risk assessment indicators should be divided into several levels according to logical levels to form an indicator system.

3. Indicator system method evaluation formula

The basic idea of the indicator system method evaluation is "weighted average", and its general evaluation formula is:

$$T = \sum_i \gamma_i X_i \tag{7.1}$$

In the formula, T is the evaluation target score; γ_i is the weight of the i-th evaluation indicator, and X_i is the score of the i-th evaluation indicator.

In the evaluation of indicator system method, the indicator score X_i and the evaluation target score T should have the same domain of definition and consistent logical mapping relationships. The scores of indicators and evaluation targets can generally adopt a percentage system, a five-point system, or a normalized score system, and it is stipulated that the higher the score, the more disadvantageous (advantageous) the status of the indicator or evaluation target. Under the above requirements, the weight of the indicator γ_i must satisfy the condition $\sum_i \gamma_i = 1$, and because the weight of the indicator represents the "relative importance" of the indicator, the weight of the indicator must be a positive number.

When the indicator system contains more than one level, the parent indicator score X_i should be calculated from the sub-indicator scores under it, the formula is:

$$\begin{cases} X_i = \sum_i \gamma_{ij} X_{ij} \\ X_{ii} = \sum_k \gamma_{ijk} X_{ijk} \\ \underbrace{X_{ij\ldots mn}}_{s\ level} = \sum_n \underbrace{\gamma_{ij\ldots mn}}_{(s+1)\ level} \underbrace{X_{ij\ldots mn}}_{(s+1)\ level} \end{cases} \tag{7.2}$$

In the formula, γ_{ij} and X_{ij} are respectively the weight and score of the j-th secondary indicator under the ith primary indicatoir. Similarly, $\underbrace{\gamma_{ij\ldots mn}}_{(s+1)\ level}$ and $\underbrace{X_{ij\ldots mn}}_{(s+1)\ level}$ are respectively the weight and score of the nth (s + 1)-level indicator under the s-level indicator $\underbrace{X_{ij\ldots m}}_{s\ level}$. The weights of the sub-indicators under each evaluation indicator also need to satisfy the conditions of summing to 1 and non-negativity.

When the indicator system has many levels, it is very cumbersome to calculate the indicator scores layer by layer according to formula (7.2), and formula (7.3) can be used at this time.

$$T = \sum_t \overline{\gamma_t \overline{X_t}} \tag{7.3}$$

In the formula, $\overline{\gamma_t}$ is the global weight of the t-th leaf indicator; $\overline{X_t}$ is the score of the t-th leaf indicator.

Here, the "leaf indicator" does not refer to the deepest level indicator in the indicator system, but refers to all evaluation indicators that do not contain sub-indicators. For example, the indicator system contains two primary indicators X_1 and X_2, and X_2 contains three secondary indicators X_{21}, X_{22}, and X_{23}, then the leaf indicators in the indicator system are X_1, X_{21}, X_{22}, and X_{23}. The global weight of the evaluation indicator refers to the weight of a certain indicator for the evaluation target. For the primary indicator, its indicator weight is the global weight; the global weight of the secondary and lower indicators can be calculated by its own indicator weight and the indicator weights of its parent indicators at all levels. Suppose the t-th leaf indicator is the nth (s + 1)-level indicator under the s-level indicator $\underbrace{X_{ij\ldots m}}_{s\ level}$, then its global weight can be calculated according to formula (7.4).

$$\overline{\gamma_t} = \gamma_i \gamma_{ij} \cdots \gamma_{ij\cdots m} \gamma_{ij\cdots mn} \tag{7.4}$$

4. Characteristics of the indicator system method

(1) Advantages of the indicator system method

① The calculation form is simple.

The basic calculation formula of the indicator system method only involves addition and multiplication, which can be realized through simple programs or manual calculations. It has a low threshold and is not prone to errors, and the analysis efficiency is high.

② The concept is clear.

The variables in the indicator system method only have indicator scores and weight coefficients. The indicator score reflects the absolute value of a certain aspect of the object, and the weight coefficient reflects the relative importance of the features. The boundary between the two is clear and they complement each other, which is in line with the analysis logic of most people.

③ High flexibility and strong operability.

The indicator system can be freely expanded or simplified according to the characteristics of the evaluation object, the coverage of the factors considered, and the requirements for the level of detail of the evaluation, and a unified, standard logic and process can be used to handle various problems.

(2) Disadvantages of the Indicator System Method

① The indicator system method requires a significant simplification of the characteristics of the evaluation object, making the evaluation results significantly influenced by the operator's subjectivity, and the dispersion between the evaluation results of different operators is large. Therefore, the application effect of the indicator system method is closely related to the quality of the indicator system, and it requires high experience, knowledge, and design level of the designer of the indicator system.

② The indicator system method uses a unified simple numerical measure for various characteristics of the evaluation object, and has significant limitations when dealing with some difficult-to-quantify and difficult-to-objectively-value characteristics. In addition, it is difficult to determine the quantification standards of indicator scores, which requires comprehensive research from multiple aspects.

③ The indicator system method assumes that the indicators are mutually independent, which is not universally appropriate in many practical problems. When there is a coupling or superposition relationship in the effect between indicators, it is difficult to get reasonable results using the indicator system method.

In summary, although the indicator system method has high requirements for the designer of the indicator system, it is very user-friendly and suitable for promotion in frontline engineering. Moreover, thanks to its flexibility, the indicator system method can consider the superposition of multiple risk sources and the coupling effect of multiple factors in the risk of urban large underground space construction by combining with other calculation methods or introducing correction factors, so it is suitable as a method for assessing the safety risk of urban large underground space construction.

7.2.2 Selection and Scoring Principles of Safety Risk Assessment Indicators for Urban Large Underground Space Construction

1. Principles for selecting safety risk assessment indicators for urban large underground space construction

The rationality of the risk assessment indicator system has a crucial impact on the accuracy of the risk assessment results and the difficulty of the risk assessment process. The following principles for selecting assessment indicators are very helpful for establishing a reasonable risk assessment indicator system.

(1) Principle of Science

In the process of establishing the indicator system, the basic scientific principles must be adhered to. The construction of urban large underground space is essentially a mechanical process, so the selection of indicators must comply with the mechanical mechanism of construction and cannot be chosen arbitrarily.

(2) Principle of Practicality

The establishment of the indicator system is not only to obtain data-based evaluation results, but also to facilitate relevant decision-making departments to take appropriate measures based on this, or to provide indicators for on-site monitoring. Therefore, the selection of the indicator system should have certain economic, social, and management practical significance.

(3) Principle of Conciseness

The purpose of selecting the indicator system is to truly reflect the risk status of the project. While ensuring the completeness of the information, the structure of the indicators should be simplified as much as possible to avoid bias in the evaluation results caused by repeated information. Therefore, when selecting indicators at different criterion levels, representative indicators should be selected purposefully.

(4) Principle of Operability

When determining the quantification standards of evaluation indicators, explanations should be made on the specific meaning of the indicator scores, the principles, basis, and methods of value determination, as far as possible to avoid the situation where the evaluator cannot make an evaluation based on objective facts due to unclear concepts.

(5) Principle of Diversity

To best reflect the actual engineering conditions, the selection of indicators should not be limited to a single evaluation indicator, but should consider the problem from a diversified perspective. When conducting risk assessments, engineering has a high degree of complexity and uncertainty. Due to the limitations of personal experience and the variability of engineering, it is difficult to use a perfect indicator to reflect multiple aspects of the situation. At this time, flexible selection should be made from multiple angle indicators.

2. Principles of scoring for urban large underground space construction safety risk assessment indicators

When using the indicator system and quantification standards given in this book to conduct safety risk assessments for urban large underground space construction, the following strategies should be adopted for the selection of assessment indicator scores.

(1) The score of the evaluation indicator can be selected according to the basic idea of "except for the upper limit of 100 points, the lower limit but not the upper limit of the range of values corresponding to the scoring basis for each grade is desirable."
(2) When the assessment indicator meets the requirements of two or more grading bases at the same time, or when it is impossible to clearly determine, the standard value of the larger grade should be taken.
(3) When the actual state of the assessment object cannot be judged based on the situation given in the scoring basis, the evaluator can refer to the scoring standards given in this study to determine the value based on their knowledge and experience.
(4) When the actual state of the assessment object is far inferior (superior) to the situation given in the lowest (highest) grade scoring basis, the indicator score can be 0 (100).
(5) When conditions permit multiple evaluators to conduct assessments, group decision-making should be used as much as possible for score selection. The final score can be the arithmetic average of the scores given by each evaluator, or the weighted average considering expert weights.

7.2.3 Method of Determining Weight Coefficients

The weight of the assessment indicators reflects the contribution of the risk factors corresponding to each risk assessment indicator to the risk. The weight of the assessment indicators is related to the specific risk formation mechanism, so the weight of the assessment indicators should be given separately for different assessment objects (risk events). The weight of the assessment indicators can be determined by various subjective or objective weighting methods. According to the characteristics of the urban large underground space construction risk assessment indicator system, it is recommended to use the Analytic Hierarchy Process (AHP) for weighting.

The Analytic Hierarchy Process belongs to the category of operations research. The basic idea is to first propose the main influencing factors of the complex system, and construct a hierarchical analysis structure model according to their interrelationships and affiliations, then judge the relative importance of each factor, and handle the problem through sorting calculations. The implementation process of the Analytic Hierarchy Process is:

(1) Establish a hierarchical structure model

The hierarchical structure model includes an evaluation goal and several evaluation indicators combined in a tree structure, and the tree structure of the evaluation indicators reflects the subordinate relationship and logical level between the indicators. For the possibility assessment of urban large underground space construction risk, the evaluation goal is the possibility of a certain risk event, and the tree structure of the indicators can be established by directly referring to the hierarchical relationship of the urban large underground space construction risk evaluation indicator system.

(2) Construct a comparison matrix

The Analytic Hierarchy Process requires the construction of a comparison matrix. The comparison matrix is based on a certain factor from the previous level as the evaluation criterion, and compares the factors at this level. All factors of this level are placed in the first row of the matrix, and then transposed to become the factors in the first column. The previous level indicator, that is, the evaluation criterion, is located in the upper left corner of the comparison matrix. The comparison matrix compares each factor at this level pairwise, not all at once, and when comparing factors at the same level, the same standard should be used for measurement.

Psychologists believe that the number of factors in pairwise comparison should not exceed 9, that is, each level should not exceed 9 factors. Use a 1–9 ratio scale evaluation set (Table 7.1), and form a comparison matrix according to the meaning given by the evaluation set.

The comparison matrix is usually described by formula (7.5).

Table 7.1 Proportional scale evaluation set

Scale	Implication
1	Indicates that two factors are equally important compared to each other
3	Indicates that compared to two factors, one factor is slightly more important than the other
5	Indicates that compared to two factors, one factor is significantly more important than the other
7	Indicates that compared to two factors, one factor is more important than the other
9	Indicates that compared to two factors, one factor is extremely important than the other
2, 4, 6, 8	The median of the two adjacent judgments mentioned above
Count backwards	Indicates the order of two factors swapping

$$A = \begin{pmatrix} a_{11} & \cdots & a_{1n} \\ \vdots & \ddots & \vdots \\ a_{n1} & \cdots & a_{nn} \end{pmatrix} \tag{7.5}$$

In the comparison matrix A, a_{ij} represents the importance of factor a_i relative to a_j. Matrix A is a positive reciprocal inverse matrix, where $a_{ii} = 1$, $a_{ij} > 0$, $a_{ij} = 1/a_{ji}$.

(3) Level-wise sorting and consistency test

The eigenvector W corresponding to the maximum eigenvalue λ_{max} of the comparison matrix A, after normalization, is the weight vector (relative weight vector) of the relative importance of the factors at this level to a certain factor at the previous level. This process is called level-wise sorting.

The consistency test is to ensure the logical consistency of the comparison matrix. The consistency index defined by the founder of the Analytic Hierarchy Process, SAATY, is:

$$CI = \frac{\lambda_{max} - n}{n - 1} \tag{7.6}$$

If CI = 0, there is complete consistency; if CI is close to 0, there is satisfactory consistency; the larger the CI, the more serious the inconsistency.

Because the satisfaction of matrices of different orders needs to be distinguished, a random consistency index RI is also introduced. SAATY randomly constructed 500 pairwise comparison matrices, calculated their consistency indices and then averaged them. The final calculation results are shown in Table 7.2.

The random consistency ratio is defined as:

$$CR = \frac{CI}{RI} \tag{7.7}$$

When $CR < 0.1$, it is considered that the comparison matrix has satisfactory consistency. Otherwise, the comparison matrix needs to be adjusted to achieve satisfactory consistency.

(4) Hierarchical total sorting and consistency test

The above are the weight vectors of a group of factors for a certain factor in the previous layer. We ultimately hope to get the sorting weights of all factors (especially

Table 7.2 Average random consistency index (RI) table

Matrix order	1	2	3	4	5	6	7	8
RI	0	0	0.52	0.89	1.12	1.26	1.36	1.41
Matrix order	9	10	11	12	13	14	15	
RI	1.46	1.49	1.52	1.54	1.56	1.58	1.59	

the factors in the lowest layer) for the target, so as to make selection judgments and choices.

For multi-level calculations, the weight calculation of each layer relative to the target layer (the top layer) uses a top-down method, layer by layer synthesis.

Assuming that the weight vector $W^{k-1} = \left(w_1^{k-1}, w_2^{k-1} \ldots, w_m^{k-1}\right)^T$ of the m factors in the k−1 layer relative to the total target has been calculated, the single sorting weight vector of the n elements in the k layer for the jth factor in the previous layer (the k−1 layer) is $P_j^k = \left(P_{1j}^k, P_{2j}^k \ldots, P_{nj}^k\right)^T$, where the weight of the factor not dominated by j is zero. Let $P^k = \left(P_1^k, P_2^k \ldots, P_m^k\right)$, then the combined sorting weight vector of the k layer factors for the total target is:

$$W^k = \left(w_1^k, w_2^k \ldots, w_n^k\right)^T = P^k W^{k-1} = \begin{pmatrix} p_{11}^k & \cdots & p_{1m}^k \\ \vdots & \ddots & \vdots \\ p_{n1}^k & \cdots & p_{nm}^k \end{pmatrix} \begin{pmatrix} w_1^{k-1} \\ w_2^{k-1} \\ w_m^{k-1} \end{pmatrix} \tag{7.8}$$

When performing the total sorting of factors, its consistency must also be tested. Assuming that the CI_j^k and RI_j^k of the k layer factors with the jth factor of the k−1 layer as the criterion have been calculated, j = 1, 2,…,m, then the comprehensive test index of the k layer is:

$$\begin{aligned} CI^k &= \left(CI_1^k, CI_2^k, \cdots, CI_m^k\right)w^{k-1} \\ RI^k &= \left(RI_1^k, RI_2^k, \cdots, RI_m^k\right)w^{k-1} \\ CR^k &= \frac{CI^k}{CR^k} \end{aligned} \tag{7.9}$$

When $CR^k < 0.1$, it is considered that the overall consistency of the comparison matrix is acceptable.

7.3 Overall Risk Assessment

7.3.1 Indicator System and Quantitative Standards

Based on the concept of overall risk level, the overall risk assessment of construction is a pre-set, global, macro, and generalized assessment, so the factors considered in the assessment are also overall and key, and generally do not need to consider the factors of construction plans.

Therefore, the overall risk assessment should mainly consider factors such as project scale, geology, environment, and technical difficulty. The indicator system and quantitative standards for the overall risk assessment of urban large underground space construction are shown in Table 7.3.

Table 7.3 Overall risk assessment indicator system

First-level indicators	Secondary and tertiary indicators	Scoring basis	Value ranges
Design conditions G_1	Engineering scale G_{11}	The depth of open excavation exceeds 35 m, the cross-sectional area of concealed excavation exceeds 650 m^2, the subway station with five floors or more, or an equivalent scale	80–100
		Open excavation depth of 30–35 m, concealed excavation section area of 500–650 m^2, four story subway station or equivalent scale	60–80
		Open excavation depth of 25–30 m, single span width of concealed excavation greater than 18 m, cross-sectional area of concealed excavation of 300–50 m^2, three story subway station or equivalent scale	40–60
Geological condition G_2	Engineering geology G21	Special rock and soil (artificial fill, soft soil, water-rich sand layer, etc.)	80–100
		poor geological condition (karst, fault, etc.)	50–80
		Complex strata (composite strata, etc.)	20–50
	Hydrogeology G22	confined water	80–100
		Phreatic water, fissure water, karst water	60–80
		perched water	40–60
Environmental condition G_3	Location of ground buildings and structures G_{31}	Located in the main affected area	80–100
		Located in the secondary affected area	60–80
		Located in the general influence area	40–60
	Location of underground structures G_{32}	Located in the main affected area	80–100
		Located in the secondary affected area	60–80
		Located in the general influence area	40–60
	Site of road G_{33}	Located in the main affected area	80–100
		Located in the secondary affected area	60–80
		Located in the general influence area	40–60

(continued)

Table 7.3 (continued)

First-level indicators	Secondary and tertiary indicators		Scoring basis	Value ranges
	Pipeline G_{34}	Pipeline position G341	Located in the main affected area	80–100
			Located in the secondary affected area	60–80
			Located in the general influence area	40–60
		Pipeline vulnerability G342	Vulnerable materials such as masonry, asbestos cement or ceramic	80–100
			Smaller strength materials such as plastics	60–80
			Grey cast iron and other materials with low reliability	40–60
			Steel tube, concrete, ball milled cast iron and other high reliability materials	20–40
		Duration of service G_{343}	Over 30 years	90–100
			5–30 years	70–90
			Less than 5 years	50–70
	Surface water body G_{35}		Distance from the project ≤ 30 m	80–100
			The distance from the project is more than 30 m and the strong permeable stratum	50–80
			The distance from the project is more than 30 m and the stratum is weakly permeable	20–50
Construction condition G_4	Construction technology maturity G_{41}		Application of new technologies, new processes, new materials and new equipment	70–100
			The process is mature and has related applications	20–70

7.3.2 Assessment Method

1. Overall risk score calculation method

The overall risk score of urban large underground space construction is calculated according to formula (7.10).

$$G = K \sum_{i=1}^{4} w_i G_i \tag{7.10}$$

In the formula, G is the overall risk score, when $G \geq 100$, it is calculated as 100; K is the coefficient of the type of underground large space, 1.0 for open excavation single underground large space, 1.1 for tunneling single underground large space, and 1.2 for large underground space by extension construction; ω_i is the weight of the first-level indicator of overall risk assessment; G_i is the score of the first-level indicator of overall risk assessment. The scores of the first-level indicators of geological conditions, design conditions, and construction conditions are calculated according to formula (7.11).

$$G_i = \sum\nolimits_j w_{ij} G_{ij} \tag{7.11}$$

In the formula, w_{ij} is the weight of the second-level indicator of overall risk assessment; G_{ij} is the score of the second-level indicator of overall risk assessment.

Each second-level indicator under the first-level indicator of environmental conditions represents the impact of different types of environmental risk sources on the overall risk. When multiple types of environmental risk sources coexist, these environmental risk sources have a superimposed effect on the overall risk, and it is not appropriate to calculate the score of the first-level indicator by the weighted sum method. Therefore, the score of the first-level indicator of environmental conditions is calculated according to formula (7.12).

$$G_3 = \max(G_{3i}) + 0.1\left[\sum_{i=1}^{5} G_{3i} - \max(G_{3i})\right]; i = 1, 2, \cdots, 5 \tag{7.12}$$

2. Precautions when scoring the overall risk indicators

(1) When scoring the second-level indicator "Project Scale G_{11}", if the project scale of the assessment object is not suitable for measurement by open excavation depth, cross-sectional area of undercuts or station layers, it can be scored according to the excavation volume (divided into three levels with 100000 m^3 and 150000 m^3 as the boundary).

(2) For the scoring of the second-level indicator "Pipeline G_{34}", one of the following methods can be chosen.

① Take the pipeline with the highest expected danger as the representative and score the subordinate third-level indicators. The determination of expected danger can be based on the proximity of the pipeline to the project, the cross-sectional size of the pipeline, the fragility of the pipeline, or a combination of these aspects.

② Score the subordinate third-level indicators for all pipelines in the project impact area separately, and calculate the second-level indicator "Pipeline G_{34}". The score is determined by the arithmetic mean or weighted average of the scores of each pipeline as the final score.

7.3.3 Overall Risk Grading Standards

The overall risk level of urban large underground space construction is divided into four levels. After obtaining the overall risk score, the overall risk level is determined according to Table 7.4.

The conclusion of the overall risk assessment of urban large underground space construction safety is the basis for judging the necessity of specialized risk assessment and formulating specialized risk assessment plans, and can provide decision support for personnel arrangement and resource allocation of related units. In principle, urban large underground spaces with an overall risk level of II and above are in a relatively unfavorable situation in one of the three aspects of project scale, geological conditions, and surrounding environment, and need to carry out more detailed specialized risk assessments. The way to carry out specialized risk assessment can refer to Table 7.5.

Table 7.4 Overall risk level standard

Overall risk level	I	II	III	IV
Risk level description	Very high risk	High risk	Intermediate risk	Low risk
Overall risk score	$75 \leq G \leq 100$	$50 \leq G < 75$	$25 \leq G < 50$	$0 \leq G < 25$

Table 7.5 The relationship between the overall risk level and the specialized risk assessment scheme

Overall risk level	Specialized risk assessment program
I	Specialized risk assessment must be carried out and risk assessment units should be divided
II	Specialized risk assessment should be carried out, and risk assessment units can be divided
III	Specialized risk assessment should be carried out
IV	Specialized risk assessment can be carried out

7.4 Specialized Risk Assessment

7.4.1 Evaluation Indicator System and Quantitative Standards for the Possibility of Open Excavation Construction Risk Events

1. Indicator system and quantitative standards

The indicator system and quantitative standards for the risk assessment of open excavation construction in urban large underground spaces are shown in Tables 7.6, 7.7, 7.8, 7.9, and 7.10.

2. Points to note when scoring

(1) The scoring basis for the secondary indicator "Type of pit retaining structure R_{12}" of the open excavation construction pit instability risk assessment only lists the types of pits retaining structures that may be used in urban large underground space projects.
(2) In the scoring basis of the secondary index "Pit support system R_{13}" of the open excavation construction pit instability risk assessment, anchor rods (cables) are not strictly part of the pit support system, but because their function is similar to that of pit support, they are listed.
(3) In the scoring basis of the secondary index "Convergence of work processes R_{51}", the tightness of the convergence of work processes refers to the timeliness of the connection between the support and excavation process, the timeliness of sealing the bottom after excavation to the bottom of the pit, and the timeliness of the permanent structure construction after the support is removed.
(4) In the scoring basis of the secondary indices "Horizontal displacement of retaining structure R_{52}", "Pipeline settlement R_{52}", "Building differential settlement R_{52}", "Building horizontal displacement R_{52}", the displacement and settlement control values can refer to the "Code for monitoring measurement of urban rail transit engineering" (GB 50,911-2013), "Code for Design of Building Foundation" (GB50007-2011), "Technical standard for monitoring of building excavation engineering" (GB50497-2019), etc., or refer to the design documents for determination.

Table 7.6 Evaluation indicator system of pit instability possibility in open-cut construction

First-level indicators	Secondary indicators	Scoring basis	Value ranges
Design conditions R_1	Depth of foundation pit R_{11}	Depth over 35 m	80–100
		The depth is 30–35 m	60–80
		The depth is 25–30 m	40–60
	Type of foundation pit retaining structure R_{12}	Row of piles	80–100
		Interlocking pile	60–80
		Diaphragm wall	40–60
	The foundation supporting system R_{13}	Steel pipe support, steel support, anchor (cable)	80–100
		Steel tube, reinforced concrete composite support	60–80
		Reinforced concrete support	40–60
Geological condition R_2	Engineering geology R_{21}	Artificial fill, soft soil, water-rich sand layer, liquefiable stratum, etc.	80–100
		Weathered rock, composite strata, karst, goaf, water sac, etc.	60–80
		Pebble layer, bedrock bulge, weathered deep groove, fault zone, ground fissure, etc.	40–60
		Collapsible loess, expansive soil, hard rock veins, etc.	20–40
	HydrogeologyR_{22}	Confined water	80–100
		Phreatic water, fissure water, karst water	60–80
		Perched water	40–60
Environmental condition R_3	Pipe diameterR_{31}	More than 2000 mm (including)	80–100
		600 (including)–2000 mm	60–80
		200 (including)–600 mm	40–60
		Under 200 mm	20–40
	Pipeline statusR_{32}	Pipeline damage, obvious leakage	80–100
		There are cracks and leakage signs in the pipeline	60–80
		The pipeline is good, no sign of leakage	40–60
	Meteorological condition R_{33}	Rainstorm is expected during the excavation period of the foundation pit	80–100
		Heavy rain is expected to occur during the excavation period of foundation pit	60–80
		Moderate rain is expected to occur during the excavation period of foundation pit	40–60
		Dry season foundation pit excavation	0–40

(continued)

Table 7.6 (continued)

First-level indicators	Secondary indicators	Scoring basis	Value ranges
Construction condition R_4	Pit bottom reinforcement R_{41}	No reinforcement measures	70–100
		There are reinforcement measures	40–70
	Ground de-watering R_{42}	No precipitation measures	70–100
		There are precipitation measures	40–70
	Soil excavation R_{43}	Whole sequential excavation	70–100
		Partition block interval excavation	20–70
	Rock excavation R_{44}	Blasting excavation	60–100
		Mechanical equivalent of light excavation	20–60
	Process joining R_{45}	Process connection is not close	80–100
		The process connection is basically close	40–80
		The process is closely connected	0–40
	Horizontal displacement of retaining structure R_{46}	$U > 0.85U_0$ or $u > 3$ mm/d	80–100
		$0.7U_0 \leq U \leq 0.85U_0$ or 2 mm/d $\leq u \leq 3$ mm/d	40–80
		$U < 0.7U_0$ or $u < 2$ mm/d	0–40
	Foundation pit soaking situation R_{47}	Long time soaking	60–100
		Short time soaking	20–60
		Not subject to immersion	0–20
	Pit edge overload R_{48}	A large amount of overload at the edge of the pit	60–100
		A small overload at the edge of the pit	20–60
		No pit-side overload	0–20
	Reducing and bailing effect R_{49}	Poor effect	80–100
		Common effect	40–80
		Good effect	20–40

Note U-actual displacement value; U_0—allowed displacement control value; u—displacement rate

7.4.2 *Risk Assessment Indicator System and Quantitative Standards for the Possibility of Occurrence of Underground Construction Events*

1. Indicator system and quantitative standards

The risk assessment indicator system and quantitative standards for urban large underground space excavation can be found in Tables 7.11, 7.12, 7.13, 7.14, and 7.15.

Table 7.7 Evaluation indicator system of possibility of base heave in open-cut construction

First-level indicators	Secondary indicators	Scoring basis	Value ranges
Geological condition R_1	Engineering geology R_{11}	Soft soil, water-rich sand layer, liquefiable stratum, etc.	80–100
		Water-rich karst, goaf, water sac, etc.	60–80
		Weathered deep troughs, fault zones, ground fissures, etc.	40–60
	Hydrogeology R_{12}	Confined water	80–100
		Karst water	60–80
		Fissure water	40–60
Environmental condition R_2	Surface water body R_{21}	Distance from the project $\leq$ 30 m	80–100
		The distance from the project is more than 30 m and the strong permeable stratum	50–80
		The distance from the project is more than 30 m and the stratum is weakly permeable	20–50
Construction condition R_3	Ground de-watering R_{31}	No precipitation measures	60–100
		There are precipitation measures	20–60
	Base water insulation R_{32}	No water-proof measures	70–100
		There are water-proof measures	40–70
	Treatment effect of confined water R_{33}	Uncut confined water	80–100
		Partially cut off the confined water	40–80
		Completely cut off confined water	20–40
	Base reinforcement treatment R_{34}	No base reinforcement treatment is done	80–100
		The effect of base reinforcement is poor	40–80
		The base reinforcement treatment effect is good	20–40

2. Points to note when scoring

(1) For the case of small pilot tunnel group excavation, in the risk assessment indicator system for underground construction roof collapse, the secondary indicator "depth-to-span ratio R_{11}" can be calculated based on the depth-to-span ratio of single tunnel; in the risk assessment indicator system for pipeline damage, surface building damage and underground building damage, the secondary indicator "span ratio R_{11}" can be calculated based on the total span of the pilot tunnel group, and adjusted downwards on this basis.

(2) In the scoring basis of the secondary indicator "advance reinforcement R_{41}" for the risk assessment of underground construction roof collapse, advance small guide pipe grouting, surface grouting, freezing, horizontal rotary spraying, pipe shed and other auxiliary methods are all considered as advance reinforcement.

Table 7.8 Evaluation indicator system of the possibility of pipeline damage in open-cut construction

First-level indicators	Secondary indicators	Scoring basis	Value ranges
Environmental condition R_1	Pipeline vulnerability R_{11}	Masonry, asbestos cement or ceramic materials	80–100
		Smaller strength materials such as plastics	60–80
		Grey cast iron and other materials with low reliability	40–60
		Steel tube, concrete, ball milled cast iron and other high reliability materials	20–40
	Connection form of pipeline R_{12}	Sockets, flat mouth, etc.	80–100
		Groove and tongue, etc.	60–80
		Flange or welding, etc.	40–60
	Pipeline position R_{13}	Located in the main affected area	80–100
		Located in the secondary affected area	60–80
		Located in the general influence area	40–60
	Service life R_{14}	Over 30 years	90–100
		5–30 years	70–90
		Less than 5 years	50–70
Design conditions R_2	Depth of foundation pit R_{21}	Depth over 35 m	80–100
		The depth is 30–35 m	60–80
		The depth is 25–30 m	40–60
Geological condition R_3	Engineering geology R_{31}	Artificial fill, soft soil, water-rich sand layer, pebble layer, etc.	80–100
		Weathered rock, composite strata, karst, goaf, etc.	60–80
		Bedrock uplift, weathered deep trough, fault zone, ground fissure, etc.	40–60
		Collapsible loess, expansive soil, hard rock veins, etc.	20–40
	Hydrogeology R_{32}	Confined water	80–100
		Phreatic water, fissure water, karst water	60–80
		Perched water	40–60
Construction condition R_4	Pipeline reinforcement R_{41}	No special reinforcement	60–100
		There is special reinforcement	20–60
	Ground de-watering R_{42}	No precipitation measures	60–100
		There are precipitation measures	20–60
	Soil excavation R_{43}	Whole sequential excavation	70–100
		Partition block interval excavation	20–70

(continued)

Table 7.8 (continued)

First-level indicators	Secondary indicators	Scoring basis	Value ranges
	Rock excavation R_{44}	Blasting excavation	60–100
		Mechanical equivalent of light excavation	20–60
	Process joining R_{45}	Process connection is not close	80–100
		The process connection is basically close	40–80
		The process is closely connected	0–40
	Pipeline settlement R_{46}	$S > 0.85S_0$ or $s > 2$ mm/d	80–100
		$0.7S_0 \leq S \leq 0.85S_0$ or 1 mm/d $\leq s \leq 2$ mm/d	40–80
		$S < 0.7S_0$ or $s < 1$ mm/d	0–40

Note S—actual settlement value; S_0—allowed settlement control value; s—settlement rate

(3) For the risk assessment of underground construction roof collapse, when the advance reinforcement method is freezing, the secondary indicator "ground de-watering R_{42}" should not be used.
(4) In the scoring basis of the secondary indicator "process connection R_{51}", the tightness of process connection refers to the timeliness of sealing or reinforcing the face after excavation, the timeliness of support construction after excavation, and the timeliness of support closure.
(5) In the scoring basis of the secondary indicators "surrounding rock deformation R_{52}", "pipeline settlement R_{52}", "building differential settlement R_{52}", "building settlement R_{52}", the deformation and settlement control values can refer to the "Code for monitoring measurement of urban rail transit engineering" (GB 50,911-2013), "Code for Design of Building Foundation" (GB50007-2011), etc., or refer to the design documents for determination.

7.4.3 Evaluation Indicator System and Quantitative Standards for the Possibility of Occurrence of Risk Events in Expansion Construction

1. Indicator system and quantitative standards

The risk assessment indicator system and quantitative standards for the construction of urban large underground spaces are shown in Tables 7.16 and 7.17.

2. Points to note when scoring

Construction of expansion projects may involve multiple excavations that repeatedly disturb environmental risk sources (pipelines, surface structures, underground structures). The following methods can be considered.

Table 7.9 Evaluation indicator system of the possibility of damage to ground structures in open-cut construction

First-level indicators	Secondary indicators	Scoring basis	Value ranges
Environmental condition R_1	Building height, bridge size, road grade R_{11}	Super high-rise, super large bridge, expressway	80–100
		High-rise, bridge, main road	60–80
		Small high-rise, middle bridge, secondary trunk road	40–60
		Multi-layer, small bridge, branch	20–40
	Foundation base R_{12}	Independent foundation, strip foundation, well format foundation	80–100
		Raft foundation, box foundation	60–80
		Pile foundation	40–60
	Location of building structures R_{13}	Located in the main affected area	80–100
		Located in the secondary affected area	60–80
		Located in the general influence area	40–60
Design condition R_2	Depth of foundation pit R_{21}	Depth over 35 m	80–100
		The depth is 30–35 m	60–80
		The depth is 25–30 m	40–60
Geological condition R_3	Engineering geology R_{31}	Artificial fill, soft soil, water-rich sand layer, pebble layer, etc.	80–100
		Weathered rock, composite strata, karst, goaf, etc.	60–80
		Bedrock uplift, weathered deep trough, fault zone, ground fissure, etc.	40–60
		Collapsible loess, expansive soil, hard rock veins, etc.	20–40
	Hydrogeology R_{32}	Confined water	80–100
		Phreatic water, fissure water, karst water	60–80
		Perched water	40–60
Construction condition R_4	Pipeline reinforcement R_{41}	No special reinforcement	60–100
		There is special reinforcement	20–60
	Ground de-watering R_{42}	No precipitation measures	60–100
		There are precipitation measures	20–60
	Soil excavation R_{43}	Whole sequential excavation	70–100
		Partition block interval excavation	20–70
	Rock excavation R_{44}	Blasting excavation	60–100
		Mechanical equivalent of light excavation	20–60

(continued)

Table 7.9 (continued)

First-level indicators	Secondary indicators	Scoring basis	Value ranges
	Process joining R_{45}	Process connection is not close	80–100
		The process connection is basically close	40–80
		The process is closely connected	0–40
	Differential settlement of structures R_{46}	$S_D > 0.002L$	80–100
		$0.001L \leq S_D \leq 0.002L$	40–80
		$S_D < 0.001L$	0–40

Note S_D -actual differential settlement value; L-horizontal distance between differential settlement measurement points

(1) If the disturbed areas are different each time, the risk of environmental risk source damage can be considered based on the maximum risk in a single disturbance.
(2) If the disturbed areas are the same each time, but the disturbances do not occur simultaneously, and the adverse effects of the first disturbance do not significantly overlap with the effects of the subsequent disturbance (such as the impact of vibration), the risk of environmental risk source damage can be considered based on the maximum risk in a single disturbance.
(3) If the disturbed areas are the same each time, but the disturbances do not occur simultaneously, and the adverse effects of the first disturbance significantly overlap with the effects of the subsequent disturbance (such as the impact of displacement), the cumulative effect of multiple disturbances can be considered in the assessment of the risk of environmental risk source damage.
(4) If the disturbed areas are the same each time, and the disturbances occur almost simultaneously, the cumulative effect of multiple disturbances can be considered in the assessment of the risk of environmental risk source damage.
(5) The cumulative effect of multiple disturbances can be considered in the following ways:

① When the same construction method is used for each disturbance source (open or closed excavation), the evaluation indicators related to the disturbance source (such as "pipeline position R_{13}", "structure position R_{13}", "soil excavation R_{43}") can be combined, and then the possibility score of environmental risk source damage risk can be calculated according to the corresponding formula.

② When different construction methods are used for multiple disturbance sources, the secondary indicators "cross-section size R_{21}", "depth-to-span ratio R_{22}", "pit depth R_{21}" cannot be directly combined. Each disturbance source must first calculate the primary indicator "excavation cross-section condition R_2" and "pit condition R_2" scores, and then combine the above primary indicators.

Table 7.10 Evaluation indicator system of the possibility of damage to underground structures in open-cut construction

First-level indicators	Secondary indicators	Scoring basis	Value ranges
Environmental conditionR_1	Dimensions of underground structures R_{11}	Large basement, large section tunnel (pipe gallery)	80–100
		Medium-sized basement, medium-section tunnel (pipe gallery)	60–80
		Small basement, small section tunnel (pipe gallery)	40–60
	Structural form R_{12}	Brick-concrete basement, prefabricated assembled tunnel	80–100
		Frame structure basement, composite lining tunnel	60–80
	Location of underground structures R_{13}	Located in the main affected area	80–100
		Located in the secondary affected area	60–80
		Located in the general influence area	40–60
Design condition R_2	Depth of foundation pit R_{21}	Depth over 35 m	80–100
		The depth is 30–35 m	60–80
		The depth is 25–30 m	40–60
Geological condition R_3	Engineering geology R_{31}	Artificial fill, soft soil, water-rich sand layer, liquefiable soil, etc.	80–100
		Weathered rock, composite strata, karst, goaf, water sac etc.	60–80
		Bedrock uplift, weathered deep trough, fault zone, ground fissure, etc.	40–60
		Collapsible loess, expansive soil, hard rock veins, etc.	20–40
	Hydrogeology R_{32}	Confined water	80–100
		Phreatic water, fissure water, karst water	60–80
		Perched water	40–60
Construction condition R_4	Reinforcement of buildings and structures R_{41}	No special reinforcement	60–100
		There is special reinforcement	20–60
	Ground de-watering R_{42}	No precipitation measures	60–100
		There are precipitation measures	20–60
	Soil excavation R_{43}	Whole sequential excavation	70–100
		Partition block interval excavation	20–70
	Rock excavation R_{44}	Blasting excavation	60–100
		Mechanical equivalent of light excavation	20–60
	Convergence of work processes R_{45}	Convergence of work processes is not close	80–100

(continued)

Table 7.10 (continued)

First-level indicators	Secondary indicators	Scoring basis	Value ranges
		Convergence of work processes is basically close	40–80
		Convergence of work processes is closely connected	0–40
	Horizontal displacement of building structures R_{46}	$U > 0.85U_0$ or $u > 3$ mm/d	80–100
		$0.7U_0 \leq U \leq 0.85U_0$ or 2 mm/d $\leq u \leq$ 3 mm/d	40–80
		$U < 0.7U_0$ or $u < 2$ mm/d	0–40

Note U—actual displacement value; U_0—allowed displacement control value; u—displacement rate

7.4.4 Risk Event Occurrence Likelihood Assessment

The likelihood score of a risk event occurring during the construction of urban large underground spaces is calculated according to formula (7.13).

$$R = \sum_{i=1}^{n} \gamma_i R_i \tag{7.13}$$

In the formula, R is the likelihood score of a risk event occurring. When $R \geq 100$, it is calculated as 100; γ_i is the weight of the primary indicator; R_i is the score of the primary indicator; n is the number of primary indicators involved in the calculation.

The primary indicator scores for the likelihood assessment of a risk event occurring is calculated according to formula (7.14).

$$R_i = \frac{\sum_{j=1}^{m_i} \gamma_{ij} R_{ij}}{\sum_{j=1}^{m_i} \gamma_{ij}} \tag{7.14}$$

In the formula, γ_{ij} is the weight of the j-th secondary indicator under the primary indicator Ri; R_{ij} is the score of the j-th secondary indicator under the primary indicator R_i; m_i is the primary indicator R_i The number of secondary indicators involved in the calculation is as follows.

The possibility of risk events occurring during the construction of urban large underground spaces can consider the amplifying or inhibiting effects of risk factors' coupling effects on risks. When considering the coupling effects of risk factors, the score of the possibility of risk events occurring during the construction of urban large underground spaces should be calculated according to formula (7.15).

$$R^C = \beta R \tag{7.15}$$

Table 7.11 Evaluation indicator system of possibility of roof collapse in underground excavation construction

First-level indicators	Secondary indicators	Scoring basis	Value ranges
Design condition R_1	Depth-to-span ratio R_{11}	Depth-to-span ratio ≤ 0.4	80–100
		$0.4 <$ depth-to-span ratio < 1	60–80
		Depth-to-span ratio ≥ 1	40–60
Geological condition R_2	Engineering geology R_{21}	Artificial fill, soft soil, water-rich sand layer, pebble layer, etc.	80–100
		Weathered rock, composite strata, karst, goaf, etc.	60–80
		Bedrock uplift, weathered deep trough, fault zone, ground fissure, etc.	40–60
		Collapsible loess, expansive soil, hard rock veins, etc.	20–40
	Hydrogeology R_{22}	The groundwater level is higher than the top surface of the excavation	80–100
		The groundwater level is located within the excavation area	60–80
		The groundwater level is lower than the bottom of the excavation	40–60
Environmental condition R_3	Pipe diameter R_{31}	More than 2000 mm (including)	80–100
		600 (including)–2000 mm	60–80
		200 (including)–600 mm	40–60
		Under 200 mm	20–40
	Pipeline status R_{32}	Pipeline damage, obvious leakage	80–100
		There are cracks and leakage signs in the pipeline	60–80
		The pipeline is good, no sign of leakage	40–60
Construction condition R_4	Pipeline reinforcement R_{41}	No special reinforcement	60–100
		There is special reinforcement	20–60
	Ground de-watering R_{42}	No precipitation measures	70–100
		There are precipitation measures	40–70
	Soil excavation R_{43}	CD method, CRD method, double side wall guide pit method	80–100
		Middle hole method, side hole method	60–80
		PBA method, multi-layer and multi-section excavation method	40–60
	Rock excavation R_{44}	Blasting excavation	60–100
		Mechanical equivalent of light excavation	20–60

(continued)

Table 7.11 (continued)

First-level indicators	Secondary indicators	Scoring basis	Value ranges
	Convergence of work processes R_{45}	Convergence of work processes is not close	80–100
		Convergence of work processes is basically close	40–80
		Convergence of work processes is closely connected	0–40
	Surrounding rock deformation R_{46}	$U > 2/3U_0$ or $u > 5$ mm/d	80–100
		$1/3U_0 \le U \le 2/3U_0$ or 2 mm/d $\le u \le 5$ mm/d	40–80
		$U < 1/3U_0$ or $u < 2$ mm/d	0–40
	Geological conformity R_{47}	The geological conditions are worse than the design documents	70–100
		The geological conditions are basically consistent with the design documents	30–70
		The geological conditions are completely consistent with the design documents	0–30

Note U—Actual displacement value; U_0—Allowed displacement control value; u—Displacement rate

In the formula, R^C is the score of the possibility of construction safety risk events occurring when considering risk factor coupling. When $R^C \ge 100$, it is calculated as 100; β is the risk factor coupling coefficient; R is the basic score of the possibility of construction safety risk events occurring, calculated according to formula (7.13).

The risk factor coupling coefficient β is calculated according to formula (7.16).

$$\beta = \frac{1}{n}\sum_{i=1}^{n}\sum_{j=1}^{n} c_{ij} \tag{7.16}$$

In the formula, c_{ij} is the relative coupling coefficient of the evaluation indicators of the i-th and j-th secondary indicators in the corresponding risk event evaluation indicator system; n is the number of secondary indicators in the corresponding risk event evaluation indicator system. The relative coupling coefficient of the evaluation indicators c_{ij} is calculated as follows.

$$c_{ij} = \begin{cases} 1, & i = j \\ \frac{k_{ij}}{n-1}, & i \ne j \end{cases} \tag{7.17}$$

$$k_{ij} = \xi_{ij}\eta_{ij}, \xi_{ij} \in \{-1,1\}, \eta_{ij} \in [0,1] \tag{7.18}$$

In the formula, κ_{ij} is the coupling coefficient of the evaluation indicators of the i-th and j-th secondary indicators; ξ_{ij} is the coupling polarity coefficient of the i-th and j-th

Table 7.12 Evaluation indicator system of possibility of mud and water inrush in underground excavation construction

first-level indicators	Secondary indicators	Scoring basis	Value ranges
Geological condition R_1	Engineering geology R11	Cavities, water sacs, water-rich sand layers, etc.	80–100
		Weathered rock, pebble layer, karst, goaf, fault zone, etc.	60–80
		Weathered deep troughs, ground fissures, composite strata, etc.	40–60
	Water abundance R_{12}	Abundant water [inflow of water > 10000 m^3/(d·10 m)]	80–100
		Medium abundant water [5000 m^3/(d·10 m) ≤ inflow of Water ≤ 10000 m^3/(d·10 m)]	60–80
		Weak abundant water [inflow of Water < 5000 m^3/(d·10 m)]	40–60
Environmental condition R_2	Surrounding water bodies R_{21}	There are lakes, reservoirs, rivers and so on above	80–100
		There are lakes, reservoirs, rivers and so on the side	60–80
		Nearby there are ponds, ditches, etc	40–60
Construction condition R_3	Advance strengthening R_{431}	No special reinforcement	60–100
		There is special reinforcement	20–60
	Ground de-watering R_{432}	No precipitation measures	60–100
		There are precipitation measures	20–60
	Geological conformity R_{33}	The geological conditions are worse than the design documents	70–100
		The geological conditions are basically consistent with the design documents	30–70
		The geological conditions are completely consistent with the design documents	0–30
	Advanced reinforcement effect R_{34}	Poor	70–100
		Medium	40–70
		Good	20–40
	Formation precipitation effect R_{35}	Poor	70–100
		Medium	40–70
		Good	20–40

Table 7.13 Evaluation indicator system of the possibility of pipeline damage in underground excavation construction

First-level indicators	Secondary Indicators	Scoring basis	Value ranges
Environmental condition R_1	Pipeline vulnerability R_{11}	Masonry, asbestos cement or ceramic materials	80–100
		Smaller strength materials such as plastics	60–80
		Grey cast iron and other materials with low reliability	40–60
		Steel tube, concrete, ball milled cast iron and other high reliability materials	20–40
	Connection form of pipeline R_{12}	Sockets, flat mouth, etc.	80–100
		Groove and tongue, etc.	60–80
		Flange or welding, etc.	40–60
	Pipeline position R_{13}	Located in the main affected area	80–100
		Located in the secondary affected area	60–80
		Located in the general influence area	40–60
	Service life R_{14}	Over 30 years	90–100
		5–30 years	70–90
		Less than 5 years	50–70
Design condition R_2	Cross-sectional size R21	Section area is greater than 650 m^2 or five or more layers of subway station or a considerable scale	80–100
		Excavation section area of 500–650 m^2 or four-story subway station or a considerable scale	60–80
		Excavation section area of 300–500 m^2 or three layers of subway station or a considerable scale	40–60
	Depth-to-span ratio R_{22}	Depth-to-span ratio ≤ 0.4	80–100
		$0.4 <$ depth-to-span ratio < 1	60–80
		Depth-to-span ratio ≥ 1	40–60
Geological condition R_3	Engineering geology R_{31}	Artificial fill, soft soil, water-rich sand layer, pebble layer, etc.	80–100
		Weathered rock, composite strata, karst, goaf, etc.	60–80
		Bedrock uplift, weathered deep trough, fault zone, ground fissure, etc.	40–60
		Collapsible loess, expansive soil, hard rock veins, etc.	20–40
	Hydrogeology R_{32}	Confined water	80–100
		Phreatic water, fissure water, karst water	60–80
		Perched water	40–60

(continued)

Table 7.13 (continued)

First-level indicators	Secondary Indicators	Scoring basis	Value ranges
Construction condition R_4	Pipeline reinforcement R_{41}	No special reinforcement	60–100
		There is special reinforcement	20–60
	Ground de-watering R_{42}	No precipitation measures	60–100
		There are precipitation measures	20–60
	Excavation method R_{43}	CD method, CRD method, double side wall guide pit method	80–100
		Middle hole method, side hole method	60–80
		PBA method, multi-layer and multi-section excavation method	40–60
	Rock excavation R_{44}	Blasting excavation	60–100
		Mechanical equivalent of light excavation	20–60
	Process joining R_{45}	Process connection is not close	80–100
		The process connection is basically close	40–80
		The process is closely connected	0–40
	Pipeline settlement R_{46}	$S > 0.85S_0$ or $s > 2$ mm/d	80–100
		$0.7S_0 \le S \le 0.85S_0$ or 1 mm/d $\le s \le$ 2 mm/d	40–80
		$S < 0.7S_0$ or $s < 1$ mm/d	0–40

Note S—Actual settlement value; S_0—Allowed settlement control value; s—Settlement rate

secondary indicators, taking 1 for positive coupling and −1 for negative coupling; η_{ij} is the coupling strength coefficient of the i-th and j-th secondary indicators, referring to Table 7.18 for values.

The coupling polarity coefficient of the evaluation indicator ξ_{ij} and the coupling strength coefficient of the evaluation indicator η_{ij} can be determined by a survey questionnaire. The survey questionnaire can adopt the forms shown in Tables 7.19 and 7.20.

7.4.5 Risk Loss Evaluation

1. Indicator system and quantification standards

Risk loss is relatively difficult to quantify, and generally can only adopt a "qualitative evaluation + quantitative adjustment" evaluation strategy. Referring to the relevant standards and guidelines for construction risk assessment, risk loss evaluation is comprehensively evaluated based on factors such as casualties, economic losses, schedule losses, and social and environmental impacts. The indicator system and quantification standards for risk loss evaluation of urban large underground space construction are shown in Table 7.21.

Table 7.14 Evaluation indicator system of the possibility of damage to ground structures in underground excavation construction

First-level indicators	Secondary indicators	Scoring basis	Value ranges
Environmental condition R_1	Building height, bridge size, road grade R_{11}	Super high-rise, super large bridge, expressway	80–100
		High-rise, bridge, main road	60–80
		Small high-rise, middle bridge, secondary trunk road	40–60
		Multi-layer, small bridge, branch	20–40
	Foundation base R_{12}	Independent foundation, strip foundation, well format foundation	80–100
		Raft foundation, box foundation	60–80
		Pile foundation	40–60
	Location of building structures R_{13}	Located in the main affected area	80–100
		Located in the secondary affected area	60–80
		Located in the general influence area	40–60
Design condition R_2	Cross-sectional size R21	Section area is greater than 650 m^2 or five or more layers of subway station or a considerable scale	80–100
		Excavation section area of 500–650 m^2 or four-story subway station or a considerable scale	60–80
		Excavation section area of 300–500 m^2 or three layers of subway station or a considerable scale	40–60
	Depth-to-span ratio R_{22}	Depth-to-span ratio ≤ 0.4	80–100
		$0.4 <$ depth-to-span ratio < 1	60–80
		Depth-to-span ratio ≥ 1	40–60
Geological condition R_3	Engineering geology R_{31}	Artificial fill, soft soil, water-rich sand layer, pebble layer, etc.	80–100
		Weathered rock, composite strata, karst, goaf, etc.	60–80
		Bedrock uplift, weathered deep trough, fault zone, ground fissure, etc.	40–60
		Collapsible loess, expansive soil, hard rock veins, etc.	20–40
	Hydrogeology R_{32}	Confined water	80–100
		Phreatic water, fissure water, karst water	60–80
		Perched water	40–60
Construction condition R_4	Reinforcement of building structures R_{41}	No special reinforcement	60–100
		There is special reinforcement	20–60

(continued)

Table 7.14 (continued)

First-level indicators	Secondary indicators	Scoring basis	Value ranges
	Ground de-watering R_{42}	No precipitation measures	60–100
		There are precipitation measures	20–60
	Excavation method R_{43}	CD method, CRD method, double side wall guide pit method	80–100
		Middle hole method, side hole method	60–80
		PBA method, multi-layer and multi-section excavation method	40–60
	Rock excavation R_{44}	Blasting excavation	60–100
		Mechanical equivalent of light excavation	20–60
	Process joining R_{45}	Process connection is not close	80–100
		The process connection is basically close	40–80
		The process is closely connected	0–40
	Differential settlement of buildings and structures R_{46}	$S_D > 0.002L$	80–100
		$0.001L \leq S_D \leq 0.002L$	40–80
		$S_D < 0.001L$	0–40

Note S_D—Actual differential settlement value; L—Horizontal distance between differential settlement measurement points

2. Evaluation method

The score of the loss of safety risk in the construction of urban large underground spaces C is calculated according to formula (7.19).

$$C = \sum_{i=1}^{5} W_i C_i \tag{7.19}$$

In the formula, C is the risk loss score; C_i is the risk loss evaluation indicator score; W_i is the weight of the risk loss evaluation indicator.

7.4.6 Specialized Risk Grading Standards

1. Risk event occurrence possibility level standards

The likelihood of risk events occurring in urban large underground space construction is represented by probability values or possibility scores, and the level is divided into 1 to 4. After obtaining the possibility score of the risk event occurrence, the likelihood level of the risk event occurrence is determined according to Table 7.22.

Table 7.15 Evaluation indicator system of the possibility of damage to underground structures in underground excavation construction

First-level indicators	Secondary indicators	Scoring basis	Value ranges
Environmental condition R_1	Building height, bridge size, road grade R_{11}	Large basement, large section tunnel (pipe gallery)	80–100
		Medium-sized basement, medium-section tunnel (pipe gallery)	60–80
		Small basement, small section tunnel (pipe gallery)	40–60
	Foundation base R_{12}	Brick-concrete basement, prefabricated assembled tunnel	80–100
		Frame structure basement, composite lining tunnel	60–80
	Location of building structures R_{13}	Located in the main affected area	80–100
		Located in the secondary affected area	60–80
		Located in the general influence area	40–60
Design condition R_2	Depth of foundation pit R_{21}	Depth over 35 m	80–100
		The depth is 30–35 m	60–80
		The depth is 25–30 m	40–60
	Depth-to-span ratio R_{22}	depth-to-span ratio ≤ 0.4	80–100
		$0.4 <$ depth-to-span ratio < 1	60–80
		depth-to-span ratio ≥ 1	40–60
Geological condition R_3	Engineering geology R_{31}	Artificial fill, soft soil, water-rich sand layer, pebble layer, etc.	80–100
		Weathered rock, composite strata, karst, goaf, etc.	60–80
		Bedrock uplift, weathered deep trough, fault zone, ground fissure, etc.	40–60
		Collapsible loess, expansive soil, hard rock veins, etc.	20–40
	Hydrogeology R_{32}	Confined water	80–100
		Phreatic water, fissure water, karst water	60–80
		Perched water	40–60
Construction condition R_4	Reinforcement of building structures R_{41}	No special reinforcement	60–100
		There is special reinforcement	20–60
	Ground de-watering R_{42}	No precipitation measures	60–100
		There are precipitation measures	20–60
	Soil excavation R_{43}	Whole sequential excavation	70–100
		Partition block interval excavation	20–70
	Rock excavation R_{44}	Blasting excavation	60–100

(continued)

Table 7.15 (continued)

First-level indicators	Secondary indicators	Scoring basis	Value ranges
		Mechanical equivalent of light excavation	20–60
	Process joining R_{45}	Process connection is not close	80–100
		The process connection is basically close	40–80
		The process is closely connected	0–40
	Settlement of building structures R_{46}	$S > 0.85S_0$ or $s > 3$ mm/d	80–100
		$0.7S_0 \le S \le 0.85S_0$ or 2 mm/d $\le s \le 3$ mm/d	40–80
		$S < 0.7S_0$ or $s < 2$ mm/d	0–40

Note S—Actual settlement value; S_0—Allowed settlement control value; s-Settlement rate

Table 7.16 Evaluation indicator system of possibility of excessive deformation or failure of the existing structure caused by the connection opening

First-level indicators	Scoring basis	Value ranges
The present situation of existing structure R_1	Serious diseases	80–100
	Minor diseases	60–80
	Basically intact	40–60
Opening size R_2	Large (total area greater than 30 m^2)	80–100
	Medium (total area 15–30 m^2)	40–80
	Small (total area less than 15 m^2)	20–40
Opening position R_3	Side wall opening	80–100
	Roof opening	60–80
	Floor opening	40–60
Opening method R_3	Blasting opening	80–100
	Static blasting opening	60–80
	Mechanical or manual cutting openings	40–60
Disturbance degree of existing structure R_3	Destroy the main stress components of the existing structure	70–100
	Not damaging the main load-bearing components of the existing structure	40–70

2. Risk Loss Level Standard

The risk loss level of urban large underground space construction is represented by the severity of the loss, divided into four levels. After determining the risk loss score, the risk loss level is determined according to Table 7.23.

Table 7.17 Evaluation indicator system of the possibility of existing structure floating caused by excavation unloading

First-level indicators	Scoring basis	Value ranges
The present situation of existing structure R_1	Serious diseases	80–100
	Minor diseases	60–80
	Basically intact	40–60
Magnitude of unloading R_2	Large (excavation volume is greater than the existing structure volume)	80–100
	Medium (excavation volume is similar to the existing structure volume)	40–80
	Small (excavation volume is less than half of the existing structure volume)	20–40
Relative position relationship R_3	Top unloading	80–100
	Side unloading	40–80
Soil layer situation R_4	Soft soil, expansive soil	70–100
	Artificial fill, water-rich sand layer	40–70

Table 7.18 The value of coupling strength coefficient of evaluation index

Description	Weak coupling	Middle coupling	Close coupling	Very strong coupling
Value	0.1	0.25	0.5	1

Table 7.19 Questionnaire on coupling of evaluation indicators

First-level indicators	R1			…	Rm
Secondary indicators	R11	…	$R_{1,n1}$	…	$R_{m,nm}$
Coupling condition	☐ Participation in coupling ☐ Not involved in coupling	☐ Participation in coupling ☐ Not involved in coupling	☐ Participation in coupling ☐ Not involved in coupling	…	☐ Participation in coupling ☐ Not involved in coupling

Note Only all secondary indicators in the corresponding risk event occurrence possibility evaluation indicator system need to be listed in the table

3. Specialized Risk Grading Standard

The grading of urban large underground space construction risks should consider the possibility of risk occurrence and the consequences of risk loss, and finally carry out comprehensive rating. According to the acceptance criteria for urban large underground space construction risks, referring to current risk assessment standards and norms, urban large underground space construction risks are divided into levels I to IV, see Table 7.24.

Table 7.20 Evaluation index coupling strength questionnaire

Coupling indicator	Coupling parameter	R_{11}	R_{13}	…	R_{42}	R44
R_{11}	Polarity of coupling		□ Positive coupling □ Inverse coupling	…	…	…
	Coupling strength		□ Weak coupling □ Middle coupling …	…	…	…
R_{13}	Polarity of coupling				…	…
	Coupling strength				…	…
…	Polarity of coupling				…	…
	Coupling strength				…	…
R_{42}	Polarity of coupling					…
	Coupling strength					…

Note Only the secondary indicators involved in the coupling in the corresponding risk event occurrence possibility evaluation indicator system need to be listed in the table

4. Specialized Risk Acceptance Criteria

(1) Concept of Risk Acceptance Criteria

Risk acceptance criteria mainly refer to the baseline values for dividing various levels and types of risk levels and formulating control levels, and are also the baseline values for risk management and risk control. It represents the acceptable risk level within a specified time or within a certain stage of system behavior, reflecting the acceptance level of risks by entities such as society, the public, or individuals. Risk acceptance criteria can directly provide reference for risk evaluation, risk response, and decision-making. The risk acceptance criteria of different countries and industries are often different, but in most cases, the following two basic principles should be followed.

① ALARP Principle

The ALARP (As Low As Reasonably Practical) principle means: every activity has risks, it is impossible to completely eliminate risks through preventive measures, and when the system's risk level is very low, it is difficult to further reduce the risk level, and its cost often rises exponentially. Therefore, a balance must be made between the system's risk level and cost, taking different risk decisions at different risk levels, but the division of risk levels and the formulation of risk countermeasures should be as reasonable and feasible as possible, and the risk cost should be as low as possible. Because the ALARP principle is widely applicable to individual, social, and environmental risk acceptance criteria, it has been widely used in risk management.

② ALARA Principle

The ALARA (As Low As Reasonably Achievable) principle means that the measures taken must be achievable when determining the measures corresponding to risk countermeasures, which is basically the same as the ALARP principle.

Table 7.21 Risk loss evaluation indicator system

First-level indicators	Scoring basis	Value ranges	Explanation
Personal injury C_1	More than 30 people died or more than 100 people were seriously injured	75–100	Casualties include engineering construction personnel and third-party casualties
	10–30 people died or 50–100 people were seriously injured	50–75	
	3 to 10 people died or 10 to 50 people were seriously injured	25–50	
	Less than 3 deaths or less than 10 serious injuries	0–25	
Economic loss C_2	10,000 million yuan or more or equivalent to the original project cost more than 100% of the economic losses	75–100	The economic loss includes the direct economic loss of the construction party and the third-party economic loss
	50 million–10 million yuan or equivalent to 50%–100% of the original project cost of economic losses	50–75	
	10 million–50 million yuan or equivalent to the original project cost 20%–50% of the economic losses	25–50	
	Economic losses below 10 million yuan or equivalent to less than 20% of the original project cost	0–25	

(continued)

Table 7.21 (continued)

First-level indicators	Scoring basis	Value ranges	Explanation
Time loss C_3	Delay of more than 12 months or more than 50% of the original construction period	75–100	The evaluation basis in the table is only applicable to the control period project, and the evaluation basis of the non-control period project should be considered as 2 times the delay time of the control period project. If the non-controlled construction period project is changed into the controlled construction period project due to delay, it should be managed according to the controlled construction period project
	6–12 months or 20%–50% of the original construction period	50–75	
	3–6 months or equivalent to 10%–20% of the original project construction period delay	25–50	
	Delays of less than 3 months or equivalent to less than 10% of the original project duration	0–25	
Social influence C_4	Poor impact, or the need to relocate more than 1000 people	75–100	Social influence includes the negative pressure of public opinion caused by risk events and the impact on normal economic and social activities
	Serious impact, or the need to relocate 500–1000 people	50–75	
	Less serious impact, or the need to relocate 100–500 people	25–50	
	Need to consider, or need to transfer the resettlement of 100 people or less	0–25	

(continued)

Table 7.21 (continued)

First-level indicators	Scoring basis	Value ranges	Explanation
Environmental influence C_5	The scope is very large, and the surrounding urban environment is seriously polluted or damaged	75–100	The surrounding urban environment includes residents' production, living environment and natural ecological environment
	The scope is large, and the surrounding urban environment is heavily polluted or damaged	50–75	
	The scope is big, and the surrounding urban environment is polluted or damaged	25–50	
	The scope is small, and the surrounding urban environment is slightly polluted or damaged	0–25	

Table 7.22 Risk event occurrence possibility level standard

Possibility level	1	2	3	4
Description	Very likely	Likely	Occasionally	Unlikely
Probit	$0.1 \le P \le 1$	$0.01 \le P < 0.1$	$0.001 \le P < 0.01$	$0 \le P < 0.001$
Score	$75 \le R \le 100$	$50 \le R < 75$	$25 \le R < 50$	$0 \le R < 25$

Table 7.23 Risk loss classification

Level	One	Two	Three	Four
The degree of risk loss	Serious	Large	General	Small
Risk loss score	$75 \le C \le 100$	$50 \le C < 75$	$25 \le C < 50$	$0 \le C < 25$

Table 7.24 Urban large underground space construction risk level matrix

Risk level		Grade of risk loss			
		One	Two	Three	Four
Risk event occurrence possibility level	1	I	I	II	II
	2	I	II	II	III
	3	II	II	III	III
	4	II	III	III	IV

The concept of general construction risk acceptance criteria is shown in Fig. 7.3.

(2) Acceptance Criteria for Urban large underground space Construction Risks

In view of the characteristics of urban large underground space construction risks and the concept of engineering construction in our country and the needs of the times, when establishing risk assessment acceptance criteria, the following two most basic principles are first considered.

① Safety First Principle: The construction of urban large underground space projects must prioritize ensuring personnel safety.

② Economic Principle: Only when safety requirements are met can the ALARP or ALARA principles be followed. The principle is to minimize the economic loss caused by engineering risks or the funds invested to avoid risks as much as possible.

Based on the above principles, the risk acceptance criteria for urban large underground space construction are shown in Table 7.25.

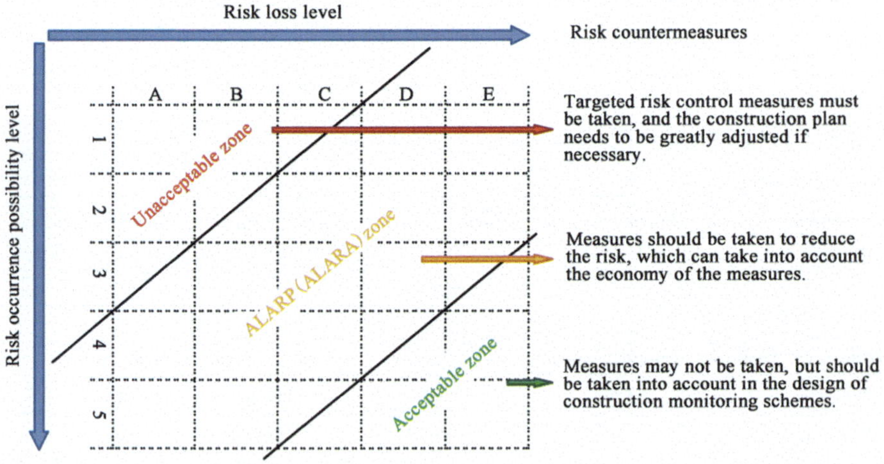

Fig. 7.3 Schematic diagram of construction risk acceptance criteria

Table 7.25 Acceptance criteria for specialized risk for urban large underground space construction

Specialized risk level	Acceptance criteria	Risk control principle
I	Unacceptable	Construction must be highly valued and effective measures must be taken to avoid or reduce risks before proceeding
II	Unwilling to accept	Effective measures should be taken to address and strengthen risk monitoring
III	Acceptable	Effective measures should be taken to address the issue and risk monitoring should be conducted
IV	Accept	No measures can be taken, but attention should be paid to prevent the risk level from rising

7.5 Urban Large Underground Space Construction Safety Risk Dynamic Assessment System

7.5.1 System Design

1. Overall system architecture

The urban large underground space construction risk dynamic assessment system mainly collects various risk factors, uses model algorithms to assess current and future risks, provides risk assessments and control measures, and synchronizes data through mobile and web platforms.

The overall framework design of the system is shown in Fig. 7.4. The system design is based on the results of system analysis, using the ideas and methods of system science to make the design results meet the objectives to the greatest extent.

2. Application system scheme

Since the urban large underground space construction risk assessment system will involve multiple business departments, such as the construction party, supervision and other third-party users, the amount of system data will gradually increase over time, and some data is shared by multiple departments, with strong cross-over. Considering the system performance issues in the later stage, it is more reasonable to adopt a three-tier B/S architecture, which is very necessary for the scalability of the system.

The first layer is the client layer: providing an information interaction interface, which is the presentation layer directly operated by the user; the second layer is the application server layer: mainly performing business processing logic; the third layer is the data/resource layer: mainly performing data reading and writing.

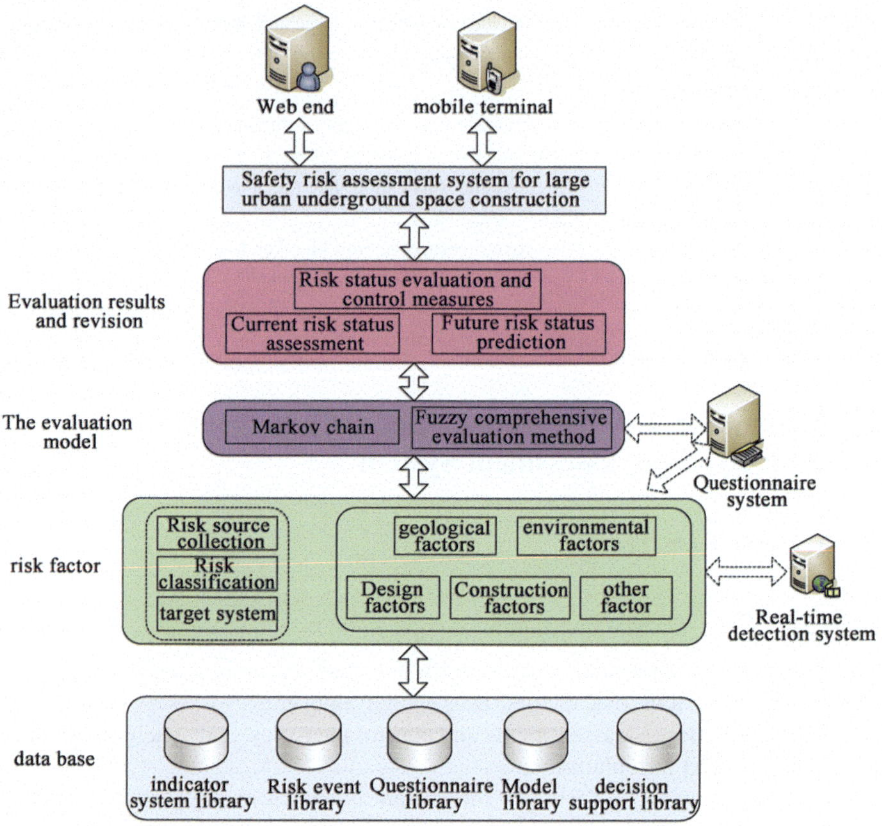

Fig. 7.4 Overall system framework planning

The advantages of the three-tier B/S architecture are as follows:

(1) Enhanced system scalability

When it is found that the speed of business processing slows down due to the increase in system clients, you can buy another middle-layer server and attach it to the original system to improve performance.

(2) Easy integration of information systems

In the three-tier architecture, the application client only needs to submit the database request to the middle layer. Through the middle layer, the interaction with various database systems is realized, and the integration of application systems is completed.

(3) High overall system performance

By connecting buffers, reducing database concurrent users; reducing network overhead, reasonably allocating system load, eliminating bottlenecks in database access through data buffering and changes in system architecture.

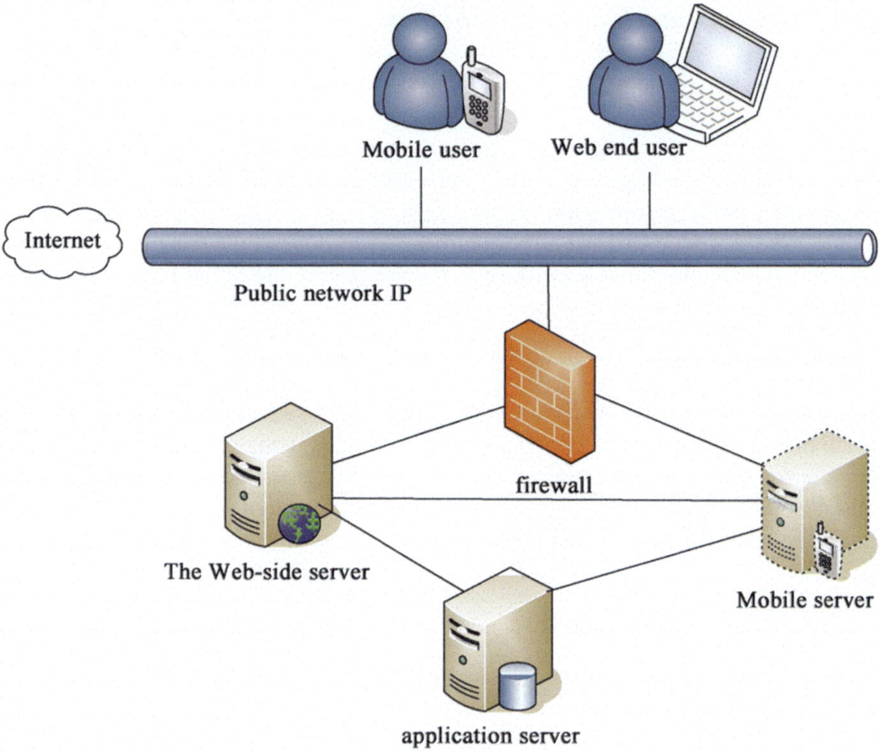

Fig. 7.5 Software application system architecture

The application system of the urban large underground space construction safety risk assessment system is shown in Fig. 7.5.

The overall system adopts the Java 2 Platform Enterprise Edition (J2EE) architecture. The multi-layer technology based on J2EE is the core of the application system. Most business systems are built using J2EE technology, providing personnel with simple and convenient WEB applications.

(1) On the database side, based on the comparison of multiple database systems, we choose MySQL as the support software for the database server.
(2) On the application server side, we choose J2EE application servers (such as: BEA WebLogic, WebSphere, JBoss, etc.) to implement business service support. By using these middleware, automatic business load balancing is achieved, ensuring business integrity.
(3) On the user end a pure browser B/S mode is used to implement the business interface.

7.5.2 System Interface and Functions

1. System Function Layout

The system comprises modules such as the homepage, project information management, monitoring management, risk engineering management, basic information management, basic information management, and permission management. Each module has possessed distinct sub-functions. According to the function classification and usage habits (later according to requirements), the menu functions can be recombined, as shown in Fig. 7.6.

2. System Main Interface

The system interface is written in HTML language, suitable for all major mainstream browsers on the market, and supports automatic recognition and switching of computer/mobile interface, making it convenient for users to access in various usage scenarios. After logging into the system, the left side is the function menu, and the right side is the function display. The main interface of the computer is shown in Fig. 7.7.

3. Project Management Interface

The project management interface is shown in Fig. 7.8, and its functions mainly include the following points.

(1) View detailed project information

Display the basic information of the participating units of the project, project overview, work surface situation (construction status of the work surface, including ongoing, temporary suspension, completion, total), risk engineering situation

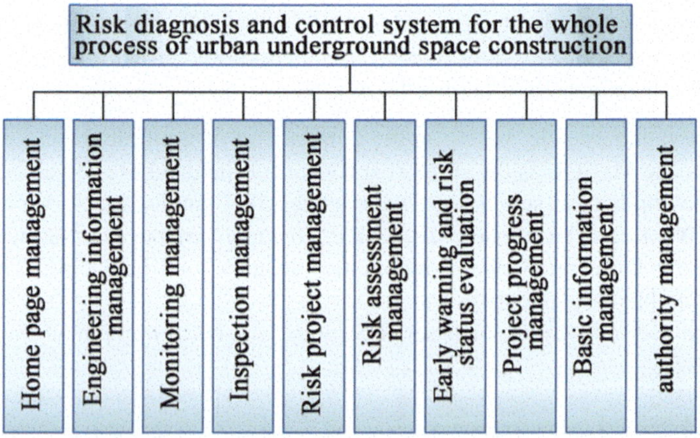

Fig. 7.6 System overall function layout

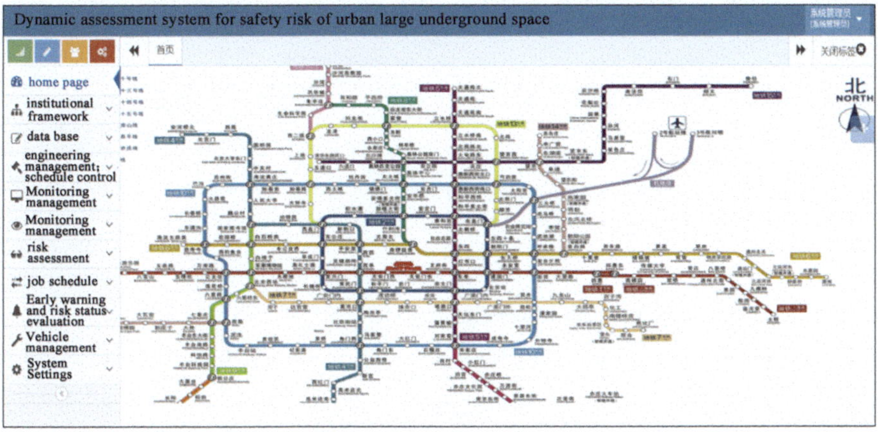

Fig. 7.7 Computer system main interface

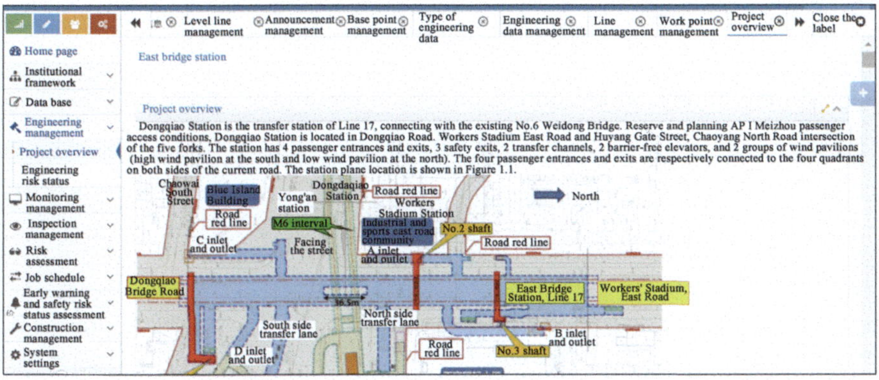

Fig. 7.8 Project overview interface

(including the number of risk projects not started, the number of risk projects started, the number of completed risk projects, total risk projects), warning situation of the measuring points, and the time tree of important nodes of the project (Fig. 7.9).

(2) Project Data Management

Classify and manage all related materials of the project, including material name, construction part, material type, preparation unit, preparation time, upload unit, upload time, associated risk project, file status. The third-party monitoring overall plan, third-party monitoring implementation plan, and GIS map materials all need to be approval.

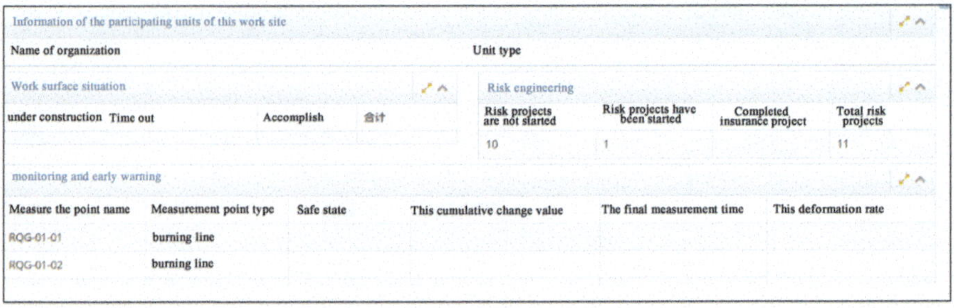

Fig. 7.9 Detailed project information interface

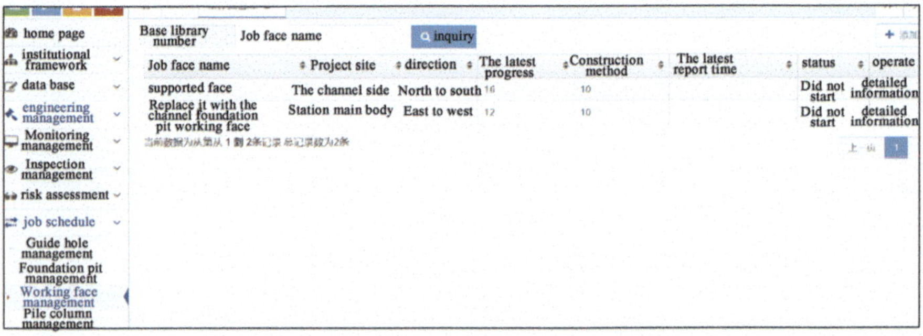

Fig. 7.10 Work surface management interface

(3) Work Surface Management

The work surface management interface can view the work surface allocation, work surface location, construction progress, construction method, etc. of the current project (Fig. 7.10).

(4) Project Instrument Management

Manage the instruments used at the project site (Fig. 7.11), including instrument equipment name, specification model, owning unit, manufacturer, factory number, calibration start date, calibration end date, calibration certificate number, calibration unit, equipment status, and the project site where the instrument is used. For instruments that need to be calibrated, record each calibration situation, including calibration unit, calibration start date, calibration end date, calibration certificate number, calibration status.

(5) User Management

The function of the user management interface is to manage all users of the platform uniformly and manage user permissions. The functions that users can access are

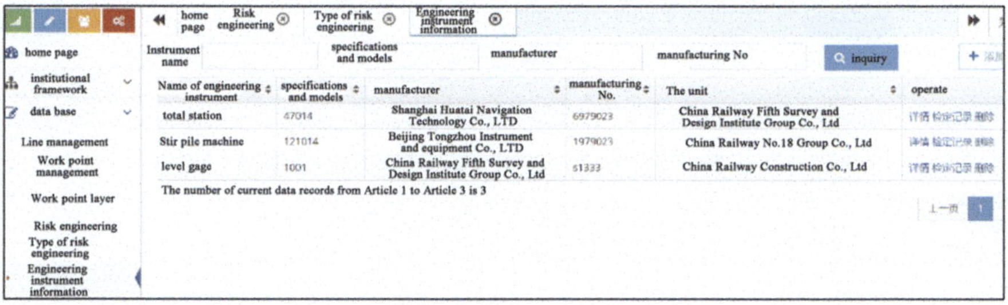

Fig. 7.11 Instrument management interface

stipulated by their roles (such as construction party, third-party monitoring, administrator, etc.), and some additional functions can also be opened to a specific user through direct authorization (Fig. 7.12).

4. Measuring Point Management Interface

The functions of the measuring point management interface mainly include the following points.

(1) View the layout of project measuring points

The system automatically obtains all measuring point information marked on the Geographic Information System (GIS) base map (CAD design drawing) and stores it in the database to generate a measuring point layout diagram (Fig. 7.13).Users can directly click on any measuring point in the picture to view the status of the measuring point.

(2) View information of each measurement point

This interface allows you to view basic information about the measurement points, including the name, category, type, corresponding risk project, coordinates, mileage, type of monitoring user, initial measurement time, latest measurement time,

Fig. 7.12 User management interface

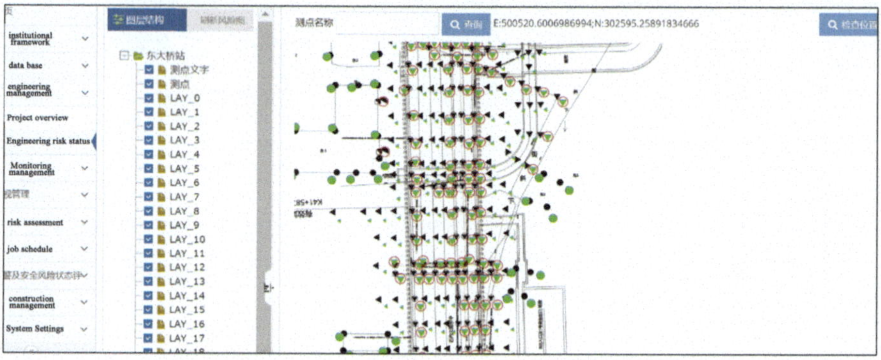

Fig. 7.13 Measurement point layout interface

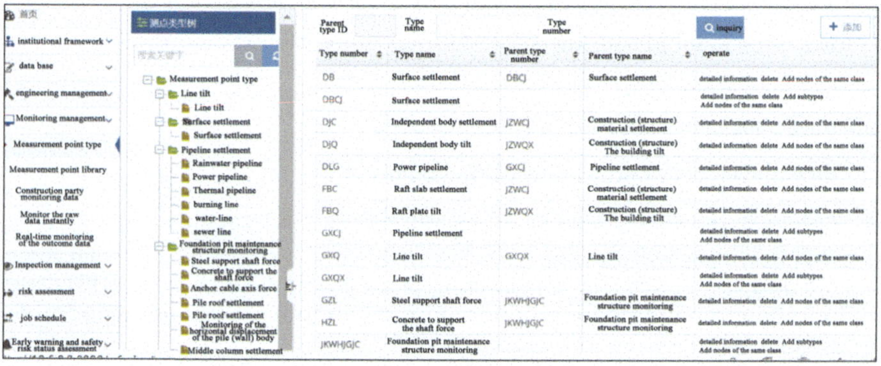

Fig. 7.14 Measurement point detailed information interface

number of measurements, data dimension, baseline value, monitoring frequency, etc. (Fig. 7.14).

(3) View monitoring warning situation

The monitoring warning management interface can view the current cumulative monitoring value, the rate of change of the monitoring value, dual control index values (cumulative value, rate of change), and the current warning level, etc., which are automatically calculated by the system based on the uploaded data (Fig. 7.15).

5. Risk Assessment Interface

The main functions of the risk assessment interface are as follows.

(1) Risk project management

The risk project management interface can enter risk project information, including the sub-items of the project corresponding to the risk project, the list of risk events, the list of risk sources, and other information related to risk assessment. It can also

Measure the point name	Measurement point type	safe state	degree of depth	This cumulative change value	The latest measurement time	This deformation rate	Cumulative alarm value	Rate alarm value	operate
WSG-02-03	污水管线	normal		-9.99	2021-04-09 16:00:40	0.16			chart
WSG-01-06	污水管线	Yellow		2.18	2021-04-09 16:00:40	0.17			chart
WSG-01-02	污水管线	Yellow		-7.47	2021-04-09 16:00:40	0.18			chart
WSG-02-13	污水管线	red		0.14	2021-04-09 16:00:40	0.3			chart
WSG-02-02	污水管线	Yellow		8.92	2021-04-09 16:00:40	0.19			chart
WSG-01-01	污水管线	red		3.03	2021-04-09 16:00:40	0.2			chart
WSG-02-01	污水管线	red		6.48	2021-04-09 16:00:40	0.21			chart
WSG-01-04	污水管线	red		-6.3	2021-04-09 16:00:40	0.22			chart
WSG-02-14	污水管线	normal		2.39	2021-04-09 16:00:40	0			chart
WSG-01-03	污水管线	red		-2.44	2021-04-09 16:00:40	0.23			chart
WSG-02-15	污水管线	normal		-4.36	2021-04-09 16:00:40	0			chart
WSG-02-11	污水管线	normal		6.4	2021-04-09 16:00:40	0			chart

Fig. 7.15 Monitoring warning interface

associate the risk project with the measurement points, so that the system can automatically carry out real-time risk assessment based on the monitoring data. Under a certain risk project, you can add, modify or delete risk assessment instances.

(2) Risk assessment management

Risk assessment management includes project information management, risk assessment indicator system management, risk assessment instance management, assessment questionnaire management, and assessment result management.

① Project information management

Project information includes project type, planar feature size, construction method, etc. The system can identify the risk event list and risk assessment indicator system of the project automatically based on the input basic information, through the risk event library and indicator database.

② Risk assessment indicator system management

The risk assessment indicator system management interface can view, modify, and delete the basic indicator system and its quantification standards. After a new project is established, the system will automatically generate the default basic indicator system. Users can add or delete risk events and corresponding basic indicator systems according to their needs (Fig. 7.16).

③ Risk assessment instance management

The risk assessment work in this system is divided into several risk assessment instances for management. A risk assessment instance represents a total risk assessment or a specialized risk assessment for a certain risk event. When establishing a risk assessment instance, the system will automatically call the current basic indicator system of the system as the assessment indicator system of this instance based on the risk event selected by the user. Users can modify the indicator system in the instance

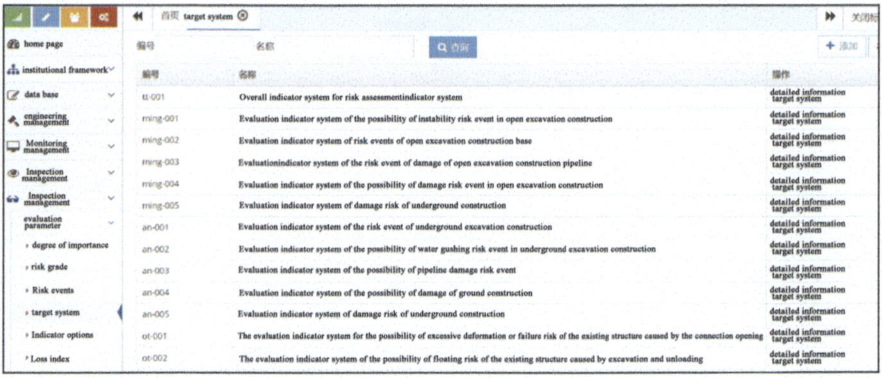

Fig. 7.16 Indicator system management interface

at will, and this modification will not affect the basic indicator system and the indicator system in other instances. After the risk assessment instance is submitted, the assessment questionnaire can be generated, issued, and the risk assessment work can be carried out (Figs. 7.17 and 7.18).

(3) Assessment questionnaire management

The function of the assessment questionnaire management interface is to generate and issue survey questionnaires. After a risk assessment instance is submitted, a questionnaire can be automatically generated through the questionnaire system. The evaluator can visit the questionnaire page through their personal dedicated link, and after filling it out, submit it. The system will automatically summarize and statistically analyze the questionnaire results, and output the corresponding questionnaire results (Figs. 7.19 and 7.20).

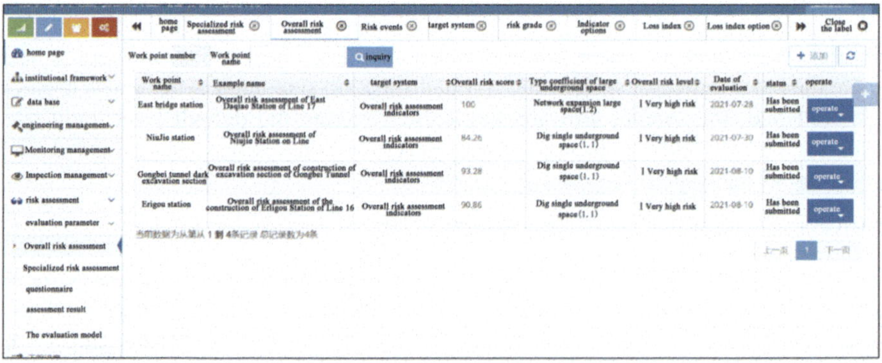

Fig. 7.17 Overall risk assessment interface

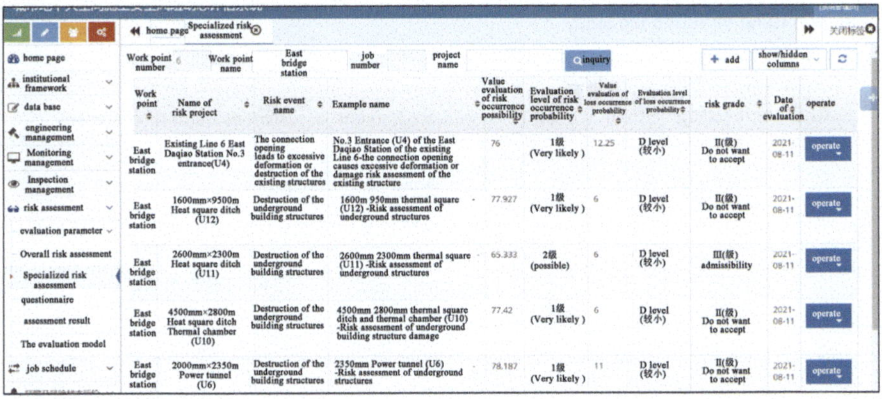

Fig. 7.18 Specialized risk assessment interface

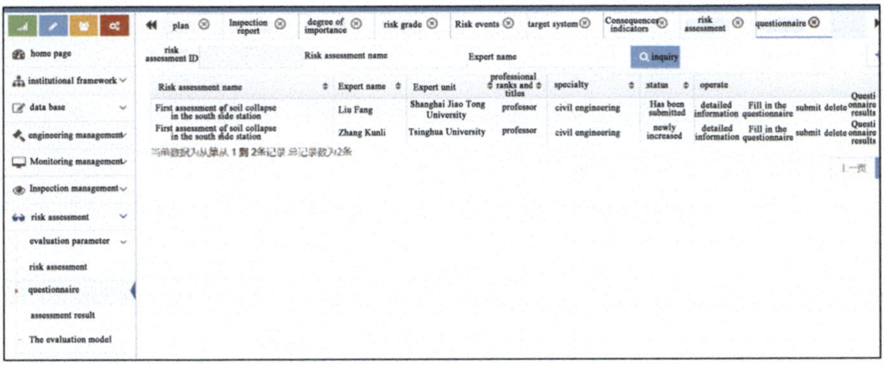

Fig. 7.19 Survey questionnaire management interface

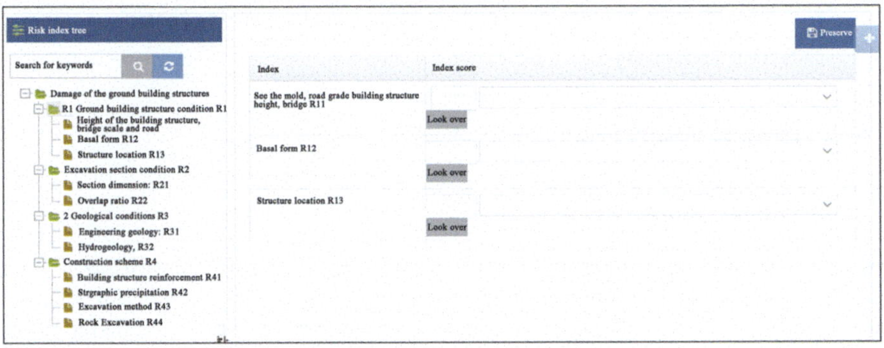

Fig. 7.20 Survey questionnaire detail interface

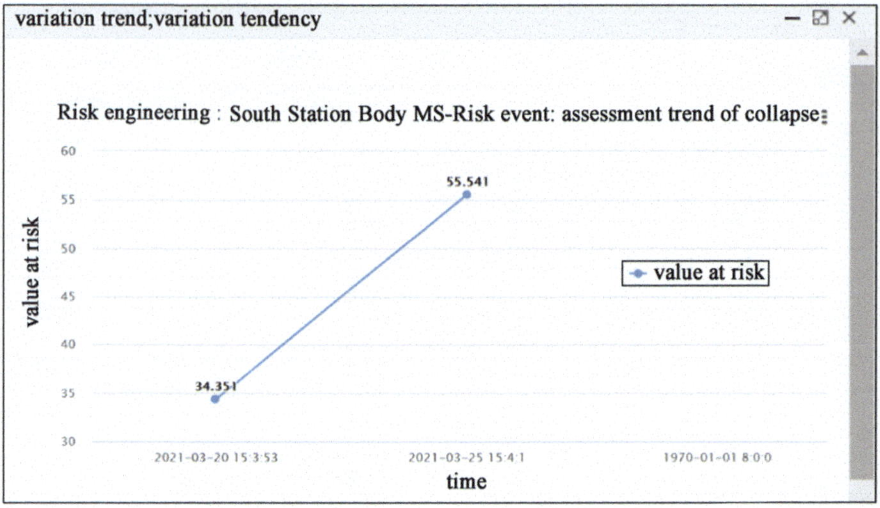

Fig. 7.21 Risk assessment result interface

(4) Evaluation Result Management

It also includes current project evaluation warnings, risk status evaluations and countermeasures, implements safety evaluations and predictions for each sub-project, views the trend graph of risk changes, and proposes corresponding control measures for existing risks, providing decision support information (Fig. 7.21).

7.6 Case Study of Construction Safety Risk Assessment for the Renovation Project of Xuanwumen Station, Beijing Subway

7.6.1 Project Overview

1. Project Background

Xuanwumen subway station is located at the intersection of Xuanwumen East and West Streets and Xuanwumen Inner and Outer Streets in the central old city area within the Second Ring of Beijing. Xuanwumen station is a transfer point between Line 4 and Line 2. The layout, structural form, and construction method of Xuanwumen station are detailed in Sect. 2.3.6.

2. Existing Structure

(1) Xuanwumen Station of Subway Line 2

Xuanwumen Station of Subway Line 2, built in 1967, is a three-span double-column rectangular frame structure. The secondary lining structure has a top plate thickness of 1000 mm, a bottom plate thickness of 900 mm, a side wall thickness of 800 mm, a center column of 1100 mm × 900 mm, a side span of 6500 mm, a middle span of 5900 mm, and a structural net height of 6050 mm. The cross-section of Xuanwumen Station on Line 2 is shown in Fig. 7.22.

① Entrance and Exit Structure of Xuanwumen Station on Subway Line 2

Xuanwumen Station on Subway Line 2 has A, B, C, D, a total of 4 entrances and exits. The cross-section form of the entrance and exit is a rectangular frame cross-section + U-shaped groove cross-section. The thickness of the secondary lining top plate is 600–700 mm, the bottom plate thickness is 500–550 mm, and the side wall thickness is 500 mm. The longitudinal section of the structure is shown in Fig. 7.23, and the cross-section is shown in Fig. 7.24.

② Southwest Air Duct of Xuanwumen Station on Subway Line 2

The southwest air duct of Xuanwumen Station on Subway Line 2 is a single-span rectangular frame structure. The standard section top plate thickness is 600 mm, the bottom plate thickness is 550 mm, the side wall thickness is 550 mm, and the structural net height is 3400 mm; the climbing section top plate thickness is 600–700 mm, the bottom plate is 550 mm, the side wall thickness is 550 mm, and the structural net height is 3400–4130 mm. The cross-section of the structure is shown in Fig. 7.25.

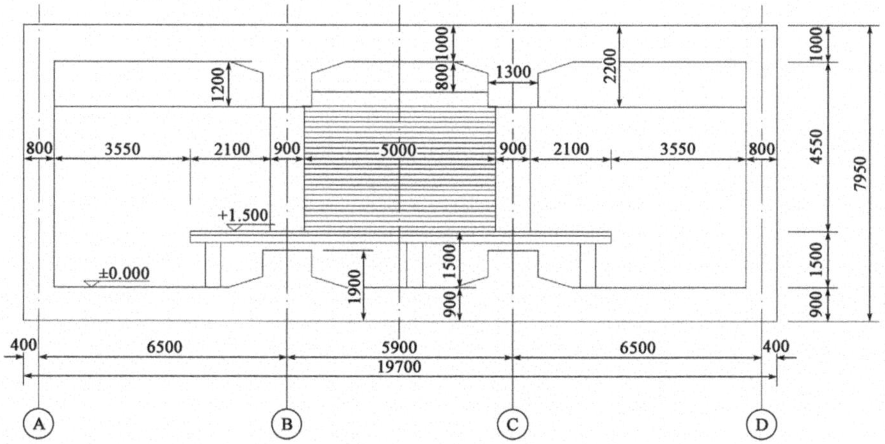

Fig7.22 Cross-section of existing line 2 Xuanwumen station (unit of measurement: mm)

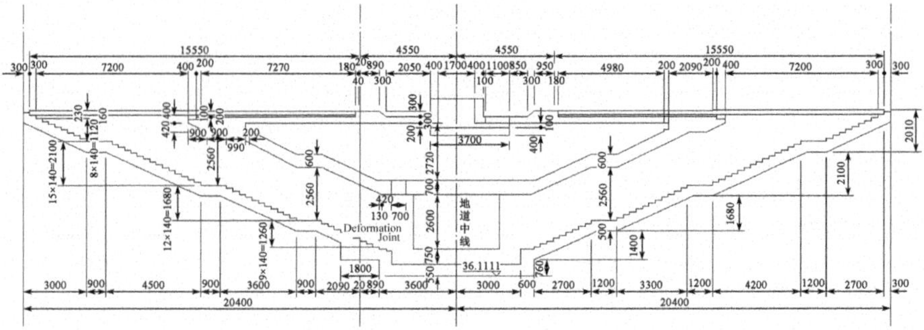

Fig. 7.23 Longitudinal section of line 2 Xuanwumen station entrance and exit (unit of measurement: mm)

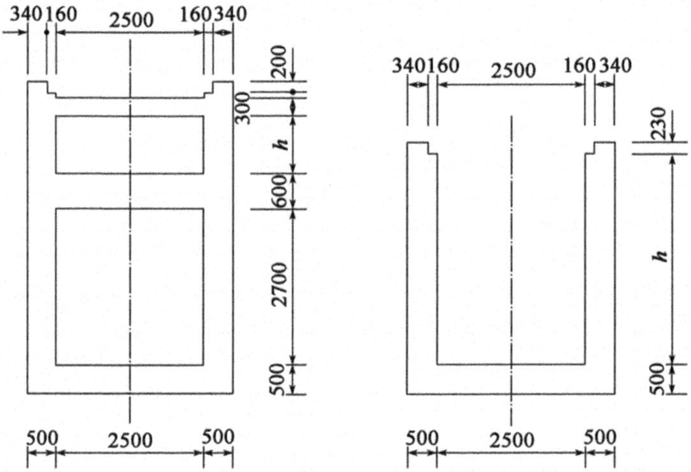

Fig. 7.24 Cross-section of line 2 Xuanwumen station entrance and exit (unit of measurement: mm)

(3) Xuanwumen Station of Subway Line 4

The existing Xuanwumen Station of Subway Line 4, built in 2006, is a three-span double-column arch-top straight-wall frame structure. The secondary lining structure has a top plate thickness of 700 mm, a bottom plate thickness of 1200 mm, a side wall thickness of 700 mm, and a center column of ϕ800 mm. Steel pipe column, side span 6750 mm, middle span 6400 mm, structural net height 12800 mm. The cross-section of Xuanwumen Station on Subway Line 4 is shown in Fig. 7.26.

① Northwest entrance and exit of Line 4 and Northwest air duct

After exiting Line 4 Xuanwumen Station from Existing Line 4 Xuanwumen Station Entrance E, climb up across the Northwest Duct to ground level in two stages. The standard cross-section of the E entrance and exit is a single-span horseshoe-shaped cross-section, with a secondary lining top plate thickness of 500 mm, bottom

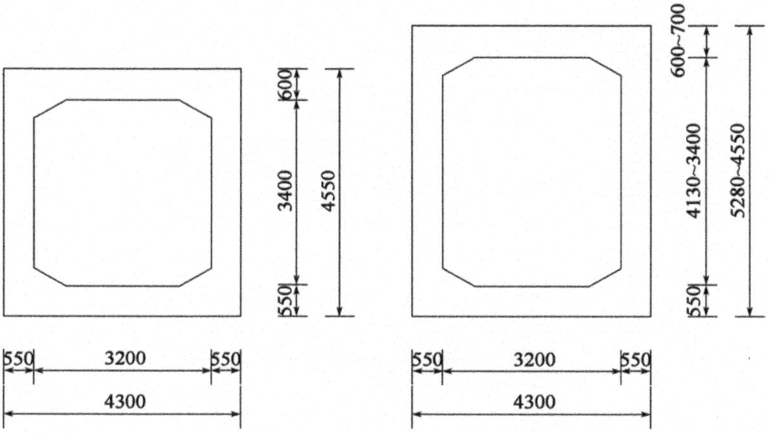

Fig. 7.25 Cross-section of Southwest air duct of line 2 Xuanwumen station (unit of measurement: mm)

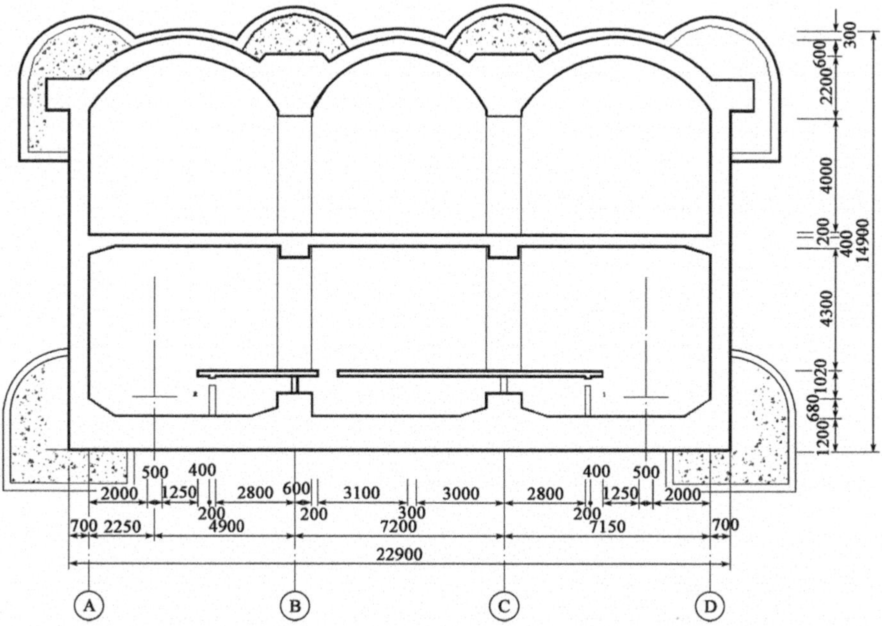

Fig. 7.26 Existing cross-section of Xuanwumen station on line 4 (unit of measurement: mm)

plate thickness of 650 mm, side wall 500 mm, structural net span 5100 mm, net height 4950 mm; the climbing section is a single-span rectangular cross-section + U shaped groove cross-section, secondary lining top plate thickness 500 mm, bottom plate thickness 500 mm, side wall 500 mm, structural net span 6800 mm, net height

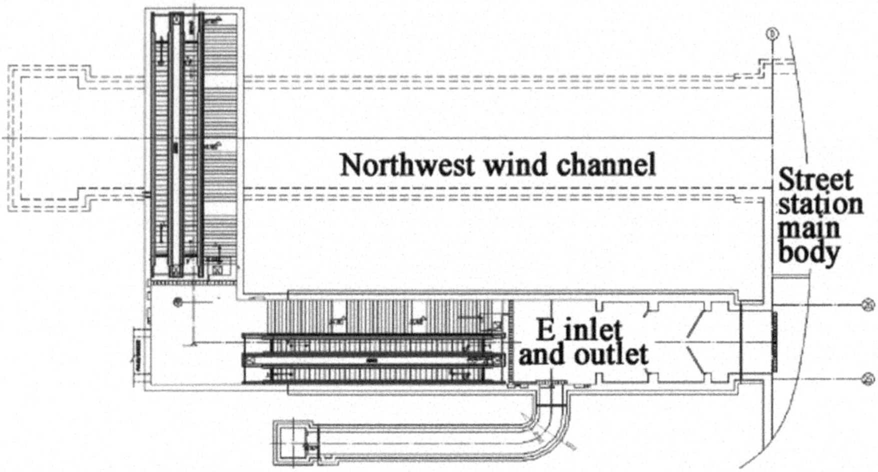

Fig. 7.27 Planar relationship between the northwest entrance and exit and the northwest air duct

1950–6450 mm; the northwest air duct is a single-span double-layer arch-top straight wall frame structure, the thickness of the secondary lining top plate is 600 mm, the thickness of the bottom plate is 650 mm, the thickness of the side wall is 600 mm, the structural net span is 8000 mm, and the net height is 11250 mm.

The planar relationship between the E entrance and exit and the northwest air duct is shown in Fig. 7.27, and the sectional relationship is shown in Fig. 7.28.

② Subway Line 4 Xuanwumen Station - Xidan Station section

After exiting from Xuanwumen Station on Subway Line 4, the Xuanwumen Station - Xidan Station section is laid from south to north, with a line spacing of about 17 m. The section structure cross-section is a single-span horseshoe-shaped cross-section, with a structure thickness of 300 mm, a top plate cover soil of 16.3 m, a bottom plate depth of 22.4 m, a structural net width of 5.20 m, a net height of 5.25 m, and the main strata through which the tunnel body passes are pebble ⑦ and loess ⑥2 layers. The structural cross-section is shown in Fig. 7.29.

3. Geological conditions of the project

(1) Regional structure

The proposed project is located in the middle and upper part of the Yongding River alluvial fan, and the micro-geomorphology is at the intersection of the ancient Jingou River and the ancient Luo River, as shown in Fig. 7.30.

According to the measured ground elevation data of the site, the overall terrain of the site is basically flat, and the ground elevation of the survey borehole is between 43.93 and 45.40 m.

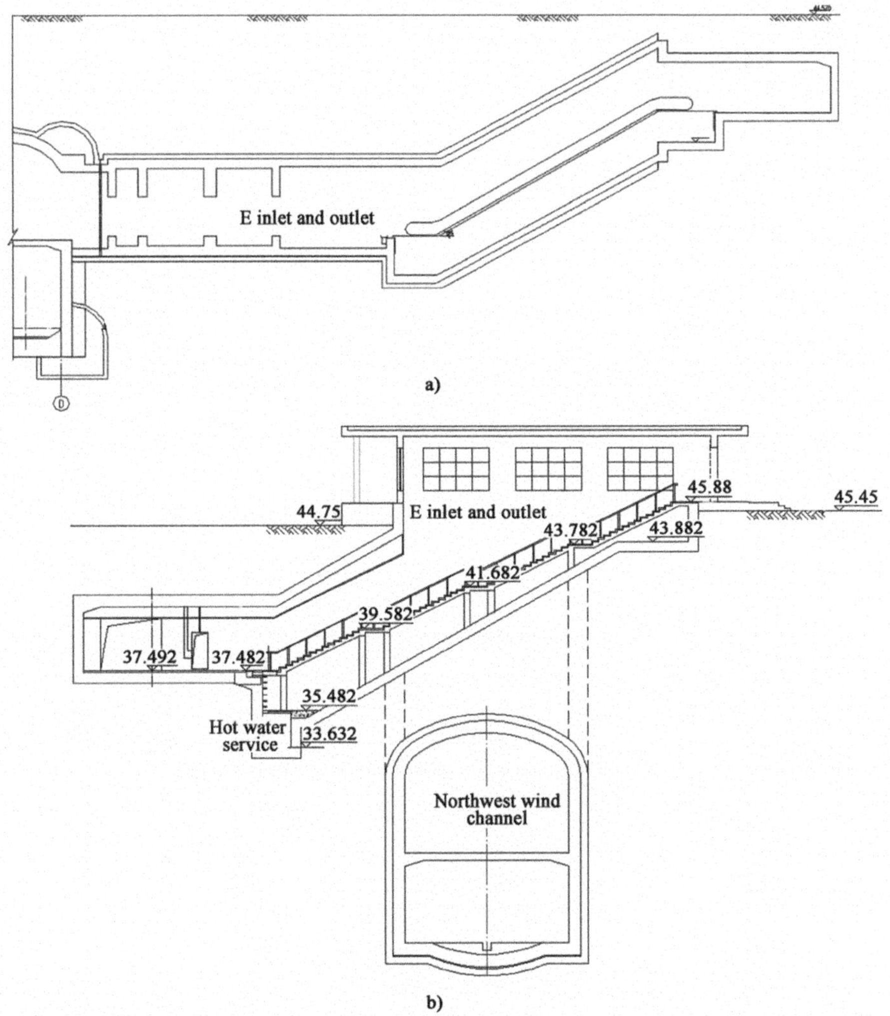

Fig. 7.28 Sectional relationship between the northwest entrance and exit and the northwest air duct (Unit of measurement: mm)

(2) Stratigraphic lithological characteristics

The maximum depth of the stratum revealed by the survey data used this time is 50 m. According to the drilling data and indoor geotechnical test results, the soil layers within the survey range of the project site are divided into artificial accumulation layer and Quaternary alluvial layer according to the sedimentary age and genesis type. The strata in the proposed project area are shown in Table 7.26 from top to bottom.

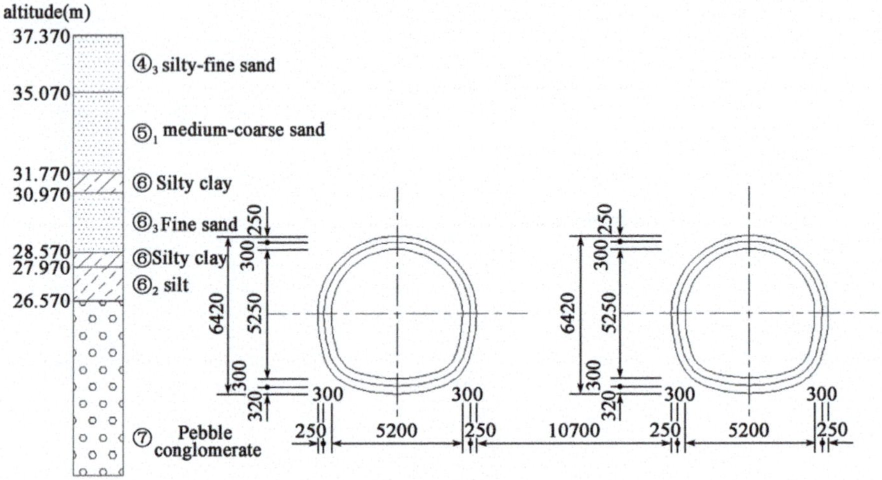

Fig. 7.29 Cross-section of Xuanwumen station - Xidan station section (unit of measurement: mm)

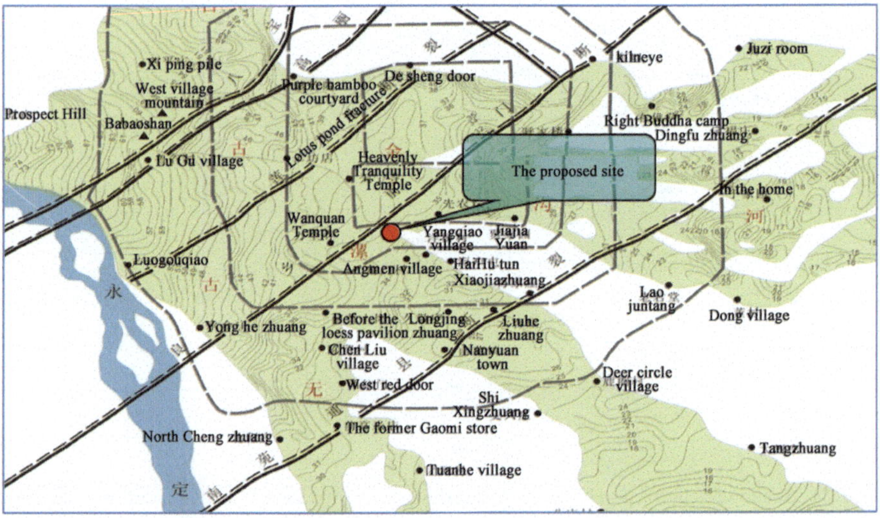

Fig. 7.30 Fault and ancient river distribution map of Beijing Plain

Table 7.26 List of stratum lithological characteristics

Deposit age	Formation code name	Lithology name	Dyestuff	Status	Closely knit linear measure	Humidity	Compressibility	Containing things	The underlying elevation(m)	Distribution
Artificial filling layer (Q^{ml})	①$_1$	Miscellaneous fill	Variegated		A little dense	A little wet		Brick slag, white ash, tree roots	37.50–42.44	Continuous distribution
	①	Powder soil filling	Isabelline		A little dense	A little wet		A small amount of ash residue, tree roots		Continuous distribution
The Quaternary Holocene floodplain (Q_4^{lala}) + ^{pl})	③	Silt	Isabelline		Closely knit	A little wet	Low compressibility	Mica, iron oxide	35.40–38.84	Continuous distribution
	③$_1$	Pink cat soil	Brown yellow–brown yellow	Soft plastic			High compressibility	Mica, iron oxide, ginger stone		Thin layer distribution
	③$_2$	Cat soil	Brown yellow–brown yellow	Can be local soft plastic			High compressibility	Mica, iron oxide, ginger stone		Lensor distribution
	③$_3$	Silty-fine sand	Isabelline		In the secret	A little wet	Low compressibility	Mica, iron oxide		Lensor distribution
	④$_3$	Silty-fine sand	Isabelline		Closely knit	Wet	Low compressibility	Mica, iron oxide	33.41–36.23	Continuous distribution

(continued)

Table 7.26 (continued)

Deposit age	Formation code name	Lithology name	Dyestuff	Status	Closely knit linear measure	Humidity	Compressibility	Containing things	The underlying elevation(m)	Distribution
	④$_4$	Medium-coarse sand	Isabelline		Closely knit	Wet	低压缩性	Mica, a small amount of ginger stone		Thin layer distribution
Quaternary Late Pleistocene alluvium(Q^{al+pl3})	⑤	Pebble conglomerate	Variegated		Closely knit	Wet	Low compressibility	The main component of pebble is diabase, sandstone, coarse sand filling	30.21–32.30	Continuous distribution, with a local lens body distribution
	⑤$_1$	Medium-coarse sand	Isabelline		Closely knit	Wet	Low compressibility	Mica, a small amount of gravel		Continuous distribution
	⑤$_2$	Silty-fine sand	Isabelline		Closely knit	Wet	Low compressibility	Mica, iron oxide		Thin layer distribution
	⑥	Pink cat soil	Brown yellow–brown yellow	Plastic for main local soft plastic			Medium compressibility	Mica, iron oxide, ginger stone	26.01–28.00	Continuous distribution

(continued)

Table 7.26 (continued)

Deposit age	Formation code name	Lithology name	Dyestuff	Status	Closely knit linear measure	Humidity	Compressibility	Containing things	The underlying elevation(m)	Distribution
	⑥$_1$	Cat soil	Brown yellow–brown yellow	Hard plastic			Medium compressibility	Mica, iron oxide, ginger stone		Thin layer or lens body distribution
	⑥$_2$	Silt	Isabelline		Closely knit	Wet	Low compressibility	Mica, iron oxide, a small amount of gravel particles		Continuous distribution
	⑥$_3$	Fine sand	Isabelline		Closely knit	Wet	Low compressibility	Mica, iron oxide		Thin layer distribution
	⑦	Pebble conglomerate	Variegated		Closely knit	Saturation	Low compressibility	The main component of pebble is diabase, sandstone, coarse sand filling	13.50–15.88	Continuous distribution
	⑦$_1$	Medium-coarse sand	isabelline		Closely knit	Saturation	Low compressibility	Mica, a small amount of gravel		Thin layer or lens body distribution

(continued)

Table 7.26 (continued)

Deposit age	Formation code name	Lithology name	Dyestuff	Status	Closely knit linear measure	Humidity	Compressibility	Containing things	The underlying elevation(m)	Distribution
	⑦$_2$	Silty-fine sand	Isabelline		Closely knit	Saturation	Low compressibility	Mica, iron oxide, and individual gravel		Thin layer or lens body distribution
	⑧	Pink cat soil	Brown yellow–brown yellow	Plastic			Medium compressibility	Mica, iron oxide	13.33–15.00	Lensor distribution
	⑧$_2$	Silt	Isabelline		Closely knit	Very wet	Low compressibility	Mica, iron oxide, a small amount of sand particles		Lensor distribution
	⑨	Pebble conglomerate	Variegated		Closely knit	Saturation	Low compressibility	The main component of pebble is diabase, sandstone, coarse sand filling	8.82–12.38	Continuous distribution
	⑨$_1$	Medium-coarse sand	Isabelline		Closely knit	Saturation	Low compressibility	Mica, a small amount of gravel		Lensor distribution

(continued)

Table 7.26 (continued)

Deposit age	Formation code name	Lithology name	Dyestuff	Status	Closely knit linear measure	Humidity	Compressibility	Containing things	The underlying elevation(m)	Distribution
	⑨$_2$	Silty-fine sand	Isabelline		Closely knit	Saturation	Low compressibility	Mica, the individual gravel		Lensor distribution
	⑩	Pink cat soil	Brown yellow–brown yellow	Plastic			Low compressibility	Mica, iron oxide	3.28–9.71	Continuous distribution
	⑩$_2$	Silt	Isabelline		Closely knit	Very wet	Low compressibility	Mica, iron oxide		Thin layer or lens body distribution
	⑪	Pebble conglomerate	Variegated		Closely knit	Saturation	Low compressibility	The main component of pebble is diabase, sandstone, coarse sand filling	–5.83 Still to this level	Continuous distribution
	⑫$_1$	Medium-coarse sand	Isabelline		Closely knit	Saturation	Low compressibility	Mica, a small amount of gravel		Lensor distribution
	⑬$_2$	Silty-fine sand	Isabelline		Closely knit	Saturation	Low compressibility	Mica, the individual gravel		Lensor distribution
	⑭$_3$	Silt	Isabelline		Closely knit	Very wet	Low compressibility	Mica, iron oxide		Lensor distribution

(3) Evaluation of soil corrosivity

The results of the soil corrosivity judgment of the project site are shown in Table 7.27.

4. Hydrogeological conditions

(1) Groundwater level overview

Combining the survey and the used boreholes, the proposed site is within a drilling depth of 30 m. Within the scope, three layers of groundwater were detected, namely the upper stagnant water (I), the confined water (II), and the interlayer water (III). Detailed information about the groundwater can be found in Table 7.28.

Due to the limitations of the initial survey, the groundwater level in this design is based on the highest groundwater level collected from this site, mainly considering the first layer of interlayer water, with the water level elevation considered at 24.41 m.

(2) Groundwater Dynamics

The dynamics of groundwater characterizes the dynamic balance of groundwater recharge and discharge over time. When the recharge of groundwater is greater than the discharge, the groundwater level rises; conversely, when the recharge of groundwater is less than the discharge, the groundwater level decreases. The dynamics of each layer of groundwater have their own characteristics

The dynamics of the upper stagnant water change with the seasons, atmospheric precipitation, and surface water recharge.

The dynamics of confined water are closely related to atmospheric precipitation. The period from July to September is the high-water period for atmospheric precipitation, and the groundwater level begins to rise in July, reaching the highest level of the year in September to October, then gradually decreases, reaching the lowest level of the year in June of the following year, with an average annual range of 2–3 m. Generally, the dynamics of phreatic water are affected by the extraction of agricultural water supply, not directly affected by urban water supply extraction, but because phreatic water has a close hydraulic connection with pressurized water, when the pressure water head decreases, the overflow recharge increases, and the phreatic water level also decreases.

The dynamics of interlayer water mainly receive lateral runoff and overflow recharge, and discharge in the form of lateral runoff. The dynamics of interlayer water lag slightly behind that of phreatic water, with the highest water level of the year appearing from September to November, and the lowest water level appearing from June to July, with an annual range of 1 to 2 m.

(3) Survey of the Highest Water Levels in Previous Years

According to the hydrogeological data near the "Survey Report", the highest water level elevation in the proposed site area in previous years can be found in Table 7.29.

It is recommended that the waterproof level for anti-floating be set at an elevation of 37.00 m.

Table 7.27 Corrosive evaluation form of the soil

Level number	Lithology name	Drill number	Sample degree of depth (m)	Humidity	Formation permeability	By environmental type		By stratum, the permeability	Cl^- (mg/kg)	For building materials Corrosion evaluation	
						SO_2^4-	Mg^{2+} mg/kg)	pH		Concrete	Concrete iron
③	Silt	IV6	8.0	Wet	Weak	89.87	36.84	8.15	144.17	Minute	Minute
$③_1$	Silty clay	IV6	6.0	Wet	Weak	124.08	154.50	8.12	40.97	Minute	Minute
$③_1$	Silty clay	IV12	5.0	Wet	Weak	59	9	8.68	27	Minute	Minute
$③_1$	Silty clay	IV5	8.0	Wet	Weak	42.02	154.50	8.59	81.61	Minute	Minute
$④_3$	Silty-fine sand	4	15.0	Wet	Stubborn	41.64	14.55	7.83	42.15	Minute	Minute
⑤	Pebble conglomerate	32–1	12.8	Wet	Stubborn	64.37	21.12	7.86	42.24	Minute	Minute
	Pebble conglomerate	34–1	15.3			59.81	14.07	7.93	37.70	Minute	Minute
$⑥_2$	Silt	IV6	15.8	Wet	Weak	17.19	6.14	8.05	54.06	Minute	Minute
$⑥_2$	Mucous silt, sandy silt	57	15.5	Wet	Weak	61	13	8.98	104	Minute	Minute
		57	16.0			64	11	8.77	59	Minute	Minute
⑥	Heavy powder clay	IV12	17.3	Wet	Weak	52	14	8.63	25	Minute	Minute

Table 7.28 Table of groundwater characteristics

Ground water type	Stable water level buried deep m)	Stabilize water level elevation (m)	Observation time (year, month)	Aquiclude	Remarks
Perched groundwater (one)	3.0	43.93	2015.10	Mucous silt fill ① layer, miscellaneous fill $①_1$ layer, the water quantity is not large	Collect the nearby engineering data
Phreatic water (two)	16.0–17.50	29.43–31.61	2015.10–2016.3	Fine medium sand ⑥ 3 layer, silty clay ⑥ layer upper water, water is not large	Collect the nearby engineering data
Interlayer water(three)	25.12	19.71	2017.7	Pebble round gravel ⑦ layer, filled with medium and coarse sand	The survey
	25.90	21.01	2017.8	Pebble round gravel ⑦ layer, filled with medium and coarse sand	The survey
	26.24	20.71	2017.8	Pebble round gravel ⑦ layer, filled with medium and coarse sand	The survey
	26.30–27.60	19.68–20.63	2015.10–2016.3	Pebble round ground ⑦ layer, filled with medium and coarse sand	Collect the nearby engineering data
	19.86–22.31	22.60–24.41	2003.7	Pebble round ground ⑦ layer, filled with medium and coarse sand	Data is collected at this site

Table 7.29 The highest water level elevation of the proposed site in the years

A particular year	Water level elevation (m)	a particular year	Water level elevation (m)
1959	44.00	2014–2019	30.00 m (go under water)
1971–1973	37.00		

(4) Evaluation of Groundwater Corrosiveness

The results of the corrosiveness assessment of the groundwater at the project site can be found in Table 7.30.

5. Evaluation of Engineering Geological Conditions

(1) Adverse Geological Effects

According to regional geological data, there are no adverse geological effects such as faults, landslides, mudslides, collapses, subsidence, sandification, liquefaction, etc. in the proposed site.

(2) Specialized Rock and Soil

The special rock and soil in the proposed site mainly consist of fill layers, with no distribution of special rock and soil such as collapsible loess, expansive soil, and residual soil.

According to the data collected from this drilling, fill layers are widely distributed in the proposed site, mainly consisting of silty clay fill layer ① and miscellaneous fill layer $①_1$, with a thickness of 2.4–5.0 m. However, due to frequent urban renovation activities in recent years, the thickness of the fill may have been increased. The consolidation of the fill is poor, and the disturbance of the construction of the subway deep foundation to the site cannot be ignored, especially the subway line 2. Open-cut slope construction, the proposed structure and the channel part of line 2 station may encounter thick fill layers. The mechanical properties of the fill vary greatly, and its stability is poor. It is not suitable to be used as a direct bearing layer without treatment.

(3) Site Stability and Suitability Evaluation

The geological data near the proposed site indicates that there are no adverse geological effects that would impact the overall stability of the site, making it generally suitable for the construction of this project.

6. Surrounding Environment

(1) Current Surrounding Building Situation

Xuanwumen Station is located at the intersection of Xuanwumen East and West Streets and Xuanwumen Inner and Outer Streets, within the second ring of the central old city area of Beijing. The site environment is described in four quadrants as follows.

The northeast quadrant is the South Church of the Catholic Patriotic Association, a national cultural heritage site. To the east of the church is the Changan Kindergarten of Xicheng District. Between Ivory Alley and Dafang Alley to the north of the church, there is a green leisure square about 50 m wide. Along Xuanwumen East Street, there is a municipal green space about 20 m wide on the south side of the church. Along Xuanwumen Inner Street, there are some street-facing shops on the west side of the church, extending into the 16 m red line of Xuanwumen Inner Street.

Table 7.30 Corrosive evaluation form of water

Underground water type	Drilling number	Sampling depth (m)	Humidity of aquifer	Soil permeability	Test index						Corrosive evaluation of building materials		
					SO^{2-}	Mg^{2+} (mg/L)	Total mine decrease degree (mg/L)	pH	HCO_3^- (mg/L)	Cl^- (mg/L)	Concrete	Concrete iron	
												Over a long period of time founder	Dry wet supersede
Perched groundwater (one)	3	3.0	Wet	Stubborn	136.7	28.8	464	7.66	3.22	73.0	Minute	Minute	Minute
Phreatic water (two)	3	17.5	Saturation	Stubborn	89.2	25.2	452	7.74	4.44	66.0	Minute	Minute	Minute
	36–1	16.5			82.7	24.7	430	7.66	4.28	63.6	Minute	Minute	Minute
Interlayer water (three)	3	26.5	Saturation	Stubborn	88.6	36.0	440	7.82	4.33	68.3	Minute	Minute	Minute
	36–1	26.3			92.9	36.6	446	7.84	4.33	68.3	Minute	Minute	Minute
	2	28.0			207.8	42.6	912	7.73	8.57	109.3	Minute	Minute	Weak
	IV12	25.12			99.09	19.01	342	8.56	3.16	73.87	Minute	Minute	Minute
	IV5	25.90			89.16	26.7	644	7.82	4.71	84.24	Minute	Minute	Minute

The northwest quadrant is the northwest entrance, pavilion, and part of the green square of Subway Line 4 station. To the west is the Xinhua News Agency and its residential area, and to the north are low-rise bungalows.

The southwest quadrant is the southwest entrance, pavilion, and part of the green square of Subway Line 4 station. To the west is the Beijing Auto Xuanwumen Gas Station, and to the south is the Global Financial Information Center Building.

The southeast quadrant is a commercial office area, including the Xuanwumen Business Hotel parking lot, Xuanwumen Business Hotel, Xuanwumen Restaurant, and Zhuangsheng Chongguang Department Store, etc.

(2) Current Ground Road and Traffic Situation

The red line width of Xuanwumen East and West Streets is 90 m, which is a city main road with 8 lanes in both directions (9 lanes at the intersection). There are non-motorized lanes and right-turn lanes on both sides of the main road. The traffic is very busy. There is a bus stop on the non-motorized isolation strip (about 60 m from the intersection), with 14 bus lines including No. 15, 22, 332, 44, 67, 7, etc.

The red line width of Xuanwumen Inner Street and Xuanwumen Outer Street is 70 m, both are city main roads with 8 lanes in both directions. There are non-motorized lanes on both sides, and the traffic is very busy. The bus lines include No. 102, 105, 109, 70, 83, etc., totaling 9 lines.

The current main form of street crossing is the ground zebra crossing. Passengers entering and exiting Subway Line 2 need to cross the motorized lane to enter the non-motorized isolation strip to take the subway. There is a pedestrian overpass on Xuanwumen Inner Street, but it is about 300 m from the intersection, and passengers are not willing to take a detour. Apart from this, there are no other crossing facilities in the surrounding area.

(3) Current Underground Pipeline Situation

This project is located at the intersection of city main roads, where underground pipelines are very dense. According to the characteristics of the project, the major control underground pipelines are described in the four quadrants of the intersection as follows.

Northeast quadrant, Xuanwumen East Street side: 2000 mm × 2350 mm power tunnel (bottom buried depth about 8.88 m), D1250 mm sewage pipe (bottom buried depth about 6.12 m), D500 mm rainwater pipe (bottom buried depth about 4.1 m), D600 mm water supply pipe (pipe top buried depth about 1.8 m); Xuanwumen Inner Street side: 2000 mm × 2350 mm power tunnel (bottom buried depth about 7.56 m), D1150 mm sewage pipe (bottom buried depth about 4.3 m), D600 mm Water supply pipe (pipe top burial depth about 2.1 m), D500 mm high-pressure gas pipe (pipe top burial depth about 1.95 m).

Northwest quadrant: 2000 mm × 2350 mm power tunnel (inner bottom burial depth about 8.1 m), 2 × D600 mm thermal pipe (pipe top burial depth 1.6 m), D600 mm water supply pipe (pipe top burial depth about 1.1 m), D1250 mm sewage pipe (inner bottom burial depth about 5.56 m), 720 mm × 520 mm telecommunications pipe (pipe top burial depth about 42.44 m), 360 mm × 250 mm wire pipe (pipe top burial depth 42.44 m), 740 mm × 790 mm wire pipe (pipe top burial depth 42.02 m), 300 mm × 200 mm (pipe top burial depth 43.69 m).

Southwest quadrant: 2600 mm × 1500 mm thermal pipe trench (inner bottom burial depth about 5.25 m), D1000 mm water supply pipe (pipe top burial depth about 3.47 m), D720 mm water supply pipe (pipe top burial depth about 1.8 m), D500 mm gas pipe (pipe top burial depth about 1.9 m), 2000 mm × 2000 mm power trench (inner bottom burial depth about 3.45 m).

Southeast quadrant: 2000 mm × 2200 mm power tunnel (inner bottom burial depth 4.6 m), D1800 mm water supply pipe (pipe top burial depth 2.37 m), 2700 mm × 1400 mm thermal pipe trench (inner bottom burial depth about 2.7 m).

D1000 mm water supply pipe top elevation 43.060 m, burial depth about 1.3 m, arranged along the auxiliary road on the west side of Xuanwumen Inner and Outer Street, pipe material, interface form unknown.

D1000 mm water supply pipe soil cover thickness 3.22 m, pipe top elevation 40.03 m, arranged along the auxiliary road on the south side of Xuanwumen East and West Street, pipe material, interface form unknown.

D720 mm water supply pipe soil cover thickness 1.52 m, pipe top elevation 42.01 m, pipeline diagonally crosses 4 The southwest entrance plaza of Xuanwumen Station on Line 4, then vertically descends under Xuanwumen Inner and Outer Street to join the main pipeline on the south side of Xuanwumen Station C on Line 2, pipe material, interface form unknown.

2000 mm × 2350 mm power tunnel soil cover is about 6.03 m, the inner bottom elevation of the tunnel is about 36.920 m, this pipeline is distributed in the southwest, northeast, southeast and southwest quadrants.

D1250 mm sewage pipeline soil cover is about 5.1 m, the inner bottom elevation of the tunnel is about 38.900 m, this pipeline is arranged along the auxiliary road on the north side of Xuanwumen East and West Street.

2000 mm × 1000 mm thermal tunnel soil cover thickness 3.22 m, tunnel inner bottom elevation 39.23 m, this tunnel is arranged in a cross pattern along the west side of Xuanwumen Inner and Outer Street and the south side of Xuanwumen East and West Street.

1800 mm × 1500 mm rainwater pipe trench soil cover thickness 2.22 m, pipe top elevation 40.02 m, arranged along the auxiliary road on the east side of Xuanwumen Inner and Outer Street.

The main pipeline layout around the new project is shown in Fig. 7.31.

7.6.2 Risk Identification

1. Risk source list

The main construction risk sources of Xuanwumen Station renovation project are shown in Tables 7.31, 7.32, 7.33, and 7.34.

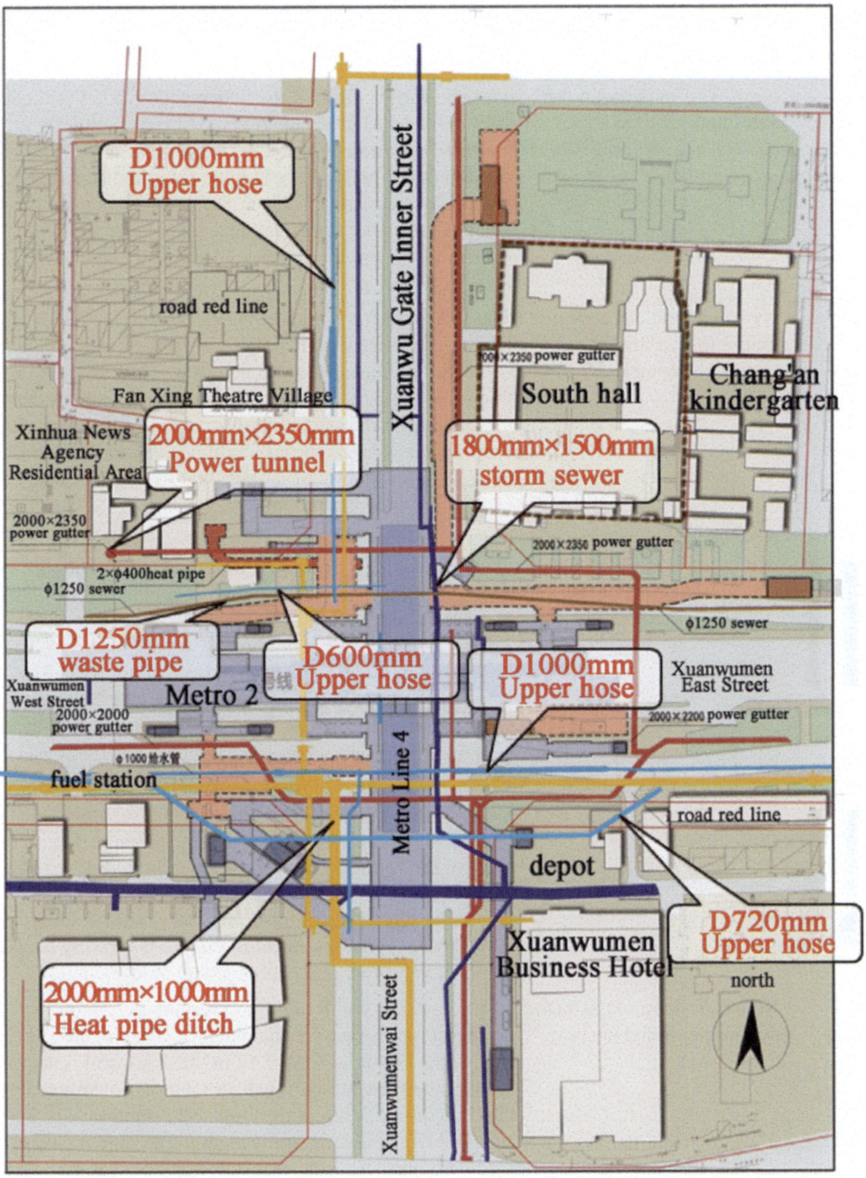

Fig. 7.31 New project surrounding main pipeline (trench, tunnel) layout diagram

Table 7.31 List of self-risk sources for Xuanwumen Station renovation project

Number	Risk source name	Risk source situation
S1	Northwest undercut ticket sales hall	The excavation width of the underground ticket office is 14.9 m, the excavation height is 7.2 m, and the soil covering thickness is about 9.7 m. The excavation is divided into three layers and three rows; the structure vault is located in the ④ layer, the structure is mainly ④ layer of fine sand, ⑥ layer, fine sand ⑥ layer and silt ⑥ layer, and the structural bearing layer is located in the ⑦ layer of pebble gravel
S2	New undercut entrance and exit in the northwest (transfer channel of line 4)	The section of the entrance channel is the channel to Line 4, the internal civil air defense, the excavation width is 6.8–8.3 m, the excavation height is 5.7 m, the soil thickness is about 11 m, is flat top straight wall structure, CRD method construction; the structure vault is located in fine sand ④ layer, the structure is ④ layer, fine sand ⑥ layer, ⑥ layer, fine sand ⑥ layer, the structure bearing layer is located in silty clay ⑥ layer
S3	New undercut entrance and exit in the northwest (transfer channel of line 2)	The entrance channel is the channel connecting the existing Line 2, is the climbing section, the climbing Angle is 30, the excavation width is 6.4–7.6 m, the excavation height is 5.93–8.87 m, the soil thickness is 3.1–11.2 m, the vault straight wall structure, using CRD method construction; the buried depth is fine sand ④ layer, the structure bearing layer is silty clay ⑥ layer; ③ layer, and the structure holding layer is fine sand ④ layer
S4	Northeast north open excavation ticket office	The excavation width is 15.9 m, Excavation length: 37.5 m, The deepest part of the foundation pit is about 17.7 m, Adopt ϕ 800 bored pile + ϕ 609 steel pipe internal support envelope system construction; The soil layer of the foundation pit from top to bottom is mainly ① layer, silty filling ① layer, silty ③ layer, silty clay ③ layer, fine sand ④ layer, medium coarse sand ⑤ layer, silty clay ⑥ layer, fine medium sand ⑥ layer, silty clay ⑥ layer, silty ⑥ layer and pebble gravel ⑦ layer
S5	New entrances and exits to the northeast and north	The excavation width of the excavation section of northeast North transfer channel is 7.2–9.9 m, the excavation height is 4.92–7.17 m, the thick soil is about 11.3 m, the vault straight wall structure is constructed by CRD method; the structure vault soil is mainly silty fine sand ④ layer, the structure is mainly fine sand ④ layer, medium coarse sand ⑤ layer, fine clay ⑥ layer, fine clay ⑥ layer, and the structural bearing layer is mainly located in the ⑦ layer of pebble round gravel

(continued)

Table 7.31 (continued)

Number	Risk source name	Risk source situation
S6	Northeast east open excavation ticket office	The open excavation pit adopts two phases: the excavation width of the first phase envelope structure is 9.4 m, Excavation length: 49.4 m, The depth of foundation pit excavation is 5.7–9.6 m, The main soil layer of the side wall of the foundation pit from top to bottom is the mixed fill layer ①, silty filling layer ①, silty fill layer ③, silty clay ③ layer, silty fine sand ④ layer, medium coarse sand ⑤ layer, The foundation pit embedded section is mainly located in the fine medium sand ⑥ layer; The excavation width of foundation pit of second II envelope structure is 5.5 m, Excavation length: 57.1 m, The excavation depth of the foundation pit is 6–10 m, The main soil layer of the side wall of the foundation pit from top to bottom is ① layer of miscellaneous filling soil, ① layer of silty filling soil, ③ layer of silty soil, ③ layer of silty clay, ④ layer of silty sand, The foundation pit insert is mainly located in the fine sand ④ layer
S7	Northeast east new undercut entrance (connect line 4 transfer channel)	The excavation width of the southeast open excavation foundation pit is 12.2 m, the excavation length is 33.3 m, the deepest foundation pit is about 8 m, and the ϕ 800 bored pile + 800 mm 1000 mm concrete support envelope system is used for construction. The main soil layer on the side wall of the foundation pit is mainly miscellaneous fill ① layer, silty filling ① layer, silty ③ layer, silty clay ③ layer, silty fine sand ④ layer, medium coarse sand ⑤ layer, silty clay ⑥ layer, fine medium sand ⑥ layer, silty clay ⑥ layer, silt ⑥ layer and pebble gravel ⑦ layer
S8	Southeast open excavation ticket office	The excavation size of no. 3 shaft is 5.3 m 10.7 m and 36.5 m deep. The shaft excavation support adopts steel grille + double mesh + anchor pipe grouting, and the inverted well wall method is adopted. In the location of the shaft, ① 1 layer, clay silt ③ layer, clay ③ 1 layer, fine sand ④ 3 layer, pebble ⑤ layer, clay ⑥ layer, clay ⑥ 2 layer, pebble ⑦ layer, medium sand ⑦ 1 layer, and groundwater is above the bottom plate
S9	Southwest undercut ticket office	The underground hall is a three-span middle column underground hall, the underground director is 21.5 m, 14.9 m wide, the height is 6.22 m, the vault straight wall, which is constructed by the middle hole method
S10	Southwest undercut entrance (line 4)	The excavation width of the southwest quadrant transfer channel (connecting line 4) is 6.1–7.2 m, and the excavation height is 4.9–8.5 m. The CRD method is adopted

(continued)

Table 7.31 (continued)

Number	Risk source name	Risk source situation
S11	Northwest construction shaft	The excavation dimension of the construction shaft is 5.1 m 7.2 m, and the excavation depth is about 20.41 m, using the "inverted well wall method" construction; the structure is mainly mixed filling soil ① layer, silty soil ③ layer ①, silty fine sand ④ layer, medium coarse sand ⑤ layer, silty clay ⑥ layer, fine medium sand ⑥ layer, silty clay ⑥ layer, silt clay ⑥ layer and pebble gravel ⑦ layer
S12	Northwest construction crossings	The excavation width of the construction transverse channel is 4.1 m, the excavation height is 7.55–10.6 m, and the soil covering thickness is 7.3–10.35 m; the structural vault is located in ④ layer of fine sand, and the structure is mainly ④ layer of fine sand, medium coarse sand ⑤ layer, silty clay ⑥ layer, fine medium sand ⑥ layer and silt ⑥ layer, and the structural bearing layer is located in the ⑦ layer of pebble round gravel
S13	Southwest construction shaft	The excavation size of the construction shaft is 7.3 m 16 m, and the excavation depth is about 15.2 m, using the "inverted well wall method" construction
S14	Southwest construction crossings	The excavation width of the transverse channel is 5.8 m, and the excavation height is 8.35–9.1 m. CRD method is adopted

2. Risk assessment unit division

The construction risk sources of Xuanwumen Station renovation project are screened, and the major risk sources are divided into risk assessment units, the results are shown in Table 7.35.

3. Division of Risk Assessment Units

The construction risk sources of the Xuanwumen Station renovation project are screened, and the major risk sources are divided into risk assessment units. The results are shown in Table 7.35.

7.6.3 Overall Risk Assessment

The overall risk assessment scores and calculation results of the Xuanwumen Station renovation project are shown in Table 7.36. The weight of the project scale G_1 indicator is 0.2, the weight of the geological condition G_2 indicator is 0.3, the weight of the environmental condition G_3 indicator is 0.4, and the weight of the technical difficulty G_4 indicator is 0.1.

According to Table 7.39, the overall risk level of the Xuanwumen Station renovation project is Level I, and a specialized risk assessment is required.

Table 7.32 List of environmental risk sources around Xuanwumen Station Reconstruction project (underground structures)

Number	Risk source name	Risk source situation	Related works	Position relation
U1	The main structure of the existing line 4 station	Three-span double column arch straight wall frame structure, side span Degree 6750 mm, medium span 6400 mm, structure net height 12800 mm	Northwest undercut ticket office	The horizontal net distance from the main structure of line 4 station is about 5.4 m
			New entrances and exits to the northeast and north	Adjacent, the horizontal net distance from the main structure of the station is about 1.7 m. Connecting, the hole form is rectangular section, the hole size is 7200 mm × 4150 mm
			Northwest transfer channel	Connecting, the hole form is rectangular section, the hole size is 6200 mm × 5000 mm
			New northeast east entrance	Connecting, the hole form is rectangular section, the hole size is 7200 mm × 4850 mm
			Southwest transfer channel	Connecting, the hole form is rectangular section, the hole size is 5500 mm × 4250 mm
			Southwest undercut ticket office	It is necessary to remove the side wall of line 4 station 5400 mm × 4500 mm range
U2	There is the main structure of line 4 interval	The roof is covered with 16.3 m, the bottom depth is 22.4 m, the structure width is 5.20 m and the net height is 5.25 m	New entrances and exits to the northeast and north	Parallel up through, and the net distance between the entrance and exit is about 2.3 m
U3	The existing northwest entrance of line 4	The standard section is a single span horseshoe section with a net span of 5100 mm and a net height of 4950 mm	Northwest construction shaft	Nearby, the net distance between the existing line 4 entrance is about 5.45 m
			Northwest undercut ticket office	Nearby, the net distance between the northwest entrance and exit structure of Xuanwumen Station of the existing line 4 is about 0.2 m. Construction shall remove the existing structure within 6200 mm × 4350 mm

(continued)

Table 7.32 (continued)

Number	Risk source name	Risk source situation	Related works	Position relation
U4	The existing southwest entrance of line 4		Southwest undercut ticket office	Adjacent, the net distance from the entrance and exit of line 4 is about 1.8 m
			New entrances and exits to the southwest	Adjacent, the horizontal net distance between the initial support skin of the entrance and exit channel is about 8.9 m
U5	The existing northwest air duct of line 4	Single-span double-layer vault straight wall frame structure with net span of 8000 mm and net height of 11250 mm	Northwest construction shaft	Adjacent, the net distance from the air well of the existing line 4 is about 8.05 m
U6	The existing line 4 transfer channel	–	Northwest undercut ticket office	Nearby, the net distance between the existing line 4 is about 0.5 m
			New northeast east entrance	Adjacent, the horizontal net distance between the civil air defense section of the entrance and exit channel is about 3.93 m
			New entrances and exits to the southwest	Adjacent, the horizontal net distance between the civil air defense section of the entrance and exit channel is about 7.9 m
U7	The main structure of the existing line 2 station	Three-span two-column rectangular frame structure, side span 6500 mm, medium span 5900 mm, structure net height 6050 mm	Southeast open excavation ticket office	Nearby, the net distance from the existing line 2 air duct and the main structure is about 1 m
			New northeast east entrance	Connecting, the hole form is rectangular section, the hole size is 5000 mm × 3950 mm
U8	The existing northwest entrance of line 2	The section is a rectangular frame section + U-shaped groove section, with a span of 3500 mm and the net height of the rectangular section of 4300 mm	Northwest transfer channel	The horizontal net distance between the climb and the entrance is 1.5–2.3 m and the vertical net distance is 3.0–5.6 m; the horizontal net distance between the entrance is 0.1–1.3 m. Connecting, the hole form is rectangular section, the hole size is 5300 mm 4200 mm

(continued)

Table 7.32 (continued)

Number	Risk source name	Risk source situation	Related works	Position relation
U9	The existing northeast entrance of line 2	The section is a rectangular frame section + U-shaped groove section, with a span of 3500 mm and the net height of the rectangular section of 4300 mm	New northeast east entrance	At the adjacent area, the horizontal net distance of the initial support skin is about 5.5 m
			Northeast winter Open excavation ticket office	Adjacent, the horizontal net distance from the entrance of Line 2 D is about 4.4 m
U10	The existing southwest entrance of line 2	The section is a rectangular frame section + U-shaped groove section, with a span of 3500 mm and the net height of the rectangular section of 4300 mm	Southwest undercut ticket office	Adjacent, the net distance from the entrance of Line 2 is about 8.3 m
			Southwest transfer channel	Connecting, the hole form is rectangular section, the hole size is 5100 mm × 4250 mm
U11	Xinhua channel		Southwest transfer hall	Cut length 15 m
U12	300 mm × 4300 mm force chamber	The bottom plate of the thermal chamber is 6.3 m deep	Northwest construction channel	Vertical down, the net distance between the bottom plate and the top of the transverse channel is about 4.1 m
U13	2600 mm × 1500 mm thermodynamic tunnel	The soil covering thickness of the thermal pipe is 3.22 m, and the bottom elevation of the pipe is 39.23 m	Southwest construction channel	Vertical down, and the construction of the horizontal channel dense paste through
			Southwest undercut ticket office	Parallel down, and the net distance from the southwest underground hall is 1.1 m
U14	2000 × 1000 mm thermodynamic tunnel	The soil covering thickness of the thermal pipe is 1.07 m, and the bottom elevation of the tunnel is 42.18 m	New entrances and exits to the southwest	Vertical down, the net distance between the southwest quadrant transfer channel is 8.11 m
U15	1900 mm × 900 mm thermodynamic tunnel	The thermal tunnel covering soil is about 5.0 m	Northwest undercut ticket office	In parallel to lower, the horizontal net distance between the entrance and exit is about 2 m, and the oblique net distance between the tunnel bottom and the initial support skin of the new structure is about 5.1 m

(continued)

Table 7.32 (continued)

Number	Risk source name	Risk source situation	Related works	Position relation
U16	2000 mm × 2350 mm power tunnel	The buried depth of the power tunnel is 6.0 m, and the bottom elevation of the tube is about 36.920 m	Northwest construction channel	In parallel to the bottom, the oblique net distance between the bottom of the groove and the transverse channel is about 3.1 m
			Northwest undercut ticket office	Vertical down, the net distance is about 1.8 m
			New entrances and exits to the northeast and north	In parallel and down, the net distance between the pipe bottom and the dark excavation structure is about 2.0 m
			Northeast north Ming dig the ticket office	Adjacent, the horizontal net distance from the foundation pit is about 5.6 m
U17	2000 mm × 2200 mm power tunnel	The soil covering thickness of the power tunnel is 7.8 m, and the pipe top elevation is 36.12 m	New northeast east entrance	Upper span, pipe top and foundation pit base close paste
			Northeast east open excavation ticket office	Upper span, the structural floor cushion is densely attached to the tunnel roof
			Southeast open excavation ticket office	Adjacent, the net distance with the open dig foundation pit is 1.89 m
U18	2000 mm × 2000 mm power tunnel	The thickness of power tunnel is 0.86 m and the bottom elevation is 41.09 m	Southwest undercut ticket office	Vertical down, and the net distance between the underground hall in the southwest quadrant is 3.14 m
			New entrance to the southwest	Vertical down, and the net distance between the construction horizontal channel is 1.54 m
U19	800 m × 1200 mm power tunnel	The soil thickness of the power tunnel is 1.4 m, and the bottom elevation of the power tunnel is 41.94 m	New northeast east entrance	Vertical down, the net distance with the dark excavation channel is 7.34 m
U20	1800 mm × 1500 mm rainwater pipe trench	The soil of the rainwater pipe trench is about 3.68 m, and the bottom elevation of the pipe is 40.050 m	New entrances and exits to the northeast and north	The parallel bottom wear, close to the initial stage of the entrance channel, the net distance between the bottom of the pipe and the dark excavation structure is about 6.3 m
			New northeast east entrance	Vertical down, the net distance with the dark excavation channel is 5.62 m

Table 7.33 List of environmental risk sources around Xuanwumen Station renovation project (ground building structures)

Number	Risk source name	Risk source situation	Related projects	Position relation
B1	The Catholic Patriotic Church, the South Hall	The building is a national cultural relics protection unit	New entrances to the northeast and north	Through parallel down, the horizontal net distance between the entrance and exit is about 15.0–17.5 m
B2	One-and two-story residences		Northwest construction shaft	Adjacent, the net distance from the bungalow is about 5.8 m

7.6.4 Specialized Risk Assessment

1. Weight of Specialized Assessment Indicators

The weights of the primary indicators used in this specialized assessment are shown in Table 7.37.

2. Risk Assessment of Roof Collapse

The risk assessment indicator system, indicator scores, and scoring instructions for the roof collapse of the Southwest Underground Ticket Inspection Hall (S_9) are shown in Table 7.38.

3. Risk Assessment of Mud and Water Inrush

The indicator system for risk assessment, indicator scores, and scoring instructions for the mud burst and water inrush of the Southwest Underground Ticket Inspection Hall (S9) are shown in Table 7.39.

4. Risk Assessment of Pipeline Damage

Taking the 2 D400 mm thermal pipelines (P1) as an example, the indicator system for risk assessment, indicator scores, and scoring instructions for pipeline damage are shown in Table 7.40. The summary of the assessment results is shown in Table 7.41. The specific scores for the major risk factors (project excavation) are shown in Table 7.42.

5. Risk Assessment of Ground Building Damage

The indicator system for risk assessment, indicator scores, and scoring instructions for the damage of the Catholic Patriotic Association South Church (B1) ground building are shown in Table 7.43.

6. Risk Assessment of Underground Building Damage

Taking the existing Line 4 section main structure (U2) as an example, the indicator system for risk assessment, indicator scores, and scoring instructions for underground

Table 7.34 List of surrounding environmental risk sources (pipeline) of Xuanwumen Station Reconstruction Project

Number	Risk source name	Risk source situation	Related projects	Position relation
P1	2 D400 mm thermal power lines	The thermal pipe is about 1.67 m and the elevation of the pipe is about 43.580 m	Northwest construction channel	Parallel to the lower wear, the net distance between the tube bottom and the top of the horizontal channel is about 8.3 m
P2	2 D600 mm thermal power lines	The thermal pipe is about 4.43 m and the elevation of the pipe is about 40.820 m	Northwest construction channel	Vertical down, the net distance between the pipe bottom and the top of the horizontal channel is about 2.3 m
			Northwest undercut ticket office	Parallel down, the net distance of about 5.33 m
P3	D600 mm Upper water pipe	The top pipe is about 1.18 m and the top elevation is about 42.04 m	New entrances and exits to the northeast and north	Through parallel down, the net distance between the pipe bottom and the dark excavation structure is about 7.8 m
			Northeast north open excavation ticket office	Adjacent, the horizontal net distance from the foundation pit is about 15.0 m
P4	D720 mm Upper water pipe	The soil thickness of the upper pipe is 1.52 m and the top elevation is 42.01 m	Southwest construction channel	Vertical down, and the net distance between the construction horizontal channel is 2.33 m
			Southwest undercut ticket office	The net distance from the southwest quadrant underground hall is 4.5 m; the pipeline diameter is greater than 0.6 m
P5	D1000 mm Upper water pipe	The thickness of the upper pipe is about 2.19 m, and the elevation of the pipe is 43.060 m	Northwest construction channel	Vertical down, the net distance between the pipe bottom and the top of the horizontal channel is about 4.1 m
P5	D1000 mm Upper water pipe	The thickness of the upper pipe is about 2.19 m, and the elevation of the pipe is 43.060 m	Northwest undercut ticket office	Parallel down, the net distance is about 6.6 m
			Northwest transfer channel	Vertical down, and the net distance between the entrance and exit vault is 5.4 m

(continued)

Table 7.34 (continued)

Number	Risk source name	Risk source situation	Related projects	Position relation
			Southwest construction channel	Vertical down, and the net distance between the construction horizontal channel is 0.35 m
			Southwest undercut ticket office	Parallel down, and the net distance from the southwest Quadrant underground hall is 2.06 m
			New entrances and exits to the southwest	Parallel down, and the net distance from the southwest Quadrant underground hall is 2.06 m
P6	D1250 mm Sewage treatment pipe	The soil covering of the sewage pipe is about 5.1 m, and the bottom elevation of the pipe is about 38.900 m	Northwest undercut ticket office	Vertical down, the net distance is about 3.4 m
			Northwest transfer channel	For line 4, the net distance between the entrance and exit vault is 0–6 m; and for line 2, the vertical net distance between the entrance and exit vault is 5.2 m
			New northeast east entrance	Adjacent, the horizontal net distance between the foundation pit and the sewage pipe is about 7.0 m
P7	D400 mm storm sewer	The soil thickness of the rainwater pipe is 2.38 m, and the bottom elevation of the pipe is 42.47 m	Northwest transfer channel	In parallel and down, the net distance between the tube bottom and the channel is 0.77 m
			New entrances and exits to the southwest	Vertical down, and the net distance with the construction horizontal channel is 2.57 m
P8	D600 mm Rainwater pipe	The thickness of rainwater pipe is 2.22 m and the top elevation is 40.02 m	New northeast east entrance	Through parallel down, the net distance between the dark excavation channel is 1.4–2.5 m

(continued)

Table 7.34 (continued)

Number	Risk source name	Risk source situation	Related projects	Position relation
P9	D500 mm Gas pipe	The gas pipe is about 0.39 m and the elevation of the pipe is about 42.9 m	New entrances and exits to the northeast and north	Through parallel down, the net distance between the pipe bottom and the dark excavation structure is about 6.3 m
			Northeast north open excavation ticket office	Adjacent, the horizontal net distance from the foundation pit is about 13.0 m
			Southwest construction channel	Vertical down, and the net distance between the construction horizontal channel is 2.84 m
			Southwest undercut ticket office	Parallel down, and the net distance from the southwest quadrant underground hall is 5.04 m
			New entrances and exits to the southwest	Vertical down, the minimum net distance between the transfer channel and the southwest quadrant is 4.48 m
P10	70 mm × 98 mm communication tube	The thickness of communication pipe is 0.96 m and the elevation of pipe top is 42.31 m	Southwest construction channel	Vertical down, and the net distance between the construction horizontal channel is 3.27 m
			Southwest undercut ticket office	Parallel down, and the net distance from the southwest quadrant underground hall is 4.63 m

building damage are shown in Table 7.44. The summary of the assessment results is shown in Table 7.45. The specific scores for the major risk factors (project excavation) are shown in Table 7.46.

7. Risk Assessment of Excessive Deformation or Damage to Existing Structures Caused by Connection Opening

Taking the existing Line 2 station main structure (U7) as an example, the risk assessment indicator system, indicator scores, and scoring instructions for excessive deformation or damage to existing structures caused by connection opening are shown in Table 7.47. The summary of the assessment results is shown in Table 7.48. The specific scores for the major risk factors (project excavation) are shown in Table 7.49.

Table 7.35 Division of risk assessment unit of Xuanwumen Station reconstruction project

Assessment unit	Risk source	Main risk factors
Project itself excavation	Southwest Undercut Ticket Office (S9)	Excavation of middle hole method
The surrounding pipeline	2 D400 mm thermal lines (P1)	Excavation of the northwest construction transverse tunnel (S12)
	2 D600 mm thermal lines (P2)	Excavation of the northwest construction transverse tunnel (S12)
		Excavation of Northwest Underground Excavation of ticket Office (S1)
	D1250 mm Sewage pipeline (P6)	Excavation of Northwest Underground Excavation of ticket Office (S1)
		Excavation of northwest transfer channel (S2, S3)
		New entrance excavation in northeast east (S7)
	D500 mm Gas line (P9)	New entrance excavation in northeast north (S5)
		Excavation for north-northeast open excavation ticket office (S4)
		Excavation of Southwest Construction Cross Corridor (S14)
		Excavation of Southwest Ticket Office (S9)
		Excavation of Southwest Transfer Channel (S10)
Ground building structures	The South Hall of the Catholic Patriotic Church (B1)	New entrance excavation in northeast north (S5)
Underground construction structures	Main structure of existing Line 4 Station (U1)	Excavation of northwest underground ticket Office (S1)
		New entrance excavation in northeast north (S5)
		Excavation of northwest transfer channel (S2, S3)
		New entrance excavation in northeast east (S7)
		Excavation of Southwest Transfer Channel (S10)
		Excavation of Southwest underground ticket Office (S9)

(continued)

Table 7.35 (continued)

Assessment unit	Risk source	Main risk factors
	Existing main structure of Line 4 (U2)	New entrance excavation in northeast north (S5)
	Existing main structure of Line 2 (U7)	Excavation of southeast open excavation ticket office (S8)
		New entrance excavation in northeast east (S7)
	3300 mm × 4300 mm thermal compartment (U12)	Excavation of the northwest construction transverse tunnel (S12)
	2600 mm × 1500 mm thermal tunnel (U13)	Excavation of Southwest Construction Cross Corridor (S14)
		Excavation of Southwest Office (S9)
	2000 mm × 1000 mm thermal tunnel (U14)	New entrance excavation in southwest (S10)
	2000 mm × 2350 mm power tunnel (U16)	Excavation of the northwest construction transverse tunnel (S12)
		Excavation of ticket Office in northwest (S1)
		New entrance excavation in northeast north (S5)
		Excavation of North Office (S4)
	2000 mm × 2200 mm power tunnel (U17)	Excavation of new entrance in northeast east (S7)
		Excavation of Northeast east open excavation Office (S6)
		Excavation of open excavation ticket Office in southeast(S8)
	2000 mm × 2000 mm power tunnel (U18)	Excavation of Southwest Office (S9)
		New entrance excavation in southwest (S10)
Expansion connection opening	Main structure of existing line 4 Station (U1)	New entrance excavation in northeast north (S5)
		Excavation of northwest transfer channel (S2, S3)
		New entrance excavation in northeast east (S7)
		Excavation of Southwest Transfer Channel (S10)

(continued)

Table 7.35 (continued)

Assessment unit	Risk source	Main risk factors
		Excavation of Southwest Office (S9)
	The existing northwest entrance and exit of line 4 (U3)	Excavation of ticket Office in northwest (S1)
	Main structure of the existing line 2 station (U7)	New entrance excavation in northeast east (S7)
	Northwest entrance and exit of the existing line 2 (U8)	Excavation of Northwest Transfer Channel (S3)
	Existing southwest entrance of line 2 (U10)	Excavation of Southwest Transfer Channel (S10)
Expansion, excavation and unloading	Main structure of existing line 4 Station (U1)	Excavation of ticket Office in northwest (S1)
		Excavation of Southwest Office (S9)
	Existing main structure of line 2 (U7)	Excavation of southeast open excavation ticket office (S8)
		New entrance excavation in northeast east (S7)

8. Risk Assessment of Existing Structure Floating Caused by Excavation Unloading

Taking the existing Line 4 Taking the main structure (U1) of Line 4 station as an example, the excavation unloading leads to the indicator system for risk assessment, index scores, and scoring instructions for the existing structure's uplift risk, as shown in Table 7.50. The assessment results summary is shown in Table 7.51. The specific scores for major risk factors (engineering excavation) are shown in Table 7.52.

7.6.5 Conclusion

(1) The overall risk assessment conclusion shows that the overall risk level of the Xuanwumen station renovation project is Level I, with numerous construction risk sources, high risk in management difficulty, and a need for specialized risk assessment.
(2) The specialized risk assessment conclusion shows that the risk levels of major risk events during the construction of the North Xuanwumen station renovation project are mostly Level II or Level III. Since this specialized assessment is based on the construction plan used in actual construction, and some lower-risk sources have been screened out in the risk identification stage, this conclusion is relatively reasonable.

Table 7.36 Overall risk assessment results of Xuanwumen Station reconstruction project

<table>
<tr><th>Primary index</th><th colspan="2">Secondary and tertiary indicators</th><th>Score</th><th colspan="2">Description</th></tr>
<tr><td>Project scale G_1</td><td colspan="2">Project scale G_{11}</td><td>80</td><td colspan="2">After the transformation, a four-story subway station will form a large underground space of the same scale</td></tr>
<tr><td rowspan="2">Geology G_2</td><td colspan="2">Engineering geology G_{21}</td><td>90</td><td colspan="2">The new structures are mainly located in silt, silty clay and silty fine sand layers</td></tr>
<tr><td colspan="2">Hydrogeology G_{22}</td><td>60</td><td colspan="2">The new structures are mainly involved in submersible and upper stagnant water</td></tr>
<tr><td rowspan="7">Environmental conditions G_3</td><td colspan="2">Ground grade position G_{31}</td><td>40</td><td colspan="2">The minimum horizontal distance between the more important ground structures and the project is about 15.0 m</td></tr>
<tr><td colspan="2">Location of the underground structure G_{32}</td><td>100</td><td colspan="2">The project involves a large number of connections and close proximity additions</td></tr>
<tr><td colspan="2">Road location G_{33}</td><td>100</td><td colspan="2">It is located below the main road in the city</td></tr>
<tr><td rowspan="3">Pipeline G_{34}</td><td>Pipeline location G_{341}</td><td>100</td><td colspan="2" rowspan="3">According to the value of D1250 mm sewage pipe, the new project passes through the pipeline vertically, and the pipeline is a concrete pipe, and the service life is inferred to be 5–30 years</td></tr>
<tr><td>Pipeline Vulnerability G_{342}</td><td>80</td></tr>
<tr><td>Years of service G_{343}</td><td>80</td></tr>
<tr><td colspan="2">Surface water body G_{35}</td><td>0</td><td colspan="2">There are no surface water bodies nearby</td></tr>
<tr><td>Technical difficulty G_4</td><td colspan="2">Construction process maturity G_{41}</td><td>40</td><td colspan="2">The new structure mostly adopts the CRD method, the upside-down shaft wall method, and the middle hole method, and the process is mature</td></tr>
<tr><td colspan="3">Factor K for the type of large underground space</td><td>1.2</td><td colspan="2">Network expansion of large space, take 1.2</td></tr>
<tr><td colspan="3">Total score G</td><td>100</td><td>Overall risk level</td><td>I</td></tr>
</table>

Table 7.37 Xuanwumen Station renovation project specialized risk assessment weight

Evaluate the object	Metric name/weight				
Cut-and-cover pipeline failure	Pipeline condition R_1	Pit condition R_2	Geological conditions R_3	Construction scheme R4	
	$\gamma_1 = 0.4$	$\gamma_2 = 0.1$	$\gamma_3 = 0.3$	$\gamma_4 = 0.2$	
Excavation of underground structures Destruction	Conditions of underground structures R_1	Foundation pit conditions R2	Geological conditions R_3	Construction plan R_4	
	$\gamma_1 = 0.4$	$\gamma_2 = 0.2$	$\gamma_3 = 0.2$	$\gamma_4 = 0.2$	
Underground excavation and collapse to the top	Coverage span ratio R_1	Geological conditions R_2	Rain in the affected area Sewer R_3	Construction scheme R_4	
	$\gamma_1 = 0.25$	$\gamma_2 = 0.35$	$\gamma_3 = 0.25$	$\gamma_4 = 0.15$	
Digging mud and gushing water	Geological conditions R_1	Environmental conditions R_2			
	$\gamma_1 = 0.8$	$\gamma_2 = 0.2$			
Underground excavation pipeline damage	Pipeline condition R_1	Excavation section condition R_2	Geological conditions R_3	Construction scheme R_4	
	$\gamma_1 = 0.4$	$\gamma_2 = 0.1$	$\gamma_3 = 0.3$	$\gamma_4 = 0.2$	
Underground excavation of ground structures Destruction	Ground Construct Condition R_1	Excavation section condition R_2	Geological conditions R_3	Construction plan R_4	
	$\gamma_1 = 0.4$	$\gamma_2 = 0.2$	$\gamma_3 = 0.2$	$\gamma_4 = 0.2$	
Underground excavation of underground structures Destruction	Conditions of underground structures R_1	Excavation section condition R_2	Geological conditions R_3	Construction plan R_4	
	$\gamma_1 = 0.4$	$\gamma_2 = 0.2$	$\gamma_3 = 0.2$	$\gamma_4 = 0.2$	
The splicing opening causes excessive deformation or destruction of the existing structure	Status quo of existing structure R_1	Opening size R_2	Opening position R_3	Opening method R_4	Degree of existing structural disturbance R_5
	$\gamma_1 = 0.2$	$\gamma_2 = 0.3$	$\gamma_3 = 0.1$	$\gamma_4 = 0.2$	$\gamma_5 = 0.2$
Excavation and unloading lead to the floating of existing structures	Status of existing structures R_1	Unload volume R_2	Relative positional relation R_3	Soil condition R_4	
	$\gamma_1 = 0.1$	$\gamma_2 = 0.35$	$\gamma_3 = 0.2$	$\gamma_4 = 0.35$	

(continued)

Table 7.37 (continued)

Evaluate the object	Metric name/weight				
Risk of loss	Casualty C_1	Economic loss C_2	Duration loss C_3	Social Impact C_4	Environmental impact C_5
	$W_1 = 0.4$	$W_2 = 0.2$	$W_3 = 0.15$	$W_4 = 0.15$	$W_5 = 0.1$

(3) The existing Line 4 station main structure (U1) has a high level of potential for deformation or damage due to multiple disturbances from expansion openings, and needs to be given special attention, with effective measures taken to reduce the cumulative adverse effects of multiple expansions.

The potential level of damage to the 2000 mm × 2200 mm power tunnel (U17) is Level 1. Special attention should be paid to the reinforcement and monitoring of these important risk sources during construction.

7.7 Case Study of Construction Safety Risk Assessment for Tianjin Metro Line 5 Siyuan Road Station

7.7.1 *Project Background*

1. Project Overview

Tianjin Metro Line 5 Siyuan Road Station is located in the plot where Siyuan Road, Qunfang Road, Baiyang Road, and Hongxing Road intersect in Hebei District. The plan location and structural form of Siyuan Road Station are detailed in Sect. 2.3.3.

2. Hydrogeological Engineering

(1) Geological Conditions

According to the survey, the strata of the proposed station from top to bottom are miscellaneous fill, silty soil, silty clay, silty clay, and a compound stratum mainly composed of silty sand. The pit is located in the $8th_1$ silty clay layer and the $8th_2$ silty soil layer, which has low permeability.

The strata in this section mainly consist of the Quaternary Holocene artificial fill layer (artificial accumulation), the Neogene sedimentary layer (Quaternary Holocene Neogene pit bottom siltation and ancient river channel, depression alluvium), the first continental layer (Quaternary Holocene upper group riverbed–river floodplain sedimentation), the first marine layer (Quaternary Holocene middle group shallow sea sedimentation), the second continental layer (Quaternary Holocene lower group swamp sedimentation), the second continental layer (Quaternary Holocene lower group riverbed–river floodplain sedimentation), the third continental layer (Quaternary Upper Pleistocene five group riverbed–river floodplain sedimentation), the

Table 7.38 Risk assessment of roof collapse results

Primary index	Primary index weight	Secondary index	Secondary index weight	Index score	Description	
Cover−span ratio R_1	$\gamma_1 = 0.25$	Coverage span ratio R_{11}	$\gamma_{11} = 1.0$	62	the southwest excavated ticket hall span is 14.9 m and the covering thickness is 6.7 m with a coverage span ratio of 0.45	
Geological conditions R_2	$\gamma_2 = 0.35$	Engineering geology R_{21}	$\gamma_{21} = 0.6$	80	There is a lack of data, and it is speculated to be located in silty clay and fine to medium sand layers	
		Hydrogeology R_{22}	$\gamma_{22} = 0.4$	70	The groundwater level is located within the excavation range	
Within the affected area storm sewer pipes R_3	$\gamma_3 = 0.15$	Pipe diameter R_{31}	$\gamma_{31} = 0.4$	0	There are no rainwater sewers in the affected area	
		Pipeline status R_{32}	$\gamma_{32} = 0.6$	0		
Construction scheme R_4	$\gamma_4 = 0.25$	Advance reinforcement R_{41}	$\gamma_{41} = 0.25$	20	Advance grouting with small catheters	
		Sedimentary precipitation R_{42}	$\gamma_{42} = 0.25$	70	No precipitation construction, due to the small amount of water, the score is lowered	
		Excavation method R_{43}	$\gamma_{43} = 0.25$	70	The southwest underground excavation ticket hall adopts the middle hole method for construction	
A score on the likelihood of risk occurring				55.4	The level of likelihood of the risk occurring	2
Personnel loss C_1		$W_1 = 0.4$		60	In the construction of the middle cave method, there are a large number of people working on the face of the tunnel in the excavation stage, and there is a possibility of causing more than 10 deaths	

(continued)

Table 7.38 (continued)

Primary index	Primary index weight	Secondary index	Secondary index weight	Index score	Description	
Economic loss C_2		$W_2 = 0.2$		40	The collapse of the ticket hall requires large-scale search and rescue, slag removal, backfilling and reinforcement, and may cause a series of secondary disasters, which is expected to cause economic losses of more than 10 million yuan	
Duration loss C_3		$W_3 = 0.15$		50	Control the construction period of the project, the collapse is expected to cause a loss of about 6 months	
Social Impact C_4		$W_4 = 0.15$		50	Above the ticket hall is below the main road in the bustling urban area, and the collapse will inevitably lead to the paralysis of the road, and the social impact is more serious	
Environmental Impact C_5		$W_5 = 0.1$		50	The collapse will lead to a wide range of overall settlement of the upper stratum, involving a large area	
Risk loss score				52.0	Risk loss level	B
Risk level					II	

second marine layer (Quaternary Upper Pleistocene four group coastal–tidal zone sedimentation), the fourth continental layer (Quaternary Upper Pleistocene three group riverbed–river floodplain sedimentation), the third marine layer (Quaternary Upper Pleistocene two group shallow sea–coastal sedimentation), the fifth continental layer (Quaternary Upper Pleistocene one riverbed–river floodplain sedimentation), and the fourth marine layer (Quaternary Middle Pleistocene upper group coastal delta sedimentation.), the VI terrestrial layer (riverbed to river floodplain sedimentation in the Middle Pleistocene of the Quaternary) is detailed in Table 7.53 and Fig. 7.32.

Table 7.39 Results of risk assessment of mud and water inrush

Primary index	Primary index weight	Secondary index	Secondary index weight	Index score	Description	
Geological conditions R_1	$\gamma_1 = 0.5$	Engineering geology R_{11}	$\gamma_{11} = 0.4$	40	There is a lack of data, and it is speculated to be located in silty clay and fine to medium sand layers	
		Degree of water abundance R_{12}	$\gamma_{12} = 0.6$	40	The results of the projection of the lack of water inflow were judged to be weak and rich according to the geological survey report	
Environmental conditions R_2	$\gamma_2 = 0.2$	Surrounding water bodies R_{21}	$\gamma_{21} = 1.0$	0	There is no surrounding water body	
Construction scheme R_3	$\gamma_3 = 0.3$	Advance reinforcement R_{31}	$\gamma_{31} = 0.4$	20	The same as the risk assessment of roof collapse	
		Sedimentary precipitation R_{32}	$\gamma_{32} = 0.6$	60	The same as the risk assessment of roof collapse	
A score on the likelihood of risk occurring				33.2	The level of likelihood of the risk occurring	3
Loss of personnel C_1		$W_4 = 0.4$		50	In the construction of the middle cave method, there are a large number of people working on the face of the tunnel in the excavation stage, and there is a possibility of causing more than 10 deaths	
Economic loss C_2		$W_2 = 0.2$		35	The mud and water inrush in the ticket hall requires large-scale dredging, backfilling and reinforcement, which is expected to cause economic losses of more than 10 million yuan	

(continued)

Table 7.39 (continued)

Primary index	Primary index weight	Secondary index	Secondary index weight	Index score	Description	
Loss of construction time C_3		$W_3 = 0.15$		30	The construction period is controlled, and the mud and water inrush is expected to cause a loss of about 3 months	
Social impact C_4		$W_4 = 0.15$		10	The social impact of the mud and water inrush is small	
Environmental impact C_5		$W_5 = 0.1$		35	The mud and water inrush will lead to local subsidence of the upper stratum, involving a large area	
Risk loss score				36.5	Risk loss level	C
Risk level					III	

(3) Hydrological conditions

The surface groundwater type in this site is Quaternary porous aquifer, which is confined within the II terrestrial layer and below in the silty sand and silty soil layers.

The silty soil and silty sand layers in and below the II terrestrial layer of the site are confined water aquifers, separated into multiple aquifers by sticky soil. The ⑧ $_2$ silty soil is the first confined aquifer, with the ⑦ clay layer and ⑧ $_1$ silty clay as the main aquiclude top; the ⑨ $_2$ silty soil layer and ⑩ $_2$ silty soil layer are the second confined aquifer, with the ⑨ $_1$ silty clay layer as the main aquiclude top. The two confined aquifers mainly receive seepage recharge from the upper phreatic water, have close hydraulic connection with the upper phreatic water, discharge through underground runoff, and recharge the lower aquifer through seepage. The confined water level is less affected by seasons, and the stable water level depth of the second confined aquifer is 3.50–4.00 m, with a water level elevation of –0.68––0.18 m.

① The first confined aquifer

The ⑧ $_2$ silty soil is the first confined aquifer, with the ⑦ clay layer and ⑧ $_1$ silty clay as the main aquiclude top, with a top burial depth of 16.9–20.4 m. The water-stop curtain in the site has cut off the first confined aquifer.

② The second confined aquifer

The ⑨ $_2$ silty soil layer and ⑩ $_2$ silty soil layer are the second confined aquifer, with the ⑨ $_1$ silty clay layer as the main aquiclude top. The top burial depth is 24.0–27.8 m,

Table 7.40 Pipeline damage risk assessment results

Primary index	Primary index weight	Secondary index	Secondary index weight	Index score	Description
Pipeline condition R_1	$\gamma_1 = 0.4$	Pipeline vulnerability R_{11}	$\gamma_{11} = 0.3$	40	Lack of information inferred that the D400 thermal line is made of stainless steel
		Pipeline fitting form R12	$\gamma_{12} = 0.2$	50	Lack of information inferred that the D400 thermal line is a flange joint
		Pipeline location R_{13}	$\gamma_{13} = 0.3$	90	The northwest construction transverse passage runs parallel to the pipeline, and the net distance between the bottom of the pipe and the underground excavation structure is about 7.8 m, the buried depth of the bottom plate of the transverse channel is about 17.9 m
		Service life R_{14}	$\gamma_{14} = 0.2$	80	There is a lack of information, and it is inferred to be 5–30 years
Excavation section conditions R_2	$\gamma_2 = 0.1$	Section size R_{21}	$\gamma_{21} = 0.5$	0	The cross-sectional area of the excavation degree of the northwest construction transverse channel is 36.72 m^2, which is much smaller than the lowest grade rating basis, take 0
		Cover−span ratio R_{22}	$\gamma_{22} = 0.5$	40	The minimum thickness of the covering soil of the northwest construction cross passage is 7.3 m with a span of 4.1 m with a span ratio of 1.78
Geological conditions R_3	$\gamma_3 = 0.3$	Engineering geology R_{31}	$\gamma_{31} = 0.6$	80	The same as the risk assessment of roof collapse
		Hydrogeology R_{32}	$\gamma_{32} = 0.4$	70	The same as the risk assessment of roof collapse

(continued)

Table 7.40 (continued)

Primary index	Primary index weight	Secondary index	Secondary index weight	Index score	Description	
Construction scheme R_4	$\gamma_4 = 0.2$	Pipeline reinforcement R_{41}	$\gamma_{41} = 0.3$	80	No pipeline reinforcement	
		Sedimentary precipitation R_{42}	$\gamma_{42} = 0.2$	20	No precipitation construction	
		Excavation method R_{43}	$\gamma_{43} = 0.25$	100	The northwest construction transverse passage is excavated by step method	
A score on the likelihood of risk occurring				64.9	The level of likelihood of the risk occurring	2
Loss of personnel C_1		$W_1 = 0.4$		0	Thermal line failures result in almost no casualties	
Economic loss C_2		$W_2 = 0.2$		10	The economic loss caused by the destruction of the thermal pipeline is expected to be no more than 10 million	
Duration loss C_3		$W_3 = 0.15$		10	The failure of the thermal pipeline has little impact on the construction period	
Social Impact C_4		$W_4 = 0.15$		30	The failure of the thermal pipeline will lead to the interruption of local district heating, which has a certain social impact	
Environmental Impact C_5		$W_5 = 0.1$		10	The failure of the thermal pipeline has less impact on the surrounding environment	
Risk loss score				9.0	Risk loss level	D
Risk level					III	

Table 7.41 Summary of pipeline failure risk assessment results

Risk sources	Pipeline conditions				Excavation section conditions	Stratigraphic conditions	
	Pipeline vulnerability	Pipeline fitting form	Pipeline location	Service life		Engineering geology	Hydrogeology
2 × D400 mm thermal lines (P1)	40	50	90.0	80	20.0	80	70
2 × D600 mm thermal lines (P2)	40	50	109.5	80	47.5	90	70
D1250 mm sewage pipeline (P6)	40	50	76.5	80	58.3	90	70
D500 mm gas pipe (P9)	40	50	129.5	80	62.0	90	70

Risk sources	Construction plan			Risk loss				
	Pipeline reinforcement	Sedimentary precipitation	Excavation method	Casualties	Economic loss	Loss of construction time	Social impact	Environmental impact
2 × D400 mm thermal lines (P1)	80	20	100.0	0	10	10	30	10
2 × D600 mm thermal lines (P2)	80	20	88.0	20	10	10	25	25
D1250 mm sewage pipeline (P6)	80	20	96.0	20	10	10	30	30

(continued)

Table 7.41 (continued)

Risk sources	Construction plan			Risk loss				
	Pipeline reinforcement	Sedimentary precipitation	Excavation method	Casualties	Economic loss	Loss of construction time	Social impact	Environmental impact
D500 mm gas pipe (P9)	80	20	131.0	20	10	10	50	50

Risk source	Total Likelihood Score	Loss of total points	Probability level	Loss level	Risk level
2 × D400 mm thermal lines (P1)	64.9	9.0	2	D	III
2 × D600 mm thermal lines (P2)	71.0	17.8	2	D	III
D1250 mm sewage pipeline (P6)	68.7	19.0	2	D	III
D500 mm gas pipe (P9)	77.7	24.0	1	D	II

Table 7.42 Scale of significant risk factors for pipeline damage

Risk sources	Risk factors	Pipeline location	Section size/pit depth	Cover-span ratio	Excavation method	Pipeline location total score	Total score of excavation section condition/foundation pit condition	Excavation total score for construction method
2 × D400 mm thermal lines (P1)	Northwest construction transverse channel excavation (S12)	90	0	40	100	90.0	20.0	100.0
2 × D600 mm thermal lines (P2)	Northwest construction transverse channel excavation (S12)	95	0	50	80	109.5	47.5	88.0
	Northwest excavation of the ticket hall excavation (S1)	100	40	50	80			
D1250 mm sewage line (P6)	Northwest excavation of the ticket hall excavation (S1)	95	40	65	80	76.5	58.3	96.0

(continued)

Table 7.42 (continued)

Risk sources	Risk factors	Pipeline location	Section size/pit depth	Cover-span ratio	Excavation method	Pipeline location total score	Total score of excavation section condition/ foundation pit condition	Excavation total score for construction method
	Northwest transfer channel excavation (S2, S3)	85	0	65	80			
	New entrances and exits were excavated in the northeast and east (S7)	88	0	50	80			
D500 mm gas pipe (P9)	New entrances and exits in the northeast and north are excavated (S5)	85	0	40	80	129.5	62.0	131.0
	The northeast north excavation ticket hall was excavated (S4)	80	40		80			

(continued)

Table 7.42 (continued)

Risk sources	Risk factors	Pipeline location	Section size/pit depth	Cover-span ratio	Excavation method	Pipeline location total score	Total score of excavation section condition/ foundation pit condition	Excavation total score for construction method
	Southwest construction transverse channel excavation (S14)	90	0	50	100			
	The southwest underground excavation ticket hall was excavated (S9)	90	40	60	70			
	Excavation of Southwest transfer channel (S10)	95	0	70	80			

Table 7.43 Damage risk assessment results of surface structures

Primary index	Primary index weight	Secondary index	Secondary index weight	Index score	Description
Ground construction Chikumono conditions R_1	$\gamma_1 = 0.4$	Building structure scaleR_{11}	$\gamma_{11} = 0.2$	0	Special low-rise buildings
		Basic form R_{12}	$\gamma_{12} = 0.3$	100	Lack of information is considered on an independent basis
		Location of the building R_{13}	$\gamma_{13} = 0.5$	50	The new entrance and exit in the northeast and north runs parallel to the building, and the clear distance from the entrance and exit level is about 15.0–17.5 m
Excavation section conditions R_2	$\gamma_2 = 0.2$	Section size R_{21}	$\gamma_{21} = 0.5$	0	The excavation section area of the new entrance and exit in the northeast is about 36 m^2, which is far smaller than the lowest grade scoring basis, and 0 is taken
		Cover−span ratio R_{22}	$\gamma_{22} = 0.5$	70	The soil cover thickness of the new inlet and outlet in the north-east is about 5.4 m, span 7.2 m with a coverage span ratio of 0.75
Geological conditions R_3	$\gamma_3 = 0.2$	Engineering geology R_{31}	$\gamma_{31} = 0.6$	90	The same as the risk assessment of roof collapse
		Hydrogeology R_{32}	$\gamma_{32} = 0.4$	70	The same as the risk assessment of roof collapse
Construction scheme R4	$\gamma_4 = 0.2$	Building structure reinforcement R_{41}	$\gamma_{41} = 0.3$	40	During the excavation process, the outside of the contour line of the primary branch of the arch 1.5 m inside 0. The soil in the range of 5 m is reinforced by deep hole grouting
		Sedimentary precipitation R_{42}	$\gamma_{42} = 0.2$	20	No precipitation construction

(continued)

Table 7.43 (continued)

Primary index	Primary index weight	Secondary index	Secondary index weight	Index score	Description	
		Excavation method R_{43}	$\gamma_{43} = 0.25$	90	The new entrances and exits in the northeast and north of the country are constructed using the CRD method	
A score on the likelihood of risk occurring				55.7	The level of likelihood of the risk occurring	2
Loss of personnel C_1		$W_1 = 0.4$		60	There is a possibility that more than 10 people will be killed in the destruction of the South Church of the Catholic Patriotic Association	
Economic loss C_2		$W_2 = 0.2$		50	In order to protect the cultural relics, the restoration work of the South Church of the Catholic Patriotic Association is expected to cost more than 10 million yuan	
Loss of construction time C_3		$W_3 = 0.15$		60	The destruction of the South Church of the Catholic Patriotic Association is expected to result in the loss of more than 6 months of construction time	
Social impact C_4		$W_4 = 0.15$		80	In order to protect cultural relics, the South Church of the Catholic Patriotic Association had a bad social impact after the destruction	
Environmental impact C_5		$W_5 = 0.1$		10	Except in extreme cases, above-ground building damage has a limited environmental impact	
Risk loss score				57.0	Risk loss level	B
Risk level					II	

Table 7.44 Damage risk assessment results of underground structures

Primary index	Primary index weight	Secondary index	Secondary index weight	Index score	Description
Underground construction building conditions R_1	$\gamma_1 = 0.4$	Dimensions of underground structuresR_{11}	$\gamma_{11} = 0.2$	50	The net width of the main structure of the existing line 4 section is 5.20 m with a clear height of 5.25 m, which belongs to the small section tunnel
		Structural form R_{12}	$\gamma_{12} = 0.3$	70	The main structure of the existing line 4 section is a prefabricated assembly tunnel
		Location of the building R_{13}	$\gamma_{13} = 0.5$	40	The new entrance and exit in the northeast and north of the northeast pass parallel to the existing line 4 section tunnel, and the net distance between the entrance and the mouth is about 2.3 m
Excavation section conditions R_2	$\gamma_2 = 0.2$	Section size R_{21}	$\gamma_{21} = 0.5$	0	The excavation section area of the new entrance and exit in the northeast is about 36 m^2, which is far smaller than the lowest grade scoring basis, and 0 is taken
		Cover−span ratio R_{22}	$\gamma_{22} = 0.5$	70	The thickness of the new entrance and exit covering soil in the northeast north is about 5.4 m, span 7.2 m with a coverage span ratio of 0.75
Geological conditions R_3	$\gamma_3 = 0.2$	Engineering geology R_{31}	$\gamma_{31} = 0.6$	90	The same as the risk assessment of roof collapse
		Hydrogeology R_{32}	$\gamma_{32} = 0.4$	70	The same as the risk assessment of roof collapse

(continued)

Table 7.44 (continued)

Primary index	Primary index weight	Secondary index	Secondary index weight	Index score	Description
Construction scheme R4	$\gamma_4 = 0.2$	Building structure reinforcement R_{41}	$\gamma_{41} = 0.3$	20	The entrance and exit structure carry out deep hole grouting on one side of the interval structure, and reinforces the soil
		Sedimentary precipitation R_{42}	$\gamma_{42} = 0.2$	20	No precipitation construction
		Excavation method R_{43}	$\gamma_{43} = 0.25$	90	The new entrances and exits in the northeast and north of the country are constructed using the CRD method
A score on the likelihood of risk occurring				51.5	The level of likelihood of the risk occurring 2
Personnel loss C_1		$W_1 = 0.4$		70	Damage to existing inter-line tunnels may result in more than 10 deaths
Economic loss C_2		$W_2 = 0.2$		30	The damage to the existing line section tunnel, affecting the operation of the subway, resulting in economic losses and repair work is expected to cost more than 10 million yuan
Duration loss C_3		$W_3 = 0.15$		50	The damage to the tunnel between the existing lines is expected to cause a loss of more than 6 months
Social Impact C_4		$W_4 = 0.15$		30	The existing line station was damaged, and the social impact was serious

(continued)

Table 7.44 (continued)

Primary index	Primary index weight	Secondary index	Secondary index weight	Index score	Description	
Environmental Impact C_5		$W_5 = 0.1$		40	The failure of the existing line section tunnel may have a serious impact on the stratigraphic structure of a certain range in the vicinity	
Risk loss score				50.0	Risk loss level	B
Risk level					II	

locally existing in lens form, and the water-stop curtain in the site has cut off the first confined aquifer.

The permeability coefficient and permeability performance of each soil layer are shown in Table 7.54.

3. Surrounding environment of the pit

(1) Surrounding buildings

The Sinopec gas station on Hongxing Branch Road (Fig. 7.33), the maximum excavation depth of the pit for the underground space engineering construction site of Siyuan Road Station is 20.15 m. The building within the range of 3 times the pit excavation depth is the Sinopec gas station on Hongxing Branch Road. This gas station is a one-story brick flat house 21 m away from the pit, and the gas station canopy is a columnar shed structure, the nearest point to the pit is 21 m. The underground oil depot of the gas station is buried 2–3 m deep and is 45 m away from the pit.

(2) Surrounding existing subway stations

The Tianjin Metro Line 5 Siyuan Road Station is located in the plot where Siyuan Road, Qunfang Road, Baiyang Road, and Hongxing Road intersect in Hebei District.

The main body of the station is located on the east side of the city's rolling mill factory three, the station platform center mileage is DK12 + 303, the starting and ending mileage is DK12 + 148.450–DK12 + 382.350, the station line spacing is 15 m, the platform width is 12 m; the total building area of the station is 14655 m^2, length 191.08 m, the total width of the standard section is 20.7 m, the depth of the foundation pit is 16.811 m, the width of the shield well section is 24.7 m, the depth of the foundation pit is 18.431 m, the bottom of the foundation pit of Siyuan Road Station is 1.99 m higher than the foundation pit of the building, combined with the structure of the upper property development, the foundation pit at the small mileage end of the station is irregular in shape.

The station structure details are shown in Table 7.55, and the plan position is shown in Fig. 7.34.

Table 7.45 Summary of damage risk assessment results of underground structures

Evaluation unit	Underground structure conditions			Excavation section conditions	Geological conditions	
	Dimensions of underground structures	Structural form	The location of the structure		Engineering geology	Hydrogeology
The main structure of the existing line 4 station (U1)	100	40	140	61.4	90	70
There is a main structure of the section of line 4 (U2)	50	70	40	16.0	90	70
The main structure of the existing line 2 station (U7)	100	40	88	51.6	90	70
3300 mm × 4300 mm thermal chamber (U12)	55	40	90	20.0	90	70
2600 mm × 1500 mm thermal tunnel (U13)	40	40	86	50.0	90	70
2000 mm × 1000 mm thermal tunnel (U14)	40	40	70	28.0	90	70
2000 mm × 2350 mm power tunnel (U16)	40	40	127	51.6	90	70
2000 mm × 2200 mm power tunnel (U17)	40	40	116	86.0	90	70
2000 mm × 2000 mm power tunnel (U18)	40	40	99	50.8	90	70

(continued)

Table 7.45 (continued)

Evaluation unit	Construction plan			Risk of Loss				
	Reinforcement of buildings	Sedimentary precipitation	Excavation method	Casualties	Economic loss	Loss of construction time	Social impact	Environmental impact
The main structure of the existing line 4 station (U1)	20	20	119	80	50	50	50	60
There is a main structure of the section of line 4 (U2)	20	20	90	70	30	50	30	40
The main structure of the existing line 2 station (U7)	20	20	88	80	50	50	50	60
3300 mm × 4300 mm thermal chamber (U12)	70	20	80	0	10	10	20	10
2600 mm × 1500 mm thermal tunnel (U13)	70	20	107	0	10	10	20	10
2000 mm × 1000 mm thermal tunnel (U14)	70	20	80	0	10	10	20	10

(continued)

Table 7.45 (continued)

Evaluation unit	Construction plan			Risk of Loss				
	Reinforcement of buildings	Sedimentary precipitation	Excavation method	Casualties	Economic loss	Loss of construction time	Social impact	Environmental impact
2000 mm × 2350 mm power tunnel (U16)	70	20	96	0	10	10	40	10
2000 mm × 2200 mm power tunnel (U17)	70	20	96	0	10	10	40	10
2000 mm × 2000 mm power tunnel (U18)	70	20	87	0	10	10	40	10

Evaluation unit	Total likelihood score	Loss of total points	Probability level	Loss level	Risk level
The main structure of the existing line 4 station (U1)	74.1	63.0	2	B	II
There is a main structure of the section of line 4 (U2)	48.7	50.0	3	C	III
The main structure of the existing line 2 station (U7)	65.7	63.0	2	B	II
3300 mm × 4300 mm thermal chamber (U12)	59.6	7.5	2	D	III
2600 mm × 1500 mm thermal tunnel (U13)	65.4	7.5	2	D	III
2000 mm × 1000 mm thermal tunnel (U14)	56.0	7.5	2	D	III
2000 mm × 2350 mm power tunnel (U16)	73.7	10.5	2	D	III
2000 mm × 2200 mm power tunnel (U17)	77.9	10.5	1	D	II
2000 mm × 2000 mm power tunnel (U18)	66.8	10.5	2	D	III

Table 7.46 Evaluation table of major risk factors for damage of underground structures

Risk sources	Key risk factors	Construct building location	Section size/pit depth	Cover-span ratio	Excavation construction method/ earthwork excavation	Structure, location, total score	Total score of excavation section condition/ foundation pit condition	Excavation method/ earthwork excavation score
There is a main station of line 4 structure (U1)	Northwest excavation of the ticket hall excavation (S1)	80	40	50	80	140.0	61.4	119.0
	New entrances and exits in the northeast and north are excavated (S5)	100	0	40	80			
	Northwest transfer channel excavation (S2, S3)	100	0	65	80			
	New entrances and exits were excavated in the northeast and east (S7)	60	0	50	80			

(continued)

Table 7.46 (continued)

Risk sources	Key risk factors	Construct building location	Section size/pit depth	Cover-span ratio	Excavation construction method/ earthwork excavation	Structure, location, total score	Total score of excavation section condition/ foundation pit condition	Excavation method/ earthwork excavation score
	Excavation of the southwest transfer channel (S10)	70	0	70	80			
	The southwest underground excavation ticket hall was excavated (S9)	90	40	60	70			
There is a section of line 4 structure (U2)	New entrances and exits in the northeast and north are excavated (S5)	40	0	40	90	40.0	16.0	90.0
There is a 2-line main structure (U7)	Southeast cut-and-cover ticket hall excavation (S8)[a]	80	50		80	88.0	51.6	88.0

(continued)

Table 7.46 (continued)

Risk sources	Key risk factors	Construct building location	Section size/pit depth	Cover-span ratio	Excavation construction method/ earthwork excavation	Structure, location, total score	Total score of excavation section condition/ foundation pit condition	Excavation method/ earthwork excavation score
	New entrances and exits were excavated in the northeast and east (S7)	80	0	40	80			
3300 mm × 4300 mm heat chamber (U12)	Northwest construction transverse channel excavation (S12)	90	0	50	80	90.0	20.0	80.0
2600 mm × 1500 mm thermal tunnel (U13)	Southwest construction transverse channel excavation (S14)	80	0	50	100	86.0	50.0	107.0

(continued)

Table 7.46 (continued)

Risk sources	Key risk factors	Construct building location	Section size/pit depth	Cover-span ratio	Excavation construction method/earthwork excavation	Structure, location, total score	Total score of excavation section condition/foundation pit condition	Excavation method/earthwork excavation score
	The southwest underground excavation ticket hall was excavated (S9)	60	40	60	70			
2000 mm × 1000 mm thermal tunnel (U14)	Excavation of new entrances and exits in the southwest (S10)	70	0	70	80	70.0	28.0	80.0
2000 mm × 2350 mm power tunnel (U16)	Northwest construction transverse channel excavation (S12)	100	0	50	80	127.0	51.6	104.0
	Northwest excavation of the ticket hall excavation (S1)	90	40	50	80			

(continued)

Table 7.46 (continued)

Risk sources	Key risk factors	Construct building location	Section size/pit depth	Cover-span ratio	Excavation construction method/ earthwork excavation	Structure, location, total score	Total score of excavation section condition/ foundation pit condition	Excavation method/ earthwork excavation score
	New entrances and exits in the northeast and north are excavated (S5)	100	0	40	80			
	Northeast North cut-and-dig ticket office excavation (S4) [a]	80	40		80			
2000 mm × 2200 mm power tunnel (U17)	Excavation of new entrances to the northeast and east (S7)	100	0	50	80	116.0	86.0	96.0
	Northeast East Cut-and-Dig Ticket Office Excavation (S6) [a]	80	40	–	80			

(continued)

Table 7.46 (continued)

Risk sources	Key risk factors	Construct building location	Section size/pit depth	Cover-span ratio	Excavation construction method/ earthwork excavation	Structure, location, total score	Total score of excavation section condition/ foundation pit condition	Excavation method/ earthwork excavation score
	Southeast cut-and-cover ticket hall excavation (S8) *	80	80		80			
2000 mm × 2000 mm power tunnel (U18)	The southwest underground excavation ticket hall was excavated (S9)	90	40	60	70	99.0	50.8	87.0
	Excavation of new entrances and exits in the southwest (S10)	90	0	70	80			

Note [a] indicates that the risk source is open-cut construction

Table 7.47 Risk assessment results of excessive deformation or failure of existing structures due to connection openings

Primary index	Primary index weight	Indicator score	Description	
Status quo of existing structure R_1	$\gamma_1 = 0.2$	50	The existing Xuanwumen Station of line 2 is structurally intact	
Opening size R_2	$\gamma_2 = 0.3$	50	The size of the opening is 5000 mm × 3950 mm, and the area is 19.75 m^2	
Opening position R_3	$\gamma_3 = 0.1$	90	The new entrance and exit in the northeast east are opened in the side wall of the existing station	
Opening method R_4	$\gamma_4 = 0.2$	50	The new entrances and exits in the northeast and east are manually or mechanically opened	
Degree of existing structural disturbance R_5	$\gamma_5 = 0.2$	70	The jointing process can disrupt the side walls of the existing structure	
A score on the likelihood of risk occurring		58.0	The level of likelihood of the risk occurring	2
Personnel loss C_1	$W_1 = 0.4$	30	Damage to existing stations due to connection openings may result in the death of more than three people	
Economic loss C_2	$W_2 = 0.2$	40	Improper connection openings may lead to damage to the existing station floor and roof, which is difficult to repair, and is expected to cause a loss of more than 10 million yuan	
Duration loss C_3	$W_3 = 0.15$	30	The delay in construction due to the destruction of the existing station due to the connection opening is expected to be more than 3 months	
Social Impact C_4	$W_4 = 0.15$	20	The destruction of the existing station due to the connection opening will affect the operation of the subway and have a certain social impact	
Environmental Impact C_5	$W_5 = 0.1$	20	The existing station is damaged due to the connection opening, which has little impact on the surrounding environment	

(continued)

Table 7.47 (continued)

Primary index	Primary index weight	Indicator score	Description	
Risk loss score		29.5	Risk loss level	C
Risk level			II	

The station's retaining structure is an underground continuous wall, with a wall thickness of 800 mm, an effective length of 28 m, a plain concrete section of 5 m, separating the second pressurized aquifer⑨ $_2$ loess layers, the outer wall thickness of the station's main underground basement is 700 mm. The first support of the station is a concrete support, except for the shield well and the excavation outlet position due to the shield construction, the remaining supports are still retained, and the broken part of the shield well and the excavation outlet position has been restored with ϕ800 mm steel support.

The station is an underground double-layer island platform station, the underground first layer is the hall layer, the underground second layer is the platform layer, the height of the underground first layer is 5.85 m, the height of the underground second layer is 6.74 m; this project is located on both sides of the main body of Siyuan Road Station, the basement structure of this project is at the same elevation as the top of the underground first layer of Siyuan Road Station, and the bottom of the underground second layer is at the same elevation as the top of the underground second layer of Siyuan Road Station, as shown in Fig. 7.35 shown.

(3) Surrounding Pipelines

① Water Pipe

There is a water pipeline near the east side of the foundation pit of Siyuan Road Station, the closest distance from this pipeline to the continuous underground wall is 6.21 m, this water pipe is a D300 mm ductile iron pipe, buried depth is 3.5 m.

② High Voltage Power Line

On the south side of the project's foundation pit, 28 m away from the pit, there is a 10 kV high voltage transmission line, running north–south along the pedestrian walkway of Hongxing Branch Road, the line height is 12 m, the closest distance from this high voltage cable to the north side of the pit is about 7 m.

7.7.2 Risk Identification

1. Risk Source List

The risk source list for the construction of the Siyuan Road foundation pit is shown in Tables 7.56, 7.57, 7.58, and 7.59.

Table 7.48 Summary of the results of the risk assessment of excessive deformation or failure of existing structures due to connection openings

Risk sources	Existing structure status quo	Opening size	Opening position	Opening method	Existing structural disturbance degree	Risk of loss					Total likelihood score	Loss of total points	Probability level	Loss level	Risk level
						Casualties	Economic loss	Duration loss	Social impact	Environmental impacts					
The main structure of the existing line 4 station (U1)	50	121	126	70	98	30	40	30	20	20	92.5	29.5	1	C	II
There is a northwest entrance and exit of line 4 (U3)	50	80	90	50	70	10	10	15	20	10	67.0	12.3	2	D	III
There is a 2-line main structure (U7)	50	50	90	50	70	30	40	30	20	20	58.0	29.5	2	C	II
There is a northwest entrance and exit of line 2 (U8)	50	50	90	50	70	10	10	15	20	10	58.0	12.3	2	D	III
There is a southwest entrance and exit of line 2 (U10)	50	50	90	50	70	10	10	15	20	10	58.0	12.3	2	D	III

Table 7.49 A rating sheet of major risk factors for excessive deformation or failure of existing structures due to connection openings

Risk sources	Key risk factors	Opening size	Opening position	Opening method	Existing structural disturbance degree	The opening size of the total score	Opening position total score	Open method total score	Existing structures Degree of perturbation score
Existing line 6 Dongdaqiao Station (U1)	Northeast North new entrance and exit excavation (S5)	75	90	50	70	11.0	16.0	70.0	98.0
	Northwest transfer channel excavation (S2, S3)	80	90	50	70				
	Northeast East new entrance and exit excavation (S7)	85	90	50	70				
	Excavation of Southwest transfer channel (S10)	70	90	50	70				

(continued)

Table 7.49 (continued)

Risk sources	Key risk factors	Opening size	Opening position	Opening method	Existing structural disturbance degree	The opening size of the total score	Opening position total score	Open method total score	Existing structures Degree of perturbation score
	Southwest excavation ticket hall excavation (S9)	90	90	50	70				
Existing line 4 Northwest Entrance (U3)	Northwest Underground excavation ticket office excavation (S1)	80	90	50	70	80.0	90.0	50.0	70.0
Existing line 2 main structure (U7)	Northeast East new entrance and exit excavation (S7)	50	90	50	70	50.0	90.0	50.0	70.0
Existing line 2 Northwest entrance (U8)	Northwest transfer channel excavation (S3)	50	90	50	70	50.0	90.0	50.0	70.0
Existing line 2 Southwest entrance (U10)	Excavation of Southwest transfer channel (S10)	50	90	50	70	50.0	90.0	50.0	70.0

Table 7.50 Risk assessment results of existing structure floating due to excavation and unloading

Primary index	Primary index weight	Indicator score		Description	
Status of existing structures R_1	$\gamma_1 = 0.2$	50		The existing Xuanwumen Station of Line 2 is structurally intact	
Unload volume R_2	$\gamma_2 = 0.3$	S1	40	S1 is the excavation of the northwest underground excavation ticket hall, and the excavation volume is less than half of the existing structure; S9 is the excavation of the southwest underground excavation ticket hall, and the excavation volume is less than half of the existing structure; The multi-risk factor composite score is based on the formula: Ri = max (Rij) + 0.1[ΣRij - max (Rij)] 计算	
		S9	40		
		Score	44.0		
Relative positional relation R_3	$\gamma_3 = 0.1$	S1	60	The S1 and S9 excavations are located on the side of the existing structure	
		S9	60		
		Score	66.0		
Soil condition R_4	$\gamma_4 = 0.2$	0		The floor of Xuanwumen Station of Line 2 is located in the pebble layer, and the resilience is low, so 0 is taken	
A score on the likelihood of risk occurring		33.6		The level of likelihood of the risk occurring	3
Personnel loss C_1	$W_1 = 0.4$	0		Excavation and unloading lead to the floating of existing stations, which will hardly lead to casualties	

(continued)

Table 7.50 (continued)

Primary index	Primary index weight	Indicator score	Description	
Economic loss C_2	$W_2 = 0.2$	45	The excavation and unloading led to the floating of the existing station, which is difficult to repair, and is expected to cause a loss of more than 30 million yuan	
Duration loss C_3	$W_3 = 0.15$	60	The excavation and unloading lead to the floating repair of the existing station takes a long time, and the resulting delay is expected to be more than 9 months	
Social Impact C_4	$W_4 = 0.15$	60	Excavation and unloading will cause the existing station to float and suspend the operation of the existing line, and the social impact will be serious	
Environmental Impact C_5	$W_5 = 0.1$	20	The excavation and unloading lead to the floating of the existing station, which has little impact on the surrounding environment	
Risk loss score		29.0	Risk loss level	C
Risk level			III	

2. Risk Assessment Unit Division

The risk sources for the construction of the Siyuan Road foundation pit are screened, and the major risk sources are divided into risk assessment units, the results are shown in Table 7.60.

7.7.3 Overall Risk Assessment

The overall risk assessment scores and calculation results for the construction of the Siyuan Road foundation pit are shown in Table 7.61. The weight of the project scale G_1 indicator is 0.2, the weight of the geological G_2 indicator is 0.3, the weight of

Table 7.51 Summary of risk assessment results of existing structure floating due to excavation unloading

Risk sources	Existing structure status quo	The amount of unloaded	Relative position relationship	Soil layer condition	Risk of Loss					Total likelihood score	Loss of total points	Probability level	Loss level	Risk level
					Casualties	Economic loss	Duration loss	Social impact	Environmental impacts					
Main structure of the existing line 4 station (U1)	50	44	66	0	0	45	60	60	20	33.6	29.0	3	C	III
Existing line 2 main structure (U7)	50	64	66	0	10	10	15	20	10	40.6	12.3	3	D	III

Table 7.52 Evaluation table of major risk factors resulting from excavation and unloading of existing structures

Risk sources	Key risk factors	The amount of unloaded	Relative positional relationships	Total unloaded score	Total score for relative positional relations
Main structure of the existing line 4 station (U1)	Northwest underground excavation ticket office excavation (S1)	40	60	44.0	66.0
	Southwest excavation ticket hall excavation (S9)	40	60		
Existing line 2 main structure (U7)	Southeast open-cut ticket hall excavation (S8)	60	60	64.0	66.0
	Northeast East new entrance and exit excavation (S7)	40	60		

the environment G_3 indicator is 0.4, the technical difficulty G_4 The weight of the indicator is 0.1.

According to the calculation results, the overall construction safety risk level of Siyuan Road pit construction is Level II, and a specialized risk assessment is required.

7.7.4 *Specialized Risk Assessment*

1. Specialized assessment indicator weight

The indicator weight used in this specialized assessment is shown in Table 7.62.

2. Pit instability risk assessment

The summary of the specialized risk assessment results for pit instability in the construction of Siyuan Road Station pit is shown in Table 7.63.

Table 7.53 Engineering geological characteristics of Siyuan Road station

Soil layer number	Geotechnical name	Soil thickness (m)	Underlying elevation (m)	Indoor penetration test		Permeability coefficient suggested values (m/d)	Permeable	Scaling bases (Hits)
				Horizontal permeability coefficient Kh (cm/s)	Vertical permeability coefficient y (cm/s)			
①$_1$	miscellaneous fill	0.50–3.60	−1.35–1.62	–	–	–	–	–
①$_2$	Plain fill	0.75–1.10	−0.85–0.82	–	–	–	–	–
④$_1$	Clay	1.10–4.90	−4.78–−0.58	1.50×10^{-7}	5.10×10^{-7}	0.1	Weak water permeability	–
④$_2$	Silt	0.00–3.70	−5.27–−4.60	3.65×10^{-5}	9.70×10^{-5}	0.5	Weak water permeability	8.5
⑤$_1$	Clay	0.50–1.00	−5.12–−4.90	1.90×10^{-7}	1.30×10^{-7}	0.05	Weak water permeability	
⑤$_2$	Silty clay	0.00–1.10	−4.4	1.00×10^{-7}	1.00×10^{-7}	0.008	Micro-permeable layer	
⑥$_1$	Silty clay	2.60–3.30	−8.00–−7.75	8.20×10^{-5}	1.50×10^{-6}	0.4	Weak water permeability	9.3
⑥$_4$	Silty clay	3.50–4.50	−12.33–−11.26	3.40×10^{-6}	1.40×10^{-6}	0.3	Weak water permeability	8.9
⑦	Silty clay	1.00–1.90	−13.89–−12.57	4.90×10^{-7}	1.00×10^{-7}	0.2	Weak water permeability	10.0
⑧$_1$	Silty clay	1.80–6.50	−20.00–−15.00	5.70×10^{-7}	3.00×10^{-7}	0.2	Weak water permeability	12.5
⑧$_2$	Silt	1.00–4.70	−20.06–−18.25	–	–	–	–	23.3

(continued)

Table 7.53 (continued)

Soil layer number	Geotechnical name	Soil thickness (m)	Underlying elevation (m)	Indoor penetration test		Permeability coefficient suggested values (m/d)	Permeable	Scaling bases (Hits)
				Horizontal permeability coefficient Kh (cm/s)	Vertical permeability coefficient y (cm/s)			
$⑨_1$	Silty clay	2.50–5.40	−25.18–−22.45	7.60×10^{-7}	6.50×10^{-7}	0.2	Weak water permeability	21.4
$⑨_2$	Silt	3.00–5.70	−28.30–−27.26	1.30×10^{-4}	6.40×10^{-5}	0.8	Weak water permeability	27.5
$⑩_1$	Silty clay	0.90–2.90	−31.00–−28.99	4.90×10^{-7}	9.00×10^{-8}	0.2	Weak water permeability	16.0
$⑩_2$	Silt	0.70–3.00	−32.51–−31.45	2.00×10^{-6}	1.00×10^{-6}	0.8	Weak water permeability	
$⑪_1$	Silty clay	1.90–9.90	−40.40–−32.75	1.00×10^{-6}	1.60×10^{-7}	0.2	Weak water permeability	
$⑪_2$	Silt	0.60–5.70	−41.56–−35.22	1.20×10^{-5}	2.50×10^{-5}	1	Moderately permeable	
$⑪_3$	Silty clay	0.80–7.50	−43.34–−39.98	7.40×10^{-7}	4.00×10^{-7}	0.2	Weak water permeability	
$⑪_4$	Silt	1.70–6.20	−47.12–−44.00	3.00×10^{-4}	7.50×10^{-6}	2	Moderately permeable	

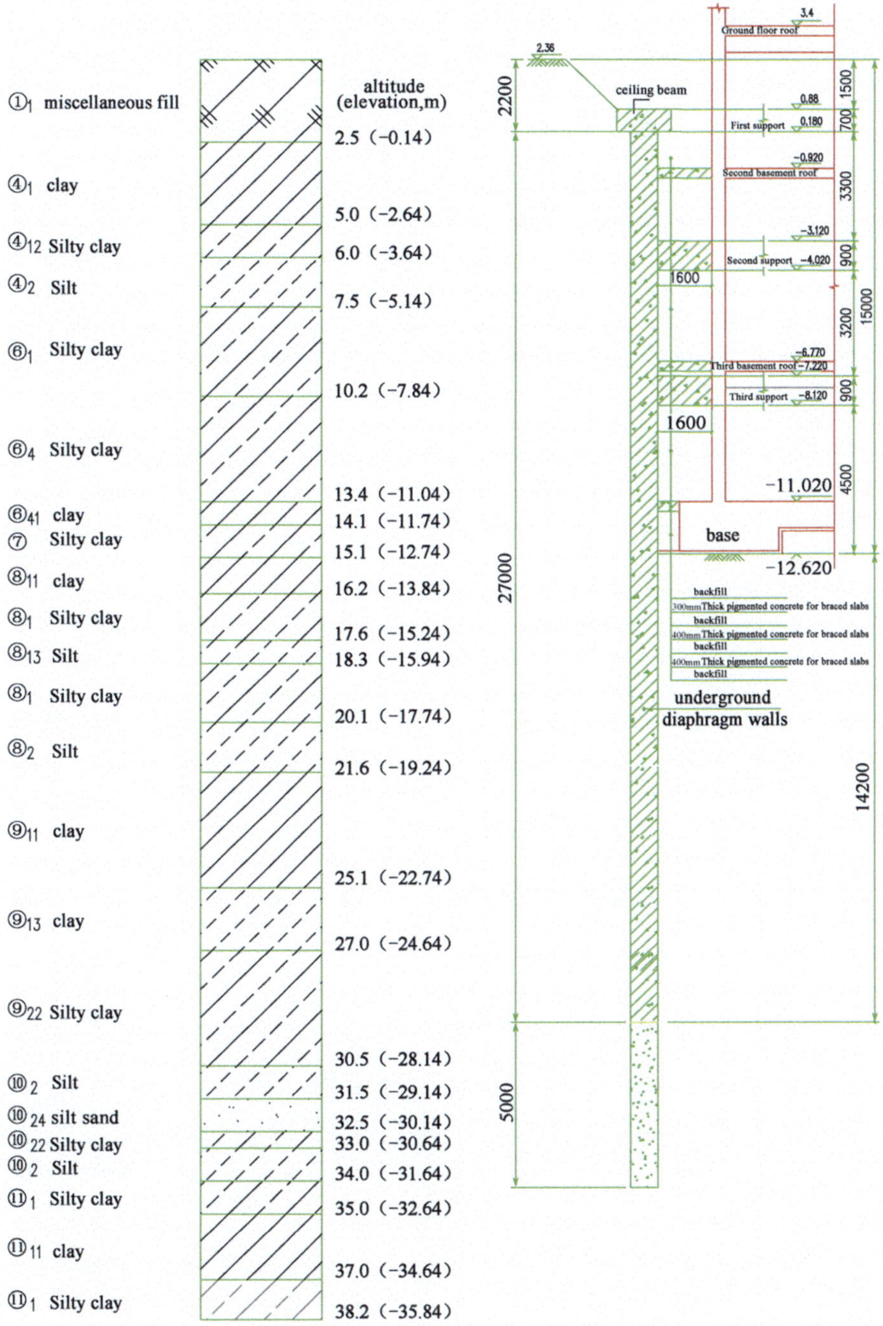

Fig. 7.32 Geological cross-section (dimension unit: mm; elevation unit: m)

Table 7.54 Permeability coefficient and permeability performance table of each soil layer

Soil layer number	Geotechnical name	Indoor penetration test		On-site pumping test permeability coefficient k (m/d)	Recommended value for permeability coefficient (m/d)	Permeable
		Horizontal permeability coefficient Kh (cm/s)	Vertical permeability coefficient Kv (cm/s)			
④$_{1}$	clay	1.50×10^{-7}	5.10×10^{-7}		0.10	Weak water permeability
④$_{12}$	Silty clay	1.15×10^{-6}	1.90×10^{-7}		0.20	Weak water permeability
④$_{2}$	Silt	3.65×10^{-5}	9.70×10^{-5}	0.29–0.36	0.50	Weak water permeability
⑤$_{1}$	clay	1.90×10^{-7}	1.30×10^{-7}		0.05	Weak water permeability
⑤$_{2}$	Silty soils	1.00×10^{-7}	1.00×10^{-7}		0.008	Slightly permeable
⑥$_{1}$	Silty clay	8.20×10^{-5}	1.50×10^{-6}		0.40	Weak water permeability
⑥$_{4}$	Silty clay	3.40×10^{-6}	1.40×10^{-6}		0.30	Weak water permeability
⑦	clay	4.90×10^{-7}	1.00×10^{-7}		0.20	Weak water permeability
⑧$_{1}$	Silty clay	5.70×10^{-7}	3.00×10 -7		0.20	Weak water permeability
⑧$_{13}$	Silt	8.20×10^{-5}	4.60×10^{-5}		0.50	Weak water permeability
⑧$_{2}$	Silt	9.40×10^{-6}	2.90×10^{-6}		0.80	Weak water permeability
⑧$_{22}$	Silty clay	4.60×10^{-7}	1.30×10^{-7}		0.50	Weak water permeability
⑧$_{24}$	Powdered sand	1.60×10^{-4}	1.35×10^{-4}		2.00	Medium permeability
⑨$_{1}$	Silty clay	7.60×10^{-7}	6.50×10^{-7}		0.20	Weak water permeability
⑨$_{13}$	Silt	2.20×10^{-5}	2.60×10^{-6}	3.87–5.53	0.80	Weak water permeability
⑨$_{2}$	Silt	1.30×10^{-4}	6.40×10^{-5}		0.80	Weak water permeability
⑨$_{22}$	Silty clay	5.00×10^{-7}	3.00×10^{-7}		0.50	Weak water permeability
⑨$_{24}$	Powdered sand	4.00×10^{-4}	1.10×10^{-4}		2.00	Medium permeability
⑩$_{1}$	Silty clay	4.90×10^{-7}	9.00×10^{-8}		0.20	Weak water permeability

(continued)

Table 7.54 (continued)

Soil layer number	Geotechnical name	Indoor penetration test		On-site pumping test permeability coefficient k (m/d)	Recommended value for permeability coefficient (m/d)	Permeable
		Horizontal permeability coefficient Kh (cm/s)	Vertical permeability coefficient Kv (cm/s)			
⑩$_2$	Silt	2.00×10^{-6}	1.00×10^{-6}		0.80	Weak water permeability
⑩$_{24}$	Powdered sand	1.70×10^{-4}	8.30×10^{-5}		2.00	Medium permeability
⑪$_1$	Silty clay	1.00×10^{-6}	1.60×10^{-7}		0.20	Weak water permeability
⑪$_1$3	Silt	1.50×10^{-5}	1.40×10^{-6}		0.80	Weak water permeability
⑪$_2$	Silt	1.20×10^{-5}	2.50×10^{-5}		1.00	Medium permeability
⑪$_2$4	Powdered sand	3.00×10^{-4}	7.80×10^{-5}		2.00	Medium permeability
⑪$_3$	Silty clay	7.40×10^{-7}	4.00×10^{-7}		0.20	Weak water permeability
⑪$_4$	Silt	7.20×10^{-5}	3.90×10^{-5}		2.00	Moderately permeable
⑪$_4$4	silt	3.00×10^{-4}	7.50×10^{-6}		1.00	Moderately permeable
⑪$_5$	Silty clay	1.20×10^{-6}	8.40×10^{-7}		0.20	Weak water permeability

Fig. 7.33 Sinopec gas station on Hongxing Branch Road

3. Base heave risk assessment

The summary of the specialized risk assessment results for base heave in the construction of Siyuan Road Station pit is shown in Table 7.64.

Table 7.55 Station structure list

The name of the building	Year of construction	Floor height	Structural form	Basic form	Relationship with foundation pits
Siyuan Road Station	In 2014	Two basement floors	frame	Cast-in-place piles	Wear it diagonally

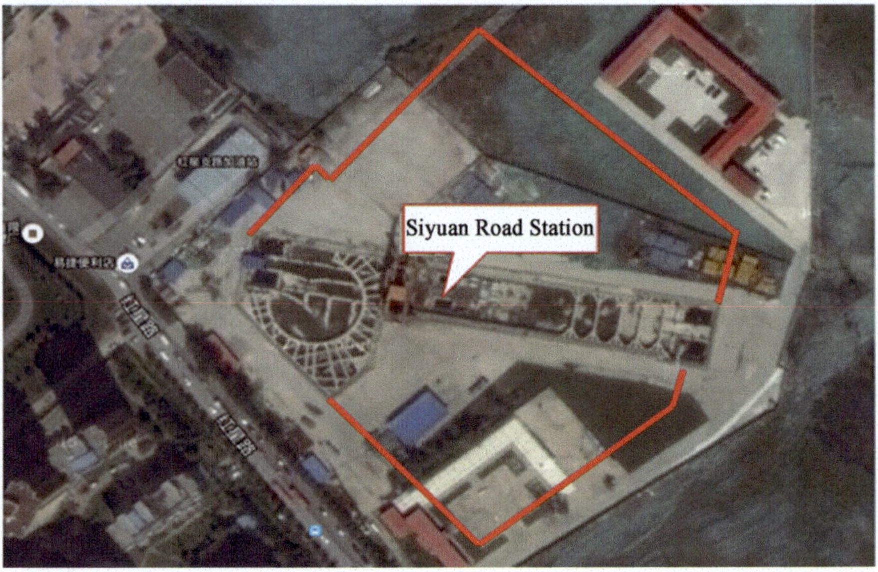

Fig. 7.34 Siyuan road station plan position diagram

4. Pipeline damage risk assessment

The risk assessment results for the damage of the water pipe near Pit No. 1 on the east side of Siyuan Road Station are shown in Table 7.65.

5. Ground building damage risk assessment

The risk assessment results for the damage of the Sinopec gas station on Hongxing Branch Road are shown in Table 7.66.

6. Underground building damage risk assessment

The impact assessment of the construction of Siyuan Road Station pit on the existing Subway Line 5 Siyuan Road Station is shown in Table 7.67, and the risk assessment results for the damage of the existing Subway Line 5 Siyuan Road Station are shown in Table 7.68.

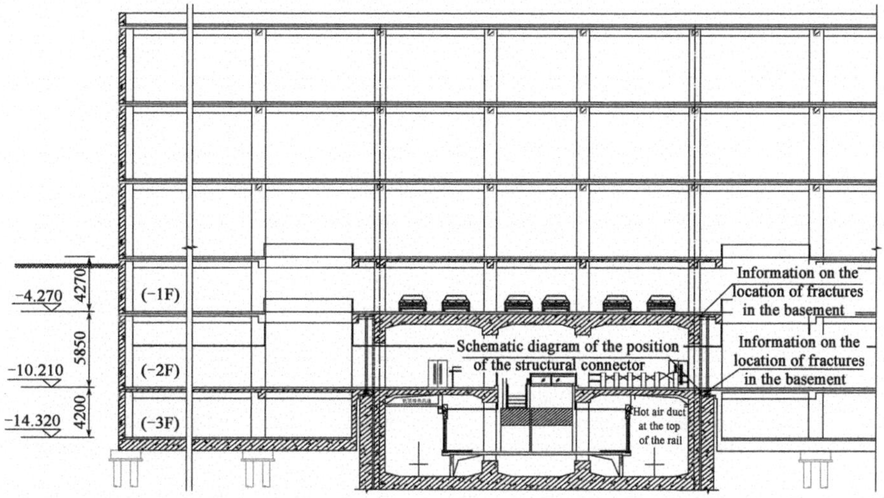

Fig. 7.35 Basement and station main section diagram (dimension unit: mm, elevation unit: m)

Table 7.56 Foundation pit itself risk source

Serial number	The name of the risk project	A description of the basic profile of the risk
S1	The foundation pit on the west side	Circular foundation pit with a radius of 36 m and an excavation depth of 19.95 m, the foundation pit is located in the (8)1 silty clay layer and the (8)2 silty layer, weak permeability
S2	No. 1 foundation pit on the east side	Circular foundation pit, radius 41 m, excavation depth 20.15 m, the foundation pit is located in the (8)1 silty clay layer and the (8)2 silt layer, weak permeability
S3	Foundation pit no. 2 on the east side	Circular foundation pit, radius of 20 m, excavation depth of 15 m, foundation pit is located in the (8)1st silty clay layer and (8)2nd silt layer, weak permeability

Table 7.57 Pipeline risk source

Numbering	Name of the source of risk	Source of risk	Related works	Location relationships
P1	Water pipes	The water pipe is a D300 mm ductile iron pipe with a buried depth of 3.5 m	No. 1 foundation pit on the east side	The pipeline is closest to the construction of the diaphragm wall 6.21 m

Table 7.58 Ground building risk source

Numbering	Name of the source of risk	Source of risk	Related works	Location relationships
B1	Sinopec gas station on the Red Star Branch Road	The gas station is a one-story brick bungalow 21 m away from the foundation pit, and the gas shed is a column shed structure	No. 1 foundation pit on the east side	The closest point to the foundation pit is 21 m

Table 7.59 Underground building risk source

Numbering	Name of the source of risk	Source of risk	Related works	Location Relationships
U1	Metro line 5 Siyuandao Station	The existing Guang-Cai section of Line 7 is a shield section, with an outer contour of 6 m, a structural thickness of 300 mm, and a distance of 15 m between lines	The foundation pit on the west side	The station is an underground double-layer island platform station, this project is located on the east and west sides of the main body of Siyuan Road Station, the basement structure of this project is the same elevation as the basement floor of the first floor of the Siyuan Road Station, and the second floor of the underground floor is the same elevation as the second floor of the Siyuan Road Station
			No. 1 foundation pit on the east side	
			Foundation pit No. 2 on the east side	

Table 7.60 Siyuan road foundation pit construction risk assessment unit division

Evaluation unit	Risk sources	Key risk factors
Foundation pit excavation	West Side Foundation Pit (S1)	Cut-and-cover construction
	East Side No. 1 Foundation Pit (S2)	Cut-and-cover construction
	East Side No. 2 Foundation Pit (S3)	Cut-and-cover construction
Perimeter pipelines	Water Mains (P1)	Construction of the No. 1 foundation pit on the east side
Ground structures	Sinopec Gas Station (B1)	Construction of the No. 1 foundation pit on the east side
Underground structures	Metro Line 5 Siyuan Road Station (U1)	Construction of foundation pit on the west side
		Construction of the No. 1 foundation pit on the east side
		Construction of No. 2 foundation pit on the east side

Table 7.61 Overall risk assessment results of foundation pit construction of Siyuan Road

Primary index	Secondary and tertiary indicators		Score	Description
Project scale G_1	Project scale G_{11}		40	The maximum excavation depth is about 20.15 m
Geology G_2	Engineering geology G_{21}		80	The strata are clay and silt
	Hydrogeology G_{22}		80	Diving, confined water
Environmental conditions G_3	Ground grade position G_{31}		50	The closest point to the foundation pit is 21 m
	Location of the underground structure G_{32}		95	It is close to Siyuan Road subway station
	Road location G_{33}		0	–
	Pipeline G_{34}	Pipeline location G_{341}	90	The cast iron water pipe is closest to the construction of the underground diaphragm wall at a distance of 6.21 m, the service life is inferred to be 5–30 years
		Fragile pipeline G_{342}	30	
		Years of service G_{343}	70	
	Surface water body G_{35}		0	There are no surface water bodies nearby
Technical difficulty G_4	Construction process maturity G_{41}		20	Cut-and-cover construction
Total score G			74	

7.7.5 Conclusion

The overall risk of the Siyuan Road pit construction project of Tianjin Subway Line 5 is Level II. The results of the special construction risk assessment are: the possibility of construction instability risk of the west pit, Pit No. 1 on the east side, and Pit No. 2 on the east side is assessed as Level 3 before construction; the possibility of base heave risk is Level 3; the possibility of water pipe damage risk is assessed as Level 3 before construction; the possibility of damage risk event of the Sinopec gas station on Hongxing Branch Road is assessed as Level 3 before construction; the possibility of damage risk event caused by pit excavation to Subway Line 5 Siyuan Road Station is assessed as Level 2 before construction.

Table 7.62 Specialized risk assessment weight of foundation pit construction of Siyuan Road

Evaluate the object	Metric name/weight			
Cut-and-cover pit instability	Pit condition R_1	Geological and Meteorological Conditions R_2	Affecting the storm sewer in the area R_3	Construction scheme R_4
	$\gamma_1 = 0.2$	$\gamma_2 = 0.5$	$\gamma_3 = 0.1$	$\gamma_4 = 0.2$
Cut-and-cover base heave	Geological conditions R_1	Environmental conditions R_2	Construction Scheme R_3	–
	$\gamma_1 = 0.4$	$\gamma_2 = 0.2$	$\gamma_3 = 0.4$	–
Cut-and-cover pipeline failure	Pipeline condition R_1	Excavation section condition R_2	Geological conditions R_3	Construction scheme R_4
	$\gamma_1 = 0.5$	$\gamma_2 = 0.2$	$\gamma_3 = 0.1$	$\gamma_4 = 0.2$
Damage to open cut ground structures and structures	Ground construct condition R_1	Excavation section condition R_2	Geological conditions R_3	Construction scheme R_4
	$\gamma_1 = 0.4$	$\gamma_2 = 0.2$	$\gamma_3 = 0.2$	$\gamma_4 = 0.2$
Damage to open cut underground structures and structures	Conditions of underground structures R_1	Excavation section condition R_2	Geological conditions R_3	Construction scheme R_4
	$\gamma_1 = 0.4$	$\gamma_2 = 0.2$	$\gamma_3 = 0.2$	$\gamma_4 = 0.2$
Risk of loss	Casualty C_1	Economic loss C_2	Duration loss C_3	Social Impact C_4
	$W_1 = 0.4$	$W_2 = 0.2$	$W_3 = 0.15$	$W_4 = 0.25$

7.8 Case Study of Construction Safety Risk Assessment of Gongbei Tunnel of Hong Kong-Zhuhai-Macao Bridge

7.8.1 Project Background

1. Project Overview

The Gongbei Tunnel is a controlling project of the Zhuhai connection line of the Hong Kong-Zhuhai-Macao Bridge. The tunneling section of the Gongbei Tunnel is a double-layer stacked tunnel with six lanes in both directions, passing under the restricted area of the Gongbei Port. The planar position, structural form, and construction method of the Gongbei Tunnel are shown in Sect. 2.3.10, and the design parameters are shown in Table 7.69.

Table 7.63 Summary of evaluation results of foundation pit instability risk

Evaluation units	Sources of risk	Foundation pit conditions			Geological and meteorological conditions			Effect area rain and dirt water pipe	
		The depth of the foundation pit	Type of foundation pit enclosure structure	Foundation pit support system	Engineering geology	Hydrogeology	Meteorological conditions	Tube diameter	Pipeline status
The foundation pit on the west side	Cut-and-cover construction	30	40	40	80	80	40	20	40
No. 1 foundation pit on the east side	Cut-and-cover construction	40	40	40	80	80	40	20	40
Foundation pit no. 2 on the east side	Cut-and-cover construction	20	40	40	80	80	40	20	40

Evaluation units	Sources of risk	Construction plan				Risk of loss			
		Reinforcement of the pit bottom	Sedimentary precipitation	Earthwork excavation	Rock excavation	Casualties	Economic loss	Loss of construction time	Social impact
The foundation pit on the west side	Cut-and-cover construction	0	40	25	0	0	15	20	10
No. 1 foundation pit on the east side	Cut-and-cover construction	0	40	30	0	0	15	20	10

(continued)

Table 7.63 (continued)

Evaluation units	Sources of risk	Construction plan				Risk of loss			
		Reinforcement of the pit bottom	Sedimentary precipitation	Earthwork excavation	Rock excavation	Casualties	Economic loss	Loss of construction time	Social impact
Foundation pit no. 2 on the east side	Cut-and-cover construction	0	40	20	0	0	15	20	10

Evaluation unit	Sources of risk	Total likelihood score	Loss of total points	Probability level	Loss level	Risk level
The foundation pit on the west side	Cut-and-cover construction	44.6	8.5	3	D	III
No. 1 foundation pit on the east side	Cut-and-cover construction	46.6	8.5	3	D	III
Foundation pit no. 2 on the east side	Cut-and-cover construction	42.6	8.5	3	D	III

Table 7.64 Summary of base heave risk assessment results

Evaluation units	Sources of risk	Geological conditions		Environmental conditions	Construction plan	
		Engineering geology	Hydrogeological	Surface bodies of water	Stratigraphic precipitation	Substrate water barrier
The foundation pit on the west side	Cut-and-cover construction	80	80	0	30	40
No. 1 foundation pit on the east side	Cut-and-cover construction	80	80	0	40	42
Foundation pit no. 2 on the east side	Cut-and-cover construction	80	80	0	35	40

Evaluation units	Sources of risk	Risk of loss				Total likelihood score	Loss of total points	Probability level	Loss level	Risk level
		Casualties	Economic loss	Duration loss	Social impact					
The foundation pit on the west side	Cut-and-cover construction	0	25	30	10	6.4	12.0	3	D	III
No. 1 foundation pit on the east side	Cut-and-cover construction	0	23	25	10	48.5	10.9	3	D	III
Foundation pit no. 2 on the east side	Cut-and-cover construction	0	25	30	10	47.2	12.0	3	D	III

Table 7.65 Pipeline damage risk assessment results

Sources of risk	Pipeline conditions				Foundation pit conditions	Stratigraphic conditions	
	Pipeline vulnerability	Pipeline fitting form	Pipeline location	Service life	Excavation depth	Engineering geology	Hydrogeology
Water pipes	20	40	80	60	40	80	80

Sources of risk	Construction plan			Risk loss			
	Pipeline reinforcement	Sedimentary precipitation	Excavation method	Casualties	Economic loss	Loss of construction time	Social impact
Water pipes	0	60	20	0	15	10	10

Sources of risk	Total likelihood score	Loss of total points	Probability level	Loss level	Risk level
Water pipes	49	7	3	D	III

2. Geological Overview of the Project

(1) Engineering Geology

The soil body traversed by the tunneling section of the tunnel mainly includes the first layer, the third layer 1, the third layer 2, and the third layer 3, followed by the fourth layer 3, the fifth layer 2, and the fifth layer 3. Layer, the $3rd_1$ layer of soil is mainly silt, as shown in Fig. 7.36. It is gray-black to dark black, plastic, saturated, slightly fishy smell, contains a large amount of humus and organic matter, contains a small amount of shells and sand, feels slippery and sticky when pinched by hand; it has low mechanical strength, high compressibility, high thixotropy, high sensitivity, high water content, large porosity, low strength and other soft soil characteristics, and the engineering geology is very poor. The $4th_3$ layer and the $5th_3$ layer of soil are mainly silt clay and silt clay, generally gray-black, plastic, feels slippery when pinched by hand, contains sand and shell debris and sand, locally relatively enriched, high compressibility, poor engineering geology. The $5th_2$ layer is mainly coarse, gravelly sand, followed by medium sand, occasionally silt, fine sand layer, contains more sticky soil, locally mixed with a small amount of silt soil, the engineering geological conditions vary greatly.

(2) Hydrogeology

The surface water in the tunnel is mainly seawater, and the groundwater is mainly stored in the $3rd_1$ layer of soft soil, the $3rd_3$ sand layer, followed by the $4th_1$ coarse, gravelly sand, and then the $3rd_2$, $4th_3$ sticky soil or sticky soil mixed with sand and Quaternary residual layer and other soil layers and bedrock fractures, among which sandy soil, especially relatively loose coarse-grained sandy soil, is a strong permeable

Table 7.66 Risk assessment results of excessive differential settlement of ground structures

Evaluation unit	Sources of risk	Ground construction conditions			Foundation pit conditions	Geology condition	
		Structures height, bridges scale, roads grade	Basic form	The location of the structure	Excavation depth	Engineering geology	Hydrogeology
Excavation of no. 1 foundation pit on the east side	Hongxing branch road in the petrochemical gas station	20	40	60	40	80	80

Evaluation unit	Sources of risk	Construction scheme			Risk of loss			
		Reinforcement of buildings	Sedimentary precipitation	Excavation method	Casualties	Economic loss	Loss of construction time	Social impact
Excavation of No. 1 foundation pit on the east side	Hongxing branch road in the petrochemical gas station	0	60	20	0	15	10	10

Evaluation unit	Sources of risk	Total likelihood score	Loss of total points	Probability level	Loss level	Risk level
Excavation of no. 1 foundation pit on the east side	Hongxing branch road in the petrochemical gas station	46.9	7	3	D	III

Table 7.67 Evaluation of impact of zone excavation of foundation pit on existing stations

Evaluation unit	Risk sources	The location of the structure	The depth of the foundation pit	Total score for the location of the building	Total score of foundation pit depth
Metro line 5 Siyuan Road station (U1)	Construction of foundation pit on the west side	95	30	118.5	45
	Construction of the No. 1 foundation pit on the east side	100	40		
	Construction of No. 2 foundation pit on the east side	90	20		

layer, followed by silt or silty soil, general sticky soil, residual soil as a relatively weak permeable layer.

3. Construction environment

The dark excavation section of the port passes through the narrow strip between Gongbei Port and Macau Gate. The distance from the tunnel curtain engineering to the structure pile foundation is only 0.46 m, and the soil thickness on the top of the curtain is 4–5 m. The port has an average of 10,000 vehicles entering and exiting every day, and the total number of people entering and exiting every day is about 300,000. It is required that the construction should not have any impact on the port clearance.

7.8.2 Risk Identification

1. Risk source list

The risk source list for the construction of the Gongbei Tunnel of the Hong Kong-Zhuhai-Macao Bridge is shown in Tables 7.70, 7.71, and 7.72.

2. Division of risk assessment units

The risk sources of the dark excavation section of the Gongbei Tunnel are screened, and the major risk sources are divided into risk assessment units. The results are shown in Table 7.73.

Table 7.68 Damage risk assessment results of underground structures

Evaluation unit	Sources of risk	Conditions of underground structures			Foundation pit conditions	Geological conditions	
		Underground structure building size	Structural form	Construct building location	Depth of foundation pit	Engineering geology	Hydrogeological
Excavation	Subway line 5 Siyuan road station (U1)	50	40	118.5	45	80	80

Evaluation unit	Sources of risk	Total likelihood score	Loss of total points	Probability level	Loss level	Risk level
Excavation	Subway line 5 Siyuan road station (U1)	63	7.0	2	C	II

Table 7.69 Tunnel design parameter table of port excavation section

Support type		Design parameters
Pipe curtain		Diameter 1620 mm, a total of 36, spacing 35.5–35.8 cm, odd tube filled with C30 micro-expanded concrete
Freeze the waterproof curtain		The odd number of tubes is the positive frozen tube, and the even number is the reinforced frozen tube; The whole length of the whole ring is actively frozen, and the segmented partition is maintained and frozen; Thickness of upper permafrost:2–2.3 m, Thickness of the underlying permafrost: 2–2.6 m
Excavation		Excavation area:336.8 m^2; Circumference of the structure: 66 m Excavation width:18.8 m; Excavation height:21.0 m
The soil in the cave is grouting and reinforcing		The internal soil of the tunnel in the underground excavation section adopts the retreating horizontal pre-grouting type, and the grouting material adopts cement slurry, and the pressure relief hole is arranged in the sandy stratum, and the specific construction parameters are determined according to the field test results
Initial support	Reinforcement mesh	Φ8 mm steel bar, double-layer arrangement
	Steel	I22b I-beam, spacing 0.40 m, welded with the pipe curtain
	Shotcrete	C25 shotcrete, 30 cm thick
	Temporary support	HN 400 × 200 × 8 × 13 sections with a spacing of 1.2 m, the vertical brace is set up in four pieces, and the first step cross brace is set up in four pieces, and the rest of the cross braces are set up in double sets
Secondary lining	Grille steel frame	It is welded with φ25 mm steel bar, and the cross-sectional size is 19.16 cm × 14.38 cm
	concrete	The vault is made of shotcrete, and the rest is molded with C35 concrete, and the thickness of the structural layer is 30 cm
Triple lining	Inverted arch	64.5–218.6 cm thick anti-corrosion reinforced concrete, concrete grade C45

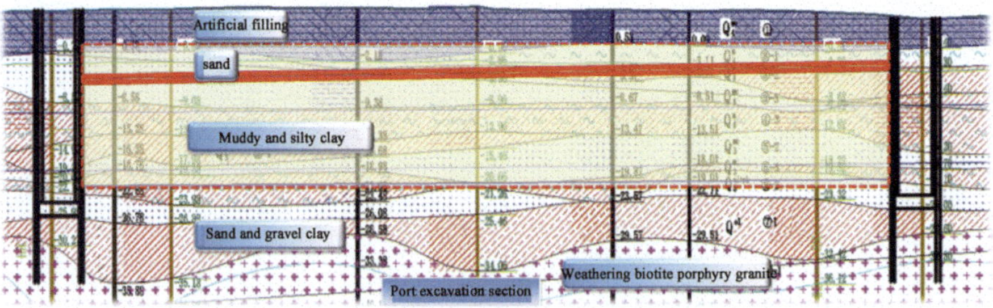

Fig. 7.36 Geological longitudinal section of the dark excavation section

Table 7.70 The tunnel itself is the source of risk

Serial number	The name of the risk project	A description of the basic profile of the risk
S1	Freeze method construction	(1) The tunnel of the underground excavation section is located at the Gongbei port, which is a highly sensitive area, and there are important buildings and underground pipelines around, and the safety and reliability of the water sealing must be ensured in the construction of the freezing method; At the same time, the amount of frost heave and thaw settlement during the freezing and thawing period should be actively reduced to protect nearby buildings (structures) (2) The engineering geological conditions are poor, the water content is abundant, and it is easy to produce large thawing and subsidence during the thawing period (3) the underground excavation section of the tunnel belongs to the offshore area, and the groundwater is mainly seawater, and the salt content in the water will reduce the freezing temperature of the soil (4) The temperature in the area is high and there is a lot of rain, which has an adverse effect on the freezing effect
S2	Underground tunnel construction	The submersible water level in the tunnel construction area is high, the soil strength of the lower part of the submersible head is low, the permeability of the sand layer gravel layer is strong, and the difficulty of underground excavation construction in the marine and continental alternating sedimentary strata is high, and the groundwater has a great influence on the formation strength. The strata are mainly silty clay, and the stability of the surrounding rock is poor. The tunnel is excavated step by step using the multi-step method, and the excavation width is 18.8 m; Excavation height 21.0 m, excavation section area 336.8 m^2

7.8.3 *Overall Risk Assessment*

The overall risk assessment index scoring and calculation results of the dark excavation section of the Gongbei Tunnel are shown in Table 7.74. The weight of the project scale G_1 index is 0.2, the weight of the geology G_2 index is 0.3, the weight of the environment G_3 index is 0.4, the weight of the technical difficulty G_4 The weight of the indicator is 0.1.

According to the calculation results, the overall construction safety risk level of the Gongbei Tunnel excavation section is Level I, and a specialized risk assessment is required.

Table 7.71 Pipeline risk source

Numbering	Name of the source of risk	Pipeline location	Related works	Location relationships
P1	D1400 mm sewage pipe	Tube insole elevation 4.7 m	Tunnel excavation	The underground excavation tunnel passes through multiple pipelines, and the multi-step method is used for step-by-step excavation, and the excavation width is 18 0.8 m; Excavation height 21.0 m, excavation section area 336.8 m^2
P2	D1000 mm water supply pipe to Macau	Pipe top elevation 5.2 m		
P3	D500 mm sewage pipe	Tube insole elevation 2.55 m		
P4	D400 mm sewage pipe	Tube insole elevation 4.12–4.25 m		
P5	D200 mm sewage pipe	Tube insole elevation 4.07 m		

Table 7.72 Ground building risk source

Numbering	Name of the source of risk	Source of risk	Related works	Location relationships	Security risk level
B1	Macau joint inspection building	Multi-layer frame structure, the foundation form is a high strength prestressed concrete pipe pile	Tunnel excavation	It is located 3 m on the south side of the tunnel	Level
B2	Immigration wind and rain corridor	Single-story structure, glass roof, enlarged foundation	Tunnel excavation	Located above the tunnel	Level 2
B3	Frontier detachment five residence halls	Multi-layer frame structure, hammered immersed pile	Tunnel excavation	The north side of the tunnel is about 14 m	Level 2

7.8.4 Specialized Risk Assessment

1. Specialized assessment indicator weight

The indicator weight used in this specialized assessment is shown in Table 7.75.

2. Risk assessment of roof collapse

The impact assessment of the pipelines around the Gongbei Tunnel of the Hong Kong-Zhuhai-Macao Bridge on the risk of construction roof collapse is shown in Table 7.76, and the results of the construction roof collapse risk assessment are shown in Table 7.77.

Table 7.73 Subdivision of risk assessment unit for underground tunneling

Evaluation unit	Risk sources	Key risk factors
Tunnel excavation	Underground excavation station body (S2)	The multi-step method is excavated step by step
		D1400 mm sewage pipe
		D1000 mm water supply pipe to Australia
		D500 mm sewage pipe
Tunnel excavation	Underground excavation station body (S2)	D400 mm sewage pipe
		D200 mm sewage pipe
Perimeter pipelines	D1400 mm sewage pipe (P1)	Freezing method construction (S1)
		Underground tunneling (S2)
	D1000 mm water supply pipe (P2)	Freezing method construction (S1)
		Underground tunneling (S2)
	D500 mm sewage pipe (P3)	Freezing method construction (S1)
		Underground tunneling (S2)
	D400 mm sewage pipe (P4)	Freezing method construction (S1)
		Underground tunneling (S2)
	D200 mm sewage pipe (P5)	Freezing method construction (S1)
		Underground tunneling (S2)
Ground structures	Macao joint inspection building	Tunnelling (S2)
	Entry and exit corridor	Tunnelling (S2)
	Dormitory building of the fifth frontier detachment	Tunnelling (S2)

Table 7.74 Overall risk assessment results of Gongbei tunnel

Primary index	Secondary and tertiary indicators		Score	Description
Project scale G1	Project scale G11		60	Excavation section area 336.8 m^2
Geology G2	Engineering geology G21		90	The tunnel is located in a silty silt layer
	Hydrogeology G22		80	The tunnel is located on the submersible level
Environmental conditions G3	Ground grade position G31		90	The minimum vertical distance between the ground structure and the project is about 3 m
	Location of the underground structure G32		0	–
	Road location G33		90	It is located below the main road in the city
	Pipeline G34	Pipeline location G341	90	D1400 mm sewage pipe invert elevation 4.7 m, the pipeline is concrete pipe, and the service life is inferred to be 5–30 years
		Fragile pipeline G342	60	
		Years of service G343	70	
	Surface water body G35		90	There is a lot of seawater on the surface
Technical difficulty G4	Construction process maturity G41		70	The pipe curtain method + freezing method is used to support, and the multi-step method is excavated step by step
Total score G			93.28	

Table 7.75 Gongbei tunnel excavation section specialized risk assessment weight

Evaluate the object	Metric name/weight			
Underground excavation and collapse to the top	Coverage span ratio R_1	Geological conditions R_2	Affecting the storm sewer in the area R_3	Construction scheme R_4
	$\gamma_1 = 0.3$	$\gamma_2 = 0.4$	$\gamma_3 = 0.1$	$\gamma_4 = 0.2$
Digging mud and gushing water	Geological conditions R_1	Environmental conditions R_2	Construction scheme R_3	
	$\gamma_1 = 0.4$	$\gamma_2 = 0.2$	$\gamma_3 = 0.4$	
Underground excavation pipeline damage	Pipeline condition R_1	Excavation section condition R_2	Geological conditions R_3	Construction scheme R_4
	$\gamma_1 = 0.4$	$\gamma_2 = 0.1$	$\gamma_3 = 0.3$	$\gamma_4 = 0.2$
Destruction of underground excavation of ground structures and structures	Ground construct condition R_1	Excavation section condition R_2	Geological conditions R_3	Construction scheme R_4
	$\gamma_1 = 0.\ 3$	$\gamma_2 = 0.\ 2$	$\gamma_3 = 0.3$	$\gamma_4 = 0.2$
Risk of loss	Casualty C_1	Economic loss C_2	Duration loss C_3	Social Impact C_4
	$W_1 = 0.4$	$W_2 = 0.2$	$W_3 = 0.15$	$W_4 = 0.25$

Table 7.76 Evaluation of pipeline factors for roof collapse

Risk sources	Risk factors	Pipe diameter	Pipeline status	Total pipe diameter	Total pipeline status score
Tunnel excavation (S1)	D1400 mm sewage pipe (P1)	65	40	80.5	52
	D1000 mm water supply pipe (P2)	60	40		
	D500 mm sewage pipe (P3)	50	40		
	D400 mm sewage pipe (P4)	45	40		

3. Risk assessment of mud and water inrush

The results of the risk assessment of mud and water inrush during the construction of the Gongbei Tunnel of the Hong Kong-Zhuhai-Macao Bridge are shown in Table 7.78.

Table 7.77 Risk assessment of roof collapse results

Assess unit	Sources of risk	Cover-span ratio	Geology condition		Storm sewer pipes in the affected area	
		Cover-span ratio	Engineering geology	Hydrogeology	Tube diameter	Pipeline status
Tunnel excavation	Underground tunneling excavation	90	90	80	80.5	52
Assess unit	Sources of risk	Likelihood total score	Loss score	Possibility grade	Loss grade	Risk grade
Tunnel excavation	Underground tunneling excavation	63.34	47.0	2	C	III

Table 7.78 Results of risk assessment of mud and water inrush

Evaluation unit	Sources of risk	Geological conditions		Environmental conditions
		Engineering geology	Water-rich degree	Surrounding water bodies
Tunnel excavation	Underground tunneling excavation	90	90	90

Evaluation unit	Sources of risk	Construction plan		Risk loss			
		Advance reinforcement	Stratigraphic precipitation	Casualties	Economic loss	Duration loss	Social impact
Tunnel excavation	Underground tunneling excavation	20	0	50	35	30	10

Evaluation unit	Sources of risk	Likelihood total score	Loss score	Possibility grade	Loss grade	Risk grade
Tunnel excavation	Underground tunneling excavation	57.2	34	2	C	II

4. Risk assessment of pipeline damage

The results of the risk assessment of pipeline damage caused by the construction of the Gongbei Tunnel of the Hong Kong-Zhuhai-Macao Bridge are shown in Table 7.79.

5. Risk assessment of damage to ground structures

The results of the risk assessment of damage to ground structures caused by the construction of the Gongbei Tunnel of the Hong Kong-Zhuhai-Macao Bridge are shown in Table 7.80.

7.8.5 Conclusion

The overall risk of the Gongbei Tunnel excavation section construction is Level I. The results of the special construction risk assessment are: the possibility of roof collapse risk during tunnel excavation is Level 2, the possibility of mud and water inrush risk is Level 3; the possibility of damage risk to the D1400 mm sewage pipe, D1000 mm water supply pipe to Macau, D500 mm sewage pipe, D400 mm sewage pipe, etc. is Level 2; the risk of excessive differential settlement deformation of the Macau Joint Inspection Building is Level 1; the risk of excessive differential settlement deformation of the Immigration Rain Shelter and Border Defense Fifth Squadron Dormitory Building is Level 2.

Table 7.79 Summary of pipeline failure risk assessment results

Risk sources	Pipeline conditions				Excavation section condition		Stratum condition	
	Pipeline vulnerability	Pipeline fitting form	Pipeline location	Service life	Section size	Cover-span ratio	Engineering geology	Hydrogeology
D1400 mm sewage pipe	40	45	90	70	60	90	90	90
D1000 mm water supply pipe to Macau	35	50	90	80	60	90	90	90
D500 mm sewage pipe	30	60	90	70	60	90	90	90
D400 mm sewage pipe	25	50	90	80	60	90	90	90
D200 mm sewage pipe	20	55	90	80	60	90	90	90

Risk sources	Construction plan			Risk loss			
	Pipeline reinforcement	Sedimentary precipitation	Excavation method	Casualties	Economic loss	Loss of construction time	Social impact
D1400 mm sewage pipe	0	0	40	20	10	10	30
D1000 mm water supply pipe to Macau	0	0	40	20	10	10	25
D500 mm sewage pipe	0	0	40	20	10	10	30
D400 mm sewage pipe	0	0	40	20	10	10	50

(continued)

Table 7.79 (continued)

Risk sources	Construction plan			Risk loss			
	Pipeline reinforcement	Sedimentary precipitation	Excavation method	Casualties	Economic loss	Loss of construction time	Social impact
D200 mm sewage pipe	0	0	40	20	10	10	40

Risk sources	Total likelihood score	Loss of total points	Probability level	Loss level	Risk level
D1400 mm sewage pipe	62.0	19.0	2	D	III
D1000 mm water supply pipe to Macau	62.6	17.8	2	D	III
D500 mm sewage pipe	62.0	19.0	2	D	III
D400 mm sewage pipe	61.4	24.0	2	D	III
D200 mm sewage pipe	61.2	21.5	2	D	III

Table 7.80 Damage risk assessment results of surface structures

Evaluation unit	Sources of risk	Condition of the ground structure			Excavation section condition		Geology condition	
		Building height, bridge scale, road grade	Basic form	The location of the structure	Section size	Cover-span ratio	Engineering geology	Hydrogeology
Tunnel excavation	Macao Joint inspection building	40	70	100	60	90	90	90
	Entry and exit corridor	20	65	90	60	90	90	90
	Dormitory building of the fifth frontier detachment	40	40	80	60	90	90	90

Evaluation unit	Sources of risk	Construction plan			Risk loss			
		Reinforcement of buildings	Sedimentary precipitation	Excavation method	Casualties	Economic loss	Loss of construction time	Social impact
Tunnel excavation	Macao joint inspection building	95	0	40	70	50	60	50
	Entry and exit corridor	75	0	40	60	35	60	25
	Dormitory building of the fifth frontier detachment	80	0	40	55	30	60	40

Evaluation unit	Sources of risk	Total likelihood score	Loss of total points	Probability level	Loss level	Risk level
Tunnel excavation	Macao joint inspection building	76.0	59.5	1	B	I
	Entry and exit corridor	71.2	46.3	2	C	II
	Dormitory building of the fifth frontier detachment	69.1	47.0	2	C	II

Chapter 8
Summary and Outlook

8.1 Summary

This book addresses the new characteristics and problems presented by contemporary urban underground space development, proposes a new concept of urban large underground space, and conducts in-depth research on its construction safety risk formation mechanism and key risk assessment theories through methods such as engineering case studies, probability statistical analysis, numerical simulation, and expert consultation. A technical system suitable for the safety risk assessment of urban large underground space construction has been established. The main contents of this book are summarized as follows:

(1) The current development status and trends of urban underground space are analyzed, pointing out that the development of urban underground space presents a trend of "up and down planning, scientific layout, functional integration, interconnection, intelligent integration, and multi-dimensional space". On this basis, the concept of urban large underground space is proposed, and according to different construction methods, urban large underground space projects are divided into three types: monolithic large space by open excavation method projects, monolithic large space by undercut method projects, and large underground space by networked extension projects. This book has statistically analyzed urban large underground space project cases in China and summarized the characteristics of urban large underground space projects through case analysis.

(2) A survey was conducted on 295 urban underground space construction safety accidents in more than twenty cities nationwide, including Beijing, Shanghai, Shenzhen, and Guangzhou, from 2004 to 2019, and several major foreign underground space construction safety accidents. Statistical analysis was conducted on the types of accidents, the geological conditions of the accident areas, and the direction of accident risk sources. The occurrence of several major urban underground space construction safety accidents was analyzed in detail, and the

S.-X. Lei et al., *Safety Risk Assessment of Urban Underground Large Space Construction*, https://doi.org/10.1007/978-981-97-7437-1_8

accident occurrence mechanism was summarized in the form of accident cause and effect diagrams.

(3) The basic theory of urban underground space construction safety risk assessment is introduced. Based on engineering case analysis and accident case analysis, the characteristics and classification of urban large underground space construction safety risks are summarized. A city underground large space construction risk system model centered on subsystem division and interactive control is established. Based on Bayesian networks, fuzzy comprehensive evaluation method, and engineering structure reliability theory, a method for analyzing the safety risk of urban large underground space construction is constructed, realizing the combination of risk cause analysis and observation measurement.

(4) The definition of risk factor coupling is clarified and reveals the formation mechanism of multi-factor coupling risk in urban large underground space construction, that is, when certain specific risk factors occur simultaneously, the coupling amplification effect between these factors will develop into larger risk factors, causing their adverse effects to exceed the system's bearing capacity, ultimately inducing the occurrence of risk events. According to the principle of independent causality, the definition of risk factor coupling coefficient is proposed, realizing the quantitative description of risk factor coupling effect, and proposing a method for calculating the risk factor coupling coefficient using numerical simulation or physical simulation combined with Bayesian network inference. Through numerical simulation, the coupling effect of key risk factors in urban large underground space construction is studied, and the grading standards and recommended values for the coupling coefficient of risk factor coupling effect are given.

(5) The mechanism of risk evolution in urban large underground space construction has been revealed, and the fundamental causes of construction risk evolution have been analyzed, including: ① natural evolution of risk factors; ② evolution of construction mechanics mechanism; ③ evolution caused by emergent phenomena in the construction management process; ④ coupled evolution caused by the coupling effect of risk factors. The pathways of action (① changing the mechanical properties of rock and soil; ② changing the stress state of the structure) and manifestations (the probability of risk events changes over time) have also been identified. A multi-factor coupling evolution model for construction risk in urban large underground spaces has been established, using the Markov-Fuzzy comprehensive prediction model to describe the natural evolution of risk factors, the dynamic Bayesian network to describe the evolution of mechanical mechanisms, the system dynamics transmission model to describe emergent evolution, and the risk factor coupling coefficient to describe coupled evolution. This model enables risk prediction throughout the entire process of urban large underground space construction, and its effectiveness has been verified with actual engineering cases.

(6) Based on the indicator system method, a risk coupling evaluation coefficient has been proposed, and a graded assessment method for safety risks in urban large underground space construction has been established. This includes a risk

assessment indicator system with 56 factors in 3 levels across 4 major categories: geology, design, construction, and environment. Quantitative standards for assessment indicators have been proposed based on surrounding rock levels, proximity to the environment, design standards, and construction methods and techniques. Referencing current regulations, methods and quantitative standards for assessing construction risk losses in urban underground spaces have been proposed, considering factors such as the number and type of casualties, economic losses, impact on the construction period, and social impact. On this basis, a dynamic safety evaluation system for urban large underground space construction has been constructed, including risk assessment methods, assessment indicator systems, assessment processes, risk acceptance criteria and grading standards, and a risk assessment system. A dynamic assessment system for safety risks in urban large underground space construction has been developed.

8.2 Outlook

As the "fourth territory" parallel to land, airspace, and territorial waters, underground space is not only an important resource needed for future urban construction, but also a necessary supplement to create a more comfortable and livable space environment on the city surface. It will become a key carrier to solve the "big city disease". In the process of urban underground space development and utilization, urban large underground spaces should be the main future development trend, playing a "pioneer" and "vanguard" role in exploring urban underground space development concepts and models, promoting the development of advanced construction technology for urban underground spaces, and improving the quality of urban underground spaces.

Urban large underground spaces should not only be spacious, comfortable, convenient, and functionally integrated underground spaces, but also safe, green, environmentally friendly, and intelligent underground spaces. In the construction process of urban large underground spaces, we must adhere to the style of being "bold and careful". We should be willing to try and innovate, but always put production safety and urban environmental protection first, and deeply implant risk awareness in our hearts. We should have a dialectical understanding of risk, seeing it both as a Damocles sword hanging over our heads and as the first line of defense for production safety. Only by recognizing the existence of risk can we stay alert at work and nip problems in the bud. Therefore, construction safety risk management methods and risk analysis theory are important guarantees for the development of urban large underground spaces. We must consistently emphasize the importance of risk research, ensuring that relevant theories and methods keep pace with the times and match the levels of planning, design, and construction.

The ultimate goal of risk research is to establish a scientific, mature, and efficient risk control system to enhance risk management capabilities. The research in this book focuses on the fundamental theory of risk assessment, which is the "first step of a long march" in constructing a safety risk management system for urban underground large-space construction. In future work, we should continue to pay attention to the research of the basic scientific theory of risk assessment, improve the theory of risk coupling and evolution, deepen the connotation and expand the extension, promote the application of the theory, and form a universal and practical risk coupling evolution analysis method. On this basis, we should constantly improve risk assessment methods, optimize assessment indicators and quantitative standards, expand the coverage of assessment content in combination with engineering practice, and actively explore the application of new uncertainty mathematical theories such as theory of fuzzy mathematics, rough set theory, and evidence theory in risk assessment. In addition, the automation and real-time nature of risk assessment is also one of the future research focuses. How to combine risk assessment with construction monitoring, establish a construction safety control system that combines local monitoring and warning with global dynamic risk warning, is a question worth considering for every risk and safety researcher. On the other hand, we should pay attention to technology development, demonstration applications, and achievement transformation related to theory, and deepen research in areas such as visualization management of construction safety risks, standardized assessment, and intelligent decision-making. The future urban large underground space will further break away from the constraints of traditional structural forms and construction methods, presenting more complex and changeable forms such as networked expansion of underground large spaces. Therefore, construction safety risk assessment should also be timely and keep up with the times, constantly refine assessment strategies and assessment content in combination with various types of underground large space characteristics, and form a multi-objective, multi-perspective, multi-level urban large underground space construction safety risk assessment technical system.

Risk represents infinite uncertainty, so the study of risk is also a difficult path of constant exploration. We look forward to future underground engineering personnel not forgetting their original intentions, keeping their mission in mind, and pushing the theoretical and technical level of urban large underground space construction safety risk research to new heights!

References

Al-BAHAR J F. Risk management in construction projects:A systematic analytical approach for contractors [D]. University of California,Berkeley. 1988.

BROOKER P. Aircraft Collision Risk in the North Atlantic Region [J]. Journal of the Operational Research Society,1984,35(8):695-703.

BU-QAMMAZ, Amani S, et al. Risk Management Model for International Public Construction Joint Venture Projects in Kuwait [D]. The Ohio State University, 2015.

BURENJIRIGALA,ALATANTUYA,GUO C. The Analysis on the Coupling Characteristics of Ecological Environment, Natural Disasters and Poverty in Inner Mongolia Autonomous Region [J]. Communications in Computer & Information Science,2013,399:454–465.

Chen Guixiang, Huang Hongwei, You Jianxin. Study on Life Cycle Risk Management of Metro [J]. Chinese Journal of Underground Space and Engineering,2006,2(1):47-51.

Chen Liping. The Macro Mesoscopic Damage Mechanism and Control of the Sandy Tunnel Surrounding Rock [D]. Beijing: Beijing Jiaotong University,2015.

Chen Mengjie, Pan Haize, He Jian, et al. The Application of Fuzzy Entropy in Metro Tunnel Water Leakage [J]. Chinese Journal of Underground Space and Engineering,2016(S1):62–65.

Chen Ronfang, Jiang Anmin, Dong Yanchen, et al. Construction Risk Assessment of Deep and Large Foundation Pit Based on Entropy Weight Extension Model [J]. Journal of Mathematics in Practice and Theory,2019,49(02):313-322.

Chen Taihong, Wang Mingyang, Xie Dongsheng, et al. Risk Identification and Control in Construction of Metro Station Foundation Pit [J]. Journal of Disaster Prevention and Mitigation Engineering,2008,28(003):375-381.

Chen Zhilong, Liu Hong. 2015 China's urban underground space development blue book [M]. Shanghai:Tongji University Press,2016.

Chen Zhong. Risk Analysis and Strategy During Construction of Chengdu Metro Shield Section Tunnel [D]. Chengdu: Southwest Jiaotong University,2008.

Cheng Xuesong. Mechanism and Redundancy Study of Some Collapse Problems in Underground Engineering [D]. Tianjin: Tianjin University,2013.

CHIDIEBERE E E . Analysis of Rework Risk Triggers in the Nigerian Construction Industry [J]. Organization Technology & Management in Construction,2018,10.

CHINKULKIJNIWAT A, HORPIBULSUK S, SEMPRICH S. Modeling of Coupled Mechanical-Hydrological Processes in Compressed-Air-Assisted Tunneling in Unconsolidated Sediments [J]. Transport in Porous Media,2015,108(1):105-129.

CHOI H H,CHO H N ,SEO J W . Risk Assessment Methodology for Underground Construction Projects [J]. Journal of Construction Engineering & Management, 2004, 130 (2):258-272.

Ding Lieyun, Wu Xianguo, Luo Hanbin, et al. Research on standard for construction safety assessment of metro engineering [J]. China Civil Engineering Journal,2011,44(11):121–127.

S.-X. Lei et al., *Safety Risk Assessment of Urban Underground Large Space Construction*, https://doi.org/10.1007/978-981-97-7437-1

Ding Lieyun, Zhou Cheng, Ye Xiaowei, et al. Study on real-time sensing and early-warning of construction safety risk for metro crossing passage under Yangtze River [J]. China Civil Engineering Journal,2013,46(07):141–150.

Du Feng. Risk assessment and practice of tunnel engineering design and construction [M]. Beijing: China Building Materials Press,2017.

EINSTEIN H H, INDERMITTE C, SINFIELD J, et al. Decision aids for tunneling [J]. Transportation research record,1999,1656(1):6-13.

EINSTEIN H H, Vick S G. Geological model for a tunnel cost model[J]. Proc RETC 2nd,II,1974:1701–1720.

EINSTEIN H H. Risk and risk analysis in rock engineering [J]. Tunnelling and Underground Space Technology,1996,11(2):141-155.

FORTEZA F J, CARRETERO-GóMEZ, JOSE M, et al. Effects of organizational complexity and resources on construction site risk [J]. Journal of Safety Research,2017,62(9):185–198.

General Office of the State Council. Guiding Opinions of The General Office of the State Council on promoting the construction of urban underground integrated pipe corridors [J]. Installation ,2015(9):10-13.

GOMATHI S, LINDA P E. Interaction Coupling: A Modern Coupling Extractor [M] / / Computational Intelligence, Cyber Security and Computational Models. 2014.

Guo Jian, Qian Jindou, Chen Jian, et al. Risk Identification and Evaluation for Foundation Pit Construction of Subway Station [J]. Journal of Civil Engineering and Management,2017,034(005):32-38.

Guo Zhongwei. Risk analysis and decision making [M]. Beijing. China Machine Press,1987.

Hangzhou Bendibao. Hangzhou planned 18 underground complex actually has underground hospital [EB/OL]. http://hz.bendibao.com/news/2015724/58195.shtm. 2015-07-24/2017-08-11.

He Meili, Liu Ji, Liu Lang, et al. Unascertained measure model of assessment tunnel collapse risk and its application in engineering [J]. Journal of Central South University: Science and Technology.2012,43(9):3665–3671.

He Zhilong. Research on the combination of urban civil air defense project and underground space development and utilization [D]. Changsha: Hunan University,2009.

HECKERMAN D. Causal independence for knowledge acquisition and inference [M] / / Uncertainty in Artificial Intelligence San Francisco: Morgan Kaufmann,1993:122–127.

Hou Yanjuan, Zhang Dingli, Li Pengfei. Analysis and Control Measures of Safety Accidents in Beijing Subway Construction [J]. Journal of Beijing Jiaotong University,2009,33(003):52–59.

Hu Xingjun, Yan Xiaoli. Construction Project Safety Assessment Based on Risk Coupling Mechanism [J]. Safety and Environmental Engineering,2015,22(6):134–138.

Huang Changqian, Zhang Jianqing, Chen Changyan. Investigation and surveying and evaluation of the artificial soil [J]. Geotechnical Investigation & Surveying,2010(S1):187–191, 371

Huang Hongwei, Hu Qunfang. Current situation and progress of engineering risk analysis and insurance research [C] / / Development of risk analysis and reliability design technology for water conservancy and hydropower projects. China Water Conservancy and Hydropower Investigation and Design Association,2010:12.

Huang Hongwei, Zeng Ming, Chen Liang, et al. Risk Management Software (TRM1.0) Based on Risk Database for Shield Tunneling [J]. Chinese Journal of Underground Space and Engineering,2001,01:36–41.

Huang Hongwei, Zhang Dongming. Recent Progress on Refined Sensing and Control of Safety and Risk of Long-and Large-scale Tunnel Lining Structures [J]. China Journal of Highway and Transport,2020,33(12):46-61.

Huang Hongwei. State-of-the-Art of the Research on Risk Management in Construction of Tunnel and Underground Works [J]. Chinese Journal of Underground Space and Engineering,2006,2(1):13-20.

Jiang Xin, Wu Yuanli. Research on evolution mechanisms of risks in hydropower project group cross operation [J]. China Safety Science Journal.2015,25(12):157–163.

KAMPMANN J, ESKESEN S, SUMMERS J. Risk assessment helps select the contractor for the Copenhagen metro System [C] / / Proceedings of the World Tunnel Congress, Sao Paulo: Cowi,1998:123–128.

Lei Shengxiang, Li Wensheng, Zhou Biao, et al. Quality Improvement and Evaluation Index System of Urban Network Underground Space [J]. Chinese Journal of Underground Space and Engineering, 2021, 17(4): 987–1007.

Lei Shengxiang, Shen Yanjun, Xiao Qing-Hua, et al. Present Situations of Development and Utilization for Underground Space in Cities and New Viewpoints for Future Development [J]. Chinese Journal of Underground Space and Engineering,2019,15(04):965-979.

Lei Shengxiang, Xiao Qinghua, Deng Yong. Four-color underground rapid transit for city [M]. Beijing : Science Press, 2017.

Lei Shengxiang. Development and utilization of underground space in the future [M]. Beijing : China Communications Press Co., Ltd., 2020.

Lei Shengxiang. Integrated pipe corridor and pipe shield [M]. Beijing : China Railway Publishing House, 2015.

Lei Shengxiang. Metro construction handbook [M]. Beijing : China Communications Press Co., Ltd., 2020.

Li Bing, Xu Mingxin, Lu Jinghui. Risk analysis and strategy for metro station construction [J]. Shanxi Architecture,2007(05):269–270.

Li Feng. Risk Management for The Construction through Zones of Weathered Rock of Extremely Poor Quality in Xiang'an Tunnel [D]. Shanghai: Tongji University,2007.

Li Shucai, Zhou Zong-Qing, Li Li-Ping, et al. RISK EVALUATION THEORY AND METHOD OF WATER INRUSH IN KARST TUNNELS AND ITS APPLICATIONS [J]. Chinese Journal of Rock Mechanics and Engineering,2013,32(9):1858–1867.

Li Yicheng, Xue Yadong, Li Yanjie. A New Construction Risk Assessment Method Based on Dynamic Weight [J]. Chinese Journal of Underground Space and Engineering,2017,13(S1):209–215.

Li Zhentuan. A dissertation on distribution of spherical weathering bodies in granite and the project influence [J]. Communications Science and Technology Heilongjiang,2017,40(04):4–5.

Liao Weiquan, Deng Siquan. Analysis of construction schedule risk [J]. Water Resources & Hydropower of Northeast China,2004,22(5):32-34.

Lin Jian, Huang Fei, Zhao Xingshuo. ACCELERATING THE LEGISLATION OF UNDERGROUND SPACE UTILIZATION, PROMOTING THE CAPABILITY OF URBAN SUSTAINABLE DEVELOPMENT [J]. City Planning Review,2015,39(3):24-28.

Liu Qing, Han Dandan, Chen Yanqing, et al. Research on navigation risk evolution in Three Gorges dam area based on system dynamics [J]. China Safety Science Journal. 2016,26(4):19-23.

Meng Xiangkun, Chen Guoming, Zhu Hongwei, et al. Complex network analysis on risk evolution of submarine pipeline leakage [J]. Journal of Safety Science and Technology. 2017,13(4):26-31.

MERCAT-ROMMENS C, CHAKHAR S, CHOJNACKI E, et al. Coupling GIS and Multi-Criteria Modeling to Support Post-Accident Nuclear Risk Evaluation [M] / / Evaluation and Decision Models with Multiple Criteria. 2015.

MOLAG M, TEIJSSENAAR-BUHRE IJM. Risk Assessment Guidelines for Tunnels [J]. Safe&Reliable Tunnels,Lausanne,2006:17–32.

Qian qihu. To Transform Way of Urban Development by Constructing Underground Utility Tunnel [J]. Tunnel Construction,2017,37(6):647–654.

Qiao Wanguan, Li Xinchun. Risk Evaluation of Coal Mine Enterprises under Role of Multi-Factor Coupling [J]. Coal Engineering,2014,46(4):145–148.

Qiao Yongkang, Zhang Mingyang, Liu Yang, et al. Strategy Research on the Master Planning for Underground Space Utilization in Historical and Cultural City of Ancient Capital Type [J]. Chinese Journal of Underground Space and Engineering,2017,13(4):859-867.

REILLY J. J. The management process for complex underground and tunneling projects [J]. Tunnelling & Underground Space Technology,2000,15(1):31–44.

Ren Qiang, Huang Hongwei. Study on Risk Evaluation and Control Technology of Shielding Construction of Beijing Underground [D]. Beijing:China University of Geosciences,2010.
Ren Zhen. Study on the coupling models for construction risk of deep excavation in subway station [D]. Wuhan: Huazhong University of Science and Technology,2013.
SAHIN O, MOHAMED S. Coastal vulnerability to Sea Level Rise: A spatio-temporal decision making tool [C] / / IEEE International Conference on Industrial Engineering & Engineering Management. IEEE,2010.
SATYAVANI Ch V V, VENKATESWARARAO B, RAJU P M. Physicochemical and Microbial Analysis of Ground Water Near Municipal Dump Site for Quality Evaluation [J]. International Journal of Bioassays,2013,2(08):1139–1144.
SEJNOHA J, JARUSKOVA J, et al. Risk Quantification for Tunnel Excavation Process [J]. World Academy of Science, Engineering and Technology,2009,(58):383–401.
Shen Yanjun, Zhang Huan, Lei Shengxiang, et al. Development Concept and Features of Civil Air Underground Defense Project in China under New Situation [J]. Chinese Journal of Underground Space and Engineering, 2019, 15(6): 1599-1608.
Shu Yu,Lu Shan,Ruan Yejing. Urban underground space planning and design [M]. Shanghai: Tongji University Press,2015.
SMITH,DAVID J . Reliability, Maintainability and Risk Practical Methods for Engineers [M]. Oxford: Butterworth-Heinemann,2017:37–49.
Sohu News. Analysis of the development status of urban rail transit in China [EB/OL]. http://www.sohu.com/a/76178744_115559. 2016-05-19/2017-08-10.
STURK R, OLSSON L, JOHANSSON J. Risk and decision analysis for large underground projects, as applied to the Stockholm Ring Road tunnels [J]. Tunnelling & Underground Space Technology,1996,11(2):157-164.
Sugijiang Gong. Construction Status & Characteristics of Common Tunnel in Japan [J]. China Municipal Engineering,2016(S1):87–90,120.
TAH J,CARR V . A proposal for construction project risk assessment using fuzzy logic [J]. Construction Management & Economics,2000,18(4):491–500.
Tan Pan, Feng Zhitao, Zhang Jie, et al. Research on evolution of navigation safety risk in Three Gorges dam area [J]. China Safety Science Journal.2018,28(12):140-147.
Tan Zhuoying. Underground space planning and design [M]. Beijing : Science Press,2015.
Wang Chengbing, Zhu Hehua. Study on the Test Method for Progressive Failure Mechanism of the Surrounding Rock of Tunnel with Model [J]. Journal of Railway Engineering Society,2009,26(3):48-53.
Wang Chengbing. Laboratory and numerical investigation on failure process of tunnel constructed in homogeneous rock [J]. Rock and Soil Mechanics,2012(1):103-108.
Wang Chengtang, Wang Hao, Qin Weimin, et al. Evaluation of collapse possibility of deep foundation pits in metro stations based on multi-state fuzzy Bayesian networks. Rock and Soil Mechanics,2020,41(05):1670–1679 +1689.
Wang Fan. Study on the modelling and Dynamics of Safety Risk during Tunnel Construction [D]. Wuhan: Huazhong University of Science and Technology,2013.
Wang Gong. Study on Accidents Statistical Analysis and Risk Assessment Methods of Urban Subway Tunnel [D]. Beijing:Beijing Jiaotong University,2018.
Wang Huawei. Study on the Construction Site Safety Management of Metro Station [D]. Chengdu: Southwest Jiaotong University,2008.
Wang Hui, Huang Hongwei. Reliability evaluation method for segment joints of metro shield tunnels in soft soils [J]. Chinese Journal of Geotechnical Engineering,2011,33(s1):278-283.
Wang Jianbo, Niu Fayang, Zhao Jia, et al. CONSTRUCTION OF METRO PIT-BASED RISK ASSESSMENT DEA-AHP-BP NEURAL NETWORKS [J]. Architecture Technology,2019(3).
Wang Mingzhuo, Huang Hongwei. Monitoring and Pre-warning of Risk Visualization For Civil Engineering [J]. Journal of Disaster Prevention and Mitigation Engineering,2015,35(05):612-616.

Wang Yongsheng, Leng Yajun. A Study of Project Group Risk Theory and Its Risk Evolvement Based-on Bayesian Nets Reasoning [J]. Journal of Northeast Electric Power University. 2011,31(z1):104-109.
WELLE T, BIRKMANN J. Measuring the Unmeasurable:CoMParative Assessment of Urban Vulnerability for Coastal Megacities-New York,London,Tokyo,Kolkata and Lagos [J]. Journal of Extreme Events,2017,03(03):33.
WIKIPEDIA. List of metro systems [EB/OL]. https://en.wikipedia.org/wiki/Listofmetrosystems, 2017-04-12/2017-7-28.
Wu Xianguo, Shen Meifang, Qin Yawei, et al. Risk evaluation of subway foundation pit based on variable-weight and matter-element theory [J]. Engineering Journal of Wuhan University(6),2016.
Wu Xianguo, Wu Gang, Luo Hanbin. Risk Research on Shield Construction of Wuhan Changjiang River Tunnel [J]. China Municipal Engineering,2007(1):51–53.
Wu Xianguo, Wu Kebao, Shen Meifang, et al. Research on coupling of safety risks in metro construction based on N-K model [J]. China Safety Science Journal,2016,26(4):96–101.
www.AsCI.com. In 2017, 31 Chinese cities opened subways, with Shanghai running the most [EB/OL]. http://www.askci.com/news/chanye/20170512/14532497916_2.shtml. 2017-05-12/2017-08-10.
Xiao Xiaochun, Yuan Jinrong, Zhu Yanfei. Causation Analysis of Singapore MRT Circle Line C824 Nicoll Highway Collapse (Part III)——Technically Breaching Back Analysis and Insufficient Instrumentation [J]. Modern Tunnelling Technology,2010,47(01):22-28.
Xiao Xiaochun, Yuan Jinrong, Zhu Yanfei. Causation Analysis of the Collapse on Singapore MRT Circle Line Lot C824 (Part I)- Project Background and Process of Collapse [J]. Modern Tunnelling Technology,2009,46(05):66-72.
Xiao Xiaochun, Yuan Jinrong, Zhu Yanfei. Causation Analysis of the Collapse on Singapore MRT Circle Line Lot C824 (Part)——Critical Design Errors in Temporary Retaining System [J]. Modern Tunnelling Technology,2009,46(06):28-34.
Xinmin Evening News. The underground space of the 20-channel hub Hongqiao is comparable to the underground city of Montreal [EB/OL]. http://shanghai.xinmin.cn/xmsq/2015/07/07/28061538.html. 2015-07-07/2017-08-10.
Xu Kui. Reseach on the Risk Control of the Engineering that a New Subway Tunnel Closely Traverses Under an Existing Subway Station [D]. Beijing: Beijing Jiaotong University,2012.
Xu Tao. Study on Safety Risk Control of Shield Construction in Underwater Tunnel with Multiple Factor Coupling [D]. Chongqing: Chongqing Jiaotong University,2016.
Xu Tian. Research on construction risk of subway deep foundation pit based on improved fuzzy comprehensive evaluation method [D]. Xi'an: Xi'an Technological University,2018.
Yang Ganhui. Research on Safety Risk Identification and Assessment in Deep Foundation of Metro Project [D]. Changsha: Central South University ,2013.
Yang Huijun, Lei Chonghong. Probing into Safety Risk Control of Working Face of Metro Works Undercut under Shallow Cover [C] / / National Academic Conference on Engineering Safety and Protection,2010.
YODZIS M. Examining the Role of Social Feedbacks and Misperception in a Model of Fish-Borne Pollution Illness [J]. Springer International Publishing,2016.
You Pengfei. Risk Mechanism Research on Subway Tunnel Construction [D]. Chengdu:Southwest Jiaotong University,2013.
Yu Chenlong, Zhang Zuohui. The development process and current situation of urban underground integrated pipe corridor at home and abroad [J]. Construction Science and Technology,2015(17):49–51.
Yu Kanzhu. Engineering geological characteristics of residual silty soils in northwest China [J]. Advances in Science and Technology of Water Resources,2000(05):41–43 + 52.
Yuan Yong, Wang Shenghui, Peng Dingchao. Fuzzy assessment of lifetime risk in water-proof of shield tunnel [J]. Journal of Natural Disasters,2005,14(2):81-88.

Zhai Zhigang. Technology Management of Safety Risk for Construction Stage in Beijing Subway [J]. Construction Technology,2011,40(355):86-88.
Zhang Bo. Research on construction risk management and application of subway shield method [D]. Changsha: Central South University,2009.
Zhang Chengping, Yue Yuejing, Wang Mengshu. Influence of pipeline leakage on ground collapse and its control during adjacent tunnelling [J]. China Civil Engineering Journal,2015(S1):351-356.
Zhang Fuqing, Hu Haisheng. The evaluation model of coupling degree of regional industry ecology and its empirical study -- taking Poyang Lake Ecological Economic Zone as an example [J]. Jiangxi Social Sciences,2010(4):219–224.
Zhang Lianwen, Guo Haipeng. Introduction to Bayesian nets [M]. Beijing : Science Press,2006.
ZHANG N L,POOLE D. Exploiting causal independence in Bayesian network inference [J]. Journal of Artificial Intelligence Research,1996,5:301–328.
Zhang Yong. Study on safety risk assessment and control of deep foundation pit construction adjacent to existing subway [D]. Xi'an: Xi'an University of Architecture and Technology,2017.
Zhang Zhiqiang, Kan Cheng, Sun Fei, et al. EXPERIMENTAL STUDY OF CATASTROPHIC BEHAVIOR FOR NATM TUNNEL IN DEBRIS FLOW STRATA [J]. Chinese Journal of Rock Mechanics and Engineering,2014(12):2451-2457.
Zhao Xianli, Luo Fan. Research on the Risks Evolution Model Based on Complex Networks in Airport Flight Area [J]. Journal of University of Electronic Science and Technology of China (Social Sciences Edition). 2013,15(4):31-34.
Zheng Gang, Dai Xuan, Zhang Xiaoshuang. EXPERIMENTAL STUDY AND NUMERICAL SIMULATION OF LEAKING PROCESS OF SAND AND WATER IN UNDERGROUND ENGINEERING [J]. Chinese Journal of Rock Mechanics and Engineering,2014(12):2458-2471.
Zheng Zhibin. Research on risk identification and control technology of urban shallow buried tunnel [D]. Beijing: Beijing Municipal Engineering Research Institute,2009.
Zhou Hongbo, YAO Hao, LU Jianhua. Construction risk assessment on deep foundation pits of a metro line in Shanghai [J]. Chinese Journal of Geotechnical Engineering,2006,28(s1):1902–1906.
Zhu Yanfei. Comment on the tunnel repair scheme of Nanpu Bridge Station - Pudong South Road Station of Shanghai Metro Line 4 [D]. Shanghai:Tongji University,2008.

The manufacturer's authorised representative in the EU is Springer Nature Customer Service Centre GmbH, Europaplatz 3, 69115 Heidelberg, Germany. If you have any concerns regarding our products, please contact ProductSafety@springernature.com

Printed and bound by CPI Group (UK) Ltd, Croydon, CR0 4YY

01/10/2026

02225962-0018